Smart Nanomaterials Technology

Nanotechnology is a rapidly growing scientific field and has attracted a great interest over the last few years because of its abundant applications in different fields like biology, physics and chemistry. This science deals with the production of minute particles called nanomaterials having dimensions between 1 and 100 nm which may serve as building blocks for various physical and biological systems. On the other hand, there is the class of smart materials where the material that can stimuli by external factors and results a new kind of functional properties. The combination of these two classes forms a new class of smart nanomaterials, which produces unique functional material properties and a great opportunity to larger span of application. Smart nanomaterials have been employed by researchers to use it effectively in agricultural production, soil improvement, disease management, energy and environment, medical science, pharmaceuticals, engineering, food, animal husbandry and forestry sectors.

This book series in Smart Nanomaterials Technology aims to comprehensively cover topics in the fabrication, synthesis and application of these materials for applications in the following fields:

- Energy Systems—Renewable energy, energy storage (supercapacitors and electrochemical cells), hydrogen storage, photocatalytic water splitting for hydrogen production
- Biomedical—controlled release of drugs, treatment of various diseases, biosensors,
- Agricultural—agricultural production, soil improvement, disease management, animal feed, egg, milk and meat production/processing,
- Forestry—wood preservation, protection, disease management
- Environment—wastewater treatment, separation of hazardous contaminants from wastewater, indoor air filters.

Azamal Husen
Editor

Enhanced Crop Production Setup Under Various Micro and Nano Contaminants

Sustainable Use, Production and Future Prospects

Springer

Editor
Azamal Husen
Wolaita Sodo University
Wolaita, Ethiopia

Graphic Era (Deemed to be University)
Dehradun, Uttarakhand, India

ISSN 3004-8273 ISSN 3004-8281 (electronic)
Smart Nanomaterials Technology
ISBN 978-981-96-6098-8 ISBN 978-981-96-6099-5 (eBook)
https://doi.org/10.1007/978-981-96-6099-5

This Springer imprint is published by the registered company Springer Nature Singapore Pte Ltd.
The registered company address is: 152 Beach Road, #21-01/04 Gateway East, Singapore 189721, Singapore

Preface

Contaminants in soil-crop plant systems are a significant challenge for sustainable agriculture, environmental health, and food security. Understanding their sources, behavior, and effects on plants and humans is essential for developing strategies to mitigate contamination and ensure the safety of agricultural products. A group of synthetic chemicals and their transformed products are being used for industrial, therapeutic, agricultural, and cosmetic purposes that are reflected as emerging contaminants in most of the environmental regimes. These emerging contaminants are now concerns of scientific communities and policy makers as they may impart severe health hazard. Many of them have been used extensively for decades. On the other hand, every year, new formulations of antibiotics, fungicides, and other drugs have been added in the list. Moreover, plastic and micro-plastics are taking place in almost all the ecosystems. The list also contains engineered nanomaterials and persistent organic pollutants. These contaminants have found a way to get introduced in the food chain through soil-crop plant system. As a part of the system, plants can tolerate the detrimental effects of contaminants or compromise their integrity. In both the cases, the persistency of such chemicals within the ecosystem remains almost constant. There has to be a good management practices for the use and disposal of emerging contaminants to restrict their dynamics. Taken together, these emerging contaminants have a negative impact on the physical, chemical, and biological health of soils. Other known contaminants in soil-crop plant systems are herbicides, insecticides, fungicides, salts, toxic metals, or metalloids like cadmium, lead, mercury, arsenic, copper, zinc, nickel, boron, and so on. Indiscriminate fertilizer use also leads to an accumulation of too many nutrients, which has various adverse impacts on soil-crop plant systems. It has been noticed that the increased contamination renders soil unfit for typical growth and interferes with the growth of crop plants. Crop plants are naturally sedentary, making them the main targets of soil toxins that stress their surroundings. Stress-related plant growth results in altered morpho-physiological and biochemical processes. Crop plants tolerance mechanisms can be significantly altered by employing signaling molecules, phytohormones, plant nutrients, bioaugmented agents or products, and nanomaterials to decipher morpho-physiological and biochemical processes. Crop plant development and their productivity are also affected by both environmental and genetic factors. The key pathways and target genes are being found using functional genomic techniques and OMICS technologies utilizing the CRISPR methodology, which may help to improve agricultural output under difficult circumstances. Moreover, numerous microbes and or PGPR strains are naturally helpful for the growth and development of plants under unfavorable conditions; thus, a sustainable solution can be ensuring for higher crop plant productivity and global food security. Additionally, innovative soil-friendly fertilizer management strategies and phytoremediation techniques may improve the health of the soil and preserve crop plants without sacrificing yield and quality.

In the present book, various general as well as emerging contaminants uptake, toxicity and tolerance in crop plants and management have been explained. Genetic and molecular methods and various biological tools have been explored for the improvement of crop plant growth under contaminants. Other specific topics such as physiological and biochemical pathways, defense mechanisms in crop plants using phytohormones, and nanomaterials under contaminants have been also examined. Moreover, safety issues associated with micro and nanocontaminants have been also explored. Thus in order to increase agricultural output, the current book elucidates the recent developments and methods that could help to improve soil quality and shield plants from various toxins and adverse situation.

Overall, this book is providing an up to date, most important, and selected information associated with the enhanced crop production setup under various micro and nanocontaminants. This book will be a good resource for upper-level students, faculty, scientists, and researchers working specially in nanobiotechnology, plant biology, plant biochemistry, botanical or crop sciences, environmental sciences, ecological sciences, and other allied subjects and or science. I am grateful to all contributors for readily accepting my invitation, sharing their knowledge in specialized areas of research, and readily adjusting the suggestions for improving the shape of their contributions. Finally, my special thanks goes to Shagufta, Zaara, Mehwish, and Huzaifa for providing their time and overall extended support to put everything together.

Wolaita, Ethiopia/Dehradun, India Azamal Husen

Contents

About the Editor

Azamal Husen is served as Full Professor and Head of the Department of Biology, University of Gondar, Ethiopia; and currently working as Foreign Delegate at Wolaita Sodo University, Wolaita, Ethiopia. He is also working as Adjunct Professor at Graphic Era (Deemed to be University), Dehradun, Uttarakhand, India. He also worked as Full Professor at Sankalchand Patel University, Visnagar, India, and Visiting Faculty of the Forest Research Institute and the Doon College of Agriculture and Forest at Dehra Dun, India. His research and teaching experience of 25 years encompasses nanobiotechnology, plant physiology, environmental botany, medicinal plants, and clonal propagation. He has conducted research sponsored by the World Bank, the National Agricultural Technology Project, the Indian Council of Agriculture Research, the Indian Council of Forest Research Education, and the Japan Bank for International Cooperation. Husen has published extensively (over 300, which includes research papers, books, book chapters, etc.) and served on the Editorial Board and as a reviewer of reputed journals published by Elsevier, Frontiers Media, Taylor & Francis, Springer Nature, RSC, Oxford University Press, Sciendo, the Royal Society, CSIRO, PLOS, MDPI, John Wiley & Sons, and UPM Journals. He is on the advisory board of Cambridge Scholars Publishing, UK. He is a fellow of the Plantae group of the American Society of Plant Biologists and a member of the International Society of Root Research, Asian Council of Science Editors and International Natural Product Sciences. He has given numerous talks and presentations in many conferences and workshops. Husen has also given various orientation courses for academic staff, prepared different modules, developed B.Sc., M.Sc., and Ph.D. curriculum. Moreover, he has received a recognition award from University of Gondar, Ethiopia, for a decade of service in teaching, research and community services. He is an editor-in-chief of the *American Journal of Plant Physiology*, and a Series Editor of *Exploring Medicinal Plants* (Taylor & Francis Group, USA); *Plant Biology, Sustainability, and Climate Change* (Elsevier, USA); and *Smart Nanomaterials Technology* (Springer Nature, Singapore). He has achieved the distinguished honor of being recognized as one of the "World's Top 2% Scientists" for the years 2022, 2023, 2024 and 2025 by Stanford University, USA. This recognition has also been prominently featured in the Elsevier Data Repository. Moreover, he has also worked as a special issue guest editor on various topics such as Advances in Plant Metabolomics: Insights, Applications, and Integration (*Plant Physiology and Biochemistry*), Exploring Nanotechnology in Medicinal Plants (*Journal of Herbal Medicine*), Integrating Nanotechnology with Plant Compounds: Enhancing Abiotic Stress Resilience and Crop Improvement via Signaling Molecules (*Frontiers in Plant Science*), etc.

Contaminants and Soil-Crop Plant Systems

Megha Barot, Arpan Modi, and Azamal Husen

Abstract Soil contaminants such as heavy metals, organic compounds, microbial pathogens, nanocomposites from various sources are interact with plant system. The translocation, absorption, adsorption, photodegradation, and phytoremediation plays important role in soil–plant interaction of these contaminants. In the current chapter, we have highlighted different soil contaminants their sources and various phenomena takes place through plant-soil system. Nanocomposites are emerging contaminants so special focus has been given to these contaminants. The main causes of soil contamination are discussed in this chapter, along with information on the different pollutants linked to each of them. In the present chapter, types and characteristics of major soil contaminants are elaborated with respect to their origin along with the characteristics of soil that governs the dynamics of the contaminants through plants within the ecosystem.

Keywords Soil contaminants · Bioremediation · Translocation · Industrial waste

M. Barot (✉)
Research and Development Cell, Parul University, Vadodara, Gujarat, India
e-mail: meghabarot1011@gmail.com

Department of Environmental Science, Parul Institute of Applied Sciences, Parul University, Vadodara, Gujarat, India

A. Modi
Axenic Agritech Private Limited, Ahmedabad, Gujarat, India

A. Husen
Wolaita Sodo University, Wolaita, Ethiopia

Department of Biotechnology, Graphic Era (Deemed to be University), Dehradun, Uttarakhand, India

A. Husen (ed.), *Enhanced Crop Production Setup Under Various Micro and Nano Contaminants*, Smart Nanomaterials Technology,
https://doi.org/10.1007/978-981-96-6099-5_1

1 Introduction

For the basic needs like food and feed, humans have created many resources like field crops cultivation [4], poultry [16], aquaculture [63], and other aquatic products like seaweeds [14]. However, the successful running of these businesses demand the production of many natural and synthetic chemicals (both organic and inorganic), and which are often discarded in the form of landfills. As a result, soil and aquatic ecosystems are polluted. A large proportion of urban and arable land in the world has been affected by inorganic and organic contaminants. These chemicals are toxic and posse a considerable health risk to humans and other environmental systems. Major source of contaminants to various ecosystems are anthropogenic activities. These include mining and smelting activities; the use of pesticides, herbicides, and fertilizers; leather tanning industries; and heavy mechanical diligences. As a result, accumulation of various organic and inorganic toxins takes place which increases the vulnerability of soils as far as safety of humans, animals, plants, and agriculturally important microbes are concerned [60].

Several chemicals have been identified or recognized recently as potential environmental pollutants. However, they are not traditionally regulated. These include endocrine-disrupting chemicals (EDCs), perfluoroalkyl substances (PFAS), painkillers, hormones, nano- and microplastics, and some antibiotics. They are known as emerging contaminants (ECs) in agriculture. These substances are becoming a concern because of their potential negative effects on crops, soil, water, and the environment. These pollutants can come from a variety of sources, such as industrial operations, personal care items, medications, urban runoff, and agricultural practices (such as the use of pesticides, herbicides, and agrochemicals) [61]. To improve crop productions, usage of new chemical-based solutions has been increased [49, 69], which leads to the addition of heavy metals like lead, cadmium, and arsenic in the soils and thereby severely harming the plant system and environmental health [59, 82]. EDCs can enter in to agricultural system through water or other mode of application and affect the reproductive and developmental processes of crop plants and animals [67, 78, 83]. Concerns over PFAS compounds' effects on crops and the food chain have been raised since they restrain at least one aliphatic perfluorocarbon fraction and have been found in soil and water close to agronomic areas as a result of human activity [2, 47, 62]. In addition to that, virulent plant contaminants, like bacteria and fungi, have spread more easily due to numerous changes in global climatic patterns and greater international trade. These pathogens deteriorate the plant's health directly or by secreting toxin chemicals that restricts the plant growth. As a result, crop productivity is severely compromised [80] (Fig. 1).

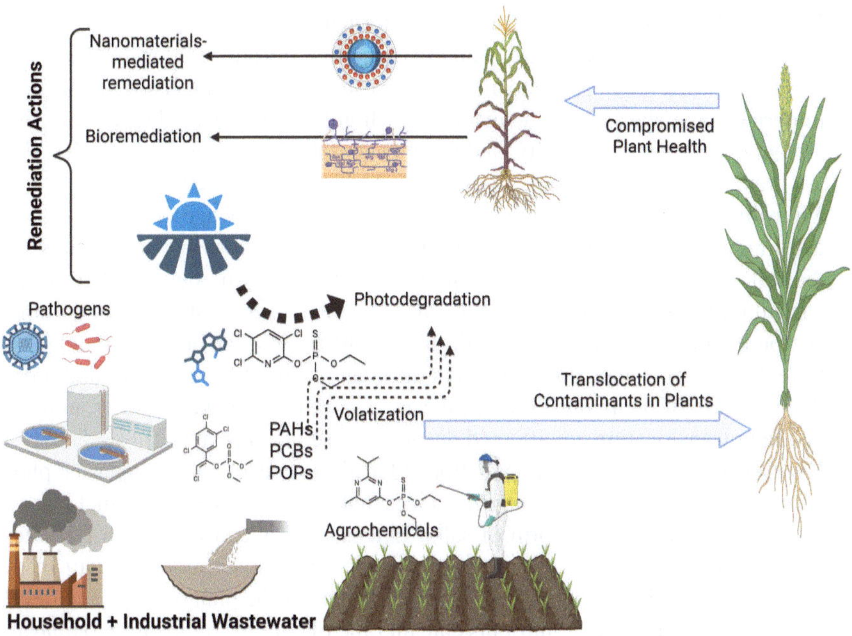

Fig. 1 Exposure of different contaminants to plants through plant-soil environment and their remediation

2 Types and Effect of Soil Contaminants on Plants

- **Heavy Metals**: Common heavy metals found in contaminated soils include lead (Pb), arsenic (As), cadmium (Cd), chromium (Cr), and mercury (Hg). These metals can accumulate in soils through industrial processes, mining, sewage sludge, and the use of certain fertilizers and pesticides. Even at smaller concentrations (less than 1 ppb), these metallic elements with high atomic weight and density—can have harmful effects [13]. Soil pollution with heavy metal is not a region-specific issue. It is now common problem for many of the developed as well as developing countries of the world. Accumulation of heavy metals in soil–plant system delivers harmful effects on plants (growth retardation), and its consequences on animals and human as part of the food chain [7].
- **Pesticides and Herbicides**: These chemicals, used to protect crops from pests and weeds, can persist in soil for extended periods and affect non-target organisms, including beneficial soil microbes. They are used to boost the productivity in the agricultural practices. However, their degradation after usage is also essential to preserve soil and plant health. Cytotoxic effects of such chemicals include necrosis, stunting, yellowing, chlorosis, epinasty, vein clearing, and hyponasty on plants [30].

- **Organic Pollutants**: These include polychlorinated biphenyls (PCBs), polycyclic aromatic hydrocarbons (PAHs), and dioxins, often introduced through industrial pollution, waste disposal, and improper waste management. They posse three important characteristics that made them highly resistant to degradation viz., aromatic ring, hydrophobicity, and thermostability. Due to these properties, they are highly toxic, carcinogenic, mutagenic, and immune-toxic [11].
- **Microbial Pathogens**: Contamination from untreated sewage or improper agricultural practices can introduce harmful bacteria, fungi, viruses, or parasites into the soil. These pathogens not only hinders the plant growth, but also disturbs the agriculturally important microorganisms from the soil ecosystem. Most common symptom of pathogen attack is the drying of particular organ or whole plant. They have the defined routes through which they can reach other vegetative and reproductive parts of the plant. Photosynthesis, respiration, uptake of water, nutrient, and their translocation are affected during disease infestation [72].
- **Excessive Nutrients**: Over-application of fertilizers can lead to nitrogen and phosphorus runoff, causing nutrient imbalances that harm soil health. Synthetic fertilizers are provided in the form of salts. Extensive application of such fertilizers causes salt accumulation in soil due to which water absorption capacity of the plant is disturbed. Soil pH is very essential factor for the plant growth and development. Excessive use of nutrients can alter the pH of the soil leading to unavailability of specific nutrients to the plant. Other physiological changes include leaf discoloration, poor plant growth, and death [65].
- **Nanomaterials**: The growing usage of NMs prompts questions regarding their toxicological impact on the environment, animals, and people [19, 26]. Tourinho [84] reviewed a significant number of research that assessed the behavior and impacts of various metallic NMs in soil. Different metallic nanoparticles affect soil microbial diversity [32]. The engineered nanoparticles plays both negative and positive effect on plants, the research reviewed by [68].

Soil contaminants are categorized based on their chemical nature. They are natural as well as synthetic one. Degraded root exudates, leached-out secondary metabolites, toxins released from pathogenic microorganisms, heavy metals, inorganic ions may be categorized as natural soil contaminants, whereas, pesticides, chemical fertilizers, pharmaceutical wastes, polycyclic aromatic hydrocarbons (PAHs), plastic material (both macro- and micro-plastics), and polychlorinated biphenyls (PCBs) are synthetic ones [71]. The study of different contaminants found in soil system and their effect on plants are listed in Table 1.

3 Uptake of Contaminants by Plants

Soil contaminants include heavy metals, organic pollutants, pesticides, trace elements, and excessive fertilizers [17]. These contaminants are natural or synthetic origin which includes pharmaceutical products, personal care products, persistent

Table 1 Various soil contaminants and their impact on plant's health

S. no	Soil contaminants	Affected plant species	Adverse effects	Key reference
1	Cadmium (Heavy metal)	*Cucumis sativus* L	• Chlorosis in leaves, damages in thylakoid and chloroplasts, reduced rate of photosynthesis • Activity of nitrogen reductase, glutamine synthetase, glutamine synthetase, glutamate synthase, and glutamate dehydrogenase	[27]
2		*Lactuca sativa* L	• Significant reduction in uptake of macro- and micro-nutrient from soil	[45]
3		Various plant species	• Chlorophyll degradation and reduced rate of photosynthesis • Cytotoxicity and cell death, decreased uptake of other nutrients • Generation of reactive oxygen species	[36]
4	Lead (Heavy metal)	*Oryza sativa* L	• Reduced levels of chlorophyll and carotenoids • Increased production of H_2O_2, malonaldehyde (MDA), leaf leachates • Deterioration in grain quality	[8]
5		*Macrotyloma uniflorum* (Lam.); *Cicer arietinum* L	• Increased rate of lipid peroxidation, concentrations of MDA • Increased activity of SOD, catalase, peroxidase	[73]
6	Zinc and lead (Heavy metals)	*Amaranthus* cultivars	• Loss of total as well as organ dry and fresh weight • Decreased root length, volume, and numbers of tips • Decreased levels of pigments and MDA in susceptible cultivars	[37]
7	Trace elements	Vegetable plants	• Accumulation in leaves (leafy vegetables), roots (tuber crops), and fruits	[34]

(continued)

Table 1 (continued)

S. no	Soil contaminants	Affected plant species	Adverse effects	Key reference
8	Copper	*Zea mays* L	• Significant reduction in root numbers, root length, and shoot length of the plant • Decreased content of carotenoid and chlorophyll • Increased content of protein bound thiols and decreased content of non-protein thiols as well as acid soluble thiols	[3]
9	Chlorpyrifos (insecticide)	*Vigna radiata* L	• Decreased root and shoot length at all the stages of flower development • Decreased content and activities of nitrate and nitrate reductase, respectively • Decreased content of soluble sugars and proteins	[66]
10	Flurochloridone (Pesticide)	*Helianthus annuus* L	• Decreased chlorophyll, carotenoid, and relative water content • Increased activities of glutathione reductase, superoxide dismutase, glutathione S-transferase, and malonaldehyde content • Decreased activities of ascorbate peroxidase and catalase	[48]
11	Glyphosate (Herbicide)	*Medicago sativa* L	• Reduced root and shoot length, fresh biomass (in a dose-dependent manner) • Increased content of H_2O_2 and MDA • Significantly increased activity of antioxidants and proline content (at maximum concentration tested)	[28]

(continued)

Table 1 (continued)

S. no	Soil contaminants	Affected plant species	Adverse effects	Key reference
12	Carbendazim, Hexaconazole (Fungicide)	*Pisum sativum* L	• Delayed germination, decreased root and shoot biomass and size • Decline in total chlorophyll, carotenoids, grain yield, grain protein, root phosphorous, shoot nitrogen • Increased levels of proline, MDA, electrolyte leakage, H_2O_2	[77]
13	*O. ariae*, *S. anemones*, *T. anemones*, *U. anemones* (Fungal pathogen)	*Anemone nemorosa*	• No significant reduction in plant growth parameters	[85]
14	*Rhizoctonia solani* (Fungal pathogen)	*Solanum lycopersicum* L	• Damping-off, root rot, stem canker • Up to 20% yield loss • Lowering fruit quality	[55]
15	*Fusarium oxysporum* (Fungal pathogen)	*Cuminum cyminum* L	• Up to 100% yield loss • Wilting and shivering of leaves, browning of the vascular system, stunning and damping-off	[1]
16	*Burkholderia glumae* (Bacterial pathogen)	*Oryza sativa* L	• Up to 75% crop loss • Reduced root and shoot length • Seedling rot, panicle browning, aborted seeds, reduced grain filling	[10]
17	Polycyclic aromatic compounds (Organic pollutants)	*Brassica napus* L	• Reduced rate of germination, shoot and root biomass, and chlorophyll content	[74]
18	Anthracene, chlorpyrifos (Organic pollutant and pesticide)	*Lolium multiflorum*	• Germination percentage reduced significantly in higher doses	[51]
19	Nitrogen (Excessive application)	Many plant species	• Soil acidification • Increased chances of soil-borne disease to plants	[94]

(continued)

Table 1 (continued)

S. no	Soil contaminants	Affected plant species	Adverse effects	Key reference
20	Cu nanoparticles	Rye grass (*Lolium perenne*) and radish (*Raphanus sativus*)	• Growth pattern • Concentrations of primary metabolites such as proteins and carbohydrates were affected	[9]
		Cilantro (*Coriandrum sativum*)	• Germination percentage decreased • Morphological parameters, such as root and shoot length also decreased	[99]
21	Multiwall carbon nanotubes	Broad bean (*Vicia faba* L.)	• Cellular oxidative damages	[86]

organic pollutants, pesticides, heavy metals, and industrial chemicals. These chemicals are less frequently degraded by soil microflora in terrestrial ecosystem as compared to aquatic ecosystem [12]. For the contaminants like pesticides and organic pollutants, manufacturers and government agencies use different models like PEARL (Pesticide Emission Assessment at Regional and Local Scales), PRZM (Pesticide Root Zone Model), PELMO (Pesticides Leaching Model), and MACRO (Model for Assessing Contaminant Fate and Transport in Groundwater) to assess their potential for groundwater contamination [20, 50, 52, 54]. In essence, these models provide a scientific basis for understanding and managing the risks associated with pesticide and contaminant use in agriculture. They help policymakers, researchers, and farmers make informed decisions to safeguard groundwater resources and ensure sustainable agricultural practices. These models also provide the information whether the edible part of the plant will be consumable or not.

Plants provide a gateway for the entry of environmental contaminants to the food chain. It has been observed that plants of the Cucurbitaceae family are hyperaccumulators of chlorinated pesticides in which organic acids released from the root system facilitates their uptake [89]. While in case of wheat and earthworm co-existence system, both the organism accumulates perfluoroalkyl and polyfluoroalkyl substances (PFAS) significantly higher than the individual [96]. Soil contaminants enter in to the plant system through two pathways. Most common route for the entry of soil contaminants in the plant is through roots. The other pathway is soil-air-plant pathway [23]. In most cases, the soil contaminants impart detrimental effects on plants [93]. However, some organic and inorganic nano-contaminants at the smaller concentrations showed positive impacts on crop growth and development [15, 18, 31, 92]. Root architecture play vital role for the entry of nano- and micro-contaminants in the plant system. Fundamental physiological processes of plants are engaged in the uptake of contaminants. Water soluble soil contaminants can actively enter into the plant system via the process of transpiration. Cross-sectional examination of root provides detailed information on how the movement of contaminants (or nutrients) takes place from soil to inner part of the plant (xylem and phloem). The transferring entity can be taken through 1. Apoplastic pathway in which the movement occurs through extracellular space; 2. Symplastic pathway in which the contaminants travel through cells using the network of plasmodesmata; or 3. Through transmembrane which is between cells with inner membrane space but without using extracellular space [57]. Before reaching to the vascular tissues, the Casperian strip (cells with cell wall containing lignin and suberin) provides strong barrier for the entry of contaminants. Due to lipophilicity of various chemicals, they are diffused easily across lipid bilayers and reached to the vascular tissues. Root hairs present on the roots also enhance the surface area to facilitate the absorption process [56, 81]. These contaminants react with soil organic matter and nutrients and gets absorbed along with them by root structures. On the other hand, commonly found root exudates such as citric acid, oxalic acid, and malonic acids are involved in the process of degradation of organic contaminants [44]. Several abiotic factors influence the uptake of these contaminants. These factors include humic content, colloidal clay, soil pH, hydraulic conductivity of soil, swelling capacity of soil, and wettability of soil [68]. Once absorbed, the contaminants are translocated to the vegetative, reproductive, and edible parts of the plants. The translocation is facilitated by the transpiration process. Pharmaceutical and personal care products were found to be actively translocated through transpirational pull in a variety of plant species [25]. In some cases, like engineered nanomaterials, the absorbed contaminants are bio-transformed and then mobilized in the plant system [76]. Translocation is not a process where the uptake mechanism ends. Translocated compounds tend to get accumulated over the period of time. Bioaccumulation of soil contaminants ends up with either cytotoxicity or phytoremediation. For the former phenomenon, most plants are susceptible and for the phytoremediation some plants have evolved with a mechanism that counteract the negative impacts of contaminants (Iqbal et al. 2015; Shmaefsky 2023) [39, 75, 90].

4 Nanomaterials as Plant Growth Promoters and Contaminants

Due to extensive use of chemicals and fertilizers in a higher dose, terrestrial ecosystem is getting disturbed as many of them are hard to mitigate. One of the proposed sustainable solutions is the use of the same chemicals and fertilizers in agriculture in a very small concentrations with increased surface area. Bulk materials are often used in a higher concentration due to relatively less surface area. On the other hand, nanomaterials imparts the same effects on plants in very minute concentrations [33]. Commonly used nanomaterials in agriculture are TiO_2 nanomaterials, Cu–Zn nanoparticles, and iron nanoparticles [6, 43, 53]. Their positive impacts on crop health includes increased germination percentage, increased chlorophyll and carotenoid content, reduced leaching, remediation of metal and metalloids from soil–plant system [98]. Nanomaterials applied to plants promote growth, soil health, nutrient uptake and consumption, and tolerance to biotic and abiotic stresses [22, 38–42, 70, 79]. Numerous studies have proved that nano-TiO_2, copper, zinc, and iron showed positive impact on germination percentage, and other growth-related parameters. Nanomaterials like Se-doped carbon quantum dots (Se-CQDs), nano zero-valent iron (nZVI), and polyacrylamide dispersed magnetite (PAM-MAG) can significantly improve soil health by enhancing diversity and relative abundance of agriculturally important microorganisms, remediate metal toxicity by absorbing, reducing, co-precipitating, and immobilizing processes [46, 91, 97]. Some nanoparticles like ZnO and FeO forms non-reactive complexes with soil nutrients. Uptake of these nutrients by plants were found to be difficult in absence of these nanoparticles. These nanoparticles not only help in the absorption process, but also modulate hormonal signaling in plants to facilitate the transportation of absorbed nutrients to the sink tissues [64]. It has been observed that yellowing of leaves due to iron deficiency was reversed using nanocellulose-iron chelates in pear. Nanomaterials are not only readily absorbed, but also converted in to simpler form to form complexes with proteins [88]. As mentioned, nanomaterials also regulate essential metabolic processes such as photosynthesis and respiration. Chlorophyll and carotenoid pigments are directly involved in the photosynthesis process and by enhancing the content of such pigments, the rate of photosynthesis increases in the plant. As a result, total biomass is increased which is an indicative of growth [35]. Moreover, nanomaterials were also found to impart biotic and abiotic stress resistance. Nanomaterials like silver and silicon nanoparticles, have antimicrobial properties that can be effectively utilized against variety of pathogenic strains of bacteria and fungi. Nanoparticles also enhance the expression of phytohormone-related genes thereby enhancing the plant's innate immune system. Broad-spectrum antimicrobial activities of nano-silver has been found in number of plant species [58, 95]. On the other hand, abiotic stress mitigation is achieved by nanomaterials through their antioxidant nature. They protect membrane-bound organelles during oxidative stress generated due to adverse environmental conditions [24].

Apart from their growth promoting activities, cytotoxic effects of nanomaterials are also well-studied. It has been observed that, certain nanomaterials beyond particular concentrations, exposure time, and very low size (<50 nm) are toxic to plants and soil ecosystems. All the growth-indicative parameters that were found to be increased due to nanomaterial applications, can be affected to show negative values. Protein, chlorophyll, and carotenoid contents were found to be decreased during toxicity [21]. Nanomaterials when used Detrimental effects of nano-silver, zinc, carbon nanotubes, cellarium, cobalt, nickel, NZVI (nanoscale zero-valent iron) have been observed in different plant species. In a field crops like wheat, it was also observed that the application of nanomaterials delayed the harvest period of the crop and nanoparticles were accumulated in the plants instead of excreted showing potential threat to food chain [5, 29, 87].

5 Conclusion

Contaminants in soil-crop systems are a significant challenge for sustainable agriculture, environmental health, and food security. Understanding their sources, behavior, and effects on plants and humans is essential for developing strategies to mitigate contamination and ensure the safety of agricultural products. A group of synthetic chemicals and their transformed products are being used for industrial, therapeutic, agricultural, and cosmetic purposes that are reflected as emerging contaminants in most of the environmental regimes. These emerging contaminants are now concerns of scientific communities and policymakers as they may impart severe health hazard. Many of them have been used extensively for decades. On the other hand, every year, new formulations of antibiotics, fungicides, and other drugs have been added in the list. Moreover, plastic and micro-plastics are taking place in almost all the ecosystems. The list also contains engineered nanomaterials and persistent organic pollutants. These contaminants have found a way to get introduced in the food chain through soil-crop plant system. As a part of the system, plants can tolerate the detrimental effects of contaminants or compromise their integrity. In both the cases, the persistency of such chemicals within the ecosystem remains almost constant. There has to be a good management practices for the use and disposal of emerging contaminants to restrict their dynamics.

Competing Interests The authors declare that they have no competing interest.

Declaration of Generative AI and AI-Assisted Technologies in the Writing Process During the preparation of this work, the book chapter author(s) have not taken any assistance from AI, and the book chapter author(s) reviewed and edited the content as needed and take(s) full responsibility for the content of the publication.

References

1. Abo-Elyousr KA, Saad MM, Al-Qurashi AD, Ibrahim OH, Mousa MA (2022) Management of cumin wilt caused by Fusarium oxysporum using native endophytic bacteria. Agronomy 12(10):2510
2. Adu O, Ma X, Sharma VK (2023) Bioavailability, phytotoxicity and plant uptake of per-and polyfluoroalkyl substances (PFAS): A review. J Hazard Mater 447:130805. https://doi.org/10.1016/j.jhazmat.2023.130805
3. Aly AA, Mohamed AA (2012) The impact of copper ion on growth, thiol compounds and lipid peroxidation in two maize cultivars (*Zea mays* L.) grown '*in vitro*'. Aust J Crop Sci 6(3):541–549
4. Anastasi U, Scavo A (2023) Cropping systems and agronomic management practices of field crops. Agronomy 13(9):2328
5. Antisari LV, Carbone S, Bosi S, Gatti A, Dinelli G (2018) Engineered nanoparticles effects in soil-plant system: Basil (*Ocimum basilicum* L.) study case. Appl Soil Ecol 123:551–560
6. Antonoglou O, Moustaka J, Adamakis IDS, Sperdouli I, Pantazaki AA, Moustakas M, Dendrinou-Samara C (2018) Nanobrass CuZn nanoparticles as foliar spray nonphytotoxic fungicides. ACS Appl Mater Interfaces 10(5):4450–4461

7. Asati A, Pichhode M, Nikhil K (2016) Effect of heavy metals on plants: an overview. Int J Appl Innov Eng Manag 5(3):56–66
8. Ashraf U, Kanu AS, Deng Q, Mo Z, Pan S, Tian H, Tang X (2017) Lead (Pb) toxicity; physio-biochemical mechanisms, grain yield, quality, and Pb distribution proportions in scented rice. Front Plant Sci 8:259
9. Atha DH, Wang H, Petersen EJ, Cleveland D, Holbrook RD, Jaruga P, Dizdaraoglu B, Xing B, Nelson BC (2012) Copper oxide nanoparticle mediated DNA damage in ter restrial plant models. Environ Sci Tech 46(3):1819–1827
10. Azzahra RSNA, Miranti M, Prismantoro D, Mohd Suhaimi NS, Alizadeh M, Mispan MS, Doni F (2024) Bacterial panicle blight caused by Burkholderia glumae: a major disease in rice cultivation. Cogent Food Agric 10(1):2381600
11. Barathi S, Gitanjali J, Rathinasamy G, Sabapathi N, Aruljothi KN, Lee J, Kandasamy S (2023) Recent trends in polycyclic aromatic hydrocarbons pollution distribution and counteracting bio-remediation strategies. Chemosphere, 139396
12. Barot M, Kumar N (2021) Remediation action on persistent organic pollutants by wonder weeds and associated microbiomes. In: Microbe mediated remediation of environmental contaminants. Woodhead Publishing, pp 355–368
13. Barot M, Modi A, Kumar A (2021) Application of plant–microbe systems in bioremediation of metalloid-contaminated soils. In: Microbe mediated remediation of environmental contaminants. Woodhead Publishing, pp 227–240
14. Barot M, Nirmal Kumar JI, Kumar RN (2019) An evaluation of the nutritional composition of seaweeds as potential source of food and feed. Natl Acad Sci Lett 42(6):459–464
15. Begum P, Ikhtiari R, Fugetsu B (2014) RETRACTED: potential impact of multi-walled carbon nanotubes exposure to the seedling stage of selected plant species. Nanomaterials 4(2):203–221
16. Bist RB, Bist K, Poudel S, Subedi D, Yang X, Paneru B, Mani S, Wang D, Chai L (2024) Sustainable poultry farming practices: a critical review of current strategies and future prospects. Poult Sci, 104295
17. Brunetti G, Kodešová R, Švecová H, Fér M, Nikodem A, Klement A, Grabic R, Šimu°nek J (2021) On the use of mechanistic soil–plant uptake models: a comprehensive experimental and numerical analysis on the translocation of carbamazepine in green pea plants. Environ Sci Technol 55(5):2991–3000
18. Burke DJ, Pietrasiak N, Situ SF, Abenojar EC, Porche M, Kraj P, Lakliang Y, Samia ACS (2015) Iron oxide and titanium dioxide nanoparticle effects on plant performance and root associated microbes. Int J Mol Sci 16(10):23630–23650
19. Bystrzejewska-Piotrowska G, Golimowski J, Urban PL (2009) Nano particles: their potential toxicity, waste and environmental management. Waste Manage 29(9):2587–2595
20. Carsel RF, Mulkey LA, Lorber MN, Baskin LB (1985) The pesticide root zone model (PRZM): a procedure for evaluating pesticide leaching threats to groundwater. Ecol Model 30(1–2):49–69
21. Chen M, Zhou S, Zhu Y, Sun Y, Zeng G, Yang C, Xu P, Yan M, Liu Z, Zhang W (2018) Toxicity of carbon nanomaterials to plants, animals and microbes: recent progress from 2015-present. Chemosphere 206:255–264
22. Choudhary A, Sharma S, Kaur H, Sharma N, Gadewar MM, Mehta S, Husen A (2024) Plant system, abiotic stress resilience, reactive oxygen species, and coordination of engineered nanomaterials: a review. S Afr J Bot 171:45–59. https://doi.org/10.1016/j.sajb.2024.05.053
23. Collins C, Fryer M, Grosso A (2006) Plant uptake of non-ionic organic chemicals. Environ Sci Technol 40(1):45–52
24. Cui B, Luo H, Yao X, Xing P, Deng S, Zhang Q, Yi W, Gu Q, Peng L, Yu X, Zuo C (2024) Nanosized-selenium-application-mediated cadmium toxicity in aromatic rice at different stages. Plants 13(16):2253
25. Dodgen LK, Ueda A, Wu X, Parker DR, Gan J (2015) Effect of transpiration on plant accumulation and translocation of PPCP/EDCs. Environ Pollut 198:144–153
26. Elsaesser A, Howard CV (2012) Toxicology of nanoparticles. Adv Drug Del Rev 64(2):129–137
27. Feng J, Shi Q, Wang X, Wei M, Yang F, Xu H (2010) Silicon supplementation ameliorated the inhibition of photosynthesis and nitrate metabolism by cadmium (Cd) toxicity in *Cucumis sativus* L. Sci Hortic 123(4):521–530

28. Fernandes B, Soares C, Braga C, Rebotim A, Ferreira R, Ferreira J, Fidalgo F, Pereira R, Cachada A (2020) Ecotoxicological assessment of a glyphosate-based herbicide in cover plants: *Medicago sativa* L. as a model species. Appl Sci 10(15):5098
29. Fernandes JP, Mucha AP, Francisco T, Gomes CR, Almeida CMR (2017) Silver nanoparticles uptake by salt marsh plants–Implications for phytoremediation processes and effects in microbial community dynamics. Mar Pollut Bull 119(1):176–183
30. Gautam H, Singh S, Prashad H, Kumar A, Choudhary A, Kaur H, Narayan OP, Lal SK, Mehta S (2023) Pesticide toxicity and their impact on plant growth, active constituents and productivity. In: Plants and their interaction to environmental pollution. Elsevier, pp 231–252
31. Ge Y, Schimel JP, Holden PA (2012) Identification of soil bacteria susceptible to TiO2 and ZnO nanoparticles. Appl Environ Microbiol 78(18):6749–6758
32. Gomes AR, Justino C, Rocha-Santos T, Freitas AC, Duarte AC, Pereira R (2017) Review of the ecotoxicological effects of emerging contaminants to soil biota. J Environ Sci Health, Part A 52(10):992–1007. https://doi.org/10.1080/10934529.2017.1328946
33. Gong X, Huang D, Liu Y, Peng Z, Zeng G, Xu P, Cheng M, Wang R, Wan J (2018) Remediation of contaminated soils by biotechnology with nanomaterials: bio-behavior, applications, and perspectives. Crit Rev Biotechnol 38(3):455–468
34. Gupta N, Yadav KK, Kumar V, Kumar S, Chadd RP, Kumar A (2019) Trace elements in soil-vegetables interface: translocation, bioaccumulation, toxicity and amelioration-a review. Sci Total Environ 651:2927–2942
35. Gupta SD, Agarwal A, Pradhan S (2018) Phytostimulatory effect of silver nanoparticles (AgNPs) on rice seedling growth: an insight from antioxidative enzyme activities and gene expression patterns. Ecotoxicol Environ Saf 161:624–633
36. Hayat MT, Nauman M, Nazir N, Ali S, Bangash N (2019) Environmental hazards of cadmium: past, present, and future. In: Cadmium toxicity and tolerance in plants. Academic Press, pp 163–183
37. Hunková J, Lisinovičová M, Lancíková V, Szabóová M, Kačírová J, Mistríková V, Hricová A (2024) A comparative analysis of heavy metal stress responses in different grain amaranth cultivars. Plant Stress 100619
38. Husen A (2022) Engineered nanomaterials for sustainable agricultural production, soil improvement and stress management. Elsevier, Cambridge, MA, USA https://doi.org/10.1016/C2021-0-00054-7
39. Husen A (2023) Nanomaterials and Nanocomposites exposures to plants (Response, Interaction, Phytotoxicity and Defense Mechanisms). Springer Nature Singapore, Singapore. https://doi.org/10.1007/978-981-99-2419-6
40. Husen A (2025a) Plant Response to Silver Nanoparticles. Springer Nature Singapore, Singapore
41. Husen A (2025b) Plant Response to Gold Nanoparticles. Springer Nature Singapore, Singapore
42. Husen A (2025c) Emerging carbon nanomaterials for sustainable agricultural practices (Synthesis, plant growth, performance, production and protection). Springer Nature, Singapore
43. Iannone MF, Groppa MD, de Sousa ME, van Raap MBF, Benavides MP (2016) Impact of magnetite iron oxide nanoparticles on wheat (*Triticum aestivum* L.) development: evaluation of oxidative damage. Environ Exp Bot 131:77–88
44. Jha P, Panwar J, Jha PN (2015) Secondary plant metabolites and root exudates: guiding tools for polychlorinated biphenyl biodegradation. Int J Environ Sci Technol 12:789–802
45. Jibril SA, Hassan SA, Ishak CF, Megat Wahab PE (2017) Cadmium toxicity affects phytochemicals and nutrient elements composition of lettuce (*Lactuca sativa* L.). Adv Agric 2017(1):1236830
46. Jin Y, Wang Y, Li X, Luo T, Ma Y, Wang B, Liang H (2023) Remediation and its biological responses to Cd (II)-Cr (VI)-Pb (II) multi-contaminated soil by supported nano zero-valent iron composites. Sci Total Environ 867:161344
47. Karamat A, Tehrani R, Foster GD, Van Aken B (2023) Plant responses to per- and polyfluoroalkyl substances (PFAS): a molecular perspective. Int J Phytoremed 18:1–9. https://doi.org/10.1080/15226514.2023.2232874

48. Kaya A, Yigit E (2014) The physiological and biochemical effects of salicylic acid on sunflowers (*Helianthus annuus*) exposed to flurochloridone. Ecotoxicol Environ Saf 106:232–238
49. Kilani-Morakchi S, Morakchi-Goudjil H, Sifi K (2021) Azadirachtin-based insecticide: overview, risk assessments, and future directions. Front Agron 3:676208. https://doi.org/10.3389/fagro.2021.676208
50. Klein M, Müller M, Dust M, Görlitz G, Gottesbüren B, Hassink J, Kloskowski R, Kubiak R, Resseler H, Schäfer H, Stein B (1997) Validation of the pesticide leaching model PELMO using lysimeter studies performed for registration. Chemosphere 35(11):2563–2587
51. Korade DL, Fulekar MH (2009) Effect of organic contaminants on seed germination of *Lolium multiflorum* in soil. Biol Med 1(1):28–34
52. Larsbo M, Jarvis N (2003) MACRO 5.0: a model of water flow and solute transport in macroporous soil: technical description, vol 1. Department of Soil Sciences, Swedish University of Agricultural Sciences, Uppsala
53. Latef AA, Srivastava AK, El-sadek MS, Kordrostami M, Tran LS (2018) Titanium dioxide nanoparticles improve growth and enhance tolerance of broad bean plants under saline soil conditions. Land Degrad Dev 29(4):1065–1073
54. Leistra M, Van der Linden AMA, Boesten JJTI, Tiktak A, Van den Berg F (2001) PEARL model for pesticide behaviour and emissions in soil-plant systems: description of the processes in FOCUS PEARL v 1.1. 1, no 13. Alterra
55. Marzouk T, Chaouachi M, Sharma A, Jallouli S, Mhamdi R, Kaushik N, Djébali N (2021) Biocontrol of Rhizoctonia solani using volatile organic compounds of solanaceae seed-borne endophytic bacteria. Postharvest Biol Technol 181:111655
56. Mc Farlane C, Trapp S (1994) Plant contamination: modeling and simulation of organic chemical processes. CRC Press
57. Miller EL, Nason SL, Karthikeyan KG, Pedersen JA (2016) Root uptake of pharmaceuticals and personal care product ingredients. Environ Sci Technol 50(2):525–541
58. Mishra S, Singh BR, Singh A, Keswani C, Naqvi AH, Singh HB (2014) Biofabricated silver nanoparticles act as a strong fungicide against Bipolaris sorokiniana causing spot blotch disease in wheat. PLoS ONE 9(5):e97881
59. Mohanty B, Das A (2023) Heavy metals in agricultural cultivated products irrigated with wastewater in India: a review. AQUA Water Infrastruct Ecosyst Soc 72(6):851–867. https://doi.org/10.2166/aqua.2023.122
60. Murtaza G, Murtaza B, Niazi NK, Sabir M (2014) Soil contaminants: sources, effects, and approaches for remediation. Improv Crop Era Clim 2:171–196
61. Naeem M, Gill R, Gill SS, Singh K, Sofo A, Tuteja N (2023) Emerging contaminants and their effect on agricultural crops. Front Plant Sci 14:1296252
62. Nassazzi W, Wu T-C, Jass J, Lai FL, Ahrens L (2023) Phytoextraction of per- and polyfluoroalkyl substances (PFAS) and the influence of supplements on the performance of short–rotation crops. Environ pollut 333:122038. https://doi.org/10.1016/j.envpol.2023.122038
63. Naylor RL, Hardy RW, Buschmann AH, Bush SR, Cao L, Klinger DH, Little DC, Lubchenco J, Shumway SE, Troell M (2021) A 20-year retrospective review of global aquaculture. Nature 591(7851):551–563
64. Oldroyd GE, Leyser O (2020) A plant's diet, surviving in a variable nutrient environment. Science 368(6486):eaba0196
65. Pandey R, Vengavasi K, Hawkesford MJ (2021) Plant adaptation to nutrient stress. Plant Physiol Rep 26(4):583–586
66. Parween T, Jan S, Mahmooduzzafar, Fatma T (2011) Alteration in nitrogen metabolism and plant growth during different developmental stages of green gram (*Vigna radiata* L.) in response to chlorpyrifos. Acta physiologiae plantarum 33:2321–2328
67. Pridemore CM (2021) The effects of endocrine disrupting chemicals on plants. Theses, Dissertations and Capstones. Available at: https://mds.marshall.edu/etd/1412

68. Pullagurala VLR, Rawat S, Adisa IO, Hernandez-Viezcas JA, Peralta-Videa JR, Gardea-Torresdey JL (2018) Plant uptake and translocation of contaminants of emerging concern in soil. Sci Total Environ 636:1585–1596
69. Rajak BK, Rani P, Mandal P, Chhokar RS, Singh N, Singh DV (2023) Emerging possibilities in the advancement of herbicides to combat acetyl-CoA carboxylase inhibitor resistance. Front Agron 5:1218824. https://doi.org/10.3389/fagro.2023.121882
70. Rajpal VR, Dhingra Y, Khungar L, Mehta S, Minkina T, Rajput VD, Husen A (2024) Metal and metal-oxide nanoparticles in nanosensing as well as managing biotic stress in plant system. Curr Res Biotechnol 7:100219. https://doi.org/10.1016/j.crbiot.2024.100219
71. Rajput P, Singh A, Agrawal S, Ghazaryan K, Rajput VD, Movsesyan H, Mandzhieva S, Minkina T, Alexiou A (2024) Effects of environmental metal and metalloid pollutants on plants and human health: exploring nano-remediation approach. Stress Biol 4(1):1–25
72. Ray RV (2024) Effects of pathogens and disease on plant physiology. In: Agrios' plant pathology. Academic Press, pp 63–92
73. Reddy AM, Kumar SG, Jyothsnakumari G, Thimmanaik S, Sudhakar C (2005) Lead induced changes in antioxidant metabolism of horsegram (*Macrotyloma uniflorum* (Lam.) Verdc.) and bengalgram (*Cicer arietinum* L.). Chemosphere 60(1):97–104
74. Ren L, Zeiler LF, Dixon DG, Greenberg BM (1996) Photoinduced effects of polycyclic aromatic hydrocarbons onbrassica napus (Canola) during germination and early seedling development. Ecotoxicol Environ Saf 33(1):73–80
75. Sabir M, Baltrėnaitė-Gedienė E, Ditta A, Ullah H, Kanwal A, Ullah S, Faraj TK (2022) Bioaccumulation of heavy metals in a soil–plant system from an open dumpsite and the associated health risks through multiple routes. Sustainability 14(20):13223
76. Servin AD, Castillo-Michel H, Hernandez-Viezcas JA, Diaz BC, Peralta-Videa JR, Gardea-Torresdey JL (2012) Synchrotron micro-XRF and micro-XANES confirmation of the uptake and translocation of TiO2 nanoparticles in cucumber (*Cucumis sativus*) plants. Environ Sci Technol 46(14):7637–7643
77. Shahid M, Ahmed B, Zaidi A, Khan MS (2018) Toxicity of fungicides to Pisum sativum: a study of oxidative damage, growth suppression, cellular death and morpho-anatomical changes. RSC Adv 8(67):38483–38498
78. Sharma BM, Scheringer M, Chakraborty P, Bharat GK, Steindal EH, Trasande L et al (2022) Unlocking India's potential in managing endocrinedisrupting chemicals (EDCs): importance, challenges, and opportunities. Health, Expo. https://doi.org/10.1007/s12403-022-00519-8
79. Siddiqi KS, Husen A (2022) Plant response to silver nanoparticles: a critical review. Crit Rev Biotechnol 42(7):773–990. https://doi.org/10.1080/07388551.2021.1975091
80. Singh BK, Delgado-Baquerizo M, Egidi E, Guirado E, Leach JE, Liu H et al (2023) Climate change impacts on plant pathogens, food security and paths forward. Nat Rev Microbiol 21:640–656. https://doi.org/10.1038/s41579-023-00900-7
81. Sterling TM (1994) Mechanisms of herbicide absorption across plant membranes and accumulation in plant cells. Weed Sci 42(2):263–276
82. Sun N, Thompson RB, Xu J, Liao S, Suo L, Peng Y et al (2021) Arsenic and cadmium accumulation in soil as affected by continuous organic fertilizer application: implications for clean production. Agronomy 11(11):2272. https://doi.org/10.3390/agronomy11112272
83. Thacharodi A, Hassan S, Hegde TA, Thacharodi DD, Brindhadevi K, Pugazhendhi A (2023) Water a major source of endocrine-disrupting chemicals: an overview on the occurrence, implications on human health and bioremediation strategies. Environ Res 231:116097. https://doi.org/10.1016/j.envres.2023.116097
84. Tourinho PS, van Gestel C, Lofts S, Svendsen C, Soares AMVM, Loureiro S (2012) Metal-based nanoparticles in soil: fate, behavior, and effects on soil invertebrates. Environ Toxicol Chem 31(8):1679–1692
85. van Dijk LJ, Ehrlén J, Tack AJ (2021) Direct and insect-mediated effects of pathogens on plant growth and fitness. J Ecol 109(7):2769–2779
86. Wang C, Liu H, Chen J, Tian Y, Shi J, Li D, Guo C, Ma Q (2014) Carboxylated multi walled carbon nanotubes aggravated biochemical and subcellular damages in leaves of broad bean

(*Vicia faba* L.) seedlings under combined stress of lead and cadmium. J Hazard Mater 274:404–412
87. Wang J, Fang Z, Cheng W, Yan X, Tsang PE, Zhao D (2016) Higher concentrations of nanoscale zero-valent iron (nZVI) in soil induced rice chlorosis due to inhibited active iron transportation. Environ Pollut 210:338–345
88. Wang Y, Feng LJ, Sun XD, Zhang M, Duan JL, Xiao F, Lin Y, Zhu FP, Kong XP, Ding Z, Yuan XZ (2023) Incorporation of selenium derived from nanoparticles into plant proteins in vivo. ACS Nano 17(16):15847–15856
89. White JC, Mattina MI, Lee WY, Eitzer BD, Iannucci-Berger W (2003) Role of organic acids in enhancing the desorption and uptake of weathered p, p′-DDE by Cucurbita pepo. Environ Pollut 124(1):71–80
90. Yan A, Wang Y, Tan SN, Mohd Yusof ML, Ghosh S, Chen Z (2020) Phytoremediation: a promising approach for revegetation of heavy metal-polluted land. Front Plant Sci 11:359
91. Yin K, Bao Q, Li J, Wang M, Wang F, Sun B, Gong Y, Lian F (2024) Molecular mechanisms of growth promotion and selenium enrichment in tomato plants by novel selenium-doped carbon quantum dots. Chemosphere 364:143175
92. Zhai Y, Hunting ER, Wouters M, Peijnenburg WJ, Vijver MG (2016) Silver nanoparticles, ions, and shape governing soil microbial functional diversity: nano shapes micro. Front Microbiol 7:1123
93. Zhang C, Yao FENG, Liu YW, Chang HQ, Li ZJ, Xue JM (2017) Uptake and translocation of organic pollutants in plants: a review. J Integr Agric 16(8):1659–1668
94. Zhang Y, Ye C, Su Y, Peng W, Lu R, Liu Y, Huang H, He X, Yang M, Zhu S (2022) Soil Acidification caused by excessive application of nitrogen fertilizer aggravates soil-borne diseases: evidence from literature review and field trials. Agr Ecosyst Environ 340:108176
95. Zhao L, Lu L, Wang A, Zhang H, Huang M, Wu H, Xing B, Wang Z, Ji R (2020) Nano-biotechnology in agriculture: use of nanomaterials to promote plant growth and stress tolerance. J Agric Food Chem 68(7):1935–1947
96. Zhao S, Fang S, Zhu L, Liu L, Liu Z, Zhang Y (2014) Mutual impacts of wheat (*Triticum aestivum* L.) and earthworms (*Eisenia fetida*) on the bioavailability of perfluoroalkyl substances (PFASs) in soil. Environ Pollut 184:495–501
97. Zheng M, Huang Z, Ji H, Qiu F, Zhao D, Bredar AR, Farnum BH (2020) Simultaneous control of soil erosion and arsenic leaching at disturbed land using polyacrylamide modified magnetite nanoparticles. Sci Total Environ 702:134997
98. Zhu Y, Xu F, Liu Q, Chen M, Liu X, Wang Y, Sun Y, Zhang L (2019) Nanomaterials and plants: positive effects, toxicity and the remediation of metal and metalloid pollution in soil. Sci Total Environ 662:414–421
99. Zuverza-Mena N, Medina-Velo IA, Barrios AC, Tan W, Peralta-Videa JR, Gardea Torresdey JL (2015) Copper nanoparticles/compounds impact agronomic and physiological parameters in cilantro (*Coriandrum sativum*). Environ Sci: Process Impacts 17(10):1783–1793

Use of Phytoremediation in Improving Soil Status

Rakesh Kumar, LalitThakur, Manmohan Jagatram Dobriyal, Saransh Kumar Gautam, Yalal Mallesh, and Thati Vignesh

Abstract Phytoremediation is an environment friendly and low-cost approach to rehabilitate contaminated soils and enhance ecological stability. It includes several distinct mechanisms viz, phytoextraction, phytostabilization, phytovolatilization, rhizofiltration, and nano-phytoremediation, each designed to treat a specific group of contaminants. Hyperaccumulator species like *Brassica juncea* (Indian mustard) and *Helianthus annuus* (sunflower) are frequently utilized for heavy metal extraction, while *Populus deltoides* (poplar) has been used to stabilize and restore disturbed soils. Recent advances in integrated soil management and biotechnology offer promising opportunities for enhancing phytoremediation efficiency. This approach not only detoxifies contaminated soils but also encourages carbon sequestration and soil restoration. This process improves microbial diversity, encourages the cycling of nutrients, improves soil structure, restores fertility, and lowers soil erosion. Though phytoremediation takes a long time to effectively decontaminate and improve the soil health, but with efficient integrated approach can provide long-term sustainable solution. Through the genetic engineering of plants, their tolerance and accumulation of contaminants present a promising pathway for efficient soil restoration. Furthermore, integrating phytoremediation with innovative green technologies like use of nanoparticles, biochar application, and soil microbial inoculation can greatly enhance its effectiveness. By combining these methods, we can develop a more holistic and

R. Kumar (✉)
Department of Environmental Science, College of Horticulture and Forestry, Rani Lakshmi Bai Central Agricultural University, Uttar Pradesh, Jhansi, India
e-mail: negirakesh632@gmail.com

LalitThakur
Department of Tree Improvement and Genetic Resource, Dr YS Parmar University of Horticulture and Forestry, Nauni, Solan, Himachal Pradesh, India

M. J. Dobriyal · S. K. Gautam · Y. Mallesh
Department of Silviculture and Agroforestry, College of Horticulture and Forestry, Rani Lakshmi Bai Central Agricultural University, Uttar Pradesh, Jhansi, India

T. Vignesh
Department of Forest Resource Management, College of Horticulture and Forestry, Rani Lakshmi Bai Central Agricultural University, Uttar Pradesh, Jhansi, India

A. Husen (ed.), *Enhanced Crop Production Setup Under Various Micro and Nano Contaminants*, Smart Nanomaterials Technology,
https://doi.org/10.1007/978-981-96-6099-5_2

expedited strategy for restoring contaminated soils, ultimately promoting both environmental sustainability and improvement of soil health. The future market phytoremediation is expected to grow at a compound annual growth rate (CAGR) of 5.1% from 2023 to 2030. The market was valued at USD 6.2 billion in 2023.

Keywords Soil health · Contamination · Phytoremediation · Genetic engineering · Nanoparticle · Carbon sequestration

1 Introduction

Exponential growth of human population in last few decades has led to a heightened demand for food, feed, water and air, wood, and other ecosystem services to support a comfortable lifestyle, resulting in severe pressure on natural resources and the environment [31, 89]. Accelerated industrialization, rapid urbanization, and intensive agricultural practices [46] are significantly contributing to environmental pollution [11, 24–26, 87] causing contaminants like heavy metals (HMs) and metalloids radionuclides [22], organic compounds [23, 53] and agro-chemicals (Malik et al. 2017, [76]), into the soil, all of which pose serious threats to soil health and overall ecosystem stability. Soil degradation advances in loss of soil fertility, soil structure destruction, and depletion of essential nutrients. Contamination of the soil makes agricultural soils unusable, reduces crop productivity, and threatens food security [48]. The polluted soils are a potential hazard to groundwater quality along with a host of global environmental issues such as biodiversity loss and disruption of the flow of ecosystem services [82].

Phyto-remediation refers to the use of plants for the in situ treatment of soil, sediment, and ground water, which implies the use of green plants and their associated microbial, soil amendments, and agronomic techniques to remove, contain or render harmless environmental contaminants [13]. Compound targeted for this includes heavy metals, chlorinated solvents, polycyclic aromatic hydrocarbons, polychlorinated biphenyls, pesticides munitions, and radio nucleoids. Plant species such as *Thlapsi caerulescens* have been shown to accumulate very high level of Zn and Cd from soils. *Brassica juncia* has also been found an excellent accumulator plants for metal in soils such as Cd, Cr, Mi, Zn, Cu. *Eichhornia crassipes, Hydrocotyle umbeellata Lemona minor,* and *Azolla pinnata* can accumulate Pb. Similarly *Lolium perenne, Festues rubra, Trifolium repens* and *Cerastium fantannum* and other grass spp. Little Agrostis, carex, aquatic macrophyte hydro-cotyle spp. also good for phytoremediation. Forest or stand act as a sink for wide variety of pollutants. Road side plantation in highways, railway track, public places, and industrial plantation (green fencing) are managed a pollution control/reduction strategy. Plantation around factories acts as in situ remediation for the air pollutants. In macro-level forests checked the global pollution by acting as natural bio-filters.

Phytoremediation addresses removal of contaminants from the soil and simultaneously restores soil health. This approach not only eliminates harmful pollutants but also improves the physical, chemical, and biological properties of soil, leading to a more sustainable and resilient ecosystem, compared to other non-biological or conventional soil remediation techniques [44] in a more economic way. It can also be operated in situ by treating the pollutants at their site of occurrence without the need for soil excavation. This methodological characteristic eases the logistical and financial burdens associated with soil remediation and mitigates environmental ecosystem impacts associated with remediation [61, 62]. Phytoremediation employs a variety of mechanisms to address pollutants, including degradation processes like rhizo-degradation and phytodegradation, accumulation techniques such as phytoextraction and rhizofiltration, as well as dissipation through phytovolatilization [29, 69]. It also involves immobilization strategies like hydraulic control and phytostabilization. The basic mechanism is to degrade, remove or immobilize contaminants, thereby mitigating environmental pollution [52]. Phytoremediation restores soil fertility by enhancing its composition, increasing organic matter, and promoting microbial activity that leads to sustainable crop production. It improves nutrient cycling, water purification, and landscape biodiversity by providing habitat for a variety of organisms. This technique not only restores natural landscapes, but also adds aesthetic and ecological value while providing long-term remedies to soil degradation. It harness natural ability of phytos to absorb toxic substances is an environmentally friendly method of ecosystem healing [17]. The future market phytoremediation is expected to grow at a compound annual growth rate (CAGR) of 5.1% from 2023 to 2030. The market was valued at USD 6.2 billion in 2023.

This chapter describes in details the phytoremediation techniques, their mechanisms, practical applications, and associated challenges. It also discusses how phytoremediation can significantly improve soil quality, promote soil health, and restore ecosystem functions, offering a sustainable approach to managing soil contamination. Figure 1 represents the translocation, sequestration, absorption, and exclusion mechanism of improving soil health status.

2 Mechanisms of Phytoremediation

Phytoremediation employs all annual or perennial plants to remediate contaminated soil and water, serving as an innovative method for reclaiming these resources by harnessing the natural capacity of plants to eliminate, stabilize, or detoxify environmental pollutants [19]. Plants chosen for phytoremediation must meet specific selection criteria, including high metal tolerance, short life cycles, wide distribution, fast growth and high biomass production, and a translocation factor (TF) exceeding 1 [40]. The phytoremediation mechanisms to remove contamination include phytoextraction, phytostabilization, phytovolatilization, rhizofiltration, phytodegradation, and nano-phytoremediation. These approaches for the restoration of contaminated areas have different degrees of applicability that would be beneficial if appropriately addressed to maximize efficacy in different contamination sites. The mechanism of phytoremediation that summarizes the procedure is shown in Fig. 2.

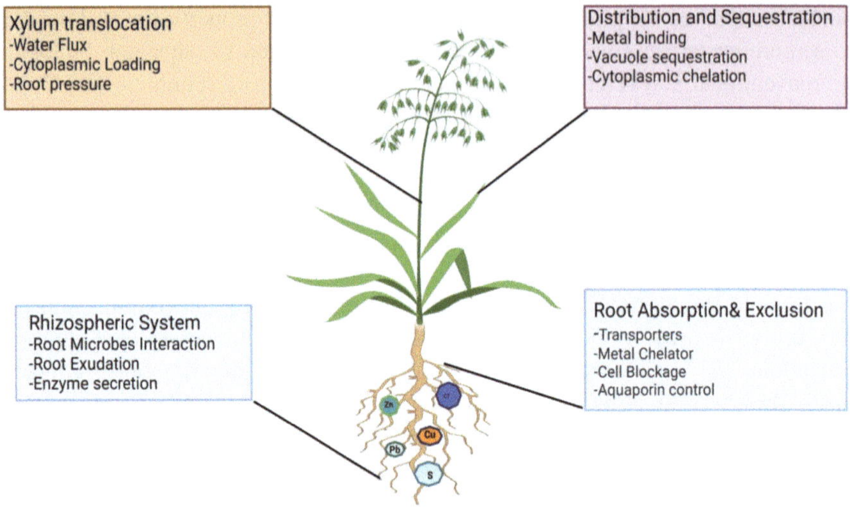

Fig. 1 Mechanisms of translocation, sequestration, absorption, and exclusion in improving soil health

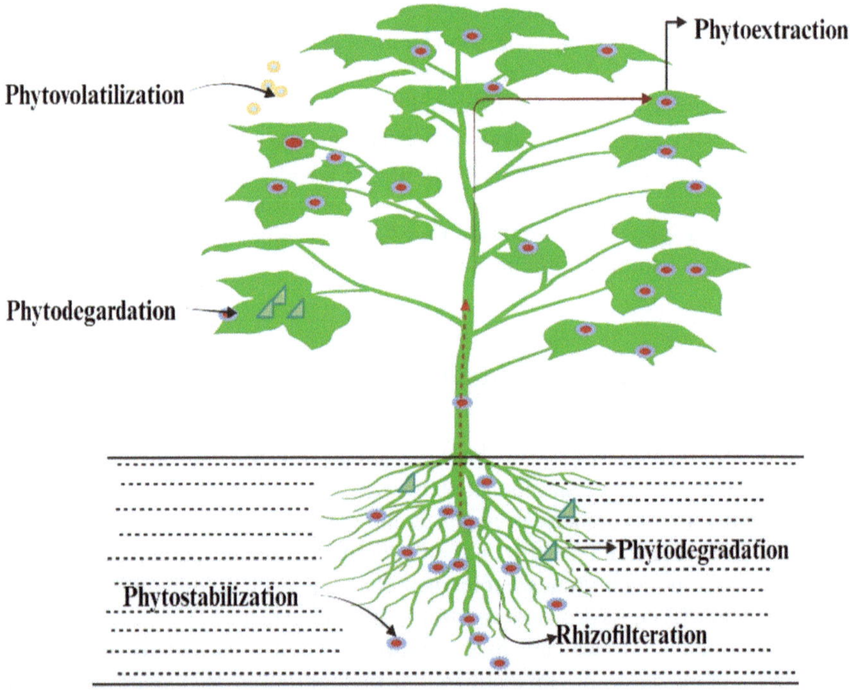

Fig. 2 Mechanism of phytoremediation

Phytoremediation approach includes multiple mechanisms through which plants interact with contaminants in soil [8, 60]. The recent omics technologies, such as genomics, transcriptomics, proteomics, and metabolomics, can also help to identify genes and pathways that can improve phytoremediation. Table 1 illustrated below provides suitable examples of the phytoremediation process.

2.1 Phytoextraction

The phytoextraction mechanism involves plants absorbing contaminants through their roots and transferring them to aerial parts, including shoots and leaves [86]. This approach is especially effective for removing heavy metals like lead (Pb), cadmium (Cd), and arsenic (As) from contaminated soil, concentrating them, and then precipitating them into the above-ground biomass. [74]. Hyperaccumulator species, such as Indian mustard (*Brassica juncea*) and sunflower (*Helianthus annuus*, are suitable for this process. Plants used for phytoextraction have traits such as tolerance to high metal concentrations in soil, high metal accumulation capacity, high biomass production, fast growth, and a profuse root system. The success of phytoextraction depends on the plant's capacity to rapidly accumulate biomass and store metals in shoot tissue [9, 47].

2.2 Phytostabilization

This mechanism involves the in activation or immobilization of toxicants within the rhizosphere of plants, where root systems limit the mobility and bioavailability of contaminants, thereby reducing their toxic effects [30]. Plants (mainly grasses) can form bound residues of toxic contaminants, rendering them unavailable in their poisonous forms or can make them unavailable to be released from the solid matrix after accumulation [39]. Heavy metals and certain organic pollutants can be converted into non-toxic compounds through conjugation with sugar, protein, and amino acid derivatives, or by forming complexes in the rhizosphere [21]. For example, glutathione secreted by the enzyme glutathione synthetase conjugates organic contaminants to detoxify their effects [28]. Furthermore, enzymes or organic compounds released from soil-root interactions, as well as the presence of pollutants, contribute to the stabilization of toxic chemicals [56, 75], while antioxidant enzymes like peroxidase play a crucial role in detoxifying the herbicide 2,4-D through polymerization [1]. Plants such as *Vetiveria zizanioides* (vetiver grass) and *Populus deltoides* (poplar) are commonly used for their strong root systems and high tolerance to contaminated soils [2]. Similarly, metal-chelating agents can be used in combination with plants to improve the plant's ability to accumulate metals, e.g., tetrasodium glutamate diacetate (GLDA) can be used in combination with *Tagetes patula* to improve the plant's ability to accumulate cadmium.

Table 1 Mechanisms of phytoremediation

Phytoremediation aspects	Plant part used	Mechanism	Processes	Example
Phytoextraction	Whole plant	Uptake and accumulation of contaminants by plant tissues	Plants absorb contaminants from soil and concentrate them in their tissues	Sunflower (*Helianthus annuus*): Extracts lead and cadmium
			Contaminants are later harvested from the plant biomass	Indian Mustard (*Brassica juncea*): Removes arsenic and lead
Phytostabilization	Root	Immobilization of contaminants in soil through plant roots, preventing their movement	Plant roots secrete compounds that stabilize contaminants in soil	Psyllium (*Plantago psyllium*): Stabilizes heavy metals
			Contaminants become less bioavailable and less likely to leach into groundwater	Willow Trees (*Salix spp.*): Stabilizes lead and cadmium
Phytodegradation	Whole plant	Breakdown of contaminants by plant enzymes or associated microorganisms	Plants generate enzymes that break down organic contaminants into less toxic compounds	Poplar trees (*Populus* spp.) fordegrading organic pollutants such as trichloroethylene (TCE)
			Microorganisms present in the rhizosphere play a crucial role in facilitating degradation processes	Alfalfa (*Medicago sativa*) for breaking down pesticides and herbicides
Phytovolatilization	Shoot	Contaminants can be released into the atmosphere via the process of plant transpiration	Plants absorb contaminants and transform them into gaseous forms, which are then released into the atmosphere	Tobacco (*Nicotiana tabacum*): Volatilizes selenium

(continued)

Table 1 (continued)

Phytoremediation aspects	Plant part used	Mechanism	Processes	Example
			Suitable for volatile contaminants, this process allows for their transformation into less toxic forms	Eucalyptus (*Eucalyptus spp.*): Volatilizes mercury
Nano-phytoremediation	Whole plant	Using nanoparticles enhance contaminant uptake through increased bioavailability, stimulating plant growth via phytohormone production, and facilitating microbial interactions	The process involves the application of nanoparticles to enhance plant uptake and degradation of environmental pollutants, improving soil remediation efficiency	Nano zerovalent iron (nZVI), magnetite nanoparticles (nFe_3O_4), and bimetallic nanoparticles (Pd/Fe)

2.3 *Phytodegradation*

In phytodegradation, plant parts play a vital role both directly and indirectly. In the direct process, plants absorb contaminants and metabolize them into smaller, less harmful compounds that are distributed throughout their tissues [77]. Plants lack active transporters; these organic contaminants are taken up through passive absorption. When degradation occurs in the rhizosphere, it is referred to as rhizodegradation and organic pollutants like pesticides can be remediated through the degradation or transformation facilitated by various plant parts [18]. Additionally, under indirect process, some microbes that live synergistically on plant root surfaces metabolize and degrade organic contaminants (Van Eerd et al. 2003). Furthermore, enzymatic degradation is performed by specific enzymes such as dehalogenase, nitro-reductase, laccase, and peroxidase, secreted by certain plants and associated microbial species that target hazardous xenobiotics [10]. Plants like *Populus* spp. (poplars) and *Salix* spp. (willows) thrive at phytodegradation owing to their ability to generate enzymes such as dehalogenase and peroxidase and catalyze that breakdown of complex organic molecules into simpler, less hazardous chemicals that enhance the transformation of pollutants [65].

2.4 Phytovolatilization

Phytovolatilization is the mechanism that involves the absorption of contaminants by plants from the soil, converting them into volatile forms, and releasing these transformed substances into the atmosphere through transpiration [78]. This method is especially effective for addressing hydrophilic pollutants as well as volatile hydrophobic pollutants like mercury (Hg) and selenium (Se),for instance, certain plants can convert selenium into dimethyl selenide, a volatile compound that is then emitted into the air. This transformation effectively lowers the concentration of selenium in the soil, thus reducing its potentially harmful effects on the environment [79].

2.5 Rhizofiltration

Rhizofiltration is a process that uses plant roots to draw out and concentrate heavy metals and other pollutants from water, making it especially effective for addressing contaminated groundwater. Plants are first cultivated in water, exposing their roots to the tainted liquid. To further remove contaminants from the groundwater, the plants are replanted into the contaminated area once their root systems have grown sufficiently [6]. The efficiency of rhizofiltration has been extensively studied in sunflower (*Helianthus annuus*) concerning its potential to eliminate uranium and other heavy metals from aquatic environments. Due to its large and strong root systems, it can absorb large amounts of contaminants, which helps clean up contaminated water and restores environmental integrity [16].

2.6 Rhizodegradation

Rhizodegradation is the biological elimination of contaminants through enhanced microbial activity in the rhizosphere of certain vascular plants, which is marked by increased microbial density and activity surrounding plant roots; this area serves a crucial part in the symbiotic relationships between plants and microorganisms, leading to insignificant microbial activity [4, 70]. Rhizoremediation, an important part of phytoremediation, can occur naturally or be induced by introducing specific pollutant-degrading bacteria or plant growth-promoting microorganisms, consequently enhancing contaminant degradation in the soil [20].

2.7 *Nano-Phytoremediation*

Nano-phytoremediation integrates nanotechnology and phytoremediation to enhance the remediation of environmental contaminants. Nanoparticles (NPs) effectively remediate soils contaminated with heavy metals and organic and inorganic pollutants due to their large surface area and reactivity, allowing them to penetrate contaminated sites more easily [71, 88]. The efficiency of these NPs is determined by their physical–chemical properties, such as composition, size, shape, and concentration, which vary from plant to plant [15]. Numerous studies have suggested that nanoparticles such as nano zerovalent iron (nZVI), magnetite nanoparticles (nFe_3O_4), and bimetallic nanoparticles (Pd/Fe) may quickly decompose a range of organic contaminants, including chlorpyrifos and trichloroethylene (TCE) pyrene, polychlorinated biphenyls (PCBs), 2,4-dinitrotoluene, and ibuprofen from polluted soil [57, 63, 72]. Nano-phytoremediation techniques allow plant species to remove toxins from the soil faster than traditional methods. Evidence suggests that nanoparticles not only encourage plant growth and development but also enhance their ability to absorb pollutants, e.g., silver nanoparticles can significantly increase phytohormone production such as ABA and GA, improving nutrient uptake and stress resilience [32], whereas salicylic acid nanoparticles can improve phytoremediation efficiency under arsenic stress [73].

3 Improving Soil Status Through Phytoremediation

Soil contamination arises from various activities, including industrial processes, agricultural practices, and improper waste disposal, leading to severe issues such as degraded soil quality, plant growth, and water quality (Raskin et al. 1997). Unlike conventional remediation methods that often involve costly excavation and chemical treatments that disrupt soil ecosystems [42], phytoremediation offers a less invasive and more environmentally acceptable alternative [13]. Phytoremediation targets both the chemical and biological aspects of remediation using plant biology as the basis of the technique for soil decontamination in an economically and ecologically sustainable manner [2]. The phytoremediation techniques restoring soil status is mentioned in Table 2.

The plants used in phytoremediation can provide some additional benefits, such as erosion control, habitat enhancement, and aesthetic improvements [90]. Regardless of its advantages, phytoremediation is constrained by certain limitations, including its gradual pace and the fact that its effectiveness is contingent upon variables like the type of soil, prevailing climatic conditions, and the concentration of the pollutant [51]. Research on plant biotechnology, nanoparticles, and soil science has made phytoremediation more efficient. Genetic engineering also holds promise for efficient soil restoration and proper treatment [67]. Phytoremediation represents a hopeful way to

Table 2 Phytoremediation technique improving soil status

Phytoremediation technique	Mechanism	Impact on soil status	Suitable plants	References
Phytoextraction	Uptake and concentration of contaminants in plant tissues	Reduces soil contamination, especially heavy metals	Indian mustard, sunflower	[2]
Phytostabilization	Immobilization of contaminants in the soil by plant roots	Prevents erosion and leaching of pollutants, enhancing soil structure	Willow trees, poplar trees	[2, 54]
Phytodegradation	Breakdown of contaminants by plant enzymes and associated microorganisms	Improves soil microbial health and breaks down organic pollutants	Grasses, *Brassica juncea*, Poplar trees, alfalfa	[50, 61, 62]
Phytovolatilization	Conversion of contaminants to less toxic volatile forms through transpiration	Reduces contaminant concentrations in soil while preventing groundwater contamination	Tobacco, Eucalyptus	[51]
Rhizofiltration	Absorption of contaminants from water or soil by plant roots	Improves water quality in the soil and reduces contaminant concentrations	Indian mustard, sunflower	[2, 60]
Rhizodegradation	Breakdown of organic contaminants in the rhizosphere by microbial action	Enhances soil nutrient cycling, microbial diversity, and pollutant breakdown	Grass species (e.g., ryegrass)	[14]
Hyperaccumulation	Plants store high concentrations of contaminants without toxicity	Improves soil health by removing toxic metals and making the soil fertile again	*Brassica juncea*, Thlaspi spp.	[2, 58]
Integration with Biochar	The use of biochar with phytoremediation enhances contaminant sorption and nutrient availability	Restores soil fertility and aids in contaminant immobilization	Multiple species, depending on the region	[7, 38]

enhance soil status and solve the issue of soil contamination. It also shows the potential of sustainable, low-cost processes restoring soil health along with the enhancement of ecosystem services and reduction in the hazardous impact of environmental contamination, utilizing the characteristic potential of plants [80].

3.1 Soil Fertility Restoration

Soils contaminated by heavy metals and organic pollutants often show weaker structures, disrupt biological activities and decreases soil fertility. These harmful substances were eliminated using phytoremediation, which helped to restore essential microbial activity and soil health. Indian mustard (*Brassica juncea*) has shown promising results in restoring important nutrient levels including phosphate and nitrogen when used for phytoremediation in cadmium-contaminated soils [43].

3.2 Soil Structure and Erosion Control

Plants used in phytoremediation, notably phytostabilization, play a pivotal role in strengthening soil structure. These plants assist enormously in soil quality restoration by reducing erosion and increasing soil aggregation [61, 62]. Short rotation species, such as poplars, willows, and Thespesia, anchor deep-rooted in the soil; reducing compaction and enhancing water infiltration resulting in reduced soil erosion. This process is essential for maintaining soil health, especially in places affected by high contamination or erosion. Their root systems provide stable conditions for soil, promoting sustainable land management practices [54].

3.3 Enhancing Soil Biological Activity

A healthy soil microbial population is essential for breaking down organic matter and cycling nutrients to maintain soil fertility. However, pollutants disrupt these microbial activities, leading to a decline in soil health. Phytoremediation offers a viable solution by reducing soil toxicity and promoting the recovery of microbial populations. For example, a study showed that the use of ryegrass (*Lolium perenne*) in the phytoremediation of a diesel-contaminated site has been shown to significantly increase microbial diversity and enzyme activity. This enhancement of microbial health and functionality underscores the potential of phytoremediation to restore damaged soils and sustain thriving soil ecosystems [45].

3.4 Carbon Sequestration

Carbon sequestration is an additional advantage of several phytoremediation species, which contributes significantly to achieving the dual objectives i.e. Soil rejuvenation and climate change mitigation. Short-rotation trees like poplars and willows effectively sequester carbon dioxide from the atmosphere and produce large amounts of biomass, through the process of photosynthesis (Rathod et al. 2024). They store this carbon in their tissues, which eventually makes its way back into the soil through root biomass and leaf litter. These trees help to restore contaminated soils reduce greenhouse gas emissions, and sequence contaminants and carbon in the atmosphere thereby storing it as above-ground biomass and below-ground biomass [36, 37].

4 Advantages of Phytoremediation

Phytoremediation is a sustainable and innovative approach that involves the deliberate use of specific plant species to address contamination in soil and water ecosystems. It has emerged as an effective and eco-friendly solution for heavy metal remediation because of its minimal environmental disruption, cost efficiency (being 60–80% less expensive), and high effectiveness [3, 49, 59]. This green technology is particularly suited for in situ treatment of contaminated soils and water, offering a sustainable method for removing toxic substances [81]. Notably, certain oil crops can be grown on metal-contaminated sites, providing the dual benefit of remediation and bioenergy production, such as biodiesel [83]. Additionally, the plants involved in phytoremediation contribute to environmental improvement by producing oxygen and sequestering pollutants from the environment. The potential advantages of phytoremediation are illustrated in Fig. 3.

5 Challenges and Limitations of Phytoremediation

Phytoremediation, though a promising and environmentally sustainable method for the decontamination of soil and water, encounters numerous challenges and limitations that hinder its broader implementation. A notable obstacle is the efficacy of phytoremediation in addressing contaminants present in elevated concentrations. Many plants have limited capacities for contaminant uptake and detoxification, making them less effective in highly polluted environments. For example, while hyperaccumulator plants can remove certain heavy metals, their ability to address a broad spectrum of pollutants or very high concentrations is constrained [60]. Another limitation is the slow growth rate of some phytoremediation species, which can prolong the time required to achieve significant decontamination results.

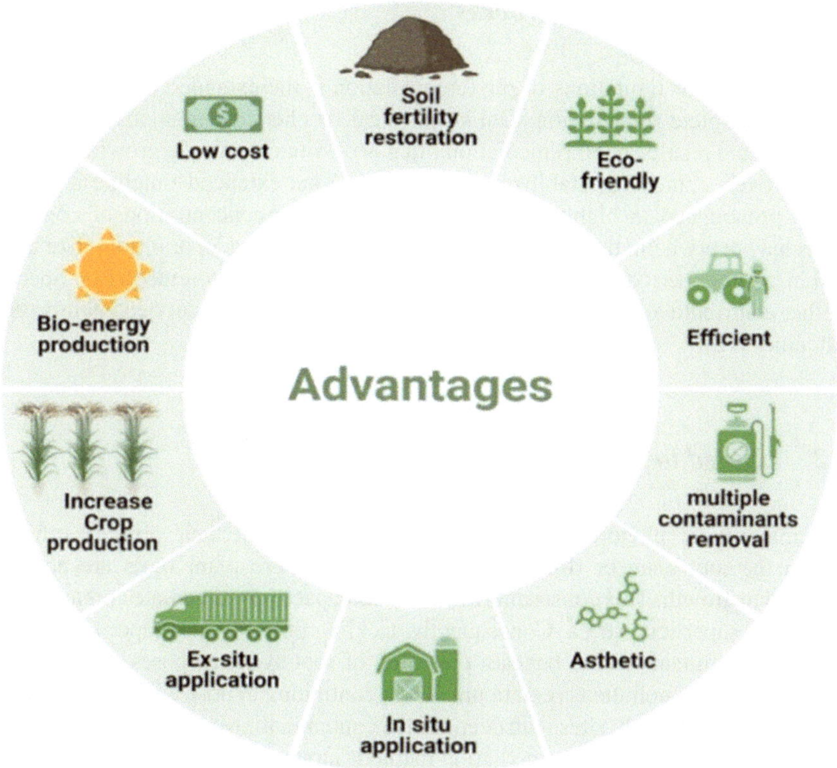

Fig. 3 Potential advantages of phytoremediation

This slow process can be a disadvantage when dealing with urgent contamination issues. Additionally, the success of phytoremediation is highly dependent on soil and climatic conditions, which can vary significantly, potentially affecting the uniformity of remediation outcomes [2]. Phytoremediation also faces challenges related to the disposal of harvested plant biomass. The removal of contaminants through plant tissues necessitates proper management of the contaminated biomass to prevent secondary pollution. Moreover, the long-term effectiveness and sustainability of phytoremediation require ongoing monitoring to ensure that contaminants are not being released back into the environment or becoming more concentrated in plant tissues [64]. Despite these challenges, phytoremediation remains a valuable tool for environmental cleanup. Continued research and development are essential to enhance its effectiveness, address limitations, and expand its applicability to various contamination scenarios [84].

5.1 Time-Consuming Process

One of the major limitations of phytoremediation is the extended time required to achieve complete remediation. Unlike physical or chemical methods, which can provide rapid results, phytoremediation often necessitates multiple growing seasons to effectively remove or stabilize contaminants. This extended timeline is particularly pronounced in highly polluted soils, where the concentration of contaminants may overwhelm the capacity of plants to absorb, degrade, or immobilize them within a short period. The slow growth rates of some phytoremediation species further contribute to the prolonged duration needed for significant decontamination outcomes [12].

5.2 Limited to Shallow Soils

The majority of phytoremediation methodologies exhibit their primary efficacy within the root zone or rhizosphere—the region where plant roots are actively engaged in growth. This constraint restricts their capacity to remediate contaminated soil to the superficial strata. Consequently, tackling pollution in deeper soil layers, where contaminants reside beneath the reach of root systems, poses a formidable challenge. These pollutants remain untreated, continuing to pose environmental and health risks. As a result, sites with deep-seated contamination cannot be fully remediated using traditional phytoremediation methods alone. This constraint necessitates exploring complementary or alternative technologies to target contaminants in deeper soil layers and achieve comprehensive site decontamination [54].

5.3 Contaminant-Specific Plants

Different plants need to treat different types of pollution, which limits the scope of phytoremediation, especially in areas with high levels of pollution. Some plants, such as hyperaccumulators, are effective in removing heavy metals but may not be suitable for degrading organic or other pollutants. This limitation requires a complex phytoremediation strategy that involves a combination of plants to deal with the various diseases present in the infected areas. Therefore, creating an effective phytoremediation program often requires careful selection and integration of different plant species to treat different diseases [55].

5.4 *Risk in the Food Chain*

Rapid industrialization, vehicle emissions, agricultural practices, and improper waste disposal are key contributors to soil contamination leading to reduced agricultural productivity and food quality. These pollutants infiltrate the food chain, posing significant risks to human health [33]. In the phytovolatilization process, plants uptake mercury (Hg) from the soil through their roots, transport it via their vascular system to above-ground parts, and release it into the atmosphere after converting it into a volatile form, with methylated Hg presenting a severe threat due to its tendency to biomagnified within the food chain (Sharma et al. 2015).

Other challenges are related to preparation of universal regulatory frameworks dovetailed with UNFCC protocols and public acceptance for the widespread adoption of phytoremediation. Developing guidelines and standards for phytoremediation and public education initiatives can foster a more favorable condition for the technology promotion, dissemination, and implementation.

6 Conclusion and Future Prospects

Phytoremediation is a promising ecologically sustainable approach for the soil restoration and the removal of contamination from the soil. This process can eliminate a wide range of contaminants, including heavy metals, organic chemicals, radionuclides, pesticides, petroleum residues, antibiotics, etc. However, the efficacy of plants in remediation varies depending on the contaminants they target, edapho-climatic condition and cost effectiveness coupled with practicality of operation. The selection of appropriate plant species based on the contaminants present and the environmental conditions at the site is critical factors in decision-making. Use of shrubs and perennials not explored compare to annuals for long-term studies in highly contaminated areas. However, significant challenges remain, including the extended timeframes required for effective remediation and the site-specific limitations approach. Despite these constraints, ongoing advancements in genetic engineering and plant biotechnology provide considerable opportunities to enhance the efficiency to broaden the applicability of phytoremediation techniques. Future innovations could facilitate the development of multi-functional plant species capable of addressing a wider range of contaminants, thereby accelerating improvements in soil health and remediation outcomes. Moreover, the integration of phytoremediation with other cutting-edge green technologies, such as the use of nanoparticles, biochar application, and soil microbial inoculation, holds great potential to significantly augment its effectiveness. Biochar, with its ability to improve soil structure and enhance the retention of nutrients and contaminants, complements the phytoremediation process by creating more favorable conditions for pollutant uptake. Concurrently, microbial inoculation can enhance soil microbial activity, thereby facilitating the degradation of organic pollutants. When combined, these integrated approaches could offer a more comprehensive, accelerated strategy for the restoration of contaminated soils. Artificial intelligence (AI) and remote sensing technologies are being used to improve the monitoring and management of phytoremediation projects.

Competing Interests The authors declare that they have no competing interest.

Declaration of Generative AI and AI-Assisted Technologies in the Writing Process During the preparation of this work, the book chapter author(s) have not taken any assistance from AI, and the book chapter author(s) reviewed and edited the content as needed and take(s) full responsibility for the content of the publication.

References

1. Agostini E, Coniglio MS, Milrad SR, Tigier HA, Giulietti AM (2003) Phytoremediation of 2, 4-dichlorophenol by Brassica napus hairy root cultures. Biotechnol Appl Biochem 37:139–144
2. Ali H, Khan E, Sajad MA (2013) Phytoremediation of heavy metals—Concepts and applications. Chemosphere 91(7):869–881
3. Al-Thani RF, Yasseen BT (2020) Phytoremediation of polluted soils and waters by native Qatari plants: future perspectives. Environ Pollut 259:113694
4. Anderson TA, Guthrie EA, Walton BT (1993) Bioremediation in the rhizosphere. Environ Sci Technol 27:2630–2636
5. Awa SH, Hadibarata T (2020) Removal of heavy metals in contaminated soil by phytoremediation mechanism: a review. Water Air Soil Pollut 231(2).
6. Bakshe P, Jugade R (2023) Phytostabilization and rhizofiltration of toxic heavy metals by heavy metal accumulator plants for sustainable management of contaminated industrial sites: a comprehensive review. J Hazard Mater Adv 10:100293
7. Beesley L, Moreno-Jiménez E, Gomez-Eyles JL (2011) A review of biochars' potential role in the remediation, revegetation, and restoration of contaminated soils. Environ Pollut 159 (12):3269–3282
8. Blaylock MJ, Huang JW (2000a) Phytoremediation of metals. In: Phytoremediation: methods and reviews. CRC Press, pp 53–70
9. Blaylock MJ, Huang JW (2000b) Phytoextraction of metals. In: Raskin I, Ensley BD (eds) Phytoremediation of toxic metals: using plants to clean up the environment. Wiley, New York, NY, pp 53–70
10. Cherian S, Oliveira MM (2005) Transgenic plants in phytoremediation: recent advances and new possibilities. Environ Sci Technol 39:9377–9390
11. Cherniwchan J (2012) Economic growth, industrialization, and the environment. Res Energy Econ 34:442–467
12. Cunningham SD, Ow DW (1996) Promises and prospects of phytoremediation. Plant Physiol 110(3):715–719
13. Cunningham SD, Berti WR, Huang JW (1996) Phytoremediation of contaminated soils. Trends Biotechnol 14(12):393–397
14. Cunningham SD, Berti WR (2000) Phytoextraction and Phytostabilization: technical, economic, and regulatory considerations of soil-lead issue. In: Terry N, Banuelos GS (eds) Phytoremediation of contaminated soil and water. CRC Press, Boca Raton, pp 359–376
15. Davis AS, Prakash P, Thamaraiselvi K (2017) Nanobioremediation technologies for sustainable environment. In: Prashanthi M (ed) Bioremediation and sustainable technologies for cleaner environment. Environmental Science. Springer International Publishing, Cham, AG, pp 13–33
16. Dushenkov S, Kumar PBAN, Motto H, Raskin I (1995) Rhizofiltration: the use of plants to remove heavy metals from aqueous streams. Environ Sci Technol 29(5):1239–1245

17. Fiorentino N, Mori M, Cenvinzo V, Duri L, Gioia L, Visconti D, Fagnano M (2018) Assisted phytoremediation for restoring soil fertility in contaminated and degraded land. Ital J Agron 13:34–44
18. Garrison AW, Nzengung VA, Avants JK, Ellington JJ, Jones WJ, Rennels D, Wolfe NL (2000) Phytodegradation of p, p '-DDT and the Enantiomers of o, p '-DDT. Environ Sci Technol 34:1663–1670
19. Gavrilescu M (2022) Enhancing phytoremediation of soils polluted with heavy metals. Curr Opin Biotechnol 74(21–31)
20. Gerhardt KE, Huang XD, Glick BR, Greenberg BM (2009) Phytoremediation and rhizoremediation of organic soil contaminants: potential and challenges. Plant Sci 176:20–30
21. Hammond CM, Root RA, Maier RM, Chorover J (2018) Mechanisms of arsenic sequestration by Prosopisjuliflora during the phytostabilization of metalliferous mine tailings. Environ Sci Technol 52:1156–1164
22. He Z, Shentu J, Yang X, Baligar VC, Zhang T, Stoffella PJ (2015) Heavy metal contamination of soils: sources, indicators and assessment. J Environ Indicat 9:17–18
23. Hu QH, Weng JQ, WangmJS (2010) Sources of anthropogenic radionuclides in the environment: a review. J Environ Radioact 101:426–437
24. Husen A (2022) Environmental pollution and medicinal plants. CRC Press, Boca Raton, FL. https://doi.org/10.1201/9781003178866
25. Husen A (2023) Plants and their interaction to environmental pollution (Damage detection, adaptation, tolerance, physiological and molecular responses). Elsevier, Cambridge, MA, USA. https://doi.org/10.1016/C2020-0-03319-0
26. Husen A, Iqbal M (2023) Medicinal Plants - Their Response to Abiotic Stress. Springer Nature Singapore, Singapore. https://doi.org/10.1007/978-981-19-5611-9
27. Hussain S, Shah AA, Rehman N (2021) Phytoremediation potential of plants for the removal of heavy metals from contaminated soils. Chemosphere 276:130090
28. Hussain S, Siddique T, Arshad M, Saleem M (2009) Bioremediation and phytoremediation of pesticides: recent advances. Crit Rev Environ Sci Technol 39:843–907
29. Iqbal M, Ahmad A, Ansari MKA, Qureshi MI, Aref IM, Khan PR, Hegazy SS, El-Atta H, Husen A, Hakeem KR (2015) Improving the phytoextraction capacity of plants to scavenge metal(loid)-contaminated sites. Environ Rev 23:44–65. https://doi.org/10.1139/er-2014-0043
30. ITRC (Interstate Technology & Regulatory Council) (2009) Phytotechnology technical and regulatory guidance and decision trees, revised. PHYTO-3. Intestate Technology & Regulatory Council, Phytotechnologies Team, Washington, DC
31. Kamran M, Danish M, Saleem M.H, Malik Z, Parveen A, Abbasi GH, Jamil M, Ali S, Afzal S, Riaz M (2021) Application of abscisic acid and 6-benzylaminopurine modulated morpho-physiological and antioxidativedefense responses of tomato (*Solanumlycopersicum* L.) by minimizing cobalt uptake. Chemosphere 263:128169
32. Khan N, Bano A (2016) Modulation of phytoremediation and plant growth by the treatment with PGPR, Ag nanoparticle and untreated municipal wastewater. Int J Phytoremediation 18(12):1258–1269
33. Kumar B, Smita K, Flores LC (2017) Plant mediated detoxification of mercury and lead. Arabian J Chem 10:S2335–S2342
34. Kumar PBAN, Dushenkov V, Motto H, Raskin I (1995) Phytoextraction: The use of plants to remove heavy metals from soils. Environ Sci Technol 29(5):1232–1238
35. Kumar PBAN, Kumar S, Shekhawat MS (1995) Heavy metal phytoremediation: a review. Environ Sci Pollut Res 1(3):133–143
36. Laureysens I, Blust R, De Temmerman L, Lemmens Cand Ceulemans R (2004) Clonal variation in heavy metal accumulation and biomass production in a poplar coppice culture: I. Seasonal variation in leaf, wood and bark concentrations. Environ Pollut 131:485–494
37. Laureysens I, Steppe K, Lemeur R, Samson R (2004) Growth and carbon sequestration of poplar and willow species in short rotation coppice systems. Biomass Bioenerg 26(1):23–32
38. Lehmann J, Rillig MC, Thies J, Masiello CA, Hockaday WC, Crowley D (2011) Biochar effects on soil biota–a review. Soil Biol Biochem 43:1812–1836

39. Li H, Sheng G, Sheng W, Xu O (2002) Uptake of trifluralin and lindane from water by ryegrass. Chemosphere 48:335–341
40. Mazumdar K, Das S (2015) Phytoremediation of Pb, Zn, Fe, and Mg with 25 wetland plant species from a paper mill contaminated site in North East India. Environ Sci pollut Res 22:701–710
41. McCutcheon SC, Schnoor JL (2003) Phytoremediation: Transformation and control of contaminants. Wiley
42. McGrath SP, Zhao FJ (2003) Phytoextraction of metals and metalloids from contaminated soils. Curr Opin Biotechnol 14(3):277–282
43. McGrath SP, Zhao FJ, Lombi E (2001) Plant and rhizosphere processes involved in phytoremediation of metal-contaminated soils. Plant Soil 232(1–2):207–214
44. Mejare M, Bulow L (2001) Metal binding proteins and peptides in bioremediation and phytoremediation of heavy metals. Trends Biotechnol 19(2):67–75
45. Merkl N, Jolliffe R, McClure C (2005) Effectiveness of ryegrass in removing diesel fuel contaminants from soil. J Environ Qual 34(5):1864–1871
46. Mozner Z, Tabi A, Csutora M (2012) Modifying the yield factor based on more efficient use of fertilizer—the environmental impacts of intensive and extensive agricultural practices. Ecol Indic 16:58–66
47. Mukhopadhyay S, Maiti SK (2010) Phytoremediation of metal enriched mine waste: a review. Glo J Environ Res 4(135–150):29
48. Oldeman LR, Hakkeling RTA, Sombroek WG (1991) World map of the status of human-induced soil degradation: an explanatory note. International Soil Reference and Information Centre
49. Olguín EJ, Sánchez-Galván G (2012) Heavy metal removal in phytofiltration and phytoremediation: the need to differentiate between biosorption and bioaccumulation. New Biotech 30(1):3–8
50. Pilon-Smits E (2005) Phytoremediation. Annu Rev Plant Biol 56:15–39
51. Pilon-Smits EA (2005) Phytoremediation. Environ Sci Technol 39(9):318A-322A
52. Pivetz BE (2001) Phytoremediation of contaminated soil and ground water at hazardous waste sites. US Environmental Protection Agency
53. Prakash D, Gabani P, Chandel AK, Ronen Z, Singh OV (2013) Bioremediation: a genuine technology to remediate radionuclides from the environment. Microb Biotechnol 6:349–360
54. Pulford ID, Watson C (2003) Phytoremediation of heavy metal-contaminated land by trees—a review. Environ Int 29(4):529–540
55. Raskin I, Ensley BD (eds) (2000) Phytoremediation of toxic metals: using plants to clean up the environment. Wiley
56. Reboreda R, Cacador I (2008) Enzymatic activity in the rhizosphere of Spartina maritima: potential contribution for phytoremediation of metals. Mar Environ Res 65:77–84
57. Reddy A, Madhavi V, Reddy KG, Madhavi G (2012) Remediation of chlorpyrifoscontaminated soils by laboratory-synthesized zero-valent nano iron particles: effect of pH and aluminium salts. J Chem 2013:7. https://doi.org/10.1155/2013/521045
58. Reeves RD, Baker AJM (2000) Metal-accumulating plants. In: Phytoremediation of toxic metals: using plants to clean up the environment, pp 193–229
59. Salido AL, Hasty KL, Lim JM, Butcher DJ (2003) Phytoremediation of arsenic and lead in contaminated soil using Chinese Brake fernsm(Pterisvittata) and Indian mustard (Brassica juncea). Intern J Phytorem 5(2):89–103
60. Salt DE, Blaylock M, Willis D (1995) Phytoremediation: a novel strategy for the removal of toxic metals from the environment using plants. Biotechnology 13:468–474
61. Salt DE, Smith RD, Raskin I (1998) Phytoremediation. Ann Rev Plant Physiol Plant Molecul Biol 49:643–668
62. Salt DE, Smith RD, Raskin I (1998) Phytoremediation. Annu Rev Plant Physiol Plant Mol Biol 49:643–668
63. Satapanajaru T, Anurakpongsatorn P, Pengthamkeerati P, Boparai H (2008) Remediation of atrazine-contaminated soil and water by nanozerovalent iron. Water Air Soil Pollut 192:349–359

64. Schnoor JL, Licht LA, McCutcheon SC, Croghan CW, Crites RW (1995) Phytoremediation of organic and nutrient contaminants. Environ Sci Technol 29(7):318A-323A
65. Schnoor JL, Licht LA, McCutcheon SC, Shea PJ, Walker JD (1995) Phytoremediation of soils, groundwater, and surface water. Crit Rev Environ Sci Technol 25(3):245–274
66. Sharma J, Kumar N, Singh N, Santal A (2023) Phytoremediation technologies and their mechanism for removal of heavy metal from contaminated soil: an approach for a sustainable environment. Front Plant Sci 14:1076876
67. Sharma P, Dubey RS, Bhatla SC (2016) Phytoremediation: role of plant growth regulators in heavy metal detoxification. Plant Growth Regul 79(1):5–13
68. Sharma P, Pandey AK, Udayan A, Kumar S (2021) Role of microbial community and metal-binding proteins in phytoremediation of heavy metals from industrial wastewater. Bioresource Technol 326:124750
69. Shmaefsky BR, Husen A (2023) Phytoremediation Potential of Medicinal Plants, *In:* Husen A, Iqbal M (Eds) Medicinal Plants - Their Response to Abiotic Stress. Springer Nature Singapore Pte Ltd. 152 Beach Road, #21–01/04 Gateway East, Singapore, pp. 443–469 https://doi.org/10.1007/978-981-19-5611-9_17
70. Siciliano SD, Germida JJ (1998) Bacterial inoculants of forage grasses enhance degradation of 2-chlorobenzoic acid in soil. Environ ToxicolChem 16:1098–1104
71. Singh BK (2009) Organophosphorus-degrading bacteria: ecology and industrial applications. Nat Rev Microbiol 7(2):156–164
72. Singh R, Misra V, Mudiam MKR, Chauhan LKS, Singh RP (2012) Degradation of γ-HCH spiked soil using stabilized Pd/Fe0 bimetallic nanoparticles: pathways, kinetics and effect of reaction conditions. J Hazard Mater 237–238:355–364
73. Souri Z, Karimi N, Sarmadi M, Rostami E (2017) Salicylic acid nanoparticle (SANPs) improves growth and phytoremediation efficiency of *IsatiscappadocicaDesv.* under arsenic stress. IET Nanobiotechnol 11(6):650–655
74. Subasic M, Samec D, Selovic A, Karalija E (2022) Phytoremediation of cadmium polluted soils: current status and approaches for enhancing. Soil Systems 6(1):3
75. Sun TR, Cang L, Wang QY, Zhou DM, Cheng JM, Xu H (2010) Roles of abiotic losses, microbes, plant roots, and root exudates on phytoremediation of PAHs in a barren soil. J Hazard Mater 176:919–925
76. Sunitha V, Reddy BM, Reddy MR (2012) Groundwater contamination from agro-chemicals in irrigated environment: field trials. Adv Appl Sci Res 3:3382–3386
77. Susarla S, Medina VF, McCutcheon SC (2002) Phytoremediation: an ecological solution to organic chemical contamination. Ecol Eng 18:647–658
78. Terry N, Zayed A, Pilon-Smits E, Hansen D (1995) Can plants solve the selenium problem? In: Proc. 14th Annual symposium focused on current topics in plant biochemistry, physiology, and molecular biology Will Plants have a role in Bioremediation? University of Missouri, Columbia, April 19–22, pp 63–64
79. Terry N, Zayed AM, De Souza MP, Tarun AS (2000) Selenium in higher plants. Annu Rev Plant Biol 51(1):401–432
80. Turrini A, Ali H, Palma S (2016) Advances in phytoremediation research. J Environ Manage 183:90–102
81. U.S. Environmental Protection Agency (USEPA) (1989) Risk assessment guidance for Superfund, Volume I: human health evaluation manual. Washington, DC, USA, Office of Emergency and Remedial Response, p 1989
82. UNEP (2016). Global environment outlook: regional assessments. United Nations Environment Programme
83. Van Ginneken L, Meers E, Guisson R (2007) Phytoremediation for heavy metal-contaminated soils combined with bioenergy production. J Enviro Engg Landscape Manag 15(4):227–236
84. Vangronsveld J, HerzigR WN, Boulet J, Adriaensen K, Ruttens A, Thewys T, Vassilev A, Meers E, Nehnevajova E, Vander Lelie D, Mench M (2009) Phytoremediation of contaminated soils and groundwater: lessons from the field. Environ Sci Pollut Res 16:765–794

85. Vijay Kumar R, Lallawmkimi MC, Gautam SK, Tiwari P, Aneesh KS, Sekhar M, Upadhyay L (2024) Agroforestry practices as a keystone for biodiversity conservation: a review
86. Wang L, Wang N, Di G, Shang Z, Zhang Y, Liu S, Wang Y (2023) Rhizobacteria helps to explain the enhanced efficiency of phytoextraction strengthened by *Streptomyces pactum*. J Environ Sci. 125:73–81
87. Wu Q, Zhou H, Tam NF, Tian Y, Tan Y, Zhou S, Li Q, Chen Y, Leung JY (2016) Contamination, toxicity and speciation of heavy metals in an industrialized urban river: implications for the dispersal of heavy metals. Mar Pollut Bull 104:153–161
88. Yadav KK, Singh JK, Gupta N, Kumar V (2017) A review of nanobioremediation technologies for environmental cleanup: a novel biological approach. J Mater Environ Sci 8(2):740–757
89. Zaheer IE, Ali S, Saleem MH, Imran M, Alnusairi GS, Alharbi BM, Riaz M, Abbas Z, Rizwan M, Soliman MH (2020) Role of iron–lysine on morpho-physiological traits and combating chromium toxicity in rapeseed (*Brassica napus* L.) plants irrigated with different levels of tannery wastewater. Plant Physiol Biochem 155:70–84
90. Zhu Y, Pilon M, Wang L (1999) Phytoremediation of metal-contaminated soils. Environ Sci Technol 33(22):3850–3858

Cadmium: Uptake, Toxicity, and Tolerance in Crop Plants and Management of Cd-Contaminated Soil

Sharfa Naaz, Ragib Husain, Vinay Kumar Singh, Swati Sachdev, Pransu Srivastava, and Mohammad Israil Ansari

Abstract The contamination of agricultural systems with cadmium (Cd) is an emerging threat that has adverse effects on plant health, food security, and environmental sustainability. Cadmium (Cd) is a non-essential toxic heavy metal that has no known biological function; nevertheless, it occurs in agricultural soils and represents a serious risk for plant health, food safety, and environmental quality. The current manuscript has been prepared in a way to provide thorough knowledge of the accumulation of Cd in plants, its tolerance mechanisms and strategies that can help in cleaning-up contaminated soil. The chapter reviews Cd absorption, translocation, and toxicity mechanisms in crops; deep insights into plant strategies to overcome the negative effect of Cd stress are also provided which includes shuttling of toxic metal to vacuoles, production of peptides having ability to bind with heavy metals such as phytochelatins and metallothioneins and activation of antioxidant defense systems against excess reactive oxygen species (ROS). It also discusses both classical and modern remediation methodologies, including soil amendments, phytoremediation along with the utilization of low-Cd fertilizers favorable for effective soil upkeep. Among the available options, phytoremediation is an eco-friendly and highly socially acceptable technology that utilizes plants to remediate Cd-contaminated soils through

S. Naaz · R. Husain
Plant Ecology and Climate Change Science Division, CSIR—National Botanical Research Institute, Lucknow, Uttar Pradesh, India

S. Naaz · M. I. Ansari (✉)
Department of Botany, University of Lucknow, Lucknow, Uttar Pradesh, India
e-mail: ansari_mi@lkouniv.ac.in

R. Husain · V. K. Singh
Department of Botany, Bareilly College, Bareilly, Uttar Pradesh, India

S. Sachdev
Department of Liberal Education, Era University, Lucknow, Uttar Pradesh, India

P. Srivastava
Department of Molecular Medicine and Biotechnology, Sanjay Gandhi Post Graduate Institute of Medical Sciences, Lucknow, Uttar Pradesh, India

A. Husen (ed.), *Enhanced Crop Production Setup Under Various Micro and Nano Contaminants*, Smart Nanomaterials Technology,
https://doi.org/10.1007/978-981-96-6099-5_3

a combination of processes such as phytovolatilization, phytoextraction, phytostabilization, rhizofiltration, and phytostimulation. All techniques that could be applied for remediation are also evaluated in terms of its effectiveness, cost, and environmental impact.

Keywords Antioxidants · Oxidative stress · Phytoremediation · Phytotoxicity · Metal uptake · And translocation

1 Introduction

Cadmium (Cd), lead (Pb), arsenic (As), and mercury (Hg) are among the most dangerous toxic metals that pose threats to human as well as plant life [94]. Although naturally present in the soil, the concentration of these metals is naturally low, but the increasing levels due to more and more anthropogenic activities have reached alarming limits like from phosphate fertilizers, industrial emission, and wastewater. This has become a world-wide environmental issue. Among these metals, cadmium is especially dangerous because of its high toxicity, environmental persistence, and potential for bio-accumulating [91]. While an important micronutrient in plant physiology, cadmium has no useful function, and low concentrations can influence plant health and agricultural production [49, 113].

Research has shown adverse effects of Cd on seed germination, plant growth reduction, root elongation inhibition, and leaf number suppression with severe cases leading to plant death [11, 120, 132]. Due to overproduction of ROS, Cd-induced oxidative stress causes more physiological disorders in plants [29]. Possibly the most significant environmental effect of Cd is its deposition and accumulation in edible parts of plants, causing serious human health problems such as kidney impairment, bone deformations, and raised cancer risk [18]. Hence, it is imperative to have a better understanding of the mechanisms of Cd uptake and toxicity as well as tolerance in plants for implementation of effective mitigation measures.

Plants cultivated in soils contaminated with Cd have developed different defense mechanisms, which can be classified into two categories: avoidance by limiting metal uptake, and tolerance involving accumulation (by roots), sequestration (by leaves), and immobilization of poisonous metals within specific tissues [63, 71, 84]. Study done by [14, 86, 125], emphasizes morphological and physiological strategies employed by plant to modulate Cd stress.

Plant immune responses are also mediated through various signaling pathways such as the calcium-calmodulin pathway, excessive ROS production, and phytohormone synthesis in response to Cd uptake [13, 70]. These responses generate various protective molecules in response to stress, including Ca^{2+} and nitric oxide (NO) [22], glutathione and proline [100, 119, 153], as well as phytohormones [13]. These molecules can bind to cadmium (Cd), and helps to reduce oxidative stress and protect the plant tissues. Additionally, modifications made by incorporating molecular mechanism are vital for regulating Cd tolerance and adaptability in plants. Genes that are activated during the Cd stress are mainly involved in processes such as Cd absorption (movement from roots to shoots), sequestration, and detoxification [33, 126, 147]. Knowing these genetic responses is important for developing new varieties of crop plants that are resistant to Cd stress and it will also minimize Cd accumulation in plants. This chapter provides a detailed overview of cadmium (Cd) toxic behavior in agriculture, focusing on its uptake by crop plants, the toxic effects it induces, and the physiological and molecular mechanisms that help plants cope Cd stress and also explores various strategies to reduce Cd contamination in soils, including phytoremediation and soil amendments uses.

2 Cadmium (Cd) Uptake in Crops

2.1 Cd Uptake by Transporter Proteins

Cadmium is found in many forms in soil, but not all of these are bioavailable [78]. Plants mainly absorb cadmium as bivalent form (Cd^{2+}), similar to calcium (Ca^{2+}) and iron (Fe^{2+}) uptake [37, 45, 56]. Bivalent Cd^{2+} enters the plant through the root cell walls by diffusion [107] and crosses the plasma membrane via active transport mechanisms, utilizing non-specific transporter proteins such as Zrt (zinc-regulated transporter), Irt (iron-regulated transporter)-related protein (ZIP), and metal-pumping ATPases [38, 118, 142]. Several other transporter proteins help in translocation of Cd within plants, including members of the natural resistance-associated macrophage protein (NRAMP), P-type ATPase, ATP-binding cassette (ABC), cation/H+ antiporters (CAX), low-affinity cation transporter (LCT), and CE families [38, 127]. Translocation of Cd takes place by xylem to aerial parts of plant by both apoplastic and symplastic pathways [38, 77, 141]. And, once the plant gets matured, apoplastic barriers restrict the direct movement of Cd into the xylem [107]. Within the apoplast and symplast, cadmium is often chelated with phosphorus and sulfur, respectively, which further affects its mobility in the plant [135]. After being absorbed into the cytosol, cadmium moves through pericycle cells, ultimately reaching the xylem and being transported to the aerial parts of the plant [107, 142] (Fig. 1).

2.2 Role of Root Systems

Root system of plants acts as primary association between the plant and soil, plays key role in the absorption of water, nutrients, and toxic elements such as cadmium (Cd). The structure of roots, including its length, density, and surface area, can determines the quantity of Cd that can be absorbed by plants [81]. Plants with extensive root systems, like those with root hairs, bears high surface area, allowing substantial Cd

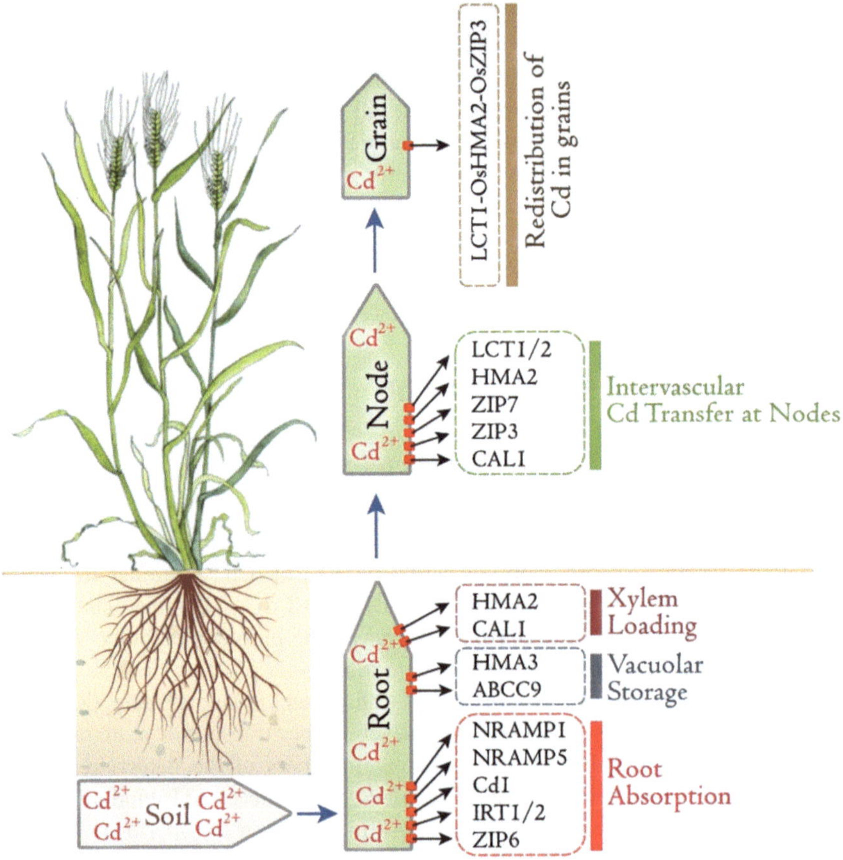

Fig. 1 Mechanism of Cd uptake and translocation in plants

absorption. This increased root-soil contact is significant in contaminated soils where Cd availability is high. The rhizosphere region of soil involved in many chemical, biological, and physical processes. Root exudates, such as phenolic compounds and organic acids can change the chemical state of Cd in the soil, and maintaining its availability for plant uptake [59]. Organic acids can chelate Cd ions, enhancing their solubility and mobility, which facilitates greater Cd absorption by the roots. The permeability of root cell membranes to Cd is another important factor [83]. For Cd to reach the xylem and be transported to above-ground plant parts, it must cross the root epidermis and cortex. The presence of casparian strip in the endodermis plays a critical role in regulating Cd entry and movement within root tissues [127].

2.3 Factors Affecting Cd Bioavailability in Soil

Cadmium has detrimental effects, causing harm to humans, plants, and animals. It is highly mobile than any other trace elements and easily enters the food chain through soil. Its presence in soil is primarily due to modern agricultural practices, industrialization, and urbanization [21]. The extent of Cd absorption by soil depends mainly on its availability, as not all forms of Cd are accessible for plants. Several factors influencing Cd uptake, includes pH of soil, temperature, organic matter content, redox potential, cation exchange capacity, and availability of other elements [48, 62, 67, 110].

2.4 Soil pH

The bioavailability of Cd in soil is mainly affected by pH levels of soil. Acidic soils enhance the Cd availability and make more easily absorption by plants [146]. Li et al. [74] explained that, low pH conditions can promote the conversion of immobile Cd into bioavailable Cd^{2+}. In acidic solutions, Cd^{2+} competes with H^+ ions for binding sites on soil particles, and as H^+ ions are more readily adsorbed, Cd desorption increases, enhancing its availability for plant uptake [148]. Conversely, in alkaline conditions, Cd tends to hydrolyze into hydroxy species such as cadmium hydroxide cation ($Cd(OH)^+$), cadmium dihydroxide ($Cd(OH)_2$), cadmium trihydroxide anion ($Cd(OH)_3$), cadmium tetrahydroxide anion ($Cd(OH)_4$), and cadmium tetrahydroxide complex ($Cd_4(OH)_3$) which have a higher affinity for adsorption to soil particles [137]. It will reduce Cd mobility and accumulation in plants [122]. As soil pH increases, more negatively charged adsorption sites are created on soil colloids, leading to greater Cd^{2+} retention [19]. Additionally, research by Rao et al. [103] demonstrated that in acidic paddy soils, Cd adsorption increased by 0.35–0.37 fold as one-unit pH decreased. Higher pH levels may cause Cd^{2+} to precipitate as $Cd(OH)_2$, making it less available to plants [17].

2.5 Soil Organic Matter

The decomposition of plant and animal material in the soil is reported to increase the bioavailability of Cd, since it forms complexes with the soil matrix. Conversely, organic matter promotes the formation of stable organometallic compounds that decrease Cd solubility and therefore limit its availability to plants [88]. In recent studies, several observations revealed that additions of biochar in the soil can have a stabilizing effect further on Cd by converting more bioavailable forms into less bioavailable forms. For instance, [6] demonstrated that the application of biochar significantly decreased the absorption of Cd by maize. Similar outcome can be seen if organic matter is combined with selenium. Up to 5.8–20.8% reduction in Cd accumulation in rice has been observed through the use of the above-mentioned mixture [79]. Amirahmadi et al. [9] in another study reported that it not only reduced the bioavailability of Cd but also the seedlings of oak forest witnessed better growth on Cd-contaminated soils after the application of the biochar. The addition of 1, 3, and 5% biochar to soil with 50 mg kg^{-1} Cd improved plant tolerance by 40.9, 56, and 60.6%, respectively. Switchgrass (*Panicum virgatum*) has also been shown to reduce Cd level in shoot and roots when grown in soil treated with a compost blend containing 5% biochar derived from beef cattle manure or poultry [93]. During rice cultivation, application of peat with iron nitrate [$Fe(NO_3)_3$] was observed to increase the immobilization of both arsenic (As) and Cd, limiting their bioavailability [139].

2.6 Reduction Potential

The term reduction potential or redox potential is defined as the ability of a substance to either receive or donate electrons. Determination significantly affects the solubility of Cd in soils, organic matter, minerals, and microorganisms in the rhizosphere [87]. In aerobic conditions, the main species present are soluble salts or cationic ions (Cd^{2+}). In anaerobic or reducing conditions, cadmium sulfide (CdS) or cadmium carbonate (CdCO3) precipitates. Changes in redox potential alter electron acceptors and chelation activity of Cd^{2+}. Study of [76] demonstrated that adding azolla (mosquito fern) to soil reduced redox potential and, in turn reduced Cd uptake by 80.3%, while boosting plant productivity by 13.4%. Moreover, addition of periphyton to rice soil lowered the redox potential by 40–120 mV, which resulted in increased Cd concentrations [82]. The reduction of metals like iron (Fe) and manganese (Mn) oxides significantly altered Cd mobility in paddy soils [139].

3 Cd Toxicity in Crop Plants

Excess Cd is toxic and can be detected by looking at different anomalies and toxicity symptoms in plants. Testing for symptoms of Cd stress is very important, as it helps identify the level of phytotoxicity and provides a means of developing proper strategies to enhance plant tolerance [3]. Cd toxicity shows both visible and invisible symptoms in plants, visible symptoms includes stunted growth, chlorosis or leaf discoloration, necrosis, wilting, reduced photosynthesis, and impaired respiration [90, 154] while non-visible symptoms are reduced biomass and alteration in mineral composition at the organelle level [117]. The number of biomass productions is significantly hampered with cadmium exposure, and at higher exposure limits, it causes the death of a plant [42]. Exposure to Cd negatively affects several physiological processes: growth, photosynthesis, enzyme activity, stomatal functioning, and free radical production with different symptoms depending on the concentration of Cd absorbed by the plant [46].

3.1 Stunted Growth

Reduced growth and vigor characterize stunting in plants. The causes can be infectious as well as non-infectious. Stunting may occur permanently through the presence of infectious sources of bacteria, fungi, viruses, or nematodes. It is typically easier to treat causes which are not infectious, such as nutritional imbalances, toxin exposure, or environmental stress. Since Cd is not a pathogen, it inhibits the absorption and transportation of major nutrients that retard the growth of the plant [128], Turgut et al. [134] demonstrated that with the elevation of Cd concentration, *Helianthus annuus* growth was significantly affected. Furthermore, [96] studied how Cd toxicity stops new plantlets of *Hydrocotyle umbellata* from being developed. In *Pisum sativum* L., Cd exposure had growth-inhibitory effects [102, 143] reported that the root and leaf growth in *Regnellidium diphyllum* were reduced by Cd and that the secondary leaves were not produced at all.

3.2 Chlorosis

Chlorosis is the removal of green pigmentation from plant leaves that turn them pale green or yellow because of the absence of chlorophyll pigment that gives leaves their green color. This condition might be due to nutrient deficiency in the necessary nutrients such as iron, phosphate, and manganese, or from a lack of exposure to enough light [72]. Chlorosis prevents the synthesis of sugars and carbohydrates in the plant photosynthesis, and if not controlled, causes further damage to the plant, which may end up dying due to death. Even with control, chlorosis may linger along with other challenges such as rust disease. With a relatively high bioavailability of cadmium, this element can build up in plants, causing interference in the uptake of nutrient substances such as calcium, phosphorus, potassium, and water. Such interference may lead to symptoms that include chlorosis of leaves and reduced photosynthesis. Earlier research had always linked Cd-induced chlorosis with the short supply of essential nutrients [150]. This form of interference was reported in many different plant species, for instance, pea [102], oilseed rape [8], and rice cultivated in Cd-contaminated soil [101].

3.3 Necrosis

Abiotic stress leads to necrosis, meaning that it causes tissue and cell death in the plant. This manifestation indicates that the plant is ailing or sick. Some of the nutrient deficiencies that include potassium, nitrogen, iron, and nickel deficiencies cause necrosis because the amount of nutrients required by plants is relatively small. Such imbalances may arise from a lack of nutrients in the soil, a pH imbalance, or an unequal application of fertilizers. In addition to the environmental factors, other pathogens such as fungal infections and nematodes can cause necrosis [66]. Cadmium also plays a role by causing antagonisms with the intake of essential nutrients by the plant. Cd usually competes with other mineral cations, especially in the root zone [55]. Similarity of the Cd ion with other mineral cations allows them to compete with the needed mineral cations in order to be taken by the plant [31]. This competition between cations leads to nutritional depletion [43]. Necrosis appear as the form of dark, wet or dry papery patches on the plant tissues, which in most cases, are dark. In other cases, they may be yellow or wilted and indicate that there has been an activation of plant cell death rather than mere discoloration. It is induced by environmental stressors such as extreme weather conditions or poor water quality.

3.4 Biomass Reduction

Phytotoxicity caused by unintentional uptake of Cd by plants causes immense damage to plant health in the form of low biomass and decreased photosynthetic efficiency, but symptoms of phytotoxicity have mainly been attributed to the toxicity of the Cd, which could be more prone to higher declines in plant biomass [121]. Rice grown under Cd-contaminated soil will significantly reduce root biomass, as stated by [25]. Turan and Esringu [133] proved that the biomass of canola and Indian mustard was reduced by above 70% Cd use. An equally significant effect was observed under hydroponic cultivation of *Pistia stratiotes*, in which although up to 20 mg/L of Cd caused no negative effects on the plant, the biomass decreased significantly with increasing concentrations of Cd [26]. Hence, Cd limits plant growth by causing significant decreases in biomass.

3.5 Photosynthesis Inhibition

For plants, photosynthesis is essential as it supplies the required energy by converting water and carbon dioxide into energy-rich compounds. It is therefore vital not to interfere with this process. The emission of Cd in the environment could severely damage the photosynthetic machinery of the plants by breaking down chloroplast structure and by reducing the synthesis level of chlorophyll [35, 97]. Cd may limit photosystem II, enzymes that take part in Calvin cycle and ATP synthetase and could also disrupt photosynthetic electron transport chain, knocking out the process of photosynthesis [114]. Treatment of *Brassica napus* with Cd showed decreased contents of chlorophyll and carotenoid, which thereby caused the discoloration of leaves [5, 8]. Exposures of Cd decreased the net photosynthetic rate, fluorescence efficiency, and stomatal conductance but increased the intercellular CO_2 level in wheat [73]. Inhibition of photosynthesis can completely cease plant growth. In *Schima superba*, Cd affected photosynthesis due to the interference with the transport of electrons from PSII to PSI, thereby affecting the distribution of energy, as well as that of CO_2 assimilation [23]. Cd inhibits carbon fixation and CO_2 assimilation, resulting in the high excitation levels that damage the photosynthetic system. Plants dissipate excess energy in the form of heat to minimize the chances of damage [51].

4 Tolerance Mechanisms in Crop Plants

4.1 Genetic and Physiological Adaptations

Plants possess the ability to sequester Cd as a mechanism of adaptation for tolerance against metal stress and survival in PTE contaminated soil [52]. This is essential to detoxify Cd and protect against toxicity. Plant roots growing in contaminated soil, first encounter with the PTE and thus are the first to be involved in metal detoxification. Plants immobilize cations to reduce metal uptake using several ways like binding to exudates, selective metal exclusion via the cell wall, and the plasma membrane, preventing further transport in plants [29, 129]. Pokeweed (*Phytolacca americana*) and *Triarrhena sacchariflora* have been reported to store Cd in their cell walls. This is assisted by its interaction with suberin and pectin and -COOH or -OH groups, resulting in Cd-COOH and Cd-OH [20, 131, 144]. Cell vacuoles serve as a primary compartment for sequestering Cd that has crossed the cell walls [80]. Sequestration in the vacuole reduces Cd content in the cytoplasm, eventually alleviating phytotoxicity [124]. Cd vacuolar sequestration is controlled by membrane transporters like the ZIP family, which carry nutrients like Zn, Fe, and Cu, as well as hazardous metals like Cd, and studies have found that Cd stress leads to increased expression of these transporters [124].

The *Arabidopsis thaliana* seedlings, transferred to Murashige and Skoog agar plates, then treated with 50 mM CdSO4 for 4 days, expressed HMA3, MTP3, IREG2, IRT2, NAS1, and NAS2 in roots where Cd is sequestered and reduced transport to the shoots [111]. In *Iris lactea* roots, Cd stress enhanced the expression of IlHMA2, which transports Cd to shoot from roots via xylem [44]. Adil et al. [1] demonstrated Cd accumulation ability of two rice genotypes, Heizhan 43 and Yinni 801, in different tissues under hydroponic condition on exposure to 10 and 15 μM L^{-1} Cd level. Heizhan 43, was found to accumulate Cd in both shoot and grain, whereas Yinni 801 resulted in Cd accumulation only in shoot. The expression of genes OsLCT1 and OsHMA2 were significantly responsible for the shoot to grain translocation of Cd in Heizhan 43.

4.2 Role of Chelating Agents and Antioxidants

The other detoxification mechanism is through phytochelatin (PCs) production. Bhuyan et al. [16] demonstrated that PCs in rice plants enhanced by 23 and 47% following the exposure of these plants with 1 and 2 mM Cd, respectively. Peco et al. [98] also established a similar rise on *Biscutella auriculata* exposed to 125 mM of Cd. Similarly, [98] reported a comparable increase in *Biscutella auriculata* when treated with 125 mM of Cd. PCs are synthesized from reduced glutathione in the presence of heavy metals with the help of the enzyme phytochelatin synthase, which acts as a precursor, with the process being driven by the overexpression of the phytochelatin synthase gene [54, 151]. In *Theobroma cacao*, Cd exposure upregulates genes associated with PC synthesis in leaves and roots [108]. PCs are peptides with the basic structure (γ-Glu-Cys) n-Gly (where $2 < n < 11$), although other forms such as (γ-Glu-Cys)n-Glu, (γ-Glu-Cys)n-Ser, and (γ-Glu-Cys)n-γ-Ala also exist [24, 36, 106].

Synthesized PCs form chelator-Cd complexes which translocates Cd from the cytosol to vacuole and is present in an inactive state (Lin and Aarts 2012) [136, 145]. Exogenously applied GSH (glutathione) treatment promoted the accumulation of increased quantities of Cd in roots but suppressed the transfer of Cd into the shoot of *Brassica campestris* when the plant was grown on 20 mM of the Cd [34, 53]. This suggests that exogenous GSH increased PCs biosynthesis, thereby increasing chelation of Cd and its sequestration within the plant roots vacuoles. Thus, PCs synthesis serves as an important step for detoxification in plants and for conferring Cd tolerance by chelating and sequestering Cd into vacuoles of plants.

The antioxidant defense system in plants combats high levels of toxic metals like Cd [65]. Cadmium exposure affects this system in such a way that detoxification and tolerance mechanisms are induced. The antioxidant defense system is composed of both enzymatic antioxidants, such as catalase, ascorbate peroxidase (APX), monodehydroascorbate reductase (MDAR), superoxide dismutase (SOD), dehydroascorbate reductase (DHAR), glutathione reductase (GR), glutathione S-transferase and glutathione peroxidase and non-enzymatic antioxidants, including glutathione, tocopherols, phenolic compounds, ascorbate, nonprotein amino acids, and alkaloids. These components assure to neutralize ROS and redox homeostasis in cells [60]. The antioxidant system safeguards the plants through scavenging free radicals against Cd or other heavy metal stress [85]. Hong et al. [50] studied that ZmWRKY4 gene influences the activity of SOD and APX under Cd stress in maize mesophyll protoplasts. Alaraidh et al. [7] reported that fenugreek plants treated with Cd, Cr, and Pb at 2.5–10 mg/L for seven days showed a higher mRNA expression of genes encoding antioxidant enzymes catalase (CAT), peroxidase (POD), and

APX. The lowest expression was of POD and CAT, whereas APX levels increased moderately, such behavior indicates the effect of Cd on gene expression in fenugreek. In transgenic tobacco plants, under Cd stress, an overexpression of TaEXPB23 led to the elevated activities of antioxidant enzymes, for example, DHAR and GR, with an increase in tolerance to Cd [39, 47]. Overexpression of the ZmVTE4 gene, encoding γ-tocopherol-methyl-transferase, in transgenic tobacco plants resulted in increased levels of accumulation of γ-tocopherol and α-tocopherol, thus enhancing Cd stress tolerance [152]. Overexpression of glutaredoxin from chickpea leaves, (CaGrx, LOC101493651), in transgenic *Arabidopsis thaliana* enhanced resistance to Cd (500 μM) through growth, enzyme activity and enhancement in physiological performance [69]. The plants have potential to tolerate the toxic effect of Cd from morphological level to genetic level. However, under high concentration of Cd and/or presence of other toxic pollutants, the efficacy of the tolerance mechanism diminishes, eventually offsetting growth and productivity. Therefore, some sustainable approaches are demanded that exhibit potential to effectively culminate the risk of Cd-induced toxicity in crop plants and human beings.

5 Plant-Based Approach for Management of Cd-Contaminated Soil

Phytoremediation is a technique of bioremediation whereby the detoxification process of the environment is achieved through the utilization of green plants. Therefore, it is one of the sustainable, economical and eco-friendly methods for detoxification and purification of contaminated environments [15]. Kafle et al. [64] further describe phytoremediation as the application of plants in reducing the impact of environmental pollution agents. Cadmium (Cd) availability in soils, streams, and rivers increases the possibility of its uptake and transportation from roots to shoots and other parts of the plant. Various approaches of phytoremediation of Cd include strategies such as phytoextraction, phytostabilization, and phytotransformation to manage and reduce the contamination with Cd efficiently [105] (Fig. 2). Different plants and strategies used to remediate Cd-contaminated environment have been listed in Table 1.

5.1 Phytoextraction

Phytoextraction is an economical phytoremediation technique that employ plants to remove heavy metals from soil [40]. This process, also referred as phytoabsorption or phytoaccumulation, involve uptake of heavy metals from water and soil by plant roots, and then transport the metals to the shoots [32, 116]. Phytoextraction is a method whereby plants absorb metals and transfer them from the roots to other

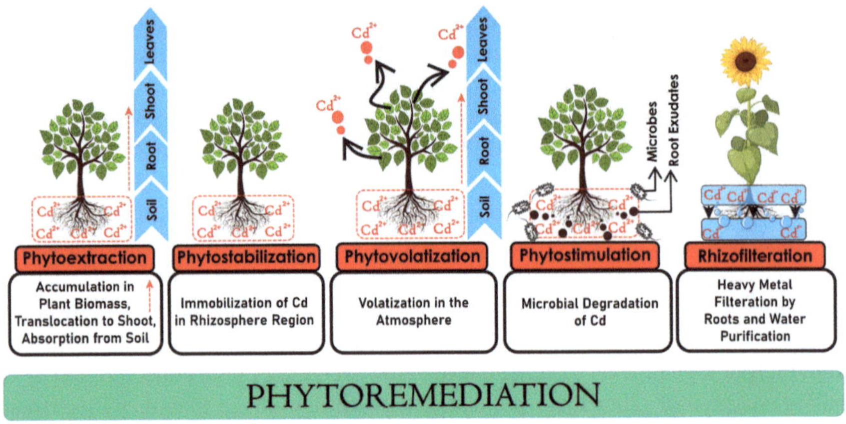

Fig. 2 Management of Cd-contaminated soil with the help of phytoremediation technology

Table 1 Potential of plants in bioremediation of Cd

Bioremediator (plant)	Action response	References
Brassica juncea and *B. carinata*	*B. juncea* was found to have better Cd accumulation potential than *B. carinata*. The process of phytoremediation was enhanced in presence of EDTA (ethylene diamine tetraacetic acid) due to increased mobilization of metal	Ahmed et al. [2]
Brassica juncea	The transgenic plant overexpressing γ-glutamylcysteine synthetase (γ-ECS) or glutathione synthetase (GS) exhibit improved tolerance as well as accumulation of Cd due to production of phytochelatin	Reisinger et al. [109]
Solanum nigrum	Cd was accumulated in range 9.92–9.66 mg/kg in aboveground parts of the plant. Result revealed that there was positive correlation between metal accumulation and plant biomass	[61]
Zea mays	Cd was accumulation after 60 days was found to be more in roots than shoot demonstrating roots were actively involved in phytoremediation process	[89]
Arundo donax	Plant efficiently removed Cd from wastewater through the mechanism of rhizofiltration	[28]
Azadirachta indica	Accumulated 0.79 ppm Cd in roots and 0.60 ppm in shoot after 2 months when grown in soil containing 5 ppm of Cd	[58]
Virola surinamensis	Cd was accumulated in different parts of the plant in order root > shoot > leaf, demonstrating that the phytostabilization was the main mechanism of remediation	[10]

plant tissues. This is an efficient contaminant-reducing technique that retains the composition of the soil as well as its fertility [130]. The effectiveness of this process lies in various factors including soil conditions, availability of pollutants, species of the plant, and also the chemical form of contaminants [99].

Toxic metals and pollutants are absorbed, concentrated, and removed from the soil by accumulating them in plant biomass [41]. Phytoextraction is particularly suitable for treating low-concentration, diffusely contaminated areas. The process of phytoextraction could be carried out either alone by phytoaccumulator or performing process in the presence of certain amendments that enhance its efficiency. These plants have a high biomass and have better tolerance capacity of absorbing heavy metals. Phytoaccumulation, the process where contaminants from the soil are taken up by the plants with their above-ground tissues accumulating these materials over multiple growth cycles in plants. The plants are harvested when there has been sufficient buildup of the metal, and this process is repeated over successive crop cycles to progressively decrease the level of metals in the soil. This is especially effective for metals like cadmium, lead, and zinc. However, amendment of chemical compounds, such as chelating agents (e.g., EDTA), to growth media enhances the bioavailability of heavy metals in the soil. By increasing the solubility of the metals, the plants can absorb larger amounts of metals than they would naturally, thereby improving phytoaccumulation efficiency. After sufficient uptake, the plants are harvested to remove the accumulated metals from the site. Both mechanisms aim to reduce metal toxicity in contaminated soils through plant growth and subsequent harvesting. The effectiveness of phytoextraction depends on factors such as plant species, metal type, and soil conditions [92, 145]. Willow (Salix viminalis) has a notable capacity for phytoextracting Cd, Zn, and Cu, with high efficiency in translocating heavy metals from roots to shoots, leading to large biomass production [140]. Similarly, Noccaea caerulescens have been reported by [57] to remediate the soil contaminated with moderate quantity of Cd and Zn by accumulating them upto 200 and 47 g/ha, respectively. Due to its ease of use and effective removal of Cd, phytoextraction is a practical method for cleaning Cd-contaminated agricultural soils, in contrast to other heavy metals.

5.2 *Phytostabilization*

Phytostabilization is an immobilization process using plants to immobilize the polluting matters from the soil surrounding the roots within the rhizosphere [12]. Thus, the roots perform a very important function in reducing metal bioavailability by establishing a zone of influence that assists metals in going into immobilization through absorption, precipitation, complexation, or reducing metal valence [4, 145]. This kind of long-term stabilization reduces the mobility and bioavailability of toxins, hence reducing the chance of toxins in the food chain. Wang et al. [138] assert that many plant species have been used for stabilizing contaminants in soil, sediments, and sludge. Phytostabilization is effective in reducing the toxicity of heavy metals within the soil by immobilizing metals thus restricting their movement to other areas [149]. Nevertheless, effectiveness depends highly on the ability of the roots of the plant.

5.3 *Phytovolatilization*

Phytovolatilization is a process wherein plants absorb pollutants with their root, transform that into volatile compounds, and releases them to the atmosphere [75]. This procedure allows heavy metals or toxic contaminants in the soil or water, then, when the plant undergoes transpiration, these contaminants will be released as volatile chemicals. Through phytotransformation, pollutants are transformed into less harmful and more volatile species. For instance, *Brassica juncea* naturally takes up selenium that often occur as extremely toxic selenocyanate, which may be transformed into a much less harmful compound dimethylselenide, having 500–700 times less toxicity [27]. This technology may be appreciated for the removal of organic pollutants from soil, yet it also has some limitations. Sabreena et al. [112] noted that phytovolatilization does not permanently exclude the pollutants from the soil, but transfers them into the atmosphere.

5.4 *Phytostimulation*

Phytostimulation, also known as rhizodegradation, boosts soil microbial activities through the encouragement of chemical exudates release from plant roots into the rhizosphere [115]. Thus, they serve as food for rhizobacteria and other useful microbes, which in turn rely on them to survive. These microorganisms break down contaminants in soils through biotransformation. This phytostimulation method is not only cost-effective but also non-toxic and environmentally friendly to remediate Cd and other organic pollutants. It is also considered one of the reliable methods for removal of organic contaminant in soil ecosystems [105].

5.5 *Rhizofiltration*

Rhizofiltration is the process of phytoremediation in which root biomass absorbs heavy metals from water involves activities like reverse osmosis, ion exchange, and precipitation, to absorb pollutants in soil and water environments using the plant roots [104]. However, these practices are often expensive and do not have widespread implementation, and rhizofiltration effectiveness hinges on the root's structure and system of the plant, with those having long, fibrous roots as well as a large surface area most suitable in the process [68]. According to the USEPA (2000), this method can partially remove heavy metals from industrial wastewater and agricultural runoff by retaining them within the roots [30]. Rhizofiltration is viewed as a promising phytoremediation technology. Plants like sunflower and mustard have the ability to extract heavy metals such as cadmium (Cd), nickel (Ni), copper (Cu), zinc (Zn), and lead (Pb) from contaminated environments [123]. Indian mustard *(Brassica juncea)* and sunflower (*Helianthus annuus*) are particularly effective for rhizofiltration, making them ideal for remediating toxic heavy metals [95]. Application of plants that exhibit different mechanisms to remove heavy metals from soil are effective approach to manage the growing challenge of pollution. However, understanding the capacity of plants to remediated the media, level and types of pollutants present, and existing environmental and physicochemical of growth media are mandatory to achieve the desired output.

6 Conclusion

Generally, the challenges of Cd contamination in agricultural systems are multidimensional and complex, requiring a holistic strategy to address impacts on crop productivity, safety of food and environment health. Mechanisms in Cd uptake, translocation, toxicity, and tolerance in crop plants have been discussed in this chapter to understand how plants interact with Cd-contaminated soils. Transporter proteins, root system adaptation, and properties of soil such as pH, organic matter, and redox potential determine the extent to which Cd is accumulated within plants. Cd introduced into the soil mainly due to anthropogenic activities such as industrial emissions, phosphate fertilizer use, and wastewater irrigation, which after uptake eventually disrupts the key physiological processes in plants, leading to reduced growth, impaired photosynthesis, and oxidative stress. The edible parts of the crop also accumulate Cd, which is posing many human health risks; therefore, its management strategies must be considered accordingly. Plants, through millions of years of evolution, have developed a number of defense mechanisms, such as Cd sequestration in vacuoles and synthesis of metal-binding peptides, and have activated antioxidant defense systems. These mechanisms not only balance the adverse effects of Cd but also open avenues for the creation of Cd-tolerant crop varieties through breeding and genetic engineering. Though plants have the mechanism to tolerate Cd stress, yet, is not sufficient to solve the challenge of Cd pollution in agricultural soil as it hampers quantitative and qualitative yield of agricultural produce.

Consequently, an approach that will focus on the alleviation of Cd content and bioavailability sustainably would help in mitigating the toxic effects. Some techniques for phytoremediation include phytoextraction, phytostabilization, phytovolatilization, phytostimulation, and rhizofiltration, which provide environmentally friendly and cost-effective solutions in mitigating Cd contamination. These methods are helpful in utilizing the natural absorption or stabilizing transformation ability of plants to utilize them as an essential part of any remediation effort. Additive approaches, phytoremediation along with other soil management practices, would then make a potential means of addressing Cd accumulation in agricultural soils as well as crop protection to ensure food security. In this regard, future research work and agricultural activities should still emphasize strategies that apply a combination of different approaches along with phytoremediation for efficiencies to be increased.

Competing Interests The authors declare that they have no competing interest.

Declaration of Generative AI and AI-Assisted Technologies in the Writing Process During the preparation of this work, the book chapter author(s) have not taken any assistance from AI, and the book chapter author(s) reviewed and edited the content as needed and take(s) full responsibility for the content of the publication.

References

1. Adil MF, Sehar S, Chen G, Chen ZH, Jilani G, Chaudhry AN, Shamsi IH (2020) Cadmium-zinc cross-talk delineates toxicity tolerance in rice via differential genes expression and physiological/ultrastructural adjustments. Ecotoxicol Environ Saf 190:110076
2. Ahmed KS, Panwar BS, Gupta SP (2001) Phytoremediation of cadmium-contaminated soil by Brassica species. Acta Agro Hung 49(4):351–360
3. Ahmad I, Akhtar MJ, Zahir ZA, Jamil A (2012) Effect of cadmium on seed germination and seedling growth of four wheat (*Triticum aestivum* L.) cultivars. Pak J Bot 44(5):1569–1574
4. Ahmad S, Sachdev S (2023) Bioremediation: an alternative tool for restoration of urban agroecosystem contaminated with harmful xenobiotics. Xenobiotics in urban ecosystems: sources, distribution and health impacts. Springer International Publishing, Cham, pp 291–319
5. Ahmed T, Masood HA, Noman M, Al-Huqail AA, Alghanem SM, Khan MM, Muhammad S, Manzoor N, Rizwan M, Qi X, Abeed AHA, Li B (2023) Biogenic silicon nanoparticles mitigate cadmium (Cd) toxicity in rapeseed (Brassica napus L.) by modulating the cellular oxidative stress metabolism and reducing Cd translocation. J Hazard Mater 459:132070
6. Alaboudi KA, Ahmed B, Brodie G (2019) Effect of biochar on Pb, Cd and Cr availability and maize growth in artificial contaminated soil. Ann Agric Sci 64(1):95–102
7. Alaraidh IA, Alsahli AA, Razik EA (2018) Alteration of antioxidant gene expression in response to heavy metal stress in *Trigonella foenum-graecum* L. S Afr J Bot 115:90–93
8. Ali E, Maodzeka A, Hussain N, Shamsi IH, Jiang L (2015) The alleviation of cadmium toxicity in oilseed rape (*Brassica napus*) by the application of salicylic acid. Plant Growth Regul 75:641–655
9. Amirahmadi E, Mohammad Hojjati S, Kammann C, Ghorbani M, Biparva P (2020) The potential effectiveness of biochar application to reduce soil Cd bioavailability and encourage oak seedling growth. Appl Sci 10(10):3410
10. Andrade Júnior WV, de Oliveira Neto CF, Santos Filho BGD, do Amarante CB, Cruz ED, Okumura RS, Barbosa AVC, Sousa DJP, Teixeira JSS, Botelho ADS (2019) Effect of cadmium on young plants of Virola surinamensis. AoB Plants 11(3):plz022
11. Bae J, Benoit DL, Watson AK (2016) Effect of heavy metals on seed germination and seedling growth of common ragweed and roadside ground cover legumes. Environ Pollut 213:112–118
12. Bakshe P, Jugade R (2023) Phytostabilization and rhizofiltration of toxic heavy metals by heavy metal accumulator plants for sustainable management of contaminated industrial sites: a comprehensive review. J Hazard Mater Adv 10:100293
13. Bali AS, Sidhu GPS, Kumar V, Bhardwaj R (2019) Mitigating cadmium toxicity in plants by phytohormones. In: Cadmium toxicity and tolerance in plants. Academic, pp 375–396
14. Benavides MP, Gallego SM, Tomaro ML (2005) Cadmium toxicity in plants. Braz J Plant Physiol 17:21–34

15. Bhat SA, Bashir O, Haq SAU, Amin T, Rafiq A, Ali M, Américo-Pinheiro JHP, Sher F (2022) Phytoremediation of heavy metals in soil and water: an eco-friendly, sustainable and multidisciplinary approach. Chemosphere 303:134788
16. Bhuyan MB, Parvin K, Mohsin SM, Mahmud JA, Hasanuzzaman M, Fujita M (2020) Modulation of cadmium tolerance in rice: insight into vanillic acid-induced upregulation of antioxidant defense and glyoxalase systems. Plants 9(2):188
17. Bolan N, Kunhikrishnan A, Thangarajan R, Kumpiene J, Park J, Makino T, Kirkham MB, Scheckel K (2014) Remediation of heavy metal (loid) s contaminated soils–to mobilize or to immobilize. J Hazard Mater 266:141–166
18. Charkiewicz AE, Omeljaniuk WJ, Nowak K, Garley M, Nikliński J (2023) Cadmium toxicity and health effects—a brief summary. Molecules 28(18):6620
19. Chen Y, Ma J, Li Y, Weng L (2019) Enhanced cadmium immobilization in saturated media by gradual stabilization of goethite in the presence of humic acid with increasing pH. Sci Total Environ 648:358–366
20. Chen Y, Zhi J, Zhang H, Li J, Zhao Q, Xu J (2017) Transcriptome analysis of *Phytolacca americana* L. in response to cadmium stress. PLoS ONE 12(9):e0184681
21. Chen Z, Zhao Y, Fan L, Xing L, Yang Y (2015) Cadmium (Cd) localization in tissues of cotton (*Gossypium hirsutum* L.), and its phytoremediation potential for Cd-contaminated soils. Bull Environ Contam Toxicol 95:784–789
22. Chmielowska-Bąk J, Gzyl J, Rucińska-Sobkowiak R, Arasimowicz-Jelonek M, Deckert J (2014) The new insights into cadmium sensing. Front Plant Sci 5:245
23. Chu J, Zhu F, Chen X, Liang H, Wang R, Wang X, Huang X (2018) Effects of cadmium on photosynthesis of *Schima superba* young plant detected by chlorophyll fluorescence. Environ Sci Pollut Res 25:10679–10687
24. Cobbett CS (2000) Phytochelatin biosynthesis and function in heavy-metal detoxification. Curr Opin Plant Biol 3(3):211–216
25. Danso OP, Acheampong A, Zhang Z, Song J, Wang Z, Dai J, Zhi T, Yin X, Zhu R (2023) The management of Cd in rice with biochar and selenium: effects, efficiency, and practices. Carbon Res 2(1):41
26. Das S, Goswami S, Talukdar AD (2014) A study on cadmium phytoremediation potential of water lettuce, *Pistia stratiotes* L. Bull Environ Contam Toxicol 92:169–174
27. de Souza MP, Lytle CM, Mulholland MM, Otte ML, Terry N (2000) Selenium assimilation and volatilization from dimethylselenoniopropionate by Indian mustard. Plant Physiol 122(4):1281–1288
28. Dürešová Z, Šuňovská A, Horník M, Pipíška M, Gubišová M, Gubiš J, Hostin S (2014) Rhizofiltration potential of Arundo donax for cadmium and zinc removal from contaminated wastewater. Chem Pap 68:1452–1462
29. El Rasafi T, Oukarroum A, Haddioui A, Song H, Kwon EE, Bolan N, Tack FM, Sebastian A, Prasad MNV, Rinklebe J (2022) Cadmium stress in plants: a critical review of the effects, mechanisms, and tolerance strategies. Crit Rev Environ Sci Technol 52(5):675–726
30. Elbehiry F, Alshaal T, Elhawat N, Elbasiouny H (2021) Environmental-friendly and cost-effective agricultural wastes for heavy metals and toxicants removal from wastewater. Cost-efficient wastewater treatment technologies: natural systems. Springer International Publishing, Cham, pp 107–127
31. Eller F, Brix H (2016) Influence of low calcium availability on cadmium uptake and translocation in a fast-growing shrub and a metal-accumulating herb. AoB Plants 8:plv143
32. Etim EE (2012) Phytoremediation and its mechanisms: a review. Int J Environ Bioenergy 2(3):120–136
33. Fan Z, Zhang Q, Li M, Niu D, Sang W, Verpoort F (2018) Investigating the sorption behavior of cadmium from aqueous solution by potassium permanganate-modified biochar: quantify mechanism and evaluate the modification method. Environ Sci Pollut Res 25:8330–8339
34. Fang Z, Hu Z, Yin X, Song G, Cai Q (2020) Exogenous glutathione alleviation of Cd toxicity in Italian ryegrass (*Lolium multiflorum*) by modulation of the Cd absorption, subcellular distribution, and chemical form. Int J Environ Res Public Health 17(21):8143

35. Farid M, Shakoor MB, Ehsan S, Ali S, Zubair M, Hanif MS (2013) Morphological, physiological and biochemical responses of different plant species to Cd stress. Int J Chem Biochem Sci 3:53–60
36. Filiz E, Saracoglu IA, Ozyigit II, Yalcin B (2019) Comparative analyses of phytochelatin synthase (PCS) genes in higher plants. Biotechnol Biotechnol Equip 33(1):178–194
37. Fontes RL, Pereira JM, Neves JC (2014) Uptake and translocation of Cd and Zn in two lettuce cultivars. An Acad Bras Ciênc 86(2):907–922
38. Gallego SM, Pena LB, Barcia RA, Azpilicueta CE, Iannone MF, Rosales EP, Zawoznik MS, Groppa MD, Benavides MP (2012) Unravelling cadmium toxicity and tolerance in plants: insight into regulatory mechanisms. Environ Exp Bot 83:33–46
39. Gao Q, Xu L, Li X, Yang W, Mi Q, Lu L, Liu X, Wang K, Lu Y, Chen Z, Li X (2022) Proteome and physiological analyses reveal tobacco (*Nicotiana tabacum*) peroxidase 7 (POD 7) functions in responses to copper stress. Transgenic Res 31(4):431–444
40. Ghori Z, Iftikhar H, Bhatti MF, Sharma I, Kazi AG, Ahmad P (2016) Phytoextraction: the use of plants to remove heavy metals from soil. In: Plant metal interaction. Elsevier, pp 385–409
41. Ghosh M, Singh SP (2005) A review on phytoremediation of heavy metals and utilization of it's by products. Asian J Energy Environ 6(4):18
42. Gill SS, Khan NA, Anjum NA, Tuteja N (2011) Amelioration of cadmium stress in crop plants by nutrients management: morphological, physiological and biochemical aspects. Plant Stress 5(1):1–23
43. Glaus MA, Frick S, Van Loon LR (2021) Competitive effects of cations on the diffusion properties of strongly sorbing trace cations in compacted illite and Opalinus Clay. ACS Earth Space Chem 5(10):2621–2625
44. Guo Q, Tian X, Mao P, Meng L (2019) Functional characterization of IlHMA2, a P1B2-ATPase in Iris lactea response to Cd. Environ Exp Bot 157:131–139
45. Gupta N, Ram H, Kumar B (2016) Mechanism of Zinc absorption in plants: uptake, transport, translocation and accumulation. Rev Environ Sci Bio/Technol 15:89–109
46. Haider FU, Liqun C, Coulter JA, Cheema SA, Wu J, Zhang R, Wenjun M, Farooq M (2021) Cadmium toxicity in plants: impacts and remediation strategies. Ecotoxicol Environ Saf 211:111887
47. Han Y, Chen Y, Yin S, Zhang M, Wang W (2015) Over-expression of TaEXPB23, a wheat expansin gene, improves oxidative stress tolerance in transgenic tobacco plants. J Plant Physiol 173:62–71
48. Helios-Rybicka E, Wójcik R (2012) Competitive sorption/desorption of Zn, Cd, Pb, Ni, Cu, and Cr by clay-bearing mining wastes. Appl Clay Sci 65:6–13
49. Hocaoğlu-Özyiğit A, Genç BN (2020) Cadmium in plants, humans and the environment. Front Life Sci Relat Technol 1(1):12–21
50. Hong C, Cheng D, Zhang G, Zhu D, Chen Y, Tan M (2017) The role of ZmWRKY4 in regulating maize antioxidant defense under cadmium stress. Biochem Biophys Res Commun 482(4):1504–1510
51. Hossain Z, Hajika M, Komatsu S (2012) Comparative proteome analysis of high and low cadmium accumulating soybeans under cadmium stress. Amino Acids 43:2393–2416
52. Hu Y, Wang J, Yang Y, Li S, Wu Q, Nepovimova E, Zhang X, Kuca K (2024) Revolutionizing soil heavy metal remediation: cutting-edge innovations in plant disposal technology. Sci Total Environ 918:170577
53. Huang Y, Chen J, Zhang D, Fang B, YangJin T, Zou J, Chen Y, Su N, Cui J (2021) Enhanced vacuole compartmentalization of cadmium in root cells contributes to glutathione-induced reduction of cadmium translocation from roots to shoots in pakchoi (Brassica chinensis L.). Ecotoxicol Environ Saf 208:111616
54. Inouhe M (2005) Phytochelatins. Braz J Plant Physiol 17:65–78
55. Irfan M, Hayat S, Ahmad A, Alyemeni MN (2013) Soil cadmium enrichment: allocation and plant physiological manifestations. Saudi J Biol Sci 20(1):1–10
56. Ismael MA, Elyamine AM, Moussa MG, Cai M, Zhao X, Hu C (2019) Cadmium in plants: uptake, toxicity, and its interactions with selenium fertilizers. Metallomics 11(2):255–277

57. Jacobs A, Drouet T, Sterckeman T, Noret N (2017) Phytoremediation of urban soils contaminated with trace metals using Noccaea caerulescens: comparing non-metallicolous populations to the metallicolous 'Ganges' in field trials. Environ Sci Pollut Res 24(9):8176–8188
58. Jaiswal N, Sachdev S, Tallapragada S, Singh RP (2018) Phytoextraction potential of neem (Azadirachtaindica) for Cddetoxification from the contaminated soil. Clim Change Environ Sustain 6(2):154–159
59. Javed MT, Akram MS, Tanwir K, Chaudhary HJ, Ali Q, Stoltz E, Lindberg S (2017) Cadmium spiked soil modulates root organic acids exudation and ionic contents of two differentially Cd tolerant maize (*Zea mays* L.) cultivars. Ecotoxicol Environ Saf 141:216–225
60. Jawad Hassan M, Ali Raza M, Ur Rehman S, Ansar M, Gitari H, Khan I, Wajid M, Ahmed M, Abbas Shah G, Peng Y, Li Z (2020) Effect of cadmium toxicity on growth, oxidative damage, antioxidant defense system and cadmium accumulation in two sorghum cultivars. Plants 9(11):1575
61. Ji P, Sun T, Song Y, Ackland ML, Liu Y (2011) Strategies for enhancing the phytoremediation of cadmium-contaminated agricultural soils by *Solanum nigrum* L. Environ Pollut 159(3):762–768
62. Jiang M, Jiang J, Li S, Li M, Tan Y, Song S, Shu Q, Huang J (2020) Glutamate alleviates cadmium toxicity in rice via suppressing cadmium uptake and translocation. J Hazard Mater 384:121319
63. Jutsz AM, Gnida A (2015) Mechanisms of stress avoidance and tolerance by plants used in phytoremediation of heavy metals. Arch Environ Prot 41(4):104
64. Kafle A, Timilsina A, Gautam A, Adhikari K, Bhattarai A, Aryal N (2022) Phytoremediation: mechanisms, plant selection and enhancement by natural and synthetic agents. Environ Adv 8:100203
65. Kaleem M, Shabir F, Hussain I, Hameed M, Ahmad MSA, Mehmood A, Ashfaq W, Riaz S, Afzaal Z, Maqsood MF, Iqbal U (2022) Alleviation of cadmium toxicity in *Zea mays* L. through up-regulation of growth, antioxidant defense system and organic osmolytes under calcium supplementation. PLoS ONE 17(6):e0269162
66. Khan NA, Singh S, Anjum NA, Nazar R (2008) Cadmium effects on carbonic anhydrase, photosynthesis, dry mass and antioxidative enzymes in wheat (*Triticum aestivum*) under low and sufficient zinc. J Plant Interact 3(1):31–37
67. Khan Z, Nisar MA, Hussain SZ, Arshad MN, Rehman A (2015) Cadmium resistance mechanism in Escherichia coli P4 and its potential use to bioremediate environmental cadmium. Appl Microbiol Biotechnol 99:10745–10757
68. Kristanti RA, Ngu WJ, Yuniarto A, Hadibarata T (2021) Rhizofiltration for removal of inorganic and organic pollutants in groundwater: a review. Biointerface Res Appl Chem 11:12326–12347
69. Kumar A, Dubey AK, Kumar V, Ansari MA, Narayan S, Kumar S, Pandey V, Shirke PA, Pande V, Sanyal I (2020) Over-expression of chickpea glutaredoxin (CaGrx) provides tolerance to heavy metals by reducing metal accumulation and improved physiological and antioxidant defence system. Ecotoxicol Environ Saf 192:110252
70. Kumar R, Kumar V, Tandon V, Kumar S, Roohi (2024) Effect and responses of cadmium in plants. In: Cadmium toxicity in water: challenges and solutions. Springer Nature Switzerland, Cham, pp 327–347
71. Kumar V, Prasad RK, Suman SN, Dash BP (2019) Uptake of cadmium by rice crop in sewage irrigated soil. ORYZA Int J Rice 56(3):325–332
72. Li J, Cao X, Jia X, Liu L, Cao H, Qin W, Li M (2021) Iron deficiency leads to chlorosis through impacting chlorophyll synthesis and nitrogen metabolism in *Areca catechu* L. Front Plant Sci 12:710093
73. Li Y, Chen Z, Xu S, Zhang L, Hou W, Yu N (2015) Effect of combined pollution of Cd and B [a] P on photosynthesis and chlorophyll fluorescence characteristics of wheat. Pol J Environ Stud 24(1)
74. Li Z, Wu L, Hu P, Luo Y, Zhang H, Christie P (2014) Repeated phytoextraction of four metal-contaminated soils using the cadmium/zinc hyperaccumulator Sedum plumbizincicola. Environ Pollut 189:176–183

75. Limmer M, Burken J (2016) Phytovolatilization of organic contaminants. Environ Sci Technol 50(13):6632–6643
76. Liu C, Guo B, Li H, Fu Q, Li N, Lin Y, Xu G (2021) Azolla incorporation under flooding reduces grain cadmium accumulation by decreasing soil redox potential. Sci Rep 11(1):6325
77. Liu J, Feng X, Qiu G, Li H, Wang Y, Chen X, Fu Q, Guo B (2023) Inhibition roles of calcium in cadmium uptake and translocation in rice: a review. Int J Mol Sci 24(14):11587
78. Liu L, Li J, Yue F, Yan X, Wang F, Bloszies S, Wang Y (2018) Effects of arbuscular mycorrhizal inoculation and biochar amendment on maize growth, cadmium uptake and soil cadmium speciation in Cd-contaminated soil. Chemosphere 194:495–503
79. Liu N, Jiang Z, Li X, Liu H, Li N, Wei S (2020) Mitigation of rice cadmium (Cd) accumulation by joint application of organic amendments and selenium (Se) in high-Cd-contaminated soils. Chemosphere 241:125106
80. Loix C, Huybrechts M, Vangronsveld J, Gielen M, Keunen E, Cuypers A (2017) Reciprocal interactions between cadmium-induced cell wall responses and oxidative stress in plants. Front Plant Sci 8:1867
81. López-Millán AF, Sagardoy R, Solanas M, Abadía A, Abadía J (2009) Cadmium toxicity in tomato (*Lycopersicon esculentum*) plants grown in hydroponics. Environ Exp Bot 65(2–3):376–385
82. Lu H, Dong Y, Feng Y, Bai Y, Tang X, Li Y, Yang L, Liu J (2020) Paddy periphyton reduced cadmium accumulation in rice (*Oryza sativa*) by removing and immobilizing cadmium from the water–soil interface. Environ Pollut 261:114103
83. Lux A, Martinka M, Vaculík M, White PJ (2011) Root responses to cadmium in the rhizosphere: a review. J Exp Bot 62(1):21–37
84. Mahmud JA, Borhannuddin Bhuyan MHM, Nahar K, Parvin K, Hasanuzzaman M (2020) Response and tolerance of Fabaceae plants to metal/metalloid toxicity. In: The plant family Fabaceae: biology and physiological responses to environmental stresses, pp 435–482
85. Mansoor S, Ali A, Kour N, Bornhorst J, AlHarbi K, Rinklebe J, Abd El Moneim D, Ahmad P, Chung YS (2023) Heavy metal induced oxidative stress mitigation and ROS scavenging in plants. Plants 12(16):3003
86. Mei X, Li Q, Wang H, Fang H, Chen H, Chen X, Yang Y, Rizwan M, Ye Z (2020) Effects of cultivars, water regimes, and growth stages on cadmium accumulation in rice with different radial oxygen loss. Plant Soil 453:529–543
87. Meng D, Li J, Liu T, Liu Y, Yan M, Hu J, Li X, Liu X, Liang Y, Liu H, Yin H (2019) Effects of redox potential on soil cadmium solubility: insight into microbial community. J Environ Sci 75:224–232
88. Mohamed I, Ahamadou B, Li M, Gong C, Cai P, Liang W, Huang Q (2010) Fractionation of copper and cadmium and their binding with soil organic matter in a contaminated soil amended with organic materials. J Soils Sediments 10:973–982
89. Mojiri A (2011) The potential of corn (Zea mays) for phytoremediation of soil contaminated with cadmium and lead. J Biol Environ Sci 5(13)
90. Navarro-León E, Oviedo-Silva J, Ruiz JM, Blasco B (2019) Possible role of HMA4a TILLING mutants of *Brassica rapa* in cadmium phytoremediation programs. Ecotoxicol Environ Saf 180:88–94
91. Nazar R, Iqbal N, Masood A, Khan MIR, Syeed S, Khan NA (2012) Cadmium toxicity in plants and role of mineral nutrients in its alleviation
92. Nedjimi B (2021) Phytoremediation: a sustainable environmental technology for heavy metals decontamination. SN Appl Sci 3(3):286
93. Novak JM, Ippolito JA, Watts DW, Sigua GC, Ducey TF, Johnson MG (2019) Biochar compost blends facilitate switchgrass growth in mine soils by reducing Cd and Zn bioavailability. Biochar 1:97–114
94. Nyiramigisha P (2021) Harmful impacts of heavy metal contamination in the soil and crops grown around dumpsites. Rev Agric Sci 9:271–282
95. Odinga CA (2018) Assessment of heavy metals and pathogens removal from municipal wastewater using a constructed rhizofiltration system. Doctoral dissertation

96. Panyakhan S, Kruatrachue M, Pokethitiyook P, Soonthornsarathoon V, Upatham S (2006) Toxicity and accumulation of cadmium and zinc in *Hydrocotyle umbellata.* Sci Asia 32(3):323–328
97. Parmar P, Kumari N, Sharma V (2013) Structural and functional alterations in photosynthetic apparatus of plants under cadmium stress. Bot Stud 54:1–6
98. Peco JD, Campos JA, Romero-Puertas MC, Olmedilla A, Higueras P, Sandalio LM (2020) Characterization of mechanisms involved in tolerance and accumulation of Cd in *Biscutella auriculata* L. Ecotoxicol Environ Saf 201:110784
99. Priya AK, Muruganandam M, Ali SS, Kornaros M (2023) Clean-up of heavy metals from contaminated soil by phytoremediation: a multidisciplinary and eco-friendly approach. Toxics 11(5):422
100. Rady MM, Elrys AS, El-Maati MFA, Desoky ESM (2019) Interplaying roles of silicon and proline effectively improve salt and cadmium stress tolerance in *Phaseolus vulgaris* plant. Plant Physiol Biochem 139:558–568
101. Rafiq MT, Aziz R, Yang X, Xiao W, Rafiq MK, Ali B, Li T (2014) Cadmium phytoavailability to rice (*Oryza sativa* L.) grown in representative Chinese soils. A model to improve soil environmental quality guidelines for food safety. Ecotoxicol Environ Saf 103:101–107
102. Rahman MF, Ghosal A, Alam MF, Kabir AH (2017) Remediation of cadmium toxicity in field peas (*Pisum sativum* L.) through exogenous silicon. Ecotoxicol Environ Saf 135:165–172
103. Rao ZX, Huang DY, Zhu QH, Liu SL, Luo ZC, Cao XL, Ren XF, Wang JY, Wang S (2013) Effects of amendments on the availability of Cd in contaminated paddy soil: a three-year field experiment. J Food Agri Environ 11:2009–2014
104. Rawat K, Fulekar MH, Pathak B (2012) Rhizofiltration: a green technology for remediation of heavy metals. Intl J Innov Biosci 2(4):193–199
105. Raza A, Habib M, Kakavand SN, Zahid Z, Zahra N, Sharif R, Hasanuzzaman M (2020) Phytoremediation of cadmium: physiological, biochemical, and molecular mechanisms. Biology 9(7):177
106. Rea PA (2006) Phytochelatin synthase, papain's cousin, in stereo. Proc Natl Acad Sci 103(3):507–508
107. Redjala T, Zelko I, Sterckeman T, Legué V, Lux A (2011) Relationship between root structure and root cadmium uptake in maize. Environ Exp Bot 71(2):241–248
108. Reis GSM, de Almeida AAF, de Almeida NM, de Castro AV, Mangabeira PAO, Pirovani CP (2015) Molecular, biochemical and ultrastructural changes induced by Pb toxicity in seedlings of Theobroma cacao L. PLoS ONE 10(7):e0129696
109. Reisinger S, Schiavon M, Terry N, Pilon-Smits EA (2008) Heavy metal tolerance and accumulation in Indian mustard (Brassica juncea L.) expressing bacterial γ- glutamylcysteine synthetase or glutathione synthetase. Int J Phytoremediat 10(5):440–454
110. Riaz U, Aslam A, uz Zaman Q, Javeid S, Gul R, Iqbal S, Javid S, Murtaza G, Jamil M (2021) Cadmium contamination, bioavailability, uptake mechanism and remediation strategies in soil-plant-environment system: a critical review. Curr Anal Chem 17(1):49–60
111. Roberts JT, McCubbin TJ, Braun DM (2023) A plate growth assay to quantify embryonic root development of Zea mays. Bio-Protocol 13(20)
112. Sabreena, Hassan S, Bhat SA, Kumar V, Ganai BA, Ameen F (2022) Phytoremediation of heavy metals: an indispensable contrivance in green remediation technology. Plants 11(9):1255
113. Sachdev S, Ansari SA, Ansari MI (2023) ROS generation in plant cells orchestrated by stress. In: Reactive oxygen species in plants: the right balance. Springer Nature Singapore, Singapore, pp 23–43
114. Sachdev S, Ansari SA, Ansari MI, Fujita M, Hasanuzzaman M (2021) Abiotic stress and reactive oxygen species: generation, signaling, and defense mechanisms. Antioxidants 10(2):277
115. Saeed Q, Xiukang W, Haider FU, Kučerik J, Mumtaz MZ, Holatko J, Naseem M, Kintl A, Ejaz M, Naveed M, Brtnicky M (2021) Rhizosphere bacteria in plant growth promotion, biocontrol, and bioremediation of contaminated sites: a comprehensive review of effects and mechanisms. Int J Mol Sci 22(19):10529

116. Salt DE, Smith RD, Raskin I (1998) Phytoremediation. Annu Rev Plant Biol 49(1):643–668
117. Sarwar N, Saifullah, Malhi SS, Zia MH, Naeem A, Bibi S, Farid G (2010) Role of mineral nutrition in minimizing cadmium accumulation by plants. J Sci Food Agric 90(6):925–937
118. Sebastian A, Prasad MNV (2018) Exogenous citrate and malate alleviate cadmium stress in *Oryza sativa* L.: probing role of cadmium localization and iron nutrition. Ecotoxicol Environ Saf 166:215–222
119. Semida WM, Hemida KA, Rady MM (2018) Sequenced ascorbate-proline-glutathione seed treatment elevates cadmium tolerance in cucumber transplants. Ecotoxicol Environ Saf 154:171–179
120. Sethy SK, Ghosh S (2013) Effect of heavy metals on germination of seeds. J Nat Sci Biol Med 4(2):272
121. Shaari NEM, Tajudin MTFM, Khandaker MM, Majrashi A, Alenazi MM, Abdullahi UA, Mohd KS (2022) Cadmium toxicity symptoms and uptake mechanism in plants: a review. Braz J Biol 84:e252143
122. Shahid M, Dumat C, Khalid S, Niazi NK, Antunes PM (2017) Cadmium bioavailability, uptake, toxicity and detoxification in soil-plant system. Rev Environ Contam Toxicol 241:73–137
123. Shakoor A, Abdullah M, Sarfraz R, Altaf MA, Batool S (2017) A comprehensive review on phytoremediation of cadmium (Cd) by mustard (*Brassica juncea* L.) and sunflower (*Helianthus annuus* L.). J Biosersity Environ Sci (JBES) 10(3):88–98
124. Sharma SS, Dietz KJ, Mimura T (2016) Vacuolar compartmentalization as indispensable component of heavy metal detoxification in plants. Plant Cell Environ 39(5):1112–1126
125. Singh PK, Wang W, Shrivastava AK (2018) Cadmium-mediated morphological, biochemical and physiological tuning in three different Anabaena species. Aquat Toxicol 202:36–45
126. Song J, Finnegan PM, Liu W, Li X, Yong JW, Xu J, Zhang Q, Wen Y, Qin K, Guo J, Li T (2019) Mechanisms underlying enhanced Cd translocation and tolerance in roots of *Populus euramericana* in response to nitrogen fertilization. Plant Sci 287:110206
127. Song Y, Jin L, Wang X (2017) Cadmium absorption and transportation pathways in plants. Int J Phytorem 19(2):133–141
128. Souza LA, Piotto FA, Dourado MN, Schmidt D, Franco MR, Boaretto LF, Tezotto T, Ferreira RR, Azevedo RA (2017) Physiological and biochemical responses of *Dolichos lablab* L. to cadmium support its potential as a cadmium phytoremediator. J Soils Sediments 17:1413–1426
129. Sterckeman T, Thomine S (2020) Mechanisms of cadmium accumulation in plants. Crit Rev Plant Sci 39(4):322–359
130. Suman J, Uhlik O, Viktorova J, Macek T (2018) Phytoextraction of heavy metals: a promising tool for clean-up of polluted environment. Front Plant Sci 9:1476
131. Sun, Q., Wang, X. R., Ding, S. M., & Yuan, X. F. (2005). Effects of exogenous organic chelators on phytochelatins production and its relationship with cadmium toxicity in wheat (*Triticum aestivum* L.) under cadmium stress. Chemosphere, 60(1), 22–31.
132. Tao L, Guo M, Ren J (2015) Effects of cadmium on seed germination, coleoptile growth, and root elongation of six pulses. Pol J Environ Stud 24(1):255–299
133. Turan M, Esringu A (2007) Phytoremediation based on canola (Brassica napus L.) and Indian mustard (*Brassica juncea* L.) planted on spiked soil by aliquot amount of Cd, Cu, Pb, and Zn. Plant Soil Environ 53(1):7
134. Fu Y, Zhatova H, Li Y, Liu Q, Trotsenko V, Li C (2022) Physiological and transcriptomic comparison of two sunflower (Helianthus annuus L.) cultivars with high/low cadmium accumulation. Front Plant Sci 13:854386
135. Van Belleghem F, Cuypers A, Semane B, Smeets K, Vangronsveld J, d'Haen J, Valcke R (2007) Subcellular localization of cadmium in roots and leaves of *Arabidopsis thaliana*. New Phytol 173(3):495–508
136. Vitelli V, Giamborino A, Bertolini A, Saba A, Andreucci A (2024) Cadmium stress signaling pathways in plants: molecular responses and mechanisms. Curr Issues Mol Biol 46(6):6052–6068

137. Wang F, Bao K, Huang C, Zhao X, Han W, Yin Z (2022) Adsorption and pH values determine the distribution of cadmium in terrestrial and marine soils in the Nansha area, Pearl River Delta. Int J Environ Res Public Health 19(2):793
138. Wang M, Zhu Y, Cheng L, Andserson B, Zhao X, Wang D, Ding A (2018) Review on utilization of biochar for metal-contaminated soil and sediment remediation. J Environ Sci 63:156–173
139. Wang X, Yu HY, Li F, Liu T, Wu W, Liu C, Liu C, Zhang X (2019) Enhanced immobilization of arsenic and cadmium in a paddy soil by combined applications of woody peat and Fe $(NO_3)_3$: possible mechanisms and environmental implications. Sci Total Environ 649:535–543
140. Wani KA, Sofi ZM, Malik JA, Wani JA (2020) Phytoremediation of heavy metals using Salix (Willows). In: Bioremediation and biotechnology, vol 2. Degradation of pesticides and heavy metals, pp 161–174
141. Wu Z, Yin X, Bañuelos GS, Lin ZQ, Liu Y, Li M, Yuan L (2016) Indications of selenium protection against cadmium and lead toxicity in oilseed rape (*Brassica napus* L.). Front Plant Sci 7:1875
142. Wu Z, Zhao X, Sun X, Tan Q, Tang Y, Nie Z, Qu C, Chen Z, Hu C (2015) Antioxidant enzyme systems and the ascorbate–glutathione cycle as contributing factors to cadmium accumulation and tolerance in two oilseed rape cultivars (*Brassica napus* L.) under moderate cadmium stress. Chemosphere 138:526–536
143. Wunder DA, Droste A, Windish PG (2009) Megaspore germination and initial development of Regnellidium diphyllum Lindman (*Pteridophyta, Marsileaceae*) sporophytes in the presence of cadmium. Braz J Bot 32:177–181
144. Xin JP, Zhang Y, Tian RN (2018) Tolerance mechanism of Triarrhena sacchariflora (Maxim.) Nakai. Seedlings to lead and cadmium: translocation, subcellular distribution, chemical forms and variations in leaf ultrastructure. Ecotoxicol Environ Saf 165:611–621
145. Yan A, Wang Y, Tan SN, Mohd Yusof ML, Ghosh S, Chen Z (2020) Phytoremediation: a promising approach for revegetation of heavy metal-polluted land. Front Plant Sci 11:359
146. Yang Y, Xiong J, Chen R, Fu G, Chen T, Tao L (2016) Excessive nitrate enhances cadmium (Cd) uptake by up-regulating the expression of OsIRT1 in rice (Oryza sativa). Environ Exp Bot 122:141–149
147. Yu R, Tang Y, Liu C, Du X, Miao C, Shi G (2017) Comparative transcriptomic analysis reveals the roles of ROS scavenging genes in response to cadmium in two Pak Choi cultivars. Sci Rep 7(1):9217
148. Zhai SW, Lu JJ, Zhao PY, Chen XH (2018) Effect of grape seed proanthocyanidins on alleviating dietary cadmium (Cd) induced growth retardation and oxidative stress in hepatopancreas of juvenile tilapia (*Oreochromis niloticus*)
149. Zhakypbek Y, Kossalbayev BD, Belkozhayev AM, Murat T, Tursbekov S, Abdalimov E, Pashkovskiy P, Kreslavski V, Kuznetsov V, Allakhverdiev SI (2024) Reducing heavy metal contamination in soil and water using phytoremediation. Plants 13(11):1534
150. Zhang Q, Wen Q, Ma T, Zhu Q, Huang D, Zhu H, Xu C, Chen H (2023) Cadmium-induced iron deficiency is a compromise strategy to reduce Cd uptake in rice. Environ Exp Bot 206:105155
151. Zhang X, Rui H, Zhang F, Hu Z, Xia Y, Shen Z (2018) Overexpression of a functional *Vicia sativa* PCS1 homolog increases cadmium tolerance and phytochelatins synthesis in Arabidopsis. Front Plant Sci 9:107
152. Zhu Q, Zhang J, Yu H, Li L, Chen X, Jiang M, Tan M (2019) Maize Cd-tolerant ZmVTE4 encoding γ-tocopherol-methyl-transferase alleviated Cd-toxicity through its product α-tocopherol. Environ Exp Bot 158:171–179
153. Zouari M, Elloumi N, Ahmed CB, Delmail D, Rouina BB, Abdallah FB, Labrousse P (2016) Exogenous proline enhances growth, mineral uptake, antioxidant defense, and reduces cadmium-induced oxidative damage in young date palm (*Phoenix dactylifera* L.). Ecol Eng 86:202–209
154. Zulfiqar U, Ayub A, Hussain S, Waraich EA, El-Esawi MA, Ishfaq M, Ahmad M, Ali N, Maqsood MF (2022) Cadmium toxicity in plants: Recent progress on morpho-physiological effects and remediation strategies. J Soil Sci Plant Nutr 22(1):212–269

Lead: Uptake, Toxicity, and Tolerance in Crop Plants and Management of Pb-Contaminated Soil

Sanika Jain, Jyoti Mathur, Arti, and Aditi Anvita

Abstract Heavy metals such as mercury (Hg), lead (Pb), cadmium (Cd), chromium (Cr), and arsenic (As), once they accumulate in living organisms, lead to heavy metal toxicity and harmful biological effects. These metals, unlike organic toxins, cannot be broken down and persist in the environment, bioaccumulating in plants, animals, and humans. One such hazardous metal that is frequently utilized in mining, metallurgy, battery, fertilizer, pesticide, and other sectors is lead (Pb). The buildup of lead in the surroundings causes unfavorable circumstances for plants, including stunted development, decreased levels of chlorophyll and cell divisions, disruptions in the absorption of nutrients, and the inhibition of enzymes. However, certain plant species sometimes survive and grow in lead-contaminated region. This tolerance possibly involves enhanced root growth and morphological adaptations, lead exclusion and sequestration, chelation and detoxification, antioxidant defense mechanisms, and phytoremediation potential. This chapter focuses on the lethal effects of lead and the mechanism of adaptive resistance in plants.

Keywords Lead toxicity · Phytoextraction · Tolerance mechanism · Hyperaccumulation

1 Introduction

In the Earth's crust, lead (Pb) is naturally found in trace amounts [3, 61]. Lead contamination levels have increased due to anthropogenic activity in various places across the world [19, 52]. Industrial effluents are a major source of lead contamination in soil [73, 76]. High levels of lead (Pb) accumulate due to the combustion of coal, the use of lead-based shotgun pellets and paint, the application of lead arsenate pesticides, and leaded gasoline. Additional contributors include explosives, the disposal of Pb-enriched municipal sewage sludge, and discarded lead-acid batteries

S. Jain · J. Mathur (✉) · Arti · A. Anvita
Department of Bioscience and Biotechnology, Banasthali Vidyapith, Banasthali, Rajasthan, India
e-mail: contact.srivastava@gmail.com

A. Husen (ed.), *Enhanced Crop Production Setup Under Various Micro and Nano Contaminants*, Smart Nanomaterials Technology,
https://doi.org/10.1007/978-981-96-6099-5_4

[7]. Groundwater, surface waterways, and soil are the main sources of lead pollution. Ionic lead (Pb^{2+}), along with lead oxides, hydroxides, and lead-metal oxyanion complexes, represents the major form of lead. Among these, the most stable species are Pb^{2+} ions and lead hydroxy complexes. The most prevalent and reactive form of lead is Pb^{2+}, which can produce both mononuclear and polynuclear hydroxides and oxides. Soil poisoned with lead from mining and firing range operations is present in vast tracts of land across the globe. Lead is one of the most prevalent, long-lasting, and hazardous contaminants found in soil. The consumption and subsequent uptake of environmental lead (Pb) have long raised concerns about the possible negative effects on human health. Pb can enter the food chain through plants and directly injure humans when it enters the soil [30, 45, 110]. Humans may experience gastrointestinal distress, brain damage, and kidney damage from exposure to high levels of lead (Pb for a short duration, while their central nervous system, blood, liver, and reproductive systems may all be impacted by long-term exposure [93, 123]. Consequently, numerous enterprises as well as governmental organizations will have to spend a substantial amount of money cleaning these Pb-contaminated soils. Plant genotype and physiological traits determine a plant's susceptibility to metals or various plant response mechanisms [21, 40, 41, 43]. Exposure to lead (Pb) can cause the production of reactive oxygen species (ROS) and increase the expression of antioxidant enzymes [18, 51]. The term "hyperaccumulator plant species" refers to those that, in contrast, naturally have increased antioxidant activity, such as that of superoxide dismutase (SOD) [72]. Compared with non-accumulator plants, accumulator plants exposed to varying amounts of lead (Pb) retain higher levels of ascorbic acid and H_2O_2 in their root cells. Furthermore, Pb exposure causes a more pronounced increase in catalase (CAT), ascorbate peroxidase (APX), and dehydroascorbate (DHA), levels, and in accumulator plants than in their non-accumulator counterparts [36, 83, 91]. The resistance of plants to lead toxicity is facilitated by this characteristic of accumulator plants [90]. Another method of tolerance involves the detoxification of Pb inside plant cells [35, 54, 77, 104, 122]. Lead detoxification and accumulation within plant cells are highlighted in this chapter. It also explores the functions and expression of phytochelatins, along with gene expression induced by lead, shedding light on the physiological and molecular mechanisms that enable plants to tolerate lead [75, 103, 105]. This study provides further insight into the mechanisms underlying plant detoxification and lead endurance.

2 Sources of Lead

Industrial effluents and leaded gasoline are the principal contributors to environmental lead contamination [56, 86]. Lead is often present as an impurity in fertilizers and is used to supply agricultural soil with essential nutrients, such as nitrogen, phosphorus, and potassium, and prolonged usage can increase Pb levels in agricultural land [28, 65]. Lead-based pesticides, such as lead arsenate, biosolids (manures,

Fig. 1 Natural and anthropogenic activities leads to environmental Pb contamination

composts, and sewage), and extensive lead ore mining and smelting also accumulate lead in the soil [9, 89]. Various sources of Pb in environment are shown in Fig. 1.

3 Uptake, Accumulation, and Toxicity of Pb on Plants

Lead initially becomes saturated by roots, where it combines with polysaccharides on the cell surface or the carboxyl groups of uronic acid. Later, it passively enters the root through channels, such as Ca-channels in the plasma membrane [66, 92, 97]. Afterward, it inhibits the Ca channel by competing with Ca and blocking the channel. Another viewpoint suggests that the uptake of Pb in plants occurs through H+/ATPases and that Pb further translocates through apoplastic and symplastic movements [59]. Once, it fully accumulates inside the root, some portion of it remains, whereas some is translocated to the above ground parts of the plant [12, 24] (Fig. 2). The order of accumulation in plant parts is as follows: roots > shoots > leaves > inflorescence > seeds with slight variation depending upon the species [6, 67, 84]. Pb residues remain in the apoplast at relatively low concentrations, however, at relatively high concentrations, Pb enters the cells. They can mostly be found in intercellular spaces, the cell wall, and vacuoles; however, some amounts can also be present in the ER, Golgi body, and vesicles [26, 29, 31, 53, 109].

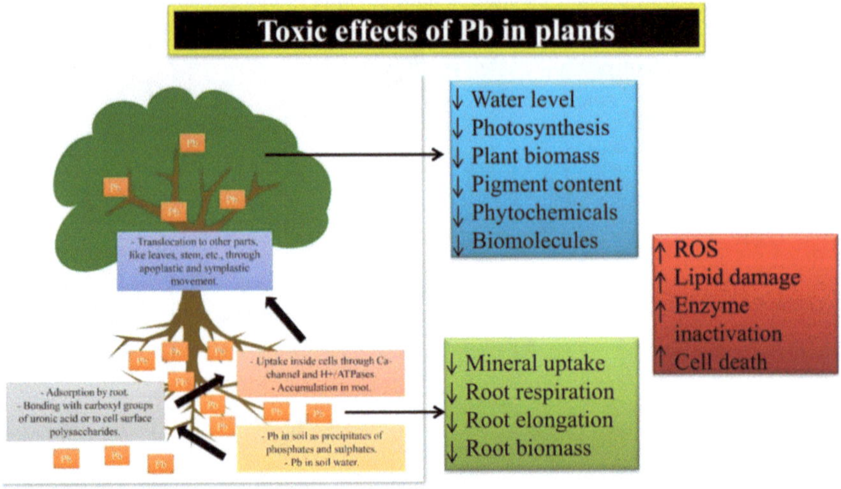

Fig. 2 The absorption, accumulation, and toxic impact of lead in plants

At very high concentrations, Pb causes plant death; however; at relatively low concentrations it induces several ultrastructural, biochemical, and physiological changes [74]. Growth inhibition is one of the most prominent effects of lead toxicity [14, 64]. This phenomenon results from decreased protein and chlorophyll synthesis, nitrate and nitrite reductase activity, a decrease in the number of dividing cells, and a decrease in the nutrients that plants absorb, including nitrate. It stops cell division, which prevents roots from growing. The synthesis of cell wall polysaccharides increases, while the amount of protein decreases and the lipid content changes [2, 48, 114]. Pb also reduces the activity of CAT (chloramphenicol acetyl transferase, an enzyme that removes toxic peroxides) by reducing its synthesis and assembly. It also inhibits other enzymes, such as ATP synthetase/ATPase and RuBisCO, which are enzymes involved in the pentose phosphate pathway [25, 96, 115]. According to several studies leafy vegetables, e.g., lettuce, and above ground root crops, e.g., carrots, are more likely to accumulate higher concentrations of lead. Fruiting plants, e.g., corn, beans, tomatoes, and strawberries, are less likely to accumulate lead [69, 95, 114].

4 Adaptive Mechanisms of Plants to Pb Toxicity

Like other living things organisms, plants also develop defense mechanisms against lead poisoning, which aids in their survival in the environment [47, 85]. When lead levels are elevated, plants respond in a variety of ways. For example, they may involve complex biochemical pathways to detoxify Pb ions or they may undergo structural

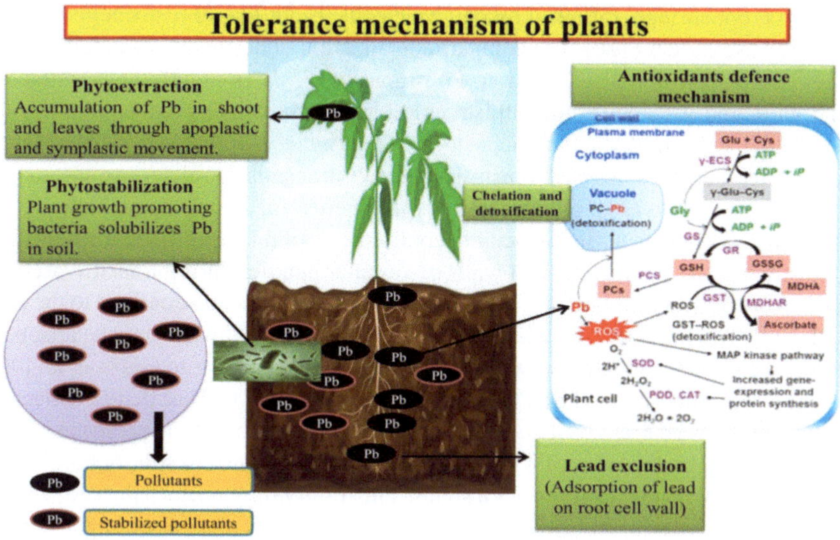

Fig. 3 Tolerance mechanism of plant against lead

changes at the cellular level. Despite this persistent environmental challenge, plants show remarkable adaptability [60] (Fig. 3).

4.1 *Lead Exclusion and Sequestration*

One strategy to prevent lead poisoning is to restrict the accumulation of toxic elements in roots. Plants absorb more Pb in their roots, limiting their movement to the shoots and resulting in limited Pb mobility within the plant [78, 101]. Numerous plant species, including *Brassica juncea, Lactuca sativa, Festuca rubra, Funaria hygrometrica, Vigna unguiculata,* and others, have been shown to bind lead to their roots [39]. Lead can passively enter the roots after being adsorbed onto the surface of the rhizoderm roots and persist in translocating water streams. The majority of Pb absorbed by the roots is held at the surface to prevent it from moving to vascular tissue through xylem loading and then being transferred to the shoots. *B.juncea* root tips exhibit a significant and primarily intracellular absorption of Pb [78]. Lead is thought to be removed from the food chain by reduced absorption and transfer to the aerial plant portions of vegetables. X-ray microanalysis revealed that root sections from *Sesbania drummondii* present a diminishing Pb gradient that runs from the axis to the root epidermis [50]. The Pb concentration in various sections of rice grains usually changes as follows: aleuronic layer < inner surface of rice glume < near the aleuronic layer < surface of the caryopsis < center of the caryopsis < outer surface of the rice glume [13]. *Dianthus carthusianorum* accumulates substantial amounts of

Pb on its root surface, followed by pericycle cells, according to histochemical lead detection techniques [120]. The compartmentalization of heavy metals is considered a key mechanism for detoxification. Pb primarily localizes in cell walls, intercellular gaps, and vacuoles at the ultrastructural level, which helps reduce Pb toxicity to essential activities [31]. Pb^{2+} deposition in *S. drummondii* root cell walls, plasma membranes, and vacuoles has been investigated via transmission electron microscopy [106]. According to [34], one of the key machineries for Pb tolerance in the grass *Brachiaria decumbens* is its capacity to confiscate insoluble lead in its cell walls. Pb is mostly held in the tracheid cell walls of vascular bundles during transit. Vascular bundles are not very mobile in *S. alfredii*, and the metal is mostly confined to the cell walls [22]. The majority of the lead that is retained in the cell walls may be attached to pectin-like substances or exist in insoluble forms. According to Gupta et al. [40, 41], chloropyromorphite ($Pb_5(PO_4)3Cl$) is a major source of lead that is collected in the cell walls of grass roots. Angulano et al. (2022) reported the presence of the same substance in the roots of Agrostis tenuis plants cultivated in soil. According to [124] lead in Arabidopsis roots has a significant affinity for galacturonic acid, which is the primary constituent of the homogalacturonan and rhamnogalacturonan pectin domains. These findings suggest that the deactivation of lead by cell wall components is complicated.

4.2 *Chelation and Detoxification*

Proteins that bind metals with low molecular weights are called phytochelatins (PCs). The bonds formed between proteins and toxic metals are called mercaptide bonds [32, 102]. As physiologically active thiols, they play crucial roles in defensive plant cells from oxidative stress [15, 121]. In lead-tolerant plants, chelated lead reduces stress in combination with PC production and complexation. It has also been observed that lead complexed by thiol groups exists, most likely as phytochelatins [125]. Therefore, Pb^{2+} accumulation directly causes PCs to accumulate in the lead-hyperaccumulator aquatic fern *Salvinia minima*, and phytochelatin production plays a role in this method as a coping mechanism [87]. *Vetiveria zizanioides* accumulates up to ~20,000 mg kg^{-1} DW lead in its roots and ~ 3,500 mg kg^{-1} DW lead in its shoots by producing phytochelatins (PCs) and forming Pb-PC complexes [70]. *Ceratophyllum demersum* detoxifies lead via induction of PCs and the antioxidant system [1]. The role of GSH, a precursor of PCs, in *S. alfredii* lead decontamination was reported by Sun et al. [113]. Nevertheless, this was accomplished without triggering PCs. These findings indicate that in high-stress environments without PCs, GSH could be essential for lead detoxification. Therefore, by capturing resolvable lead in the cytoplasm and moving it to vacuoles and chloroplasts, phytochelatin lessens the harmful effects of Pb^{2+} in cells [117].

4.3 Antioxidant Defense Mechanism (ADM)

One of the main tactics used to combat the dangerous levels of Pb-induced oxidative stress is the activation of antioxidants. Antioxidants play a decisive role in preventing oxidative damage to cellular components by scavenging reactive oxygen species (ROS) produced under lead stress [80, 111]. Superoxide dismutase (SOD), peroxidase (POD), and catalase (CAT) are enzymes that help plants absorb iron from the environment by altering (ROS) into more benign compounds. Apart from this particular method, a number of enzymatic antioxidants such as carotenoids, glutathione, ascorbic acid, and tocopherols are also important. Furthermore, mitochondria and peroxisomes are important organelles in the metabolism of ROS. ADM is regulated by the transcription factors (TFs) WRKY and NAC as well as the mitogen-activated protein kinase (MAPK) signaling pathway [57, 107, 126]. Figure 3 shows various tolerance mechanisms against lead.

4.3.1 Enzymatic Antioxidants (EA)

Enzymatic antioxidants are assumed to be the first line of protection against Pb-induced oxidative stress in plants. Superoxide radicals (O_2•−) are catalyzed by SOD into O_2 and hydrogen peroxide (H_2O_2).Catalases and peroxidases work together to detoxify H_2O_2 and convert it into H_2O and O_2, preventing oxidative damage to biological components. Together with the critical role of TFs such as AP2/ERF and NAC in altering the expression of antioxidant genes, the activity and expression of EA are tightly regulated in response to lead exposure [49].

4.3.2 Nonenzymatic Antioxidants

Plants are shielded from oxidative stress and maintain redox equilibrium via the collective action of enzymatic and nonenzymatic antioxidants. As a critical redox buffer, GSH is essential for both the renaissance of other antioxidants and the scavenging of reactive oxygen species (ROS). Ascorbic acid is a powerful antioxidant that both directly scavenges (ROS) and regenerates the oxidized forms of other antioxidants. Tocopherols, or vitamin E, are another type of nonenzymatic antioxidant that plays a substantial role in shielding cell membranes from ROS-induced lipid peroxidation [33, 55, 100].

4.4 Protective Role of Cations and Related Genes in Lead Tolerance

It is commonly believed that cations, including Ca^{2+} and Mg^{2+}, reduce the toxicitiy of heavy metals by competing with them at certain locations. Lead phosphate precipitation is likely the cause of Pb-induced P shortages, which are somewhat linked to Pb toxicity [10, 23]. Lead and cadmium may participate with bivalent cations for conveyance into the roots of rice plants, which is one of the possible ways by which Ca^{2+} protects against Pb^{2+} toxicity [58]. Lead enters roots mostly through Ca^{2+} permeable channels, as several authors have shown [39]. Several cation transporters found in the plasma membrane of plants provide phytotoxic metals with possible entrance points into plant cells. Plant cells are unlikely to have specialized Pb transporters since lead (Pb) is a highly harmful and unnecessary metal for plants [116]. Expanding understanding of the mechanisms underlying lead absorption, transportation, and aggregation by plants is essential for their effective removal of lead from contaminated soils. NtCBP4 (*Nicotiana tabacum* calmodulin-binding protein) membrane protein channels are present in *N. tabacum*, according to [11], which resemble the ring nucleotide non-selective cation channels found in mammals and facilitate the transportation of lead ions cross the plasma membrane into plant cells. This study revealed that Pb can pass through the membrane of calcium ion channels and that NtCBP4 can increase tobacco's capacity to absorb lead. Consequently, NtCBP4 was the first plant protein to be able to control Pb accumulation and tolerance in plants. Using the model plant Arabidopsis, [112] documented the initial effective vacuolar compartmentalization of lead in plant shoots by overexpressing the yeast cadmium factor (YCF1) protein. The function of YCF1 is demonstrated by increased tolerance of the plants and the sequestration of Pb^{2+} bound to glutathione into vacuoles from the cytoplasm. Agrobacterium-mediated genetic transformation was used to introduce YCF1 into *B. juncea* plants to produce high biomass plant species. Research by Bhuiyan et al. [16, 17] revealed that the Pb content in YCF1-transgenic plants is 1.4–1.6 times greater than that in wild-type (WT) plants with lead accumulation in transgenic lines reaching 1.7–2.2 times greater than that in WT plants. These findings suggest that YCF1-expressing transgenic plants could be valuable for lead phytoremediation. The tolerance of the shrub *Nicotiana glauca* R. Graham to lead (Pb) significantly improved because of the overexpression of phytochelatin synthase (TaPCS1). This results in seedling roots that are 150% longer than those of WT plants. Additionally, seedlings of modified plants grown in mining soils with 1572 mg kg^{-1} lead accumulate twice as much lead as do wild-type plants. According to [37], these findings suggest that modified *N. glauca* plants are a very promising novel instrument for use in phytoremediation operations. In *B. juncea*, the overexpression of ATM3, a member of the *Arabidopsis thaliana* ATP binding cassette transporter family, resulted in increased tolerance to lead (Pb^{2+}) stress. Crucially, the shoot tissues of the transgenic seedlings presented Pb^{2+} concentrations that were between 1.3 and 1.7 times greater than those of the WT seedlings, indicating that the transgenic plants accumulated Pb^{2+} at a considerably greater rate [16, 17].

4.5 Phytoextraction of Lead by Plants

Plant species such as *Tagetes erecta*, *Medicago sativa*, *Helianthus annuus*, and others have been identified as possible hyperaccumulators of lead (Pb), which could help remove Pb from contaminated soils [20]. Unique to these plant species, they are able to absorb exceptionally high Pb concentrations from the soil without harming themselves. Because these plants can remove precious metals from polluted soil or clean contaminated soils, they are very interesting in sectors such as environmental science, bioremediation, and phytomining. According to [98], *Sesbania drummondii*, a leguminous shrub cultivated in a Pb-contaminated nutrient solution, can accumulate 1% Pb in its young shoots. According to field surveys, *Buddleja asiatica* exhibits remarkable resistance to lead (Pb); when cultivated in contaminated soils, the plant can accumulate up to 4335 mg kg^{-1} DW Pb in its shoots [118]. Moreover, common buckwheat cultivated in Pb-contaminated soil may acquire a high quantity of Pb without suffering major harm. Therefore, to ensure that phytoremediation is efficient, methods that encourage metal buildup in plants are needed. Plants with a transfer factor above one are usually categorized as hyperaccumulators, whereas those with a transfer factor below one are categorized as nonaccumulators [63].

4.6 Improving Phytoextraction Through the Use of Lead-Resistant Microorganisms

Bacteria linked with plants are essential for the host's environmental adaptation. Thus, rhizospheric microbial communities may improve plant resistance to heavy metal toxicity by generating compounds that stimulate plant development, such as siderophores, IAA, and 1-aminocyclopropane-1-carboxylate (ACC) deaminase [38, 81]. According to recent studies, plant endophytes may play a role in the phytoremediation of heavy metals [44, 68, 94]. *Burkholderia* sp., a species of bacteria resistant to heavy metals, dramatically increases the biomass of tomato and maize plants and increases the Pb level in tissue by 38–192% [46]. In contrast, inoculation of maize with multiple diazotrophs seems to be a more effective method than inoculation with a single culture to remove lead while simultaneously increasing biomass and plant growth. According to previous reports, the Pb accumulation and tolerance of the male plants are increased by arbuscular mycorrhiza (AM) fungi [99]. The isolated strain NBT of *Bacillus edaphicus* from the rhizosphere of cotton (*Gossypium hirsutum*) when incubated with *B. juncea* (Indian mustard) increases Pb uptake from 18 to 46% in soil treated with 400 and 800 mg kg^{-1} of Pb. Indole-3-acetic acid (IAA), siderophores, and ACC deaminase can all be produced by strain NBT [108]. Rape plant tissues and rhizosphere soils are invaded by endophytic bacteria that may aid in plant development and lead to uptake by generating IAA, siderophore, or ACC deaminase or by solubilizing lead in the soil [42, 127]. The addition of active mycelia from

CBRF59 appears to significantly increase the availability of soil Pb from contaminated Pb soils by 77%, according to [27]. *Bacillus* sp. isolated from the cadmium hyperaccumulator *Solanum nigrum* was able to preferentially absorb 80% of Pb^{2+} at a starting dosage of 10 mg^{L-1}. Compared with the uninoculated control inoculation with *Ochrobactrum intermedium* BN-3 significantly increased the biomass and Pb accumulation of *Eucalyptus camaldulensis* in hydroponic experiments. [119], reported that naturally occurring rhizospheric bacteria located on the root surfaces of *E. camaldulensis* play critical role in fostering Pb accumulation and plant growth. Bacteria that are capable of solubilizing lead and encouraging plant growth in polluted soils could be used to effectively remediate lead-affected soils. Different plant shows different toxic effects and exhibit different tolerance mechanism (Table 1).

5 Conclusion and Future Prospective

Plants employ various defense mechanisms to combat lead poisoning. Their primary strategy is to exclude or sequester the metal within the cell wall, thereby preventing its entry into the cell. Antioxidants that are both enzymatic and nonenzymatic are found in several subcellular compartments as part of a secondary defense mechanism, which helps to mitigate oxidative damage caused by reactive oxygen species (ROS). Consequently, enhancing a plant's resilience to Pb-induced oxidative stress could be pivotal in selecting species for the remediation of Pb-contaminated soils. Given that plants respond to lead toxicity through diverse mechanisms, the most effective strategy to augment their phytoremediation potential lies in integrating these mechanisms. This includes selecting lead hyperaccumulator species, stimulating beneficial microorganisms that facilitate lead uptake, utilizing phytochelatins for lead complexation, promoting lead accumulation in cell walls and vacuolar compartmentalization, and identifying plant genes that increase lead tolerance. Additionally, the application of exogenous amendments such as chemicals, auxins, or nitric oxide may further improve lead resilience. However, the potential ramifications of phytoremediation on the food chain raise significant concerns, particularly regarding herbivores and omnivores that consume foliage contaminated with heavy metals. Furthermore, the implications of introducing non-native plant species and employing synthetic chemical chelates for Pb phytoextraction, which increase lead bioavailability and uptake in plants, highlight urgent areas for future research.

Table 1 List of various plants with their toxicity/tolerance mechanism against lead

S. No.	Species	Toxicity/Tolerance mechanism	Concentration of lead accumulated in plant tissues	Key references
1	*Boehmeria nivea* L.	Metal accumulation and supply	9.41 mg kg^{-1}	[128]
2	*Jatropha curcas* L.	Limiting metal transport to the aerial parts and maintaining antioxidant enzymes activity	21 mg kg^{-1}	[82]
3	*Gomphrena celosoides* Mart.	Enhanced antioxidant enzyme activities and increased level of proline	155.791 mg kg^{-1}	[4]
4	*Biscutella auriculata* L.	Increased activities of antioxidant enzyme activities	3686.10 mg kg^{-1}	[88]
5	*Glycine max* L.	Restricted translocation of Pb from roots to shoots in presence of Arbuscular Mycorrhizal Fungi	Reduced translocation of from root to shoot by 57.6%	[5]
6	*Medicago sativa* L.	Pb accumulation in shoot	87.95 mg kg^{-1}	[74]
7	*Tagetes erecta* L.	Pb accumulation	54.8 mg kg^{-1}in root and 10.6 mg kg^{-1}in shoot	[71]
8	*Triticum aestivum* L.	Increased activities of antioxidants and restrict root to shoot translocation	0.027 mg kg^{-1}	[62]
9	*Helianthus annuus* L.	Accumulation of Pb in upper and lower biomass, suppression of lipid peroxidation, Enhancement in antioxidant property and chlorophylls and carotenes amount	693.69 mg kg^{-1}	[20]
10	*Solidago canadensis* L.	Exclusion or reduction in the uptake of Pb	7.14 mg kg^{-1}	[79]

Competing Interests The authors declare that they have no competing interest.

Declaration of Generative AI and AI-Assisted Technologies in the Writing Process During the preparation of this work, the book chapter author(s) have not taken any assistance from AI, and the book chapter author(s) reviewed and edited the content as needed and take(s) full responsibility for the content of the publication.

References

1. Aasim M, Akgur O, Mustafa Z (2023) (*Ceratophyllum demersum* L.): an aquatic macrophyte for phytoremediation. In: Environmental pollution impact on plants. Apple Academic, pp 205–225
2. Abedi T, Gavanji S, Mojiri A (2022) Lead and zinc uptake and toxicity in maize and their management. Plants 11(15):1922
3. Abubakar US (2024) Copper and lead as environmental pollutants and their elemental toxicity – a review. Sci World J 19(2):342–346
4. Adejumo SA, Tiwari S, Thul S, Sarangi BK (2019) Evaluation of lead and chromium tolerance and accumulation level in *Gomphrena celosoides* Mart.: a novel metal accumulator from lead acid battery waste contaminated site in Nigeria. Int J Phytorem 21(13):1341–1355
5. Adeyemi NO, Atayese MO, Sakariyawo OS, Azeez JO, Sobowale SPA, Olubode A, Adeoye S (2021) Alleviation of heavy metal stress by arbuscular mycorrhizal symbiosis in *Glycine max* L. grown in copper, lead, and zinc contaminated soils. Rhizosphere 18:100325
6. Akay A (2022) Lead tolerance and accumulation characteristics of Cubana Kordes rose in lead-contaminated soil. Environ Monit Assess 194(4):307
7. Ali S, Naseer S, Rehman M, Wei Z (2024) Recent trends and sources of lead toxicity: a review of state-of-the-art nano-remediation strategies. J Nanopart Res 26(7):1–29
9. Ankush, Lamba S, Ritambhara, Diwedi A, Kumar S, Singh V (2023) Source and distribution of lead in soil and plant—a review. In: Lead toxicity: challenges and solution, pp 3–16
10. Antosiewicz DM (2005) Study of calcium-dependent lead-tolerance on plants differing in their level of Ca-deficiency tolerance. Environ Pollut 134(1):23–34
11. Arazi T, Kaplan B, Fromm H (2000) A high-affinity calmodulin-binding site in a tobacco plasma-membrane channel protein coincides with a characteristic element of cyclic nucleotide-binding domains. Plant Mol Biol 42:591–601
12. Ashraf A, Bhardwaj S, Ishtiaq H, Devi YK, Kapoor D (2021) Lead uptake, toxicity and mitigation strategies in plants. Plant Arch 21(1):09725210
13. Ashraf U, Mahmood MHUR, Hussain S, Abbas F, Anjum SA, Tang X (2020) Lead (Pb) distribution and accumulation in different plant parts and its associations with grain Pb contents in fragrant rice. Chemosphere 248:126003
14. Aslam M, Aslam A, Sheraz M, Ali B, Ulhassan Z, Najeeb U, Gill RA (2021) Lead toxicity in cereals: mechanistic insight into toxicity, mode of action, and management. Front Plant Sci 11:587785
15. Bagchi R, Arifuzzaman MD, Rahman MA, Kabir AH, Parvez MS (2024) Sulfur mitigates lead toxicity through phytochelatin-driven vacuolar sequestration and antioxidant defense in mustard (*Brassica juncea* L.). Biologia 1–9
16. Bhuiyan MSU, Min SR, Jeong WJ, Sultana S, Choi KS, Lim YP, Liu JR (2011) An improved method for agrobacterium-mediated genetic transformation from cotyledon explants of *Brassica juncea* L. Plant Biotechnol 28(1):17–23
17. Bhuiyan MSU, Min SR, Jeong WJ, Sultana S, Choi KS, Lee Y, Liu JR (2011) Overexpression of AtATM3 in *Brassica juncea* L. confers enhanced heavy metal tolerance and accumulation. Plant Cell Tissue Organ Cult (PCTOC) 107:69–77
18. Chauhan P, Mathur J (2018) Potential of *Helianthus annuus* L. for phytoremediation of multiple pollutants in the environment: a review. J Med Biol Sci 4(3):5–16
19. Chauhan P, Mathur J (2020) Phytoremediation efficiency of *Helianthus annuus* L. for reclamation of heavy metals-contaminated industrial soil. Environ Sci Pollut Res 27(24):29954–29966
20. Chauhan P, Rajguru AB, Dudhe MY, Mathur J (2020) Efficacy of lead (Pb) phytoextraction of five varieties of *Helianthus annuus* L. from contaminated soil. Environ Technol & Innov 18:100718
21. Chen F, Aqeel M, Maqsood MF, Khalid N, Irshad MK, Ibrahim M, Lam SS (2022) Mitigation of lead toxicity in *Vigna radiata* L. R. Wilczek genotypes by silver nanoparticles. Environ Pollut 308:119606

22. Chen Z, Liu Q, Chen D, Wu Y, Hamid Y, Lin Q, Yang X (2024) Enhancing the phytoextraction efficiency of heavy metals in acidic and alkaline soils by *Sedum alfredii* L. Hance: a study on the synergistic effect of plant growth regulator and plant growth-promoting bacteria. Sci Total Environ 932:173029
23. Cheyns K, Peeters S, Delcourt D, Smolders E (2012) Lead phytotoxicity in soils and nutrient solutions is related to lead induced phosphorus deficiency. Environ Pollut 164:242–247
24. Collin S, Baskar A, Geevarghese DM, Ali MNV, Bahubali P, Choudhary R, Swamiappan S (2022) Bioaccumulation of lead (Pb) and its effects in plants: a review. J Hazard Mater Lett 3:100064
25. Dalyan E, Yüzbaşıoğlu E, Akpınar I (2020) Physiological and biochemical changes in plant growth and different plant enzymes in response to lead stress. In: Lead in plants and the environment, pp 129–147
26. de Francisco P, Amaro F, Martín-González A, Serrano A, Gutiérrez JC (2023) Quantitative proteomic analyses of a Pb-adapted Tetrahymena thermophila strain reveal the cellular strategy to Pb (II) stress including lead biomineralization to chloropyromorphite. Sci Total Environ 891:164252
27. Deng Z, Cao L, Huang H, Jiang X, Wang W, Shi Y, Zhang R (2011) Characterization of Cd- and Pb-resistant fungal endophyte Mucor sp. CBRF59 isolated from rapes (*Brassica chinensis* L.) in a metal-contaminated soil. J Hazard Mater 185(2–3):717–724
28. Dewi T, Martono E, Hanudin E, Harini R (2022) Impact of agrochemicals application on lead and cadmium concentrations in shallot fields and their remediation with biochar, compost, and botanical pesticides. IOP Conf Ser: Earth Environ Sci 1109(1):012050. IOP Publishing
29. Djavaheri M, Clothier L, Kindrachuk C, Borhan MH (2023) Plasticity of host cell organelles in response to Pb infection. bioRxiv, 2023-04
30. Dong J, Li X (2023) Lead pollution-related health of children in China: disparity, challenge, and policy. Sci Total Environ 882:163383
31. El Khattabi O, Lamwati Y, Henkrar F, Collin B, Levard C, Colin F, Fahr M (2024) Lead-induced changes in plant cell ultrastructure: an overview. BioMetals 1–19
32. Faizan M, Alam P, Hussain A, Karabulut F, Tonny SH, Cheng SH, Hayat S (2024) Phytochelatins: a key regulator against heavy metal toxicity in plants. Plant Stress 100355
33. Fatemi H, Esmaielpour B (2021) Investigation of biological and non-biological interaction on enzymatic and non-enzymatic antioxidant of coriander (*Coriandrum sativum* L.) under Pb stress condition. J Plant Res (Iran J Biol) 34(3):687–705
34. Ferreira DA, Gaião LM, Kozovits AR, Messias MC (2022) Evaluation of metal accumulation in the forage grass (*Brachiaria decumbens* L.) Stapf grown in contaminated soils with iron tailings. Integr Environ Assess Manag 18(2):528–538
35. Ghuge SA, Nikalje GC, Kadam US, Suprasanna P, Hong JC (2023) Comprehensive mechanisms of heavy metal toxicity in plants, detoxification, and remediation. J Hazard Mater 450:131039
36. Giannakoula A, Therios I, Chatzissavvidis C (2021) Effect of lead and copper on photosynthetic apparatus in citrus (*Citrus aurantium* L.) plants. The role of antioxidants in oxidative damage as a response to heavy metal stress. Plants 10(1):155
37. Gisbert C, Ros R, De Haro A, Walker DJ, Bernal MP, Serrano R, Navarro-Aviñó J (2003) A plant genetically modified that accumulates Pb is especially promising for phytoremediation. Biochem Biophys Res Commun 303(2):440–445
38. Gul I, Manzoor M, Hashim N, Shah GM, Waani SPT, Shahid M, Arshad M (2021) Challenges in microbially and chelate-assisted phytoextraction of cadmium and lead–a review. Environ Pollut 287:117667
39. Gupta DK, Huang HG, Corpas FJ (2013) Lead tolerance in plants: strategies for phytoremediation. Environ Sci Pollut Res 20:2150–2161
40. Gupta M, Dwivedi V, Kumar S, Patel A, Niazi P, Yadav VK (2024) Lead toxicity in plants: mechanistic insights into toxicity, physiological responses of plants and mitigation strategies. Plant Signal Behav 19(1):2365576

41. Gupta S, Kaur M, Kaur A, Kaur A, Kumar R, Singh V, Prasad B (2024) Microbial tolerance strategies against lead toxicity. In: Lead toxicity mitigation: sustainable nexus approaches. Springer Nature Switzerland, Cham, pp 183–210
42. Harun FA, Yusuf MR, Usman S, Shehu D, Babagana K, Sufyanu AJ, Yakasai HM (2023) Bioremediation of lead-contaminated environment by *Bacillus cereus* strain BUK_BCH_BTE2: isolation and characterization of the bacterium. Case Stud Chem Environ Eng 8:100540
43. Hassan SH, Chafik Y, Lebrun M, Sferra G, Bourgerie S, Scippa GS, Trupiano D (2023) Lead toxicity and tolerance in plants: insights from omics studies. In: Heavy metal toxicity and tolerance in plants: a biological, omics, and genetic engineering approach, pp 373–405
44. He X, Xu M, Wei Q, Tang M, Guan L, Lou L, Xia Y (2020) Promotion of growth and phytoextraction of cadmium and lead in *Solanum nigrum* L. mediated by plant-growth-promoting rhizobacteria. Ecotoxicol Environ Saf 205:111333
45. Howard JA, David L, Lux F, Tillement O (2024) Low-level, chronic ingestion of lead and cadmium: the unspoken danger for at-risk populations. J Hazard Mater 135361
46. Huang GH, Tian HH, Liu HY, Fan XW, Liang Y, Li YZ (2013) Characterization of plant-growth-promoting effects and concurrent promotion of heavy metal accumulation in the tissues of the plants grown in the polluted soil by *Burkholderia* strain LD-11. Int J Phytorem 15(10):991–1009
47. Huihui Z, Xin L, Zisong X, Yue W, Zhiyuan T, Meijun A, Guangyu S (2020) Toxic effects of heavy metals Pb and Cd on mulberry (*Morus alba* L.) seedling leaves: photosynthetic function and reactive oxygen species (ROS) metabolism responses. Ecotoxicol Environ Saf 195:110469
48. Ilyas MZ, Sa KJ, Ali MW, Lee JK (2024) Toxic effects of lead on plants: integrating multi-omics with bioinformatics to develop Pb-tolerant crops. Planta 259(1):18
49. Iqbal N, Tanzeem-ul-Haq HS, Turan V, Iqbal M (2023) Soil amendments and foliar melatonin reduced Pb uptake, and oxidative stress, and improved spinach quality in Pb-contaminated soil. Plants 12(9):1829
50. Israr M, Jewell A, Kumar D, Sahi SV (2011) Interactive effects of lead, copper, nickel and zinc on growth, metal uptake and antioxidative metabolism of *Sesbania drummondii*. J Hazard Mater 186(2–3):1520–1526
51. Jagota N, Singh S, Kaur H, Kaur R, Sharma A (2024) Oxidative stress in lead toxicity in plants and its amelioration. In: Lead toxicity mitigation: sustainable nexus approaches. Springer Nature Switzerland, Cham, pp 299–333
52. Jain S, Panwar R, Mathur J (2023) Design and development of colorimetric, whole-cell based, electrochemical biosensors for arsenic detection. Inorg Chem Commun 153:110730
53. Jogawat A, Yadav B, Chhaya, Narayan OP (2021) Metal transporters in organelles and their roles in heavy metal transportation and sequestration mechanisms in plants. Physiol Plant 173(1):259–275
54. Joshi S, Gangola S, Bhandari G, Bhandari NS, Nainwal D, Rani A, Slama P (2023) Rhizospheric bacteria: the key to sustainable heavy metal detoxification strategies. Front Microbiol 14:1229828
55. Juárez-Maldonado A (2023) Influence of nanomaterials on non-enzymatic antioxidant defense activities in plants. Nanomaterial interactions with plant cellular mechanisms and macromolecules and agricultural implications. Springer International Publishing, Cham, pp 273–298
56. Jyothi NR (2020) Heavy metal sources and their effects on human health. In: Heavy metals-their environmental impacts and mitigation, pp 1–12
57. Kaya C, Akram NA, Sürücü A, Ashraf M (2019) Alleviating effect of nitric oxide on oxidative stress and antioxidant defence system in pepper (*Capsicum annuum* L.) plants exposed to cadmium and lead toxicity applied separately or in combination. Sci Hortic 255:52–60
58. Kim YY, Yang YY, Lee Y (2002) Pb and Cd uptake in rice roots. Physiol Plant 116(3):368–372
59. Kohli SK, Handa N, Bali S, Khanna K, Arora S, Sharma A, Bhardwaj R (2020) Current scenario of Pb toxicity in plants: unraveling plethora of physiological responses. Rev Environ Contam Toxicol 249:153–197

60. Kosakivska IV, Babenko LM, Romanenko KO, Korotka IY, Potters G (2021) Molecular mechanisms of plant adaptive responses to heavy metals stress. Cell Biol Int 45(2):258–272
61. Kumar K, Singh D (2024) Toxicity and bioremediation of the lead: a critical review. Int J Environ Health Res 34(4):1879–1909
62. Lamhamdi M, El Galiou O, Bakrim A, Novoa-Munoz JC, Arias-Estevez M, Aarab A, Lafont R (2013) Effect of lead stress on mineral content and growth of wheat (*Triticum aestivum* L.) and spinach (*Spinacia oleracea* L.) seedlings. Saudi J Biol Sci 20(1):29–36
63. Laptiev V, Apori SO, Giltrap M, Tian F, Ryzhenko N (2024) Bioaccumulation of Cr, Zn, Pb and Cu in *Ambrosia artemisiifolia* L. and *Erigeron canadensis* L. Resources 13(3):43
64. Li Y, Lv H, Xue C, Dong N, Bi C, Shan A (2021) Plant polyphenols: potential antidotes for lead exposure. Biol Trace Elem Res 199:3960–3976
65. Liu Z, Bai Y, Gao J, Li J (2023) Driving factors on accumulation of cadmium, lead, copper, zinc in agricultural soil and products of the North China Plain. Sci Rep 13(1):7429
66. Ma C, Lin L, Yang J, Liu F, Berrettoni M, Zhang K, Zhang H (2023) Mechanisms of lead uptake and accumulation in wheat grains based on atmospheric deposition-soil sources. Sci Total Environ 885:163845
67. Ma W, Zhao B, Lv X, Feng X (2022) Lead tolerance and accumulation characteristics of three Hydrangea cultivars representing potential lead-contaminated phytoremediation plants. Hortic Environ Biotechnol 63(1):23–38
68. Manzoor M, Gul I, Ahmed I, Zeeshan M, Hashmi I, Amin BAZ, Arshad M (2019) Metal tolerant bacteria enhanced phytoextraction of lead by two accumulator ornamental species. Chemosphere 227:561–569
69. Martín-León V, Rubio C, Rodríguez-Hernández Á, Zumbado M, Acosta-Dacal A, Henríquez-Hernández LA, Luzardo OP (2023) Evaluation of essential, toxic and potentially toxic elements in leafy vegetables grown in the Canary Islands. Toxics 11(5):442
70. Masinire F (2022) Phytoremediation of Cr (VI) and Pd (II) using *vetiver grass* (Chrysopogon zizanioides). Master's thesis, University of Pretoria, South Africa
71. Madanan MT, Shah IK, Varghese GK, Kaushal RK (2021) Application of Aztec Marigold (*Tagetes erecta* L.) for phytoremediation of heavy metal polluted lateritic soil. Environ Chem Ecotoxicol 3:17–22
72. Mathur J, Chauhan P (2020) Mechanism of toxic metal uptake and transport in plants. In: Sustainable solutions for elemental deficiency and excess in crop plants, pp 335–349
73. Mathur J, Chauhan P, Srivastava S (2023) Comparative evaluation of cadmium phytoremediation potential of five varieties of *Helianthus annuus* L. Int J Phytorem 25(6):799–810
74. Mathur J, Panwar R (2024) Synergistic effect of pyrene and heavy metals (Zn, Pb, and Cd) on phytoremediation potential of *Medicago sativa* L. (alfalfa) in multi-contaminated soil. Environ Sci Pollut Res 31(14):21012–21027
75. Mathur J, Sharma M, Jain S (2024) Arsenic tolerance in *Tagetes erecta* L.: phytoaccumulation, physicochemical and anatomical studies through electron microscopy. Environ Exp Bot 226:105910
76. Mathur J, Yadav S, Srivastava S (2023) Evaluation of morpho-physiological and antioxidant responses of *Moringa oleifera* Lam. to lead exposure. J Trace Elem Miner 6:100093
77. Methela NJ, Islam MS, Lee DS, Yun BW, Mun BG (2023) S-nitrosoglutathione (GSNO)-mediated lead detoxification in soybean through the regulation of ROS and metal-related transcripts. Int J Mol Sci 24(12):9901
78. Meyers DE, Auchterlonie GJ, Webb RI, Wood B (2008) Uptake and localisation of lead in the root system of *Brassica juncea* L. Environ Pollut 153(2):323–332
79. Miletic Z, Jaric S, Jonjev M, Mitrovic M, Pavlovic D, Matic M, Pavlovic P (2024) Phytoremediation potential of invasive plant species for potentially toxic elements along the Sava River upstream. Heliyon 10(13)
80. Mishra N, Jiang C, Chen L, Paul A, Chatterjee A, Shen G (2023) Achieving abiotic stress tolerance in plants through antioxidative defense mechanisms. Front Plant Sci 14:1110622
81. Mishra R, Datta SP, Annapurna K, Meena MC, Dwivedi BS, Golui D, Bandyopadhyay K (2019) Enhancing the effectiveness of zinc, cadmium, and lead phytoextraction in polluted soils by using amendments and microorganisms. Environ Sci Pollut Res 26:17224–17235

82. Mohamed AA, Dardiry MH, Samad A, Abdelrady E (2020) Exposure to lead (Pb) induced changes in the metabolite content, antioxidant activity and growth of *Jatropha curcas* L. Trop Plant Biol 13:150–161
83. Natasha Shahid M, Khalid S, Saleem M (2022) Unrevealing arsenic and lead toxicity and antioxidant response in spinach: a human health perspective. Environ Geochem Health 1–10
84. Niu Z, Li X, Mahamood M (2023) Accumulation potential cadmium and lead by sunflower (*Helianthus annuus* L.) under citric and glutaric acid-assisted phytoextraction. Int J Environ Res Public Health 20(5):4107
85. Noor I, Sohail H, Sun J, Nawaz MA, Li G, Hasanuzzaman M, Liu J (2022) Heavy metal and metalloid toxicity in horticultural plants: tolerance mechanism and remediation strategies. Chemosphere 303:135196
86. Obeng-Gyasi E (2019) Sources of lead exposure in various countries. Rev Environ Health 34(1):25–34
87. Pang YL, Quek YY, Lim S, Shuit SH (2023) Review on phytoremediation potential of floating aquatic plants for heavy metals: a promising approach. Sustainability 15(2):1290
88. Peco JD, Higueras P, Campos JA, Olmedilla A, Romero-Puertas MC, Sandalio LM (2020) Deciphering lead tolerance mechanisms in a population of the plant species *Biscutella auriculata* L. from a mining area: accumulation strategies and antioxidant defenses. Chemosphere 261:127721
89. Porterfield K, Hore P, Whittaker SG, Fellows KM, Mohllajee A, Azimi-Gaylon S, Fuller R (2024) A snapshot of lead in consumer products across four US jurisdictions. Environ Health Perspect 132(7):075002
90. Priya AK, Muruganandam M, Ali SS, Kornaros M (2023) Clean-up of heavy metals from contaminated soil by phytoremediation: a multidisciplinary and eco-friendly approach. Toxics 11(5):422
91. Rady MM, Salama MM, Kuşvuran S, Kuşvuran A, Ahmed AF, Ali EF, Mahmoud AE (2023) Exploring the role of novel biostimulators in suppressing oxidative stress and reinforcing the antioxidant defense systems in *Cucurbita pepo* plants exposed to cadmium and lead toxicity. Agronomy 13(7):1916
92. Rahman S U, Qin A, Zain M, Mushtaq Z, Mehmood F, Riaz L, Shehzad M (2024) Pb uptake, accumulation, and translocation in plants: plant physiological, biochemical, and molecular response: a review. Heliyon.
93. Raj K, Das AP (2023) Lead pollution: Impact on environment and human health and approach for a sustainable solution. Environ Chem Ecotoxicol 5:79–85
94. Rajkumar M, Ae N, Freitas H (2009) Endophytic bacteria and their potential to enhance heavy metal phytoextraction. Chemosphere 77(2):153–160
95. Rani M, Vikas Kumar R, Lathwal M, Kamboj A (2024) Effect and responses of lead toxicity in plants. In Lead toxicity mitigation: sustainable Nexus approaches. Springer Nature Switzerland, Cham, pp 211–241
96. Rodriguez E, da Conceição SM, Azevedo R, Correia C, Moutinho-Pereira J, Ferreira de Oliveira JMP, Dias MC (2015) Photosynthesis light-independent reactions are sensitive biomarkers to monitor lead phytotoxicity in a Pb-tolerant *Pisum sativum* L. cultivar. Environ Sci Pollut Res 22:574–585
97. Rosca M, Cozma P, Minut M, Hlihor RM, Beţianu C, Diaconu M, Gavrilescu M (2021) New evidence of model crop Brassica napus L. in soil clean-up: comparison of tolerance and accumulation of lead and cadmium. Plants 10(10):2051
98. Sahi SV, Bryant NL, Sharma NC, Singh SR (2002) Characterization of a lead hyperaccumulator shrub, *Sesbania drummondii* L. Environ Sci Technol 36(21):4676–4680
99. Salazar MJ, Menoyo E, Faggioli V, Geml J, Cabello M, Rodriguez JH, Becerra AG (2018) Pb accumulation in spores of arbuscular mycorrhizal fungi. Sci Total Environ 643:238–246
100. Salbitani G, Maresca V, Cianciullo P, Bossa R, Carfagna S, Basile A (2023) Non-protein thiol compounds and antioxidant responses involved in bryophyte heavy-metal tolerance. Int J Mol Sci 24(6):5302

101. Schmidt F, Ledermann L, Schäffer A, Snaith HJ, Lenz M (2022) Rapid sequestration of perovskite solar cell-derived lead in soil. J Hazard Mater 436:128995
102. Seregin IV, Kozhevnikova AD (2023) Phytochelatins: Sulfur-containing metal (loid)-chelating ligands in plants. Int J Mol Sci 24(3):2430
103. Sharma J, Kumar S, Kumar V, Singh P, Khyalia P, Verma S, Sharma A (2023) Foliar application of *Glycine betaine* L. to ameliorate lead toxicity in barley plants by modulating antioxidant enzyme activity and biochemical parameters. Environ Res Commun 5(7):075002
104. Sharma M, Mathur J (2023) Phytoaccumulation of zinc from contaminated soil using ornamental plant species *Helianthus annuus* L. and *Tagetes erecta* L. Int J Phytorem 25(10):1289–1305
105. Sharma M, Mathur J, Goswami P (2023) Ultrastructural analysis and biochemical evaluation provide evidence for arsenic reclamation from contaminated soil through *Helianthus annuus* L. Acta Physiol Plant 45(7):84
106. Sharma NC, Gardea-Torresdey JL, Parsons J, Sahi SV (2004) Chemical speciation and cellular deposition of lead in *Sesbania drummondii* L. Environ Toxicol Chem: Int J 23(9):2068–2073
107. Shehzad J, Mustafa G, Arshad H, Ali A, Naveed NH, Riaz Z, Khan I (2023) Morpho-physiological and biochemical responses of *Brassica* species toward lead (Pb) stress. Acta Physiol Plant 45(1):8
108. Sheng XF, Jiang CY, He LY (2008) Characterization of plant growth-promoting *Bacillus edaphicus* NBT and its effect on lead uptake by Indian mustard in a lead-amended soil. Can J Microbiol 54(5):417–422
109. Shi Y, Tang L, Shao Q, Jiang Y, Wang Z, Peng C, Li Z (2024) The dynamic roles of intracellular vacuoles in heavy metal detoxification by *Rhodotorula mucilaginosa*. J Appl Microbiol 135(9):lxae241
110. Simatupang MM, Veronika E, Irfandi A, Azteria V (2024) Potential impacts of lead on health: a review of environmental exposure, population at risk, and toxic effects. J Environ Health 16(3)
111. Singh C, Singh R, Shekhar A (2023) Antioxidant defense: a key mechanism of lead tolerance. In: Lead toxicity: challenges and solution. Springer Nature Switzerland, Cham, pp 127–144
112. Song WY, Ju Sohn E, Martinoia E, Jik Lee Y, Yang YY, Jasinski M, Lee Y (2003) Engineering tolerance and accumulation of lead and cadmium in transgenic plants. Nat Biotechnol 21(8):914–919
113. Sun Q, Ye ZH, Wang XR, Wong MH (2005) Increase of glutathione in mine population of *Sedum alfredii* L.: a Zn hyperaccumulator and Pb accumulator. Phytochemistry 66(21):2549–2556
114. Tan Z, Wu C, Xuan Z, Cheng Y, Xiong R, Su Z, Wang D (2022) Lead exposure dose-dependently affects oxidative stress, AsA-GSH, photosynthesis, and mineral content in pakchoi (*Brassica chinensis* L.). Front Plant Sci 13:1007276
115. Tokarz K, Piwowarczyk B, Makowski W (2020) Mechanisms involved in photosynthetic apparatus protection against lead toxicity. In: Lead in plants and the environment, pp 117–128
116. Vatansever R, Ozyigit II, Filiz E (2017) Essential and beneficial trace elements in plants, and their transport in roots: a review. Appl Biochem Biotechnol 181:464–482
117. Verma MK, Paul A, Roy M (2024) Antioxidant defense: key mechanism of lead intolerance. In: Lead toxicity mitigation: sustainable nexus approaches. Springer Nature Switzerland, Cham, pp 263–275
118. Waranusantigul P, Kruatrachue M, Pokethitiyook P, Auesukaree C (2008) Evaluation of Pb phytoremediation potential in *Buddleja asiatica* Lour. and *B. paniculata* Lour. Water Air Soil Pollut 193:79–90
119. Waranusantigul P, Lee H, Kruatrachue M, Pokethitiyook P, Auesukaree C (2011) Isolation and characterization of lead-tolerant *Ochrobactrum intermedium* and its role in enhancing lead accumulation by *Eucalyptus camaldulensis* Dehnh. Chemosphere 85(4):584–590
120. Wójcik M, Tukiendorf A (2014) Accumulation and tolerance of lead in two contrasting ecotypes of *Dianthus carthusianorum* L. Phytochemistry 100:60–65

121. Xing Q, Hasan MK, Li Z, Yang T, Jin W, Qi Z, Zhou J (2023) Melatonin-induced plant adaptation to cadmium stress involves enhanced phytochelatin synthesis and nutrient homeostasis in *Solanum lycopersicum* L. J Hazard Mater 456:131670
122. Yaashikaa PR, Kumar PS, Jeevanantham S, Saravanan R (2022) A review on bioremediation approach for heavy metal detoxification and accumulation in plants. Environ Pollut 301:119035
123. Yu YL, Yang WY, Hara A, Asayama K, Roels HA, Nawrot TS, Staessen JA (2023) Public and occupational health risks related to lead exposure updated according to present-day blood lead levels. Hypertens Res 46(2):395–407
124. Zhan J, Huang H, Yu H, Zhang X, Zheng Z, Wang Y, Li T (2020) The combined effects of Cd and Pb enhanced metal binding by root cell walls of the phytostabilizer *Athyrium wardii* Hook. Environ Pollut 258:113663
125. Zhang LS, Yang L, Li L, Feng X, Cao D J, Liu ZW (2023) Phytochelatins of *Cladophora rupestris* in Pb^{2+} absorption and its detoxification. Environ Prot Eng 49(3)
126. Zhang L, Liu Z, Song Y, Sui J, Hua X (2024) Advances in the involvement of metals and metalloids in plant defense response to external stress. Plants 13(2):313
127. Zhang YF, He LY, Chen ZJ, Zhang WH, Wang QY, Qian M, Sheng XF (2011) Characterization of lead-resistant and ACC deaminase-producing endophytic bacteria and their potential in promoting lead accumulation of rape. J Hazard Mater 186(2–3):1720–1725
128. Zhu S, Shi W, Zhang J (2022) Effect of different ramie (*Boehmeria nivea* L. Gaud) cultivars on the adsorption of heavy metal ions cadmium and lead in the remediation of contaminated farmland soils. Open Chem 20(1):444–454

Arsenic: Uptake, Toxicity, and Tolerance in Crop Plants and Management of Arsenic-Contaminated Soil

Nidhi Panchal, Dhruvi Patel, Dharmendra Prajapati, Anilkumar Patani, and Chinmayi Joshi

Abstract Arsenic, a naturally occurring mineral, occurs in a wide range of natural environments. A thorough review of arsenic uptake, toxicity, and tolerance processes in agricultural plants is given in this chapter, along with management techniques for soils contaminated by arsenic. Arsenic poses serious threats to both agricultural productivity and human health. The first section covers the arsenic uptake mechanisms of plants and their effects on plant physiology. The chapter also includes plants' genetic and biochemical processes for arsenic resistance. Additionally, the chapter highlights the use of phytoremediation to reduce arsenic toxicity. The chapter also looks at soil management techniques like applying amendments and using bioremediation techniques to improve soil health and reduce arsenic bioavailability. By integrating data from toxicology, plant physiology, and soil science, this chapter aims to provide a thorough understanding of arsenic dynamics in agriculture and practical solutions for farmers and land managers dealing with arsenic contamination.

Keywords Arsenic · Toxicity · Tolerance · Phytoremediation · Soil management · Crop plants

1 Arsenic, a Toxic Metalloid: An Environmental Contaminant Harmful to Agriculture

Arsenic, the most abundant element in our environment, may travel great distances, adhere to minuscule airborne particles, and remain in the atmosphere for a long time. Arsenic is mostly used for various applications due to its pesticidal and herbicidal activity. Moreover, it can be used as wood preservative due to its resistance to rotting

N. Panchal · D. Patel · D. Prajapati · A. Patani · C. Joshi (✉)
Smt. S.S.Patel Nootan Science and Commerce College, Sankalchand Patel University, Mehsana, Gujarat, India
e-mail: ckjoshi.fsh@spu.ac.in

A. Husen (ed.), *Enhanced Crop Production Setup Under Various Micro and Nano Contaminants*, Smart Nanomaterials Technology,
https://doi.org/10.1007/978-981-96-6099-5_5

and degradation. Another key use of arsenic includes industrial manufacturing, electronics, and medicine. Arsenic and its compounds can be highly mobile and long-lived in the environment, but they may change the form of arsenic; bind to new particles, or leave existing ones. Contamination from various industrial/agricultural activities such as power plants, mining, and hydraulic fracturing, and arsenic-containing herbicides can raise arsenic levels in certain locations. Staple foods like grains and some seafood contain arsenic because it is found in soil or water. Arsenic can contaminate natural water reservoirs due to its property, i.e., water solubility [14, 20, 39, 45, 79]. Drinking water is the most common mode through which arsenic compounds enter living cells. Both the forms of arsenic, i.e., organic and inorganic, are extremely toxic.

The other name of arsenic is the "king of poisons." It is in the number one position in WHO's 2001 priority list of harmful substances and disease registries [34]. The maximum allowable arsenic (As) concentration in drinking water has been set at 10 μg/L since 1993. Before 1993, however, it was at 50 μg/L. Shaji et al. [70] stated that it is classified as a carcinogen, mutagen, and teratogen. Three West Bengali families, for the first time identified in 1983, were affected by groundwater arsenic contamination in India. Among the recent reports, some of the states most contaminated with groundwater above the permissible level are Jharkhand, Bihar, Uttar Pradesh, Assam, Manipur, and Chhattisgarh. Around the globe, more than 500 million people are affected by groundwater contamination of arsenic, which has detrimental effects on human health, such as liver, kidney, heart, and lung diseases, hyperkeratosis, and skin cancer [34, 53].

Given the preceding context, we may state that arsenic (As) is a significant environmental threat due to its high toxicity, contamination, non-biodegradability, and high environmental concentrations and accumulation. It is also an extremely expensive element that is challenging to recover and remove from the environment [2] because of its wide distribution in air, water, soil, and plants, which raises questions concerning public health and food safety. In addition to potential health risks, arsenic also affects the quality of crops, agricultural yields, and soil health. As arsenic accumulation is common in water and soil, As can easily accumulate in crop plants and can easily affect the crop yield (Table 1). Therefore, agriculture must be conducted in an environmentally friendly manner. The farmers and scientists need to know about the arsenic pollution system and how to handle it properly [20, 38]. This chapter also includes the arsenic problem in agriculture and how it gets into plants, its relationship with human health, and how it can hinder agricultural production and soil health. In such a case, this particular chapter will cover soil management as one of the measures to address this emerging problem. Combining scientific knowledge with actionable measures makes it feasible to sustain human health, maintain food safety, and develop appropriate agricultural practices, despite the presence of arsenic.

Thus, it is essential to ensure that agriculture is practiced in an environmentally friendly manner. Farmers and scientists have a role to play in understanding the complexities of arsenic pollution and how best to mitigate it [20, 38]. This chapter also includes the arsenic problem in agriculture and how it gets into plants, its relationship with human health, and how it can hinder agricultural production and soil health.

Table 1 Key concepts of arsenic toxicity in plants

Aspect	Toxic effects on plants	References
Mechanisms of toxicity	Induction of Reactive oxygen species (ROS); inhibition of activity of vital enzymes, and interruption in energy metabolism	[23]
Cellular effects	Oxidative damage to lipids, proteins, nucleic acids, cell dysfunction and death	[72]
Physiological effects	Inhibition of growth: reduced height and biomass, yellowing of leaves due to inhibition of chlorophyll biosynthesis, and necrosis	[1]
Impact on plant metabolism	Interruption in photosynthesis, electron transport chain, mitochondrial function, ATP production	[86]
Agricultural implications	Decrease crop quality and quantity, affecting food security, poses a health risks	[32]

In such a case, this particular chapter will cover soil management as one of the measures to address this emerging problem. Combining scientific knowledge with actionable measures makes it feasible to sustain human health, maintain food safety, and develop appropriate agricultural practices, despite the presence of arsenic.

2 Arsenic Uptake by Crop Plants

Arsenic present in the soil can be absorbed into the root system of crop plants; the uptake might result in toxic effects and disruption of their physiological and metabolic functions, which are important to their survival [26]. Chemical form of arsenic, pH of the soil, and amount of other essential nutrients available in the soil are the factors that affect the absorption of arsenic by agricultural crops [1]. Figure 1 describes the uptake of arsenic and its toxicity to plants. Molecular and metabolic processes of arsenic absorption have been studied widely [1, 27, 48, 68].

There are two natural types of arsenic: these include arsenite As(III), which is in the reduced state, mainly under anaerobic conditions and absorbed into living organisms through aquaporins or hexose transporters. On the other hand, arsenate is a more oxidized form of arsenic, which occurs very often under aerobic conditions, thus being transported via phosphate transporters. After being absorbed by cells, arsenate reductases will rapidly convert As(V) to As(III). After conjugation with thiols, the further detoxification of As(III) will result in extrusion or vacuolar sequestration [88, 90]. Plants have various mechanisms that reduce arsenate, that have been primarily mediated by arsenate reductases, of which ArsC, ACR2, and HACs are involved. Though reduction in arsenate is mediated by a set of independent processes carried by different processes initiated by these sets of enzymes, the result is reduction of arsenite from the process of arsenate reduction via a cysteine cascade [89].

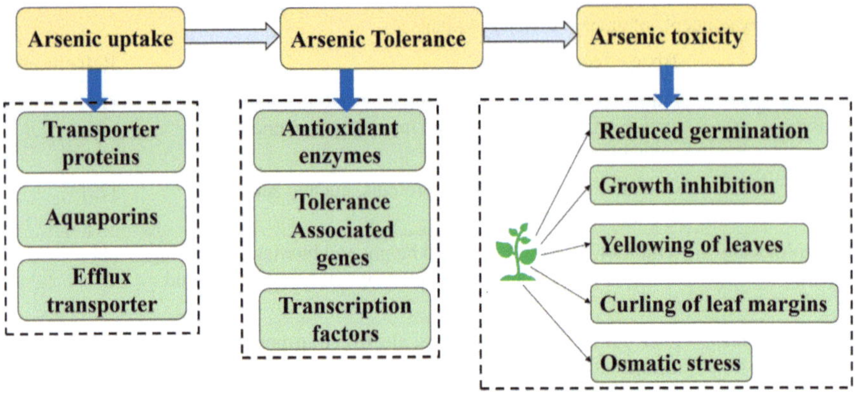

Fig. 1 Mechanisms of arsenic uptake, tolerance, and toxicity in plants

Arsenic uptake by plants is mainly regulated by specific transport proteins. The chemical speciation of As modifies the mechanisms of As uptake by plants. Since P and arsenate are chemically similar, recent evidence has surfaced that indicates that As(V) enters plant cells via several Pi entry routes. The transfer of both As(V) and Pi is increased when As(V) is present in the growth environment and phosphorus is absent. P channels are major parts of different Pi transport proteins that may be involved in the uptake of As(V) by plants. Both high- and low-affinity P transporters exist in plants. High-affinity transport is contributed to by the PHT1 proteins. Low-affinity transport proteins are yet unknown to date [1]. As, mostly in its trivalent form, As(III), enters the plants through nodulin-26-like intrinsic proteins, or NIPs. Unlike PHT, or phosphate transporters, NIP transporters enable the bidirectional movement of As(III) from plant cells into growth media and vice versa in concentration gradients [1, 47, 72, 89].

As(III) can also use silicon transporters because of structural similarities. Si deficiency increases the expression of the influx Si transporter Lsi1, which enhances Si and As transfer. Lsi1 and efflux Si transporters Lsi2 are positioned on either side of epidermal and endodermal cells, allowing for As transfer across plant tissues [11]. Besides the chemical characteristics and specific forms of arsenic present in the growth medium, physiological and tolerance processes play a very important role in determining the extent of arsenic absorbed by the plants under stress. Whenever this stress might cause undesirable effects at the cellular and molecular levels, the plants would involve a sequence of mechanisms of tolerance, including both physiological and biochemical adjustments. Notably, plants tolerate metals by reducing metal uptake, and this is considered a mechanism of tolerance. The internal physiological changes affect arsenic absorption and translocation from roots to shoots [1, 88]. Along with transport proteins, aquaporins also mediate the transport of small molecules, i.e., arsenic species. Aquaporins are membrane channels that transport water and small neutral molecules across biological membranes and are involved in the transport of arsenic. Several isoforms of these channels exist, which reflect a high diversity of cellular localizations, transport selectivity, and regulation properties [56].

Li et al. [43] reviewed that specific aquaporins in rice enable the uptake of arsenic from contaminated water. This finding represents the dual role of these proteins in nutrient and toxin transport. It refers to characterizing the proteins participating in arsenic uptake and translocation processes, particularly in the rice plant. Such molecular-level knowledge is underlined as relevant for improving breeding programs toward producing crop types that limit arsenic accumulation in edible sections, hence ensuring safer food. Faizan et al. [22] also reported that phytochelatins and their transporters are significant constituents in detoxifying arsenic in the plant. They suggest an increase in the expression of specific transporters for an improvement in the tolerance to arsenic in soils that have been treated with arsenic. This is in agreement with the findings of [1], who discuss how plants physiologically and biochemically react to arsenic exposure, specifically highlighting the generation of ROS and the function of antioxidant systems under arsenic stress.

3 Arsenic Toxicity in Plants

Arsenic has effects on plant physiology that go beyond just growth inhibition to intricate biochemical changes that compromise overall health. Arsenic has been shown to disrupt root development big time, resulting in shorter adventitious roots and hindered lateral root formation, both of which are important for nutrient uptake and stability. High concentration of arsenic results in increased lignin deposition in the root tissues that can ultimately retard water and nutrient transport [51].

Inhibition of photosynthesis in plants: Arsenic is a known inhibitor of photosynthesis in plants. When absorbed by the plant, there is interference in the light-harvesting complex of the plants; the chlorophyll content decreases, and photosynthesis is reduced. High concentrations of arsenic can severely affect the adaptive changes of photosystem I and II; synthesis of chlorophyll pigments is significantly decreased. This has been reported in several species, including *Zea mays*, *Trifolium pratense,* and *Lactuca sativa,* where arsenic stress was associated with chlorophyll biosynthesis [50, 51]. The toxic effects of arsenic on photosynthesis can occur at photochemical and (biochemical levels). Arsenic damages chloroplast membranes and disrupts fundamental photosynthetic functions. Under arsenic exposure, the rate of carbon-dioxide fixation and photosystem II (PSII) function is reduced. Arsenic also affects photochemical efficiency and heat dissipation capacity, resulting in changes in gas exchange rates and fluorescence emission [1, 46].

Plants generally accumulate and metabolize arsenate (As(V)) through phosphate (Pi) transport channels because of the chemical similarity between As(V) and Pi. This similarity allows As(V) to disrupt phosphate-dependent metabolic processes, especially during ATP synthesis. Although As(V) can compete with Pi for binding to transport proteins on root surfaces, these proteins have higher affinity for Pi, so increased Pi availability in the rhizosphere can reduce As(V) uptake. But if As(V) in the soil is higher than Pi, As(V) can bind to root surfaces and be transported into plant cells. This will form unstable As(V) adducts, which will autohydrolyze to form As(V)-ADP. This will create a futile cycle that uncouples oxidative phosphorylation and photophosphorylation, severely impeding the plant's ability to produce ATP and maintain normal metabolic functions [24, 25, 72].

Cell membranes are the most sensitive targets for damage under arsenic (As) stress conditions, and this can serve as an indicator of plant stress tolerance capacity. Membrane damage may be induced by nutrient and water imbalances that decrease stomatal conductance. The interference caused by arsenic with plant transpiration is expected to be through disrupted nutrient uptake and water transport [1, 51, 63]. For example, after exposing *Avena sativa* seedlings to the onset of arsenic toxicity, the rate of transpiration declined dramatically [17].

Another effect of arsenic toxicity is interference with nutrient uptake by displacing essential nutrients such as phosphate. Nutrient deficiencies, therefore, result in changed uptake mechanisms of macronutrients and micronutrients, inducing stunted growth, poor root development, and reduced resistance to environmental stressors [62, 72]. Symptoms of arsenic toxicity in crops can be very obvious. Some of the most common visible symptoms include chlorosis, or yellowing of the leaves, necrosis, and wilting with time as arsenic accumulation continues, plants start to show reduced leaf size, poor root development, and in extreme cases, dieback of shoots. These symptoms vary between species and depending on the environment but altogether indicate that a plant is suffering from the inability to overcome the effects of arsenic toxicity. Long-term exposure to arsenic can severely impact the health and yield of plants. Chronic exposure to arsenic reduces the general vigor of the plants, lowers reproductive success, and decreases crop yields. As this arsenic accumulates within the edible parts of these plants, it presents an important risk not only for plant health but also in terms of food safety. It can enter the food chain for humans, gradually reducing agricultural productivity, eroding revenue for farmers, and causing health hazards among consumers. That means the knowledge would be indispensable in developing strategies on ways of mitigating exposure while ensuring plant health and protecting food safety in agricultural systems [61, 64].

4 Tolerance Mechanisms in Crop Plants

The mechanisms toward the tolerance of crop plants toward arsenic toxicity will be derived from the expression of both genetic factors as well as biochemical pathways. It is very basic in nature that there be variance among genes since variance and differences found within the species and its variety owe their origin to specified alleles that bring specificity upon such species or their respective varieties. The next thing, when introduced to As, they perform complex molecular response events involved with various signaling pathways in these ways: through the mediums of calcium and mitogen-activated protein kinases, plus several other associated plant hormones, such as auxin, followed by methyl jasmonate along with cytokinin. These pathways activate more of the enzymes in the antioxidant system, which fight against the ROS and RNS produced as a result of As-induced stress. The response of plant genomics toward As exposure is extremely elaborate, involving up-regulated genes associated with uptake, chelation, and sequestration. Important aquaglyceroporin as well as ABC transporter gene families are key for facilitating As uptake and its transportation within plant tissues resistance [30].

Most plants have unique means of absorption, and absorption of the required nutrients phosphorus (P), sharing transport proteins with arsenate (As(V)). Evidence has been demonstrated that a rhizosphere with a supply of higher Pi diminishes arsenic uptake in plants [1, 4]. Pooam et al. [59] studied the functional role of SNAC3 in promoting As stress tolerance and grain productivity in rice (Oryza sativa L.). Two SNAC3-overexpressing (SNAC3-OX) and two SNAC3-RNAi transgenic lines were produced and confirmed. Wild-type and transgenic rice plants were treated with various concentrations of As stress (0, 25, and 50 μM). Results: The SNAC3 overexpression was noted to improve grain yield traits, coupled with tolerance toward As in rice. SNAC3-OX rice lines at both the levels of As stress, 25 and 50 μM, showed severe reductions in levels of biomarkers of oxidative stress and OsCRY1b expression whereas they contained high levels of gas exchange characters, chlorophyll, osmolytes, comprising soluble sugars, proteins, proline, phenols, flavonoids, antioxidant enzymes, including Superoxide dismutase (SOD), catalase (CAT), Ascorbate peroxidase (APX), and Peroxidase (POD) and stress-tolerant genes expression than wild-type plants. In contrast, suppression of SNAC3 with RNAi lowered grain yield components and reversed the measured physio-biochemical and molecular traits mentioned above. In brief, this is the first study indicating that SNAC3 is essential in enhancing As stress resistance and grain productivity of rice through regulation of antioxidants, photosynthesis, osmolyte accumulation, and expression of stress-related genes, which makes it a potential candidate for further genetic enhancement of stress resistance in many crops.

Once absorbed, plants deal with the toxicity by using transport proteins and vacuolar sequestration. There are specific transporters for internal movement that allow it to be either sequestered in the vacuoles or transferred to less sensitive tissues such as the shoot [25]. This compartmentalization aids in detoxification and the avoidance of interference with essential metabolic functions. Specific transport proteins play a crucial role in this sequestration process. For example, the Nodulin-26-like Intrinsic Proteins (NIPs) and Phosphate Transporters (PHTs) are known to facilitate the uptake and movement of arsenic within plant tissues. These transporters can lead the arsenic to vacuoles or less sensitive tissues like the shoot, which reduces its toxicity in the actively metabolizing parts of the plant [58].

It serves as a storage organelle for several ions and compounds, including heavy metals and arsenic. Research studies have demonstrated that vacuolar sequestration detoxifies arsenic while at the same time elevating the tolerance of the plant to oxidative stress from exposure to heavy metals. For instance, arsenic, when it is sequestered in vacuoles, can be transformed into a less toxic form, thus reducing its toxic effect on growth and development [88]. It has also been shown that the induction of phytochelatins—the small peptides that bind to heavy metals—is increased following exposure to arsenic. These complexes may promote the transport of arsenic into vacuoles where it is sequestered, thereby playing a critical role in the detoxification mechanism of the plant [22].

Besides keeping the plants alive, plant arsenic sequestration is critical to safety in food. Once such arsenic accumulates in the edible parts of these plants, it directly attacks human health. It goes without saying that this implies that the mechanisms by which plants sequester their arsenic will influence practices in agriculture to ensure the least uptake of arsenic from the soil by crops with the aim of food and agricultural productivity enhancement [32].

This means that exposure to arsenic also causes the production of reactive oxygen species through oxidative stress, which could damage their cellular components. Plants can counter this by increasing an antioxidant defense mechanism in the form of increased production of enzymes such as superoxide dismutase and glutathione peroxidase [44]. These antioxidants basically scavenge ROS while reducing oxidative damage, which means that the plant is more stress-tolerant.

Physical responses in plants under stress from arsenic include numerous responses. Changes in stomatal conductance, photosynthetic rates, and transpiration balance nutrient and water uptake with minimal stress. For example, most plants adjust their photosynthetic machinery when exposed to arsenic to cope with the reduced efficiency caused by oxidative stress [8]. Some species also enhance the uptake of nutrients that could counteract the toxicity of arsenic, such as selenium (Se) or silicon (Si) [60]. The rate of water–vapor escape and carbon-dioxide entrance through the leaf through its stomata can be defined as stomatal conductance. Generally, plants under stress have been known to show some variations in stomatal conductance in an attempt to ensure that they either prevent more water loss to external forces or maintain gas flow inside [80].

Arsenic exposure typically causes oxidative stress that affects the photosynthetic machinery of plants. Consequently, most species adapt by modifying their photosynthetic activities to cope with the lower efficiency caused by this toxic substance. For instance, they increase the production of antioxidants, which help eliminate ROS resulting from arsenic stress. Plants also induce protective pigments, like carotenoids, which minimize the effects of oxidative damage [69]. Such physiological changes allow plants to continue with at least a level of photosynthetic activity even though they may contain arsenic; their efficiency, however, is often compromised in such stressed conditions. This capability is necessary for continued growth and productivity under stress. Some plant species absorb some specific nutrients that may mitigate arsenic toxicity. For example, some of the protective nutrients against arsenic are selenium (Se) and silicon (Si). These are nutrients that may be used to provide protection in plants subjected to arsenic stress. Selenium, for example, has been known to increase antioxidant activity and also enhance the synthesis of stress-related proteins, which are used as a buffer to oxidative damage caused by arsenic [24, 71]. It is also known to harden plant cell walls and make plants more resistant to stress conditions. Silicon has been proven to increase the sequestration of arsenic in vacuoles, which decreases its availability to metabolically active tissues and facilitates detoxification. In this manner, silicon supports plant health while alleviating the toxicity of arsenic.

A critical mechanism of arsenic tolerance involves the metabolic transformation of inorganic arsenic into the less toxic organic forms. Methylation processes can change inorganic arsenic to monomethylarsenic and dimethylarsinic acids, thereby reducing its toxicity and increasing storage within the plant tissues [49]. This detoxification is not only necessary for the organism but also minimizes the mobility of arsenic within the plant.

Recent studies have established that exposure to arsenic may promote epigenetic modifications through the regulation of stress-response and detoxification-related genes. These effects are conducive to improving a tolerance of plants to arsenic for generations, and possibilities are being opened through breeding as well as biotechnology for improving tolerance to this metalloid in crops [18, 78, 87].

Different crop species have been proven to possess some level of tolerance to arsenic (As), thus allowing them to survive and grow in As-contaminated soils. Such tolerant crops possess unique mechanisms aimed at reducing uptake or ameliorating toxic effects. Among the crops studied are rice (*Oryza sativa*). It accumulates arsenic in its shoots with lower amounts within grains, which is also important for food safety purposes. Some research has indicated the existence of certain rice varieties that take up silicon from the soil more efficiently to reduce arsenic toxicity levels [21, 65]. Barley (*Hordeum vulgare*) is also known to resist arsenic due to strong exclusion and vacuolar sequestration mechanisms. Some cultivars have shown the capacity to survive in arsenic-contaminated soils and yet proved physiologically stable with growth. *Brassica juncea* (mustard) belongs to the arsenic-tolerant species. It accumulates arsenic in tissues but without causing extreme phytotoxic symptoms. The antioxidant system of this species was also well-equipped in handling the oxidative stress caused by arsenic [10]. For instance, *Pteris vittata* and *Pteris ensiformis* are well-known fern species that hyperaccumulate arsenic. These plants, which can tolerate high amounts of arsenic, do more than just survive well at such levels; rather, they can sequester it in their fronds efficiently, making these useful for phytoremediation purposes [31, 81]. Varieties of lettuce, *Lactuca sativa*, have been found to be resistant to arsenic stresses, characterized by limited uptake of arsenic into edible parts, with the ability for adequate growth and development in contaminated soils. Such varieties, therefore, become key targets in breeding programs that would enhance their tolerance to arsenic [41]. Proper regulation of nutrient uptake in certain varieties of wheat, *Triticum aestivum*, has shown resistance to arsenic and also an ability to counteract oxidative stress [67].

5 The Management of Arsenic-Contaminated Soils

It is integral to reducing its uptake in crops. Sun et al. [75] provide insights into how water management practices, such as intermittent flooding, can significantly lower arsenic concentrations in rice grains. This finding highlights the importance of agronomic practices in mitigating arsenic uptake and suggests that effective irrigation strategies could be employed in arsenic-affected regions. The use of ameliorating agents, such as zinc oxide nanoparticles, has also been shown to improve crop tolerance to arsenic by enhancing growth factors and reducing arsenic levels in plants [6, 35]. These innovative approaches are promising for enhancing crop productivity in contaminated environments.

5.1 Remediation of Arsenic for Agricultural Sustainability

The problem of arsenic contamination in agricultural soils is significant in the global scenario. Arsenic contamination can cause severe impacts on crop production, food safety, and human health [9]. Different remediation strategies have been proposed to mitigate the negative impacts of arsenic and to improve soil quality. Such remediation strategies may lead to sustainable agricultural practices [15].

5.1.1 Bioremediation

Bioremediation utilizes microbes that degrade or transform pollutants. This includes arsenic. Some bacteria and fungi can biotransform arsenic to less toxic forms or help immobilize it in the soil [5, 91]. It can enhance the natural degradation of arsenic and, in turn, improve the health of the soil. Bioaugmentation, with the addition of specific strains of microbes that degrade arsenic, has been able to augment bioremediation processes effectively [57]. The efficacy, however, depends on the local microbial community and environment. Among the few bioremediation methods, microbial remediation and phytoremediation are considered as cost-effective, eco-friendly methods.

Phytoremediation

Phytoremediation is the removal of toxic compounds, which are heavy metals, pesticides, and metalloids, from the soil and water of contaminated environments with the help of green plants. Use of phytoremediation is considered one of the promising removal techniques of arsenic in contaminated soils. Several have identified hyperaccumulator plant species that tolerate and extract the arsenic, thus minimizing its bioavailability. There are various pathways of the phytoremediation process, such as rhizofiltration, phytoaccumulation or phytoextraction, phytostabilization, phytodegradation or phytotransformation, phytovolatilization, etc. In the process of rhizofiltration, plants absorb toxic substances taken by their roots with the help of an adsorption or absorption process and sequestrated in their root system. Aquatic plants mainly showed this process, whereas in phytoaccumulation or phytoextraction, toxic substances are absorbed by the plants through their root system, but it transferred to some other part of the plants, like stem and leaves, or any other modified part of plants. It appeared mainly in the well-adapted plants for remediation of contaminated soil [36, 40]. Phytostabilization limits the migration of harmful substances within the soil or water. Thus, less is absorbed by the plant. Here, the toxic substances in the environment are not taken up by the plants. Some root exudates or photochemicals emitted by the plants react with toxic substances, thus forming a stable chemical bond with them and enhancing its stability within the environment [66]. Phytodegradation or Phytotransformation is the process in which the plant uptakes toxic substances from the soil or water that then degrade to form nontoxic substances. A lot of metabolic and physiological factors have to intervene in this process [92]. In Phytovolatilization, plants absorb toxic components by their root system, transport them to their aerial part of the plant especially into the leaves, and emit the toxic components in vapor form which may or may not be toxic depending on their primary source of the component [85].

Cantamessa et al. [13] demonstrated the phytoremediation technique with *Pteris vittata* L. and Arbuscular Mycorrhizal Fungi in the remediation of the highly arsenic-contaminated site. Chen et al. [16] studied arsenic-enriched farmland with maize, *Rotala rotundifolia*, and the arsenic super-accumulator *Pteris vittata* as an effort to determine how phytoremediation may influence the arsenic in the farmland. After these explorations, *Pteris vittata* has to be seeded first followed by rapid consumption of the residual As concentration in soil and thus further planting is made by maize for concurrent decontamination of the soils along with corn yield facilitate the reuse of the land safely also. Srivastava et al. in their work in 2021 published on discussing the phytoremediation techniques came up with various advantages and disadvantages together with it toward the remediation of As within and out of the water soils. High biomass and high As accumulation and economical utilization can be applied for phytoremediation.

Microbial Remediation

Microbial bioremediation is a non-destructive remediation process environment-friendly, which utilizes microorganisms to degrade, transform, or eliminate contaminants in soil, water, and other environments. A wide variety of contaminants like heavy metals, hydrocarbons, pesticides, and arsenic have been successfully remediated using this method [29]. Microbial bioremediation is reported as a feasible method for removal of arsenic. Xue et al. [83] used biochar-supported controllable genetically modified bacteria for Arsenic bioremediation in wastewater. Nam et al. [52] applied Talaromyces sp. KM-31 isolated from As-Contaminated Mine Soil and demonstrates the potential of fungi for As bioremediation. William and Magpantay [82] have pointed out the interactions of microorganisms with arsenic, mechanism adopted by microorganisms for the detoxification of arsenic, and new trends in microbial-based bioremediation of arsenic. Ghosh and Sarkar [28] have mainly focused on the removal of arsenic using genetically engineered bacterial species like *Agrobacterium tumefaciens, Bacillus subtilis, Pseudomonas aureofaciens, P. putida,* etc.

5.1.2 Soil Amendments

Soil amendment is the technique which can be used to reduce bioavailabilty and mobility of arsenic from the soil. All materials like organic, calcium compounds, and silicon can be used, and thus effectiveness can be analyzed. Application of organic amendments like compost and biochar to soil has been observed to improve the soil structure and enhance the microbial activity, thus reducing the arsenic mobility [84]. These also have a positive effect on availability of nutrients and overall condition of the soil. Both calcium carbonate and gypsum also form insoluble complexes thereby restricting the uptake of As by plants [77]. This process is very helpful in soils with a high content of arsenic. Silicon is increasingly becoming known for its role in arsenic remediation since it helps in enhancing the resilience of plants against the toxicity of arsenic and also healthy growth [3]. Silicon amendments have been very effective in rice cultivation.

Nguyen Quoc et al. [54] performed research on the impact of some soil amendments on metals' bioavailability and its outcome on the phytotoxicity and microbial diversity in a polluted farmland. The findings of that study reported that the composition of soil amendments and the subsequent composites greatly influenced both the soil, plant, and microbial metabolic functions. Suda et al. [74] researched the effects of various amendments on As concentration in flooded anaerobic soil solutions. The authors proved a promising effect of a polysilicate-iron solution precipitate as a novel As adsorbent. Sun et al. [76] tested the effect of inorganic materials in contaminated arsenic soil and achieved reduction in arsenic bioavailability using seven inorganic amendments.

5.1.3 Chemical Remediation

Chemical remediation techniques involve applying certain agents to detoxify or immobilize arsenic in soils. Manipulation of redox soil potential can allow As(V) to be changed into As(III). This has the effect on its bioavailability [44]. Those conversions are able to provide increased tolerance in some crops to the metalloid while needing a balance to the increase the toxicity. Solubilization of As by the use of a compound has been done, one is EDTA through which the arsenic being solubilized enables it to be absorbed in plants then harvested later for consumption. The mobility attached to increasing solubility increases its risk of the environmental release [12]. Soil washing is a process that involves the removal of contaminants in the soil using water or solvents physical and chemical action. The contaminated soil is mixed with a washing solution and agitated to remove the arsenic. The washing agent is a key component of the process, and the type of agent used depends on the soil properties [7, 55]. Citric acid, organic acid that dissolves iron oxides to which arsenic is bound has shown much potential in its removal ability [2, 37]. However, the removal efficiency was inconsistent in previous investigations. Other considered washing agents are inorganic and organic acids and surfactants, and chelating agents, the alkaline solutions used for washing in previous approaches. Inorganic acids have been known to destroy the soil, whereas surfactants are relatively poor at acting as wash agents [19]. Tannic acid, a biodegradable organic acid found in many plants, has a potential positive effect because it can reduce $Fe3^+$ to $Fe2^+$ and form stable $Fe2^+$-TA complexes, which help to dissolve iron oxides, thereby helping to release the arsenic. However, TA alone does not show high removal of arsenic owing to properties of soil and hardness of releasing crystalline iron oxide-bound arsenic. This has been proposed for improvement in the remediation of arsenic based on a new composite washing agent which combines TA with dithionite. Dithionite dissolves the iron oxides bound with arsenic, thereby allowing a high extraction efficiency to be possibly achieved. It is therefore a novel synergy which hitherto had not been studied before; this system will offer improved solutions against the contamination by arsenic in soils [7, 33, 61].

5.1.4 Agronomic Practices

Sustainable agronomic practices at levels of arsenic will maintain agricultural systems. Crop rotations and planting cover crops could disrupt the cycles of accumulating arsenic and provide healthier soils [15]. These help create nutrient cycling and reduce erodibility that ultimately supports agricultural sustainability. Ongoing soil testing is important to monitor the presence levels of arsenic that inform remediation efforts so the implemented management practices are working while responsive to changes.

An indispensable step for the improvement in sustainability and food safety concerning agricultural soils is the remediation of arsenic contamination. Some of the feasible remediation strategies identified during the study include phytoremediation, amendment to soil, chemical remediation, soil washing, bioremediation, and sustainable agronomic practices. Further research into the former methods will be needed before developing integrated approaches that should enhance soil health, resilience of crops, and the productivity of agriculture in those regions affected by arsenic [15, 42, 73].

6 Conclusions and Perspectives

In conclusion, serious issues exist with arsenic pollution in agricultural systems, posing threats to human health and food safety. Mechanisms of arsenic uptake, toxicity, and tolerance in plants will be key in developing efficient management plans that improve agricultural output and protect public health in affected areas. The mechanisms of arsenic absorption, which may be influenced by competitive nutrient interactions such as phosphate ions, also determine arsenic control. Knowledge of these physiological and biochemical reactions can be incorporated into effective management strategies that place emphasis upon the development of cultivars that are more tolerant and able to accumulate less in their edible sections.

Improving soil health and lowering arsenic bioavailability requires a multifaceted strategy that incorporates agronomic best practices, including fertilizer management, organic farming, and soil amendments. In order to achieve balanced management that promotes healthy plant growth and optimal yields, research should concentrate on the interaction between arsenic and important nutrients. In order to provide context-specific solutions, integrated solutions also need to take socioeconomic aspects into account and prioritize stakeholder and community involvement.

In order to integrate innovative agronomic methods with community mobilization in the future, a comprehensive understanding of arsenic dynamics and consequences is essential. Addressing the issues raised by arsenic in agriculture would require sustained investment in community engagement, education, and research. We can guarantee the sustainability of agricultural systems in impacted areas and safeguard the welfare of individuals who rely on them by promoting safer crops, healthier soils, and better public health outcomes.

Competing Interests The authors declare that they have no competing interests.

Declaration of Generative AI and AI-Assisted Technologies in the Writing Process During the preparation of this work, the book chapter author(s) have not taken any assistance from AI, and the book chapter author(s) reviewed and edited the content as needed and take(s) full responsibility for the content of the publication.

References

1. Abbas G, Murtaza B, Bibi I, Shahid M, Niazi NK, Khan MI, Amjad M, Hussain M, Natasha N (2018) Arsenic uptake, toxicity, detoxification, and speciation in plants: physiological, biochemical, and molecular aspects. Int J Environ Res Public Health 15(1):59
2. Ahmed SF, Kumar PS, Rozbu MR, Chowdhury AT, Nuzhat S, Rafa N, Mahlia TM, Ong HC, Mofijur M (2022) Heavy metal toxicity, sources, and remediation techniques for contaminated water and soil. Environ Technol Innov 25:102114
3. Ahmed T, Guo J, Noman M, Lv L, Manzoor N, Qi X, Li B (2024) Metagenomic and biochemical analyses reveal the potential of silicon to alleviate arsenic toxicity in rice (Oryza sativa L.). Environ Pollut 345:123537

4. Anawar HM, Rengel Z, Damon P, Tibbett M (2018) Arsenic-phosphorus interactions in the soil-plant-microbe system: dynamics of uptake, suppression and toxicity to plants. Environ Pollut 233:1003–1012
5. Ayilara MS, Babalola OO (2023) Bioremediation of environmental wastes: the role of microorganisms. Front Agron 5:1183691
6. Banerjee S, Mondal S, Islam J, Sarkar R, Saha B, Sen A (2024) Rhizospheric nano-remediation salvages arsenic genotoxicity: Zinc-oxide nanoparticles articulate better oxidative stress management, reduce arsenic uptake, and increase yield in Pisum sativum (L.). Sci Total Environ 913:169493
7. Bari AF, Lamb D, MacFarlane GR, Rahman MM (2022) Soil washing of arsenic from mixed contaminated abandoned mine soils and fate of arsenic after washing. Chemosphere 296:134053
8. Beniwal R, Yadav R, Ramakrishna W (2023) Multifarious effects of arsenic on plants and strategies for mitigation. Agriculture 13(2):401
9. Bhat A, Ravi K, Tian F, Singh B (2024) Arsenic contamination needs serious attention: an opinion and global scenario. Pollutants 4(2):196–211
10. Bhat JA, Faizan M, Bhat MA, Sharma G, Rinklebe J, Alyemeni MN, Bajguz A, Ahmad P (2023) Mitigation of arsenic stress in Brassica juncea L. using zinc oxide-nanoparticles produced by novel hydrothermal synthesis. S Afr J Bot 163:389–400
11. Bienert GP, Jahn TP (2010) Major intrinsic proteins and arsenic transport in plants: new players and their potential role. In: MIPs and their role in the exchange of metalloids, pp 111–125
12. Briffa J, Sinagra E, Blundell R (2020) Heavy metal pollution in the environment and their toxicological effects on humans. Heliyon 6(9)
13. Cantamessa S, Massa N, Gamalero E, Berta G (2020) Phytoremediation of a highly arsenic polluted site, using *Pteris vittata* L. and arbuscular mycorrhizal fungi. Plants 9(9):1211
14. Carlin DJ, Naujokas MF, Bradham KD, Cowden J, Heacock M, Henry HF, Lee JS, Thomas DJ, Thompson C, Tokar EJ, Waalkes MP, Birnbaum LS, Suk WA (2016) Arsenic and environmental health: state of the science and future research opportunities. Environ Health Perspect 124(7):890–899
15. Chauhan R, Awasthi S, Tiwari P, Upadhyay MK, Srivastava S, Dwivedi S, Dhankher OP, Tripathi RD (2024) Biotechnological strategies for remediation of arsenic-contaminated soils to improve soil health and sustainable agriculture. Soil & Environ Health 2(1):100061
16. Chen CC, Lin MS, Cheng PC, Huang CY, Cheng SF (2024) Strategy of phytoremediation for sustainable use of arsenic-rich farmland. Eng Proc 74(1):8
17. Chen L, Yang S, Liu Y, Mo M, Guan X, Huang L, Sun C, Yang ST, Chang XL (2018) Toxicity of graphene oxide to naked oats (Avena sativa L.) in hydroponic and soil cultures. RSC Adv 8(28):15336–15343
18. Chmielowska-Bąk J, Searle IR, Wakai TN, Arasimowicz-Jelonek M (2023) The role of epigenetic and epitranscriptomic modifications in plants exposed to non-essential metals. Front Plant Sci 14:1278185
19. Cho K, Myung E, Kim H, Park C, Choi N, Park C (2020) Effect of soil washing solutions on simultaneous removal of heavy metals and arsenic from contaminated soil. Int J Environ Res Public Health 17(9):3133
20. Chung JY, Yu SD, Hong YS (2014) Environmental source of arsenic exposure. J Prev Med Public Health 47(5):253
21. Etesami H, Jeong BR, Maathuis FJ, Schaller J (2023) Exploring the potential: Can arsenic (As) resistant silicate-solubilizing bacteria manage the dual effects of silicon on As accumulation in rice? Sci Total Environ 903:166870
22. Faizan M, Alam P, Hussain A, Karabulut F, Tonny SH, Cheng SH, Yusuf M, Adil MF, Sehar S, Alomrani SO, Albalawi T, Hayat S (2024) Phytochelatins: a key regulator against heavy metal toxicity in plants. Plant Stress 11:100355
23. Farooq MA, Islam F, Ali B, Najeeb U, Mao B, Gill RA, Zhou W (2016) Arsenic toxicity in plants: cellular and molecular mechanisms of its transport and metabolism. Environ Exp Bot 132:42–52

24. Finnegan PM, Chen W (2012) Arsenic toxicity: the effects on plant metabolism. Front Physiol 3:182
25. Garbinski LD, Rosen BP, Chen J (2019) Pathways of arsenic uptake and efflux. Environ Int 126:585–597
26. Gautam A, Pandey AK, Dubey RS (2020) Unravelling molecular mechanisms for enhancing arsenic tolerance in plants: a review. Plant Gene 23:100240
27. Ghosh D, Ghosh A, Bhadury P (2022) Arsenic through aquatic trophic levels: effects, transformations and biomagnification—a concise review. Geosci Lett 9(1):20
28. Ghosh S, Sarkar B (2023) Genetically modified bacteria for arsenic bioremediation. In: Genomics approach to bioremediation: principles, tools, and emerging technologies, pp 467–483
29. González-Benítez N, Durante-Rodríguez G, Kumar M, Carmona M (2021) Biotechnology for arsenic detection and bioremediation. Front Microbiol 12:743109
30. Gracia-Rodriguez C, Lopez-Ortiz C, Flores-Iga G, Ibarra-Muñoz L, Nimmakayala P, Reddy UK, Balagurusamy N (2024) From genes to ecosystems: decoding plant tolerance mechanisms to arsenic stress. Heliyon
31. Gumaelius L, Lahner B, Salt DE, Banks JA (2004) Arsenic hyperaccumulation in gametophytes of Pteris vittata: a new model system for analysis of arsenic hyperaccumulation. Plant Physiol 136(2):3198–3208
32. Gupta A, Dubey P, Kumar M, Roy A, Sharma D, Khan MM, Bajpai AB, Shukla RP, Pathak N, Hasanuzzaman M (2022) Consequences of arsenic contamination on plants and mycoremediation-mediated arsenic stress tolerance for sustainable agriculture. Plants 11(23):3220
33. He J, Lin Q, Luo Y, Liu Y, Fan X, Zheng J, Xu K, Ma Y (2023) Removal of arsenic from contaminated soils by combining tartaric acid with dithionite: an efficient composite washing agent. J Environ Chem Eng 11(3):109877
34. Hughes MF, Beck BD, Chen Y, Lewis AS, Thomas DJ (2011) Arsenic exposure and toxicology: a historical perspective. Toxicol Sci 123(2):305–332
35. Jalil S, Alghanem SM, Al-Huqail AA, Nazir MM, Zulfiqar F, Ahmed T, Ali S, Abeed AH, Siddique KH, Jin X (2023) Zinc oxide nanoparticles mitigated the arsenic induced oxidative stress through modulation of physio-biochemical aspects and nutritional ions homeostasis in rice (Oryza sativa L.). Chemosphere 338:139566
36. Kafle A, Timilsina A, Gautam A, Adhikari K, Bhattarai A, Aryal N (2022) Phytoremediation: mechanisms, plant selection and enhancement by natural and synthetic agents. Environ Adv 8:100203
37. Kanel SR, Das TK, Varma RS, Kurwadkar S, Chakraborty S, Joshi TP, Bezbaruah AN, Nadagouda MN (2023) Arsenic contamination in groundwater: geochemical basis of treatment technologies. ACS Environ Au 3(3):135–152
38. Khan MU, Rai N (2023) Arsenic enrichment in the north Gangetic Plains of Laksar, Uttarakhand, India. Groundw Sustain Dev 21:100913
39. Khosravi-Darani K, Rehman Y, Katsoyiannis IA, Kokkinos E, Zouboulis AI (2022) Arsenic exposure via contaminated water and food sources. Water 14(12):1884
40. Kristanti RA, Ngu WJ, Yuniarto A, Hadibarata T (2021) Rhizofiltration for removal of inorganic and organic pollutants in groundwater: a review. Biointerface Res Appl Chem 11:12326–12347
41. Kumwimba MN, Xibai Z, Lingyu B (2013) Uptake kinetics of arsenic by lettuce cultivars under hydroponics. Afr J Environ Sci Technol 7(5):321–328
42. Lavanya MB, Viswanath DS, Sivapullaiah PV (2024) Phytoremediation: an eco-friendly approach for remediation of heavy metal-contaminated soils—a comprehensive review. Environ Nanotechnol Monit & Manag 22:100975
43. Li N, Wang J, Song WY (2016) Arsenic uptake and translocation in plants. Plant Cell Physiol 57(1):4–13
44. Liao X, Li Y, Miranda-Avilés R, Zha X, Anguiano JHH, Sánchez CDM, Puy-Alquiza MJ, González VP, Garzon LFR (2022) In situ remediation and ex situ treatment practices of arsenic-contaminated soil: an overview on recent advances. J Hazard Mater Adv 8:100157

45. Madhukar M, Murthy S, Udayashankara TH (2016) Sources of arsenic in groundwater and its health significance—a review. Nat Environ & Pollut Technol 15(3):971–979
46. Martínez-Castillo JI, Saldaña-Robles A, Ozuna C (2022) Arsenic stress in plants: a metabolomic perspective. Plant Stress 3:100055
47. Mawia AM, Hui S, Zhou L, Li H, Tabassum J, Lai C, Wang J, Shao G, Wei X, Tang S, Luo J, Hu S, Hu P (2021) Inorganic arsenic toxicity and alleviation strategies in rice. J Hazard Mater 408:124751
48. Mitra A, Chatterjee S, Moogouei R, Gupta DK (2017) Arsenic accumulation in rice and probable mitigation approaches: a review. Agronomy 7(4):67
49. Mlangeni AT (2023) Methylation of arsenic in rice: mechanisms, factors, and mitigation strategies. Toxicology reports
50. Muhammad I, Shalmani A, Ali M, Yang QH, Ahmad H, Li FB (2021) Mechanisms regulating the dynamics of photosynthesis under abiotic stresses. Front Plant Sci 11:615942
51. Nahar K, Rhaman MS, Parvin K, Bardhan K, Marques DN, García-Caparrós P, Hasanuzzaman M (2022) Arsenic-induced oxidative stress and antioxidant defense in plants. Stresses 2(2):179–209
52. Nam IH, Murugesan K, Ryu J, Kim JH (2019) Arsenic (As) removal using Talaromyces sp. KM-31 isolated from as-contaminated mine soil. Minerals 9(10):568
53. Naujokas MF, Anderson B, Ahsan H, Aposhian HV, Graziano JH, Thompson C, Suk WA (2013) The broad scope of health effects from chronic arsenic exposure: update on a worldwide public health problem. Environ Health Perspect 121(3):295–302
54. Nguyen Quoc T, Nejad ZD, Jung MC (2021) Effect of commercial amendments on immobilization of arsenic, copper, and zinc in contaminated soil: comprehensive assessing to plant uptake combined with a microbial community approach. Minerals 11(10):1143
55. Olasanmi IO, Thring RW (2020) Evaluating rhamnolipid-enhanced washing as a first step in remediation of drill cuttings and petroleum-contaminated soils. J Adv Res 21:79–90
56. Ozu M, Galizia L, Alvear-Arias JJ, Fernández M, Caviglia A, Zimmermann R, Garate JA (2023) Mechanosensitive aquaporins. Biophys Rev 15(4):497–513
57. Pande V, Pandey SC, Sati D, Bhatt P, Samant M (2022) Microbial interventions in bioremediation of heavy metal contaminants in agroecosystem. Front Microbiol 13:824084
58. Pommerrenig B, Diehn TA, Bernhardt N, Bienert MD, Mitani-Ueno N, Fuge J, Bienert GP (2020) Functional evolution of nodulin 26-like intrinsic proteins: from bacterial arsenic detoxification to plant nutrient transport. New Phytol 225(3):1383–1396
59. Pooam M, El-Ballat EM, Jourdan N, Ali HM, Hano C, Ahmad M, El-Esawi MA (2023) SNAC3 transcription factor enhances arsenic stress tolerance and grain yield in rice (Oryza sativa L.) through regulating physio-biochemical mechanisms, stress-responsive genes, and cryptochrome 1b. Plants 12(14):2731
60. Qin C, Lian H, Zhang B, He Z, Alsahli AA, Ahanger MA (2024) Synergistic influence of selenium and silicon supplementation prevents the oxidative effects of arsenic stress in wheat. J Hazard Mater 465:133304
61. Rahman S, Rahman IM, Ni S, Harada Y, Kasai S, Nakakubo K, Hasegawa H (2022) Enhanced remediation of arsenic-contaminated excavated soil using a binary blend of biodegradable surfactant and chelator. J Hazard Mater 431:128562
62. Rashid A, Ayub M, Bundschuh J, Gao X, Ullah Z, Ali L, Ahmad P (2023) Geochemical control, water quality indexing, source distribution, and potential health risk of fluoride and arsenic in groundwater: Occurrence, sources apportionment, and positive matrix factorization model. J Hazard Mater 460:132443
63. Raza MB, Datta SP, Golui D, Barman M, Das TK, Sahoo RN, Naveenkumar A (2023) Synthesis and performance evaluation of novel bentonite-supported nanoscale zero valent iron for remediation of arsenic contaminated water and soil. Molecules 28(5):2168
64. Rokonuzzaman M, Li WC, Man YB, Tsang YF, Ye Z (2022) Arsenic accumulation in rice: sources, human health impact and probable mitigation approaches. Rice Sci 29(4):309–327
65. Roy A, Datta SP, Barman M, Golui D, Bhattacharyya S, Meena MC, Rahman MM (2023) Co-application of silicate and low-arsenic-accumulating rice cultivars efficiently reduces human exposure to arsenic—a case study from West Bengal, India. Toxics 11(1):64

66. Roy D, Sreekanth D, Pawar D, Mahawar H, Barman K (2021) Phytoremediation of arsenic contaminated water using aquatic, semi-aquatic and submerged weeds. In: Biodegradation technology of organic and inorganic pollutants. IntechOpen, London
67. Saeed M, Quraishi UM, Malik RN (2021) Arsenic uptake and toxicity in wheat (Triticum aestivum L.): a review of multi-omics approaches to identify tolerance mechanisms. Food Chem 355:129607
68. Sandil S, Záray G, Endrédi A, Füzy A, Takács T, Óvári M, Dobosy P (2023) Arsenic uptake and accumulation in bean and lettuce plants at different developmental stages. Environ Sci Pollut Res 30(56):118724–118735
69. Savchenko T, Tikhonov K (2021) Oxidative stress-induced alteration of plant central metabolism. Life 11(4):304
70. Shaji E, Santosh M, Sarath KV, Prakash P, Deepchand V, Divya BV (2021) Arsenic contamination of groundwater: a global synopsis with focus on the Indian Peninsula. Geosci Front 12(3):101079
71. Sherin G, Aswathi KR, Puthur JT (2022) Photosynthetic functions in plants subjected to stresses are positively influenced by priming. Plant Stress 4:100079
72. Sinha D, Datta S, Mishra R, Agarwal P, Kumari T, Adeyemi SB, Chen JT (2023) Negative impacts of arsenic on plants and mitigation strategies. Plants 12(9):1815
73. Srivastava S, Shukla A, Rajput VD, Kumar K, Minkina T, Mandzhieva S, Suprasanna P (2021) Arsenic remediation through sustainable phytoremediation approaches. Minerals 11(9):936
74. Suda A, Baba K, Yamaguchi N, Akahane I, Makino T (2015) The effects of soil amendments on arsenic concentrations in soil solutions after long-term flooded incubation. Soil Sci Plant Nutr 61(4):592–602
75. Sun L, Zheng M, Liu H, Peng S, Huang J, Cui K, Nie L (2014) Water management practices affect arsenic and cadmium accumulation in rice grains. Sci World J 2014(1):596438
76. Sun YY, Liu RL, Zeng XB, Lin QM, Bai LY, Li LF, Wang YN (2015) Reduction of arsenic bioavailability by amending seven inorganic materials in arsenic contaminated soil. J Integr Agric 14(7):1414–1422
77. Sundha P, Basak N, Rai AK, Yadav RK, Sharma PC, Sharma DK (2020) Can conjunctive use of gypsum, city waste composts and marginal quality water rehabilitate saline-sodic soils? Soil Tillage Res 200:104608
78. Talarico E, Zambelli A, Araniti F, Greco E, Chiappetta A, Bruno L (2024) Unravelling the epigenetic code: DNA methylation in plants and its role in stress response. Epigenomes 8(3):30
79. Thankachan MB, Kamath GS, Sasikumar G, Panicker SP (2023) Exploring the interplay between arsenic and cutaneous physiology, pathology, and regeneration. In: Arsenic in the environment-sources, impacts and remedies. IntechOpen.
80. Vezza ME, Llanes A, Travaglia C, Agostini E, Talano MA (2018) Arsenic stress effects on root water absorption in soybean plants: physiological and morphological aspects. Plant Physiol Biochem 123:8–17
81. Wang J, Zhao FJ, Meharg AA, Raab A, Feldmann J, McGrath SP (2002) Mechanisms of arsenic hyperaccumulation in Pteris vittata: uptake kinetics, interactions with phosphate, and arsenic speciation. Plant Physiol 130(3):1552–1561
82. William VU, Magpantay HD (2023) Arsenic and microorganisms: genes, molecular mechanisms, and recent advances in microbial arsenic bioremediation. Microorganisms 12(1):74
83. Xue Y, Li Y, Li X, Zheng J, Hua D, Jiang C, Yu B (2024) Arsenic bioremediation in mining wastewater by controllable genetically modified bacteria with biochar. Environ Technol Innov 33:103514
84. Yadav SPS, Bhandari S, Bhatta D, Poudel A, Bhattarai S, Yadav P, Oli B (2023) Biochar application: a sustainable approach to improve soil health. J Agric Food Res 11:100498
85. Yan A, Wang Y, Tan SN, Mohd Yusof ML, Ghosh S, Chen Z (2020) Phytoremediation: a promising approach for revegetation of heavy metal-polluted land. Front Plant Sci 11:359
86. Zemanová V, Pavlíková D, Hnilička F, Pavlík M (2021) Arsenic toxicity-induced physiological and metabolic changes in the shoots of Pteris cretica and Spinacia oleracea. Plants 10(10):2009

87. Zemanová V, Popov M, Pavlíková D, Kotrba P, Hnilička F, Česká J, Pavlík M (2020) Effect of arsenic stress on 5-methylcytosine, photosynthetic parameters and nutrient content in arsenic hyperaccumulator Pteris cretica (L.) var. Albo-lineata. BMC Plant Biol 20:1–10
88. Zhang J, Hamza A, Xie Z, Hussain S, Brestic M, Tahir MA, Shabala S (2021) Arsenic transport and interaction with plant metabolism: clues for improving agricultural productivity and food safety. Environ Pollut 290:117987
89. Zhang J, Liu J, Zheng F, Yu M, Shabala S, Song WY (2022) Comparative analysis of arsenic transport and tolerance mechanisms: evolution from prokaryote to higher plants. Cells 11(17):2741
90. Zhao FJ, McGrath SP, Meharg AA (2010) Arsenic as a food chain contaminant: mechanisms of plant uptake and metabolism and mitigation strategies. Annu Rev Plant Biol 61(1):535–559
91. Zheng X, Lin H, Du D, Li G, Alam O, Cheng Z, Li J (2024) Remediation of heavy metals polluted soil environment: a critical review on biological approaches. Ecotoxicol Environ Saf 284:116883
92. Zhou YQ, Li SY, Shi YD, Lv W, Shen TB, Huang QL, Li YK, Wu ZL (2013) Phytoremediation of chromium and lead using water lettuce pistia stratiotes L. In: Applied mechanics and materials, vol 401, pp 2071–2075. Trans Tech Publications Ltd.

Copper: Uptake, Toxicity, Tolerance in Crops, and Strategies for Managing Cu-Contaminated Soils

Mohamed Zouari, Ameni Ben Hassena, Lina Trabelsi, Nacim Zouari, Bechir Ben Rouina, and Pascal Labrousse

Abstract Although copper is a vital micronutrient for plants, playing key roles in photosynthesis, enzyme activity, hormone regulation, stress response mechanisms, and nutrient uptake efficiency, its excessive accumulation can lead to toxicity. Elevated copper levels interfere with plant biochemical and physiological processes, resulting in oxidative stress and damage to cellular structures, ultimately impairing plant growth and productivity. This review provides a comprehensive analysis of copper uptake and transport in plants, highlighting the toxic effects on seed germination, growth, and photosynthesis. Additionally, it explores plant detoxification mechanisms, including both enzymatic and non-enzymatic antioxidant systems, which help mitigate copper-induced stress. Sustainable soil management practices, particularly phytoremediation, are also discussed, focusing on the use of plants to tolerate, accumulate, and immobilize copper. Techniques like phytoextraction and phytostabilization emerge as cost-effective, eco-friendly solutions for soil rehabilitation. This review underscores the importance of selecting plant species with enhanced copper tolerance and understanding their detoxification strategies to optimize phytoremediation outcomes.

Keywords Copper toxicity · Nutrient uptake · Transporter · Tolerance · Phytoremediation

M. Zouari (✉) · A. B. Hassena · L. Trabelsi · B. B. Rouina
Laboratory of Improvement of Olive Productivity and Fruit Trees, Olive Tree Institute of Sfax, University of Sfax, Sfax, Tunisia
e-mail: mohamedzouari2@gmail.com

M. Zouari · P. Labrousse
University of Limoges, E2Lim UR 24133, Limoges, France

N. Zouari
Laboratory of Biodiversity, Molecules, and Applications (LR22ES02), Higher Institute of Applied Biology of Medenine (ISBAM), University of Gabes, Medenine, 4119, Tunisia

A. Husen (ed.), *Enhanced Crop Production Setup Under Various Micro and Nano Contaminants*, Smart Nanomaterials Technology,
https://doi.org/10.1007/978-981-96-6099-5_6

1 Introduction

Soil contamination by copper (Cu) has become a growing concern due to the intensification of industrial activities, such as mineral extraction and chemical production, as well as agricultural practices that rely on Cu-based agrochemicals [42, 72]. Plants have primarily absorbed Cu through their root systems. However, various transporter protein families contribute to maintaining Cu homeostasis in plants. Key Cu transporters include copper transporter proteins family (COPT), copper-transporting P-type ATPases, the ZIP (Zrt, Irt-like protein) family, yellow stripe-like (YSL) transporters, and Cu chaperones. At the cellular level, these proteins mediate Cu movement, directing it to specific target sites within the plant. Additionally, other transport proteins are believed to play a role in channeling Cu into specific organelles [34, 51, 63] .

At optimal levels, Cu contributes to key processes, such as photosynthesis, enzyme activity, hormone regulation, stress response mechanisms, and nutrient uptake efficiency. It serves as a co-factor for several enzymes, including, Cu/Zn superoxide dismutase (SOD) and polyphenol oxidase. Copper also provides protection against certain abiotic stresses by mitigating the damaging effects of reactive oxygen species (ROS) [24, 44, 79]. Nevertheless, Cu becomes highly toxic when present in excessive amounts. Cu disrupts critical biochemical and physiological pathways, ultimately impairing plant growth and productivity. High levels of Cu lead to increased production of ROS and free radicals, resulting in oxidative stress. This oxidative stress damages cellular components, including lipids, proteins, and nucleic acids. Common symptoms of Cu toxicity include chlorosis, necrosis, and inhibited root and shoot growth, which reflect the disruption of photosynthesis and mineral uptake. Plant detoxification mechanisms against Cu stress include various types of enzymatic antioxidants such as ascorbate peroxidase (APX), glutathione peroxidase (GPX), catalase (CAT) and SOD and non-enzymatic antioxidants such as phytochelatins, metallothioneins, glutathione, ascorbic acid, and proline [39, 44, 81].

Over the past few decades, designing and implementing diverse approaches to remove toxic contaminants from soil has been a major area of focus. Among them, phytoremediation, which involves employing plants to uptake and sequester heavy metals from the soil, has emerged as an environmentally friendly and cost-effective strategy for the rehabilitation of soil contaminated with Cu. The methods of phytoextraction and phytostabilization are mainly used [18, 21]. The tolerance, accumulation, and translocation of heavy metals to the aerial part of plants are fundamental aspects to be followed in the selection of species in phytoremediation programs. Previous literature has indicated that plant species vary in their capacity to withstand and adapt to copper-induced stress, and that tolerant species have adapted various means of detoxification [17, 27, 55]. The current review examines uptake, toxicity, tolerance mechanisms in plants, and management strategies for Cu-contaminated soil by the application of phytoremediation techniques.

2 Copper Uptake and Transport in Plants

Plants require Cu in trace amounts, with concentrations typically ranging between 5 and 20 $\mu g\ g^{-1}$ dry weight in plant tissues across various species [24]. The accumulation of Cu within plant tissues is influenced by species characteristics, developmental stages, and environmental conditions. Cu is absorbed and transported by roots in its reduced form. Isotopic studies of Cu absorption in dicotyledon roots reveal the uptake of trace amounts of the Cu isotope (^{63}Cu), confirming the reductive uptake of Cu from the soil by the roots [9]. These observations suggest that the reduction of Cu(II) to Cu(I) at the root surface is a critical step that enhances Cu solubility and thereby facilitates more effective uptake from the soil.

The mechanisms involved in Cu uptake and transport in plant cells remain less understood compared to other essential nutrients. However, significant progress has been made in recent years, especially by applying findings from yeast studies to other eukaryotic organisms. The cell wall acts as the initial barrier to Cu ions entering plant root cells and contains various macromolecular substances, including amino groups, phosphoric acid, hydroxyl, and carboxyl groups [51, 79]. These functional groups can engage in diverse physical or chemical interactions with positively charged Cu ions, facilitating the adsorption of excess Cu onto the cell wall and thus restricting its entry into the protoplast. This mechanism serves to mitigate the toxic effects of excessive Cu on plant cells. Once Cu traverses the cell wall, it forms complexes with specific amino acids and proteins within the cell, allowing for its subsequent transport to other tissues and organs. The Cu transport pathway is generally divided into four main steps: (1) uptake by roots, (2) sequestration in vacuoles, (3) loading into the xylem and phloem, and (4) distribution and redistribution across nodes [13].

Cu transporters are essential to the regulation of Cu within plants, participating in processes such as absorption, chelation, transport, and compartmentalization. At the cellular level, Cu transporters identified in plants can be divided into two categories: those responsible for transporting extracellular Cu to the intracellular space (absorption transporters) and those responsible for transporting intracellular Cu to the extracellular space or organelles (efflux transporters) [42, 72].

Various transporters are essential for Cu uptake and regulation within plant cells. Identified Cu transporters include COPT, copper-transporting P-type ATPases, ZIP family transporters, YSL transporters, and Cu chaperones [42, 72, 79]. Research indicates that these transporters, along with associated pumps and channels, facilitate the absorption, distribution, and intracellular regulation of Cu, maintaining a stable Cu ion balance. This homeostasis is important not only for supporting normal plant growth and development but also for enabling adaptive responses to environmental fluctuations.

2.1 Copper Transporter Proteins Family (COPT)

COPT transporters, also known as Ctr in fungi and animals, are present in all eukaryotes. Numerous studies on yeasts, mammals, insects, algae, and plants have revealed that eukaryotes utilize the conserved CTR/COPT family of proteins to facilitate high-affinity cellular Cu acquisition at the plasma membrane and Cu mobilization from intracellular storage organelles, when Cu bioavailability decreases. COPT protein family is highly specific for Cu^{+} transport, but not for Cu^{2+}. Consequently, they function in coordination with membrane metalloreductases that catalyze Cu^{2+} reduction to Cu^{+} before transport [13, 72].

A several genome-wide studies of the COPT protein family was identified in plant species such as Arabidopsis (*Arabidopsis thaliana* L.) [63, 64], rice (*Oryza sativa* L.) [80], maize (*Zea mays* L.) [73], alfalfa (*Medicago sativa* L.) [74], tomato (*Solanum lycopersicum* L.) [60] and mangrove (*Kandelia obovata*) [34].

The COPT family of transporters consists of three transmembrane domains (TMDs) with an amino-terminal region rich in methionine (Met) and/or histidine (His) residues. According to the number of domains, the six COPT members reported in *Arabidopsis* can be divided into three categories [63, 64]. The first class consists of three Met- and His-rich domains: AtCOPT1, AtCOPT2, and AtCOPT6. The proteins of this group have a high affinity for Cu and a strong transport capacity, which is hypothesized to be responsible for the ability of Cu ions to flow in and out of cells. These proteins are located at the plasma membrane and are implicated in the Cu uptake into the cytoplasm. The second subgroup consists of AtCOPT3 and AtCOPT5, which are characterized by a single Met- and His-rich domain. These transporters have a low affinity for Cu and are less efficient in transporting Cu ions. Additionally, these transporters are found in intracellular membranes, where they are involved in the recycling of Cu within the cell. Interestingly, the third class is AtCOPT4, which lacks the Met residues. This protein can assist in Cu transport in plants by interacting with other COPT proteins [43, 63, 64, 79]. According to these authors and by characterizing lack-of-function Arabidopsis mutant lines, these authors conclude that COPT1 mediates root Cu acquisition, COPT6 facilitates shoot Cu distribution, COPT5 mobilizes Cu from storage organelles, and COPT2 functions at the intersection between Cu and Fe homeostasis. COPT4 is the only member of the gene family in *A. thaliana* that could not complement yeast mutants deficient in Cu uptake, and therefore, is considered nonfunctional [63].

In tomato, six SlCOPT genes were identified [60]. These authors reported that SlCOPT1 and SlCOPT2 were probably located in the plasma membrane and are the only SlCOPTs enabled to mediate Cu transport themselves, while SlCOPT3 and SlCOPT5 were located in intracellular organelles and are only partially functional.

2.2 Copper-Transporting P-type ATPases

P-type ATPases are a family of transporters that transport heavy metal ions across the membranes by hydrolyzing ATP, and they are also known as heavy metal transporting ATPase (HMA). The role of P-type ATPase in the transport of various metals (such as Cu) across the cell membrane has been reported in various organisms [23]. During plant growth, HMAs can selectively uptake and transport not only essential metal ions such as Cu^{2+} and Zn^{2+}, but also non-essential ions such as lead (Pb^{2+}) and cadmium (Cd^{2+}). These metal transporters use ATP as an energy source to pump Cu and other metals and nutrients across the biological membranes [19].

To date, a varying number of heavy metal ATPase (HMA) genes have been identified in several plant species, including 12 in black cottonwood (*Populus trichocarpa* Nutt.) [59], 9 in rice [19], 8 in Arabidopsis [10], and 31 in mulberry (*Morus alba* L.) [23].

In *A. thaliana,* AtHMA1, AtHMA6, and AtHMA8 are high-affinity Cu transporters located in the chloroplast. They are responsible for the transport of Cu in the chloroplast. In addition, they transport Cu to the chloroplast stroma and thylakoid to synthesize plastocyanin and also provide co-factors for Cu/Zn-SOD in the stroma [33],Memon and Meraklı., 2022). AtHMA5 is regulated by higher Cu concentrations and eliminates excess Cu by symplast transport for xylem loading [3]. AtHMA7 regulates ethylene signaling by reacting with trans-cyclooctene (an ethylene receptor), expressing the characteristic feature of Cu in the ethylene signaling pathway. AtHMA7 additionally delivers Cu to the secretory pathway, which is essential for the organization of ethylene receptors [32].

In rice, OsHMA6 is expressed in all organs at different growth stages, including spikelet, and abundant in leaf blades, however, OsHMA9 is most strongly expressed in roots, but very low in spikelet. Excessive Cu^{2+} can up-regulate the expression of OsHMA6 and OsHMA9 in rice seedlings. These results suggested that OsHMA6, similar to OsHMA9, is likely a copper efflux protein located in the plasma membrane [77]. Furthermore, another HMA in rice (OsHMA5) was found to function in the transport of Cu to the xylem [19]. These authors reported that OsHMA4 is located in the vacuole membrane, which can store Cu in the vacuoles of root cells and reduce the accumulation of Cu in rice seeds.

2.3 ZIP (Zrt, Irt-Like Protein) Family Transporters

The ZIP family is composed of two types of members—zinc-regulated transporters (ZRTs) and iron-regulated transporters (IRTs). Members of the ZIP family primarily maintain the intracellular ion balance by transporting various metal cations, such as zinc, iron (Fe), Cu, and Cd, into the cytoplasm [48]. ZIP proteins are widely found in fungi, bacteria, animals, and plants, with more than 100 types identified, including

15 in *A. thaliana* (AtZIP1–AtZIP11) [30], 16 in rice (OsZIP1–OsZIP16) [14] and 10 in sorghum (*Sorghum bicolor* L.) [48].

In *A. thaliana* under Cu deficiency, expression of ZIP transporters (ZIP2 and ZIP4) was upregulated, whereas down-regulation in gene expression was reported under conditions of excess Cu (Milner et al. 2013). According to [48], ten SbZIP family transporter genes were identified in shoot and root tissues of sorghum under deficient and sufficient Zn. SbZIP2, 5, 6, 7, and 8 were expressed in all tissues under deficient and sufficient Zn. SbZIP2, 4, 5, 6, 7, 8, and 10 were highly induced in shoot tissues by deficient Zn. The expression level of SbZIP6, 7, 8, and 9 was higher in root tissues under deficient Zn. In rice, Cu stress induces the expression of OsZIP1, and complementation assays in yeast (*Saccharomyces cerevisiae*) showed that OsZIP1 reduces the accumulation of metals [14]. All these studies reveal that the expression level of ZIP family genes is marked changed and is essential to involved in root metal uptake, as well as metal transport and distribution.

2.4 *Yellow Stripe-Like (YSL) Transporters*

The YSL family of transporters is involved in heavy metal transport, as these proteins transport metals bound to phytosiderophores and nicotianamine (NA). However, only a few YSL proteins were shown to transport Cu. *A. thaliana* has been found to have eight YSL family members (AtYSL1–AtYSL8) [20, 76], and rice has 18 members (OsYSL1–OsYSL18) [83].

In *A. thaliana*, AtYSL2 is involved in the transport of Fe (II)-NA and Cu-NA, and its transcriptional level is regulated by Fe and Cu. Therefore, although it seems clear that YSL proteins in *A. thaliana* have at least an indirect role in Cu homeostasis, it is not yet clear whether they can transport Cu and what their physiological functions regarding Cu distribution [20, 76].

In rice, OsYSL16 is expressed in roots, leaves, and nodes, with higher expression in vascular tissues of the leaves [83]. According to [82] OsYSL16 is primarily responsible for transporting the Cu-NA complexes through the phloem to nascent tissues and seeds. Plants lacking OsYSL16 function also showed reduced Cu concentration in seeds and reduced fertility [82, 83].

2.5 *Copper Chaperones*

Cu chaperones are a class of metal receptor proteins with low molecular mass, which are widely present in the cells of various organisms and contain a Cu ion-binding domain. This domain is responsible for binding the Cu ions and delivering them to specific Cu transporters. Cu chaperones prevent Cu from interacting with other intracellular components, thus avoiding the toxicity of free Cu ions [52].

To date, the reported Cu chaperones in *A. thaliana* include cytochrome C Oxidase17 (AtCOX17), Cu chaperone for SOD (AtCCS), antioxidant protein1 (AtATX1), and ATX1- like Cu chaperone (AtCCH) [16]. According to these authors, the expression level of antioxidant protein 1 (AtATX1) in *A. thaliana* is upregulated by Cu-induced stress, and overexpression of AtATX1 significantly enhances the accumulation and tolerance of *A. thaliana* to Cu.

3 Copper Toxicity in Plants

Copper is an essential trace element and co-factor for various proteins, playing an essential role in many physiological and metabolic processes in plants, such as photosynthesis, respiration, antioxidative regulation, electron transport, and nitrogen and protein metabolism. Additionally, copper supports hormonal signaling and overall plant growth and development by acting as a structural and metabolic component of proteins and enzymes. However, while an optimal amount of Cu is necessary for these cellular functions, excessive concentrations can be highly toxic. High copper levels can disrupt plant physiology, causing oxidative stress, accumulation of ROS, and imbalances in nutrient uptake, which together lead to inhibited growth and compromised productivity due to alterations in root architecture and cellular metabolism [13, 45] (Fig. 1 and Table 1).

3.1 Copper Toxicity on Photosynthesis and Pigment Contents

Cu is an essential micronutrient element in the transport of electrons between photosystems II (PSII) and photosystems I (PSI) because it enters the structural constitution of plastocyanin, an important component of the electron transport chain along photosystems. However, under toxic levels of Cu, PSII is the most sensitive site of Cu action. Toxic Cu concentrations vary significantly depending on plant species and genotypes, leading to significant differences in Cu tolerance within and among species [38].

Jin et al. [39] reported that under Cu stress, photosynthesis rate, stomatal conductance, and chlorophyll content of the water hyacinth plant (*Eichhornia crassipes* Mart.) were significantly decreased as compared to the control.

In another study, Cu stress significantly ($p \leq 0.05$) disrupted the photosynthetic process of castor seedlings (*Ricinus communis* L.), affecting both photosynthesis and the chlorophyll stability index (CSI) compared to non-stressed plants [62]. According to these authors, Cu-induced ROS resulted in the denaturation of key enzymes and polypeptides involved in the photosynthetic processes, which are needed for the long-term stability of PSII, thereby adversely affecting the oxygen-evolving complex.

The significant reduction in contents of photosynthetic parameters in leaves of orange (*Citrus aurantium* L.) subjected to Cu stress indicates the ROS-mediated

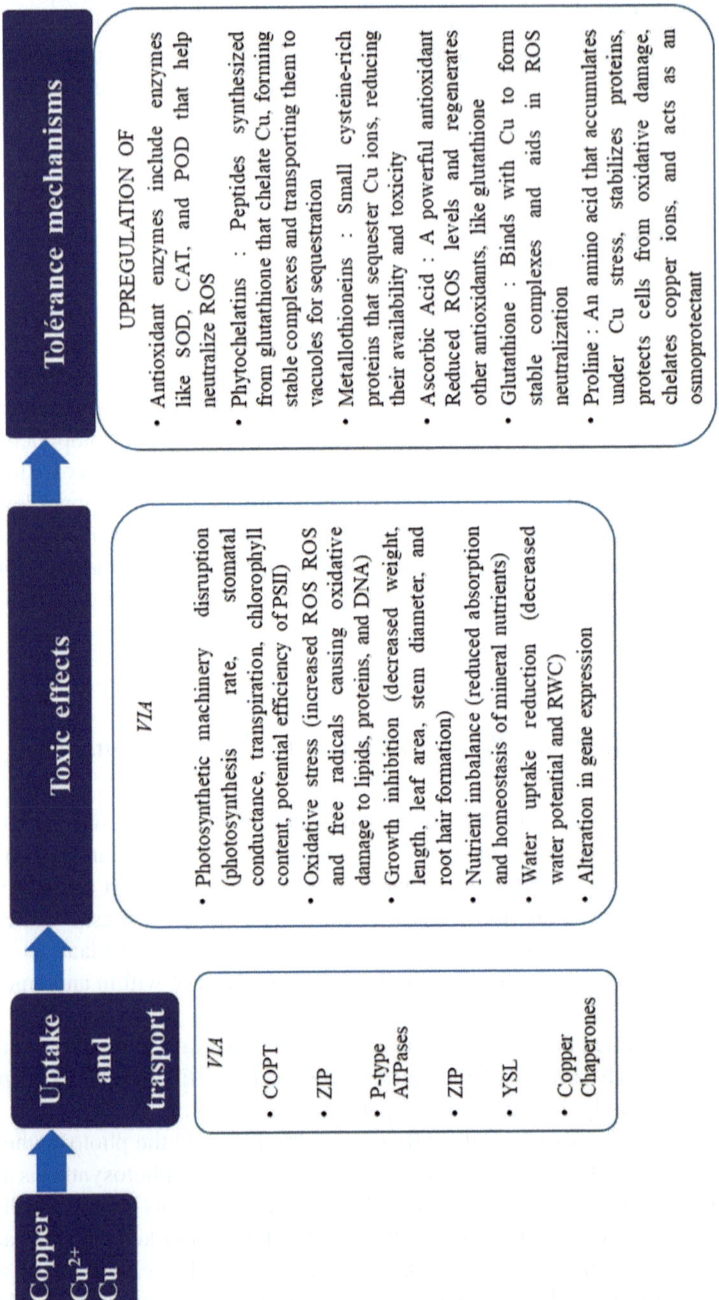

Fig. 1 Plant responses and defense mechanisms against copper stress

Table 1 Copper toxicity and tolerance in plants

Plant species	Copper concentration	Growth medium	Period	Effect on plant traits	Tolerance factors	Key references
Pisum sativum	2000 mg kg^{-1} $CuSO_4$	Soil	90 days	-Decreased growth, chlorophyll contents, photosynthetic rate, MDA, H_2O_2 and EL -Reduced activity of antioxidants (SOD, POD, and CAT)	Gibberellic acid foliar applications increased growth, activity of antioxidant enzymes and alleviated the oxidative stress induced by Cu	[38]
Ricinus communis	0.2 mM $CuSO_4$	Hydroponic	6 days	-Decreased photosystem activities and chlorophyll stability index, RWC -Increased reactive oxygen species the leaves and roots	-Increased activities of antioxidant enzymes like APX, CAT, POD, and SOD -Cell wall compounds act as principal Cu binding sites and have significant roles in Cu sequestration	[62]
Helianthus annuus	(2.5, 5, 10, 20, 50 and 100 µM)	hydroponic	7 days	Cu at the highest doses (50 and 100 µM): -Decreased contents of K, Mg, Zn and Mn -Increased oxidative stress biomarkers (MDA and H_2O_2) -inhibited antioxidative enzymes (SOD, CAT and APX)	-Stimulation of guaiacol peroxidase activity correlated with its role in membrane lignification -Tolerance threshold of a sunfower plant to Cu stress is 10 µM. When Cu exceeds this concentration in the culture medium, toxicity symptoms may occur in the plant	[61]

(continued)

Table 1 (continued)

Plant species	Copper concentration	Growth medium	Period	Effect on plant traits	Tolerance factors	Key references
Triticum aestivum	100, 300, and 500 μM Cu	Petri dishes containing Hoagland nutrient solution	6 days	Reduced growth parameter, photosynthetic pigment content, gas exchange parameters	-Increased amount of cell wall lignin in leaves and roots -Up-regulation of enzymes related to lignin synthesis, such as phenylalanine ammonia-lyase, cinnamyl alcohol dehydrogenase	[45]
Eichhornia crassipes	1, 3, 5, 7 and 9 mg L^{-1} $CuSO_4$	hydroponic	7, 14 days	Reduced growth, gas exchange, chlorophyll fluorescence	The physiological tolerance of *E. crassipes* to Cu relies on osmotic regulation, anti-lipid peroxidation and improved antioxidant properties	[39]
Pistia stratiotes	0.2, 0.4, 0.8, and 1 mg L^{-1} $CuSO_4$	Hoagland solution	20 days	-Decreased growth rates, relative water contents, chlorophyll contents -Reduced root exudates included amino acids, alkanes, aldehydes, ketones and phenols		[75]

(continued)

Table 1 (continued)

Plant species	Copper concentration	Growth medium	Period	Effect on plant traits	Tolerance factors	Key references
Vicia pannonica	Cu NPs 50, 150 and 300 ppm	Petri dishes	Just seeds germination	-Reduced the mitotic index of meristem cells -Abnormalities in cell division, such as C-metaphase, polar aberration, deconstructed anaphase, chromosome breakage, loss of genetic material, failure to assemble in the metaphase plate, deconstructed metaphase		[70]
Capsicum annuum	0.1, 0.5, 50 mM Cu	Hydroponics system		-Decreased growth parameters and increase of oxidative stress biomarkers (MDA and H_2O_2)	-Increase in proline, sucrose content, CAT, POD and SOD enzyme activity -Exogenous chrysin enhanced plant nutrient uptake	[81]
Citrus sinensis and Citrus grandis	0.5 and 300 μM $CuCl_4$	Hydroponics system	18 weeks	-Decreased shoot biomass, chlorophyll content -Destroyed the ultrastructure of the leaves and roots	Under Cu toxicity, increased root flocculent precipitate and thickened root cell wall reduce the Cu translocation from citrus roots to the shoots	[46]
Cucumis sativus	0.2, 5, 25 and 50 μM		7 days	-Altered mass of root and shoot tissues, mineral status -Limitation in the root acquisition of P	Plants exposed to Cu toxicity upregulate the expression of Pi transporters	[25]

(continued)

Table 1 (continued)

Plant species	Copper concentration	Growth medium	Period	Effect on plant traits	Tolerance factors	Key references
Solanum lycopersicum	96, 144, and 192 ppb in the form of Copper EDTA 15%	Soil	4 months	-Decreased growth, productivity, mineral status and chlorophyll content -Increased oxidative stress biomarkers (MDA and H_2O_2)	-Increased CAT and SOD enzyme activity in roots and leaves under Cu stress -Exogenous application of proline and melatonin reduce the harmful effects of excess copper	[11]
Pontederia cordota	1, 2.5 and 5 mg L^{-1}	Hydroponic	7, 14, 21, and 28 days	- With 1 to 5 mg L^{-1} Cu exposure for 7 d and 14 d, the plant normally grew by stable chlorophyll content,	-Increased peroxidase, ascorbate peroxidase, dehydroascorbate reductase activities -Increased GSH, non-protein thiol total peptide (NPT), and phytochela-tins (PCs)	[78]

destruction of chlorophyll stability. The same authors indicated that the disruption in the status of the photosynthetic apparatus could also be due to the degradation of the thylakoid membrane, leading to reduced activities of PSI and PSII, and subsequently, the inhibition of PSII electron transfer activity. High levels of Cu can affect plant photosynthesis-related factors such as PSII efficiency, carbon dioxide diffusion, ROS generation, and electron transport [28].

In wheat seedlings (*Triticum aestivum* L.) subjected to Cu stress, reduction of photosynthetic pigment content, gas exchange parameters, and chlorophyll fluorescence parameters, including the maximum photosynthetic efficiency, potential efficiency of PSII, photochemical efficiency of PSII in light, photochemical quenching, actual photochemical efficiency, quantum yield of PSII electron transport, and electron transport rate were observed [46]. This decrease in photosynthetic activity can be generated by stomatal factors, including restrictions for CO_2 diffusion, and by non-stomatal limitations, such as decreased rubisco (ribulose-1,5-bisphosphate carboxylase/oxygenase) activity or damage to photosynthetic apparatus by PSII performance impairment [46]. It has been reported that metals, including Cu^{2+}, compete for binding to -SH groups and can replace Mg^{2+}, Zn^{2+}, or Fe^{2+} in chloroplast proteins. In this way, Cu alters the ionic composition of these proteins, making them unstable or inactive, leading to their degradation [66].

3.2 *Copper-Induced Oxidative Stress*

Under normal metabolic conditions, ROS play essential roles in plants, including pathogen defense, cell wall lignification, stress adaptation, and signaling processes. They regulate redox balance, development, fruit ripening, systemic responses, and are involved in gene expression for apoptosis, senescence, and cell death [22, 53, 61]. At lower levels, Cu may not induce the overproduction of ROS. However, at toxic levels, Cu generates oxidative stress that damages cellular structures and molecules, such as DNA, proteins, and lipids. Therefore, oxidative stress is a naturally occurring physiochemical process in the biological metabolism where the level of ROS overpowers their scavenging system causing an imbalance between the ROS and antioxidants. Continued exposure of plants to excess Cu accelerates ROS generation. Cu toxicity promotes the production of superoxide radicals ($O_2{\bullet}^-$) through various chain reactions. This $O_2{\bullet}^-$ is then converted into hydrogen peroxide (H_2O_2) or hydroxyl radical (•OH) by different antioxidant enzymes under specific redox conditions. ROS generated by Cu stress can attack hydrogen atoms in fatty acid chains (such as linoleic acid) in plasma and organelle membranes, leading to the generation of aldehydes and lipid free radicals. These reactions can cause membrane lipid peroxidation, increased membrane permeability, massive extravasation of cellular contents, and cell death [28, 66].

Copper-induced ROS overproduction has been reported in various plant species (Table 1). For example, in pea plants (*Pisum sativum* L.), Cu stress influenced ROS generation by enhancing malondialdehyde (MDA), H_2O_2, and electrolyte leakage

(EL) in leaf cells [38]. These results reflected the lipid peroxidation and disruption of membrane integrity in case of Cu-treated pea plants. In castor seedlings grown under 0.2 mM $CuSO_4$, [62] reported that Cu stress induced a significant decrease in membrane stability index and an increase in EL. Similar findings were reported in sunflower (*Helianthus annuus* L.) plants cultivated under Cu stress [61]. Referring to these authors, the lipoperoxidation could be due either to the stimulation of lipoxygenase or to the ability of Cu to catalyze the formation of free radicals via the Haber–Weiss reaction.

3.3 *Effect of Copper Toxicity on Mineral Status*

Another common effect of Cu toxicity on plants is the reduced absorption and homeostasis of other mineral nutrients. The degree of nutrient disproportion is based upon plant species, Cu level in the rhizosphere, treatment time, and growth environs.

High concentrations of Cu in orange (*Citrus sinensis* L.) reduced the levels of phosphorus (P), nitrogen (N), calcium, magnesium, potassium (K), sulfur, boron and Fe in leaves and roots [15]. Cu-toxicity-induced a decrease in P level might also affect the uptake of the other nutrients because P is necessary for the biosynthesis of ATP, that is usually required for the transport of several nutrient elements against the H^+-electrochemical gradient across the plasma membrane of roots [13, 25].

In cucumber plants (*Cucumis sativus* L.), Cu toxicity (25 and 50 μM) led to decreased uptake of P, K in both shoots and roots as compared to control. Concerning Mn, no significant alterations were recorded in the root tissue, while concentration reductions by about 25 and 61% were observed in plants exposed to 25 and 50 μM Cu, respectively [25]. [61] indicated that excess Cu disrupted the mineral composition of the sunflower plant. This effect was reflected in decreased contents of K, Mg, Zn, and Mn in roots, stems, and leaves of sunflower seedlings treated with 100 μM $CuSO_4$. According to these authors, this perturbation of mineral nutrient status could be due to: (1) competition between Cu and essential elements via common transporters, altered gene expression required for nutrient uptake, and modifications in permeability of the plasma membrane, and (2) morphological and anatomical changes that disrupt root structure. Excess Cu damages the root cuticle, reduces root hair proliferation, and deforms root architecture, impairing the plant's ability to absorb nutrients. Additionally, Cu toxicity harms cell membranes in root tissues, further reducing the root surface area available for nutrient uptake from the soil.

3.4 Effect of Copper Toxicity on Plant Growth and Morphology

Cu is an essential nutrient for normal plant growth and development, but it is toxic when it exceeds the permissible limit, leading to a decrease in plant growth performance and loss of production. Thus, the reduction is a final consequence of changes that initially occur at the biochemical, physiological, morphological, and mineral levels of plants [35, 75].

Several previous studies have reported inhibition in plant growth under Cu stress. The growth properties (e.g., fresh mass, root length, height, leaf length, and width) of water hyacinth plant were negatively affected by Cu^{2+} treatments, especially at higher Cu^{2+} concentrations (9 mg/l Cu^{2+}) [39]. Excess accumulation of Cu decreased height, fresh weight, dry weight, and number of leaves in pea plants [38]. Similarly, root dry mass was significantly reduced in castor seedlings exposed to elevated Cu levels, in comparison with control plants [62].

Different mechanisms behind the Cu-induced inhibition of growth were revealed. Cu toxicity has a negative impact on gas exchange (photosynthesis, transpiration, stomatal conductance, and internal cell concentration), chloroplast pigments (chlorophylls *a* and *b*, and carotenoids), and photochemical parameters. These changes promoted by Cu contribute decisively to the reduction of vegetative growth. All these disorders in the photochemical apparatus of photosynthesis ultimately lead to a reduction in shoot growth, development, and plant production [35, 39].

Morphologically, Cu toxicity manifests as leaf chlorosis, necrosis, wilting, and stunted overall growth. Anatomically, high Cu levels (800 mg kg^{-1}) alter leaf structure by distorting leaf blade shape, reducing stomatal size, decreasing intracellular space within mesophyll cells, and increasing hair density on leaf surfaces, which collectively impair transpiration. In roots, Cu toxicity disrupts the root cuticle, damages cell membranes, and reduces root hair formation, leading to severe root deformities [49].

3.5 Effect of Copper Toxicity on Seed Germination

Excess Cu in soil significantly reduces seed germination in many plant species (Table 1). Shamas et al. [67] observed that high concentrations of copper (100, 200, and 300 μM) had a significant effect on the germination percentage, germination rate, mean period of ultimate germination, and seed vigor index germination of lettuce (*Lactuca sativa* L.) seeds. Iqbal et al. [35] observed a decline in germination of lentils (*Lens culinaris* Medik.) seeds as Cu level increased up to 50 ppm. Astuti and Nugroho [5] reported that high Cu concentration (100 mg l^{-1}) significantly reduced seed germination percentage in mung bean (*Vigna radiata* L.).

Germination and seedling establishment are vulnerable stages in the plant life cycle. Studies have shown that Cu toxicity inhibits seed germination by down-regulating key enzymes like α-amylase and invertase, which are essential for converting starch and sucrose into simpler sugars. Additionally, Cu toxicity disrupts the seed's metabolic processes and hinders water transport. Cu ions exert a toxic effect by creating an external osmotic potential that inhibits water uptake, similar to the effect of Na^+ and Cl^- ions [5, 35, 67].

3.6 Copper-Induced Genotoxicity

Genotoxicity serves as a marker of cellular dysfunction, reflecting genome toxicity through mutagenic effects, such as chromosomal aberrations. Numerous studies have highlighted the genotoxic effects of Cu in plants.

Tasar [70] demonstrated that all doses of copper nanoparticles (Cu NPs) in Hungarian vetch (*Vicia pannonica* Crantz) caused significant decreases in the Mitotic index of meristem cells compared to the control. Multiple analyzes carried out in this study showed that Cu NPs caused abnormalities in cell division such as C-metaphase, polar aberration, deconstructed anaphase, chromosome breakage, loss of genetic material, failure to assemble in the metaphase plate, deconstructed metaphase, asynchronous division remaining chromosome in metaphase, adhesion, deconstructed prophase, advanced chromosomes, and chromosome bridges in anaphase. Mosa et al. [53] indicated that Cu NPs at high exposure concentrations induced genotoxicty in cucumber plants. Referring to [1], random amplified polymorphic DNA (RAPD) analysis confirmed the genotoxic effect of Cu NPs, which altered the coriander (*Coriandrum sativum* L.) genome. This was shown by the different banding pattern of RAPD.

Additionally, Bagdonas et al. [7] suggested that Cu genotoxicity may arise from two primary mechanisms: the generation of ROS that damage DNA and the inhibition of DNA repair enzymes. The latter effect likely occurs due to nonspecific binding of Cu^{2+} ions to essential sites on enzyme molecules, hindering their DNA repair functionality.

4 Copper Tolerance in Plants

To assure protection from the damage inflicted on different metabolic processes by oxidative stress, plants adopt different strategies, such as the modulation of the antioxidative system to an adequate level that guarantees optimal ROS scavenging (Fig. 1 and Table 1).

4.1 Role of Antioxidant Enzymes

Under Cu stress, antioxidant enzymes play a vital role in detoxifying ROS to prevent their accumulation. Key antioxidant enzymes involved include SOD, CAT, as well as glutathione peroxidase (GPX) and APX. These enzymes help mitigate the impact of elevated ROS by converting them into non-toxic products, thereby reducing oxidative damage. The activation and inactivation of antioxidant enzymes to respond to Cu-induced oxidative stress depend on ROS and the plant species [54, 71, 78].

Referring to [66], continued exposure of plants to excess Cu accelerates ROS generation. Superoxide radicals ($O_2^{\bullet-}$) are produced, which are considered the primary ROS. This ROS is relatively short-lived and moderately reactive, possessing both oxidizing and reducing characteristics. Once formed in plant cells, $O_2^{\bullet-}$ is converted into secondary ROS, a process facilitated by various antioxidant enzymes. SOD catalyzes the conversion of $O_2^{\bullet-}$ into less harmful H_2O_2 and O_2, acting as a primary defense mechanism against ROS in plants. Key enzymes for detoxifying H_2O_2 include CAT, GPX, and APX. While CAT directly decomposes H_2O_2 without a reducing agent, APX and GPX require specific molecules ascorbate and glutathione, respectively for reduction [62].

Several reports have described enhanced antioxidant enzyme activity due to Cu stress. According to [57], the APX and GPX activities in root, shoot, and leaf organs of wheat was increased by copper stress. These authors reported that APX plays a critical role in chloroplasts, where it reduces H_2O_2 to oxygen and water, helping to protect the photosynthetic machinery from oxidative stress, especially during intense photosynthetic activity. GPX, on the other hand, not only detoxifies H_2O_2 but also prevents lipid peroxidation by reducing organic peroxides, thus preserving cellular membrane stability.

El-Beltagi et al. [22] reported that antioxidant defense system, through various antioxidant enzymes such as SOD, POD, APX, and CAT, enhances the resistance of flax plants (*Linum usitatissimum* L.) to Cu toxicity. Sakouhi and El Ferjani [61] reported that Cu stress contributes to the increase of SOD, CAT and APX on sunflower plants grown under Cu conditions. In this way, plants can tolerate Cu by modifying the activities of antioxidant enzymes. Yuce et al. [81] stated that excess Cu stimulates genes, which boost defensive antioxidant enzymes like SOD, CAT, GPX, and APX to reduce toxicity.

Sameena and Puthur [62] demonstrated that in castor seedlings grown under 0.2 mM $CuSO_4$, the activities of antioxidant enzymes such as APX, CAT, POD, and SOD significantly increased in roots, cotyledonary leaves, and primary leaves, with the exception of POD activity in roots. In another study, GPX was increased significantly under Cu stress whereas SOD, POD and APX activities were reduced under high treatment (100 μM) of Cu-treated sunflower plants [61]. According to these authors, the increase in GPX activity may be attributed to its role in maintaining membrane rigidity, as it catalyzes the lignification of the cell wall. This process ultimately leads to the inhibition of cell elongation.

4.2 Role of Non-Enzymatic Antioxidants

4.2.1 Glutathione and Ascorbate

Maintaining an adequate pool of ascorbate (AsA) and glutathione (GSH) under Cu stress conditions is essential for cellular defense in plants. AsA serves both as a unique substrate for APX and as a direct scavenger of ROS, reducing ROS accumulation within the cell. Similarly, GSH functions not only as a substrate for enzymes such as GPX and glutathione transferases (GST) but also plays an indirect role in ROS detoxification through its ability to biosynthesize other antioxidants such as phytochelatins (PCs). Together, AsA and GSH bolster antioxidant defense, protecting plants from oxidative stress induced by heavy metals [54, 78].

Xin et al. [78] reported that GSH and AsA contents were significantly increased in leaves of pickerel weed plant (*Pontederia cordota* L.) under 5.0 mg L^{-1} Cu^{2+} exposure for 7 days, compared to control plants. These authors reported that GSH is used as a substrate by dehydroascorbate reductase (DHAR) to convert oxidized DHA into AsA. This can alleviate oxidative damage and maintain the balance of antioxidant system in cells. Similar findings were reported by [71], who noticed that leaves of maize plants treated with 2 mM Cu showed higher AsA and GSH activity than control plants and then began to decrease at 4 and 8 mM Cu. According to these authors, increased AsA and GSH may enhance the ability of plants to cope with oxidative stress at low Cu concentration. Mostafa et al. [54] showed that AsA content decreased while GSH and PCs content increased significantly in Cu stressed rice seedlings. In addition, exogenous GSH decreased toxicity symptoms and diminished a Cu-induced increase in $O_2^{\bullet-}$, H_2O_2, MDA, and proline content.

4.2.2 Phytochelatins and Metallothioneins

The neutralization and sequestration of excess free metal ions are essential mechanisms for sustaining cellular homeostasis in plants. This process is facilitated by chelation, where free metal ions such as Cu are captured and bound by specialized molecules known as chelators. This binding reduces both the reactivity and solubility of the metal ions, thus decreasing their potential toxicity [58]. Among various trace metal chelators, PCs are a family of enzymatically synthesized cysteine-rich peptides. The biosynthesis of PCs is catalyzed by the enzyme phytochelatin synthase (PCS) using GSH as the precursor [29]. Song et al. [68] reported that PCs play a crucial role by forming stable, non-toxic metal-PC complexes. These complexes are then transported to and stored within vacuoles or integrated into the cell wall compartment, further isolating the metals from the cellular environment (Navarrete et al. 2018) [58].

Several studies have reported an increase in PCs content in plants exposed to Cu stress. According to Navarrete et al. (2018), the initial response to increased intracellular Cu concentrations in the green seaweed (*Ulva compressa* L.) involves elevated levels of GSH and PCs, followed by an increase in metallothionein (MT) expression. This suggests that thiol-containing peptides and proteins participate in Cu accumulation and detoxification in a coordinated and complementary manner.

Moreover, [69] investigated the response of Hydrilla plants (*Hydrilla verticillate* L.f.) induced by a range of Cu doses (1–5 μM). High PCs activity was observed at both levels of Cu exposure in comparison to control. In contrast, no PCs were detected in the Cu-treated alfalfa plants [26] possibly due to the binding of Cu ions to metallothionein or other chelating agents, which resulted in plant's Cu tolerance.

MTs are cysteine-rich proteins characterized by thiol groups that enable them to bind effectively to various metals, including Cu [56]. Their strong affinity for Cu allows MTs to play a critical role in maintaining cellular metal homeostasis and detoxifying excess Cu and other trace metals within the cytosol. Additionally, recent studies indicate that MTs contribute to the sequestration of Cu in senescing leaves [40]. Responses to copper exposure vary across plant species, with some studies showing up-regulation or down-regulation of MT expression, while others report no change. This variability is influenced by factors such as species differences, exposure duration, Cu concentration, and the specific MT isoform [8, 40, 43].

4.2.3 Proline

Under heavy metal stress, such as in copper (Cu) conditions, proline serves multiple functions: (i) as an osmolyte, aiding in osmotic regulation, and (ii) as a non-enzymatic antioxidant, supporting the stability of subcellular structures like membranes and proteins by neutralizing free radicals [11, 41, 57]. Additionally, it may act as a hydrotrope compatible with proteins, helping to alleviate cytoplasmic acidosis and maintain the appropriate $NADP^+/NADPH$ ratios essential for metabolism [6, 47].

Many studies have shown the reasons behind the accumulation of proline under heavy metals and environmental stress. Noreen et al. [57] reported that proline content increased in leaves and root tissues of wheat grown under copper stress conditions. The same authors showed that the exogenous application of proline at the rate of 80 mM resulted in the reduction of ROS generation and enhanced accumulation of proline and protein contents in wheat varieties under copper stress environment. Ku et al. [41] showed that application of Cu increased the proline content in the leaves of tobacco (*Nicotiana benthamiana* Domin) plants. Referring to these authors, excess Cu increased the expression of two proline synthesis genes, pyrroline5-carboxylate synthetase and ornithine aminotransferase and suppressed proline catabolism gene, proline dehydrogenase. Exogenous proline in rice seedlings under excess copper stress not only reduced the Cu absorption in the roots but also the Cu exclusion, suggesting supplement of proline accompanied by Cu exposure induces a barrier of Cu influx and efflux in rice roots [12]. Badiaa et al. [6] revealed that the proline content was increased by copper treatment in leaves and roots of tomato (*Lycopersicon esculentum* Mill.). This increase was directly proportional to the different concentrations of copper stress. Brengi [11] reported that exogenous proline led to the tomato plant's leaves having more N, P, K, and chlorophyll, which was reflected in improved plant growth characteristics compared with the control under different Cu levels. It also enhanced the antioxidant content in the roots and leaves, which contributed to an increase in the number of fruits per plant, average fruit mass, and overall plant productivity.

5 Management of Copper-Contaminated Soil

Soil pollution with heavy metals, particularly Cu, represents a pressing environmental issue with far-reaching implications for ecosystems, agricultural land, and human health. Unlike organic pollutants, heavy metals are non-biodegradable and accumulate in the environment, disrupting natural ecosystem services and contaminating food chains [13]. The major sources of Cu contamination include industrial discharges, mining activities, waste disposal, and agricultural practices such as the application of copper-based fungicides and fertilizers. Agrochemicals, including those widely used in organic farming, further contribute to the accumulation of heavy metals in soils, posing risks to food security [36, 55] (Fig. 2). According [65], the persistent nature and toxicity of Cu stress not only reduce the availability of arable land but also threaten biodiversity and compromise the safety of food and water resources, underlining the urgent need for sustainable management practices to mitigate soil Cu pollution.

Cu is a critical micronutrient for plants, essential in small quantities to support their growth and survival. However, elevated Cu levels, often resulting from extensive use of agrochemicals, can lead to toxicity, impairing plant health and reducing agricultural productivity. As already mentioned, numerous studies have demonstrated that excessive Cu levels in soil can damage chlorophyll, resulting in chlorosis, and increase ROS production, which can damage cellular components, such as lipids, proteins, and DNA, potentially leading to cell death [6, 22, 57]. Furthermore, Cu toxicity can affect soil health by harming soil microbial communities, thereby reducing nutrient availability and overall soil fertility [31].

Several research have focused on developing effective methods to decontaminate soils contaminated with toxic heavy metals, such as Cu. According to [31], traditional methods, like excavation and landfilling, are largely impractical for large-scale operations, underscoring the need for innovative, sustainable remediation strategies to protect soil health and agricultural productivity. Phytoremediation has been advanced as a promising method to remove toxic trace elements such as Cu from the soil, as well as to control or immobilize them in soils by the use of green plants (Table 2).

Phytostabilization involves plants that, while accumulating metals in their roots, restrict their transfer to the shoots, effectively immobilizing contaminants within the root zone and reducing soil erosion and leaching. Ghazaryan et al. [27] reported that for remediation of copper-contaminated soils by phytostabilization method, wormwood plant (*Artemisia absinthium* L.) is able to accumulate relatively high concentrations of copper in its root and does not transfer it to the above-ground part at the same time. The same authors reported that for the cleaning of copper-contaminated

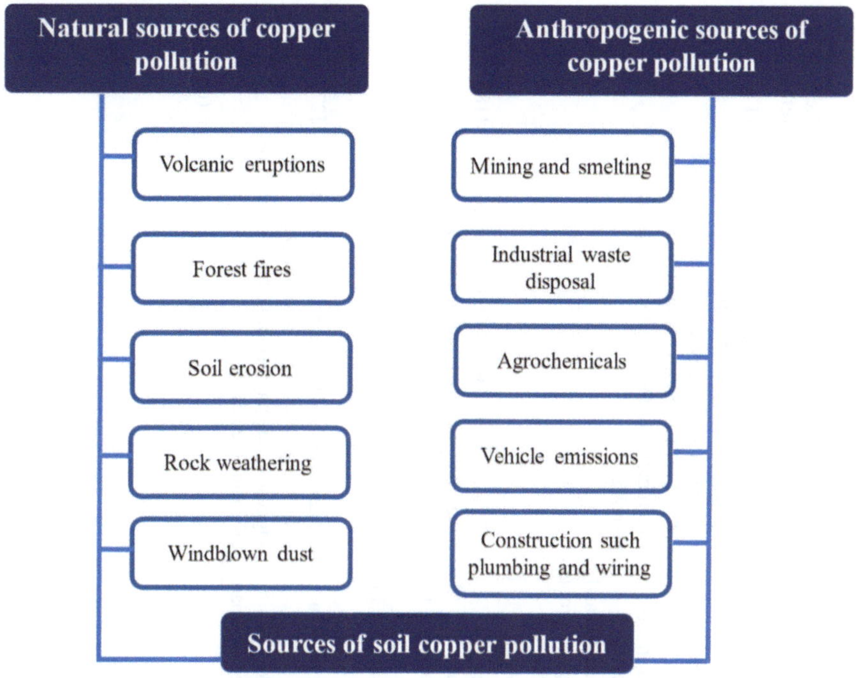

Fig. 2 Sources of soil copper pollution

soils through phytoextraction method by *A. absinthium*, the application of chemical amendments is needed for the enhancement of copper bioavailability and for its intensive transportation to the above-ground part of the plant. Daryabeigi Zand and Mühling [18] studied the effects of additions of 0, 50, 100, 200, and 300 mg kg^{-1} of Cu to soil on phytoremediation potential of maize plants. Referring to these author's roots accumulated significantly higher levels of Cu than shoots in all treatments, suggesting dominancy of phytostabilization mechanism in remediation of Cu-polluted soil by *Z. mays*. [17] studied the phytoremediation behavior of African mahogany (*Khaya ivorensis* A. Juss.) and cedar (*Cedrela fissilis* Vell.) in Cu-contaminated soil (60, 200, 400, and 600 mg kg^{-1}). The results indicated that the tested tropical species have potential to phytoremediate Cu-contaminated soils. The species present low translocation (TF < 1) of Cu, immobilizing and accumulating this metal in the roots, which indicates their phytostabilization potential.

In phytoextraction, hyperaccumulator plants uptake and transfer metals from the soil to their above-ground tissues, where concentrations can reach extraordinary levels up to 1000 μg g^{-1} of dry weight in certain species, making these metals accessible for harvesting and safe disposal [4, 21]. These authors showed that hyperaccumulators can concentrate metals in their shoot tissues, with levels often 10–500 times higher than those found in ordinary plants. This capability highlights their unique adaptation for metal absorption, transport, and accumulation, particularly

Table 2 Plant species utilized for phytoremediation of copper-contaminated environments

Plant species	Copper concentration	Growth medium	Period	Cu accumulation	Phytoremediation traits	Key references
Cedrela fissilis *Khaya ivorensis*	60, 200, 400, and 600 mg kg^{-1}	Soil	30 days	At highest Cu treatment: -*K. ivorensis*: 329 and 52 mg kg^{-1} in roots and shoots respectively; -*C. fissilis*: 207 and 25 mg kg^{-1} in roots and shoots respectively	- The plant species have potential to phytoremediate Cu-contaminated soils - The plant species present low translocation (TF < 1) of Cu, accumulating Cu in the roots, which indicates their phytostabilization potential	[17]
Zea mays	0, 50, 100, 200 and 300 mg kg^{-1}	Soil	42 days	At highest Cu treatment: -435.8 mg kg^{-1} in roots; -831.5 mg kg^{-1} in shoots	- Greater accumulation of copper in roots rather than the shoots suggesting the phytostabilization potential of this species	[18]
Artemisia absinthium	3480.3 mg kg^{-1}	Soil	5 months	-363 mg kg^{-1} in roots -57 mg kg^{-1} in Above-ground part	Accumulation of high concentrations of Cu in its root than the above-ground part suggesting the phytostabilization potential of this plant species	[27]

(continued)

Table 2 (continued)

Plant species	Copper concentration	Growth medium	Period	Cu accumulation	Phytoremediation traits	Key references
Elsholtzia splendens	586 mg kg^{-1}	Soil	Plants collected from natural contaminated sol	357, 219, and 269 mg kg^{-1} in root, shoot and leaf, respectively	The plant species is a potential candidate for the phytoextraction of Cu-contaminated soils, indicated by higher accumulation and translocation capacities	[21]
Myracrodruon urundeuva	125, 1525, 3050 mg kg^{-1}	Soil	90 days	At highest Cu treatment: 12.4, 3241 and 2600 mg kg^{-1} in leaves, stem and roots, respectively	The plant species showed the potential to stabilise and phytoextract Cu from Cu-contaminated soils	[4]

within the vacuoles of cells in their aerial parts. Numerous plant species have been recognized to possess enhanced capabilities for accumulation of excess Cu. Amin et al. [2] reported that among four plant species, significant highest Cu accumulation was found in the roots and shoots of oats (*Avena sativa* L.). This plant was a species that grows well in metal stress conditions with short cycle, high tolerance, and accumulation potential. According to these authors, the high values of bioconcentration factor, bioaccumulation coefficient, translocation factor, and phytoremediation ratio recommended that *A. sativa* was a suitable plant for effective Cu phytoextraction and remediated the Cu-contaminated soil in quick and successive flushes than other tested plant species. The harvested plant biomass was biodegradable, making it easy to dispose of and utilize as an alternative source of biofuel energy or as raw material for large-scale composting or phytomining processes. Finally, it is recommended that if the soil is deficient in Cu, Cu-encompassing fertilizers can be applied, whereas if the Cu level is high, it can be managed using phytoremediation plant species.

6 Conclusion and Future Remarks

This paper reviews the uptake, toxicity, and tolerance mechanisms of copper in crop plants. In addition, it discusses the management of Cu-contaminated soil through plant-based phytoremediation strategies. Soil contamination by Cu has increasingly become a significant environmental issue due to the intensification of industrial and agricultural activities. These activities have introduced significant amounts of Cu into the soil, which, although essential for plant growth, must be carefully regulated to maintain a proper balance. It is concluded that plants primarily absorbs Cu through their root systems, relying on various transporter proteins like COPT, copper-transporting P-type ATPases, and the ZIP family to maintain Cu homeostasis. These proteins directed Cu within the plant to support essential processes, such as mitochondrial respiration, cell wall metabolism, and photosynthesis, while acting as co-factors for important enzymes like cytochrome c oxidase and Cu/Zn SOD. However, at excessive levels, Cu disrupted these processes, leading to oxidative stress through the accumulation of ROS. Symptoms such as chlorosis, necrosis, and inhibited growth emerged, reflecting the disruption of photosynthesis and other critical pathways. To counteract this stress, plants activated various detoxification mechanisms, including enzymatic and non-enzymatic antioxidants like phytochelatins and glutathione.

Recognizing these challenges, researchers have increasingly focused on phytoremediation techniques as a sustainable solution. This approach utilizes specific plant species to remove or stabilize Cu in contaminated soils through methods such as phytoextraction and phytostabilization. The selection of suitable plant species, based on their tolerance, accumulation, and translocation abilities, is essential for the success of phytoremediation programs aimed at soil rehabilitation.

Future research should focus on understanding how Cu is taken up and transported in plants, especially the role of transport proteins. It will also be important to explore how Cu interacts with other soil nutrients and how plants manage Cu toxicity. Developing crops with better Cu tolerance through genetic improvements or biotechnology could enhance agricultural productivity. In addition, further work is needed to identify plant species that are efficient at absorbing and removing Cu from contaminated soils, which would improve phytoremediation techniques. Long-term studies on phytoremediation under real-world conditions are essential for identifying the most effective strategies for remediating Cu-contaminated soils. Furthermore, gaining a deeper understanding of how environmental factors influence Cu uptake in plants will be critical for optimizing these solutions.

Competing Interests The authors declare that they have no competing interests.

Declaration of Generative AI and AI-Assisted Technologies in the Writing Process During the preparation of this work, the book chapter author (s) have not taken any assistance from AI, and the book chapter author (s) reviewed and edited the content as needed and take (s) full responsibility for the content of the publication.

References

1. AlQuraidi AO, Mosa KA, Ramamoorthy K (2019) Phytotoxic and genotoxic effects of copper nanoparticles in coriander (*Coriandrum sativum*—Apiaceae). Plants 8(1):19
2. Amin H, Arain BA, Jahangir TM, Abbasi AR, Mangi J, Abbasi MS, Amin F (2019) Copper (Cu) tolerance and accumulation potential in four native plant species: a comparative study for effective phytoextraction technique. Geol Ecol Landsc 5(1):53–64
3. Andrés-Colás N, Sancenón V, Rodríguez-Navarro S, Mayo S, Thiele DJ, Ecker JR, Puig S (2006) The *Arabidopsis* heavy metal P-type ATPase HMA5 interacts with metallochaperones and functions in copper detoxification of roots. Plant J 45:225–236
4. Asensio V, Flórido FG, Ruiz F, Perlatti F, Otero XL, Oliveira DP, Ferreira TO (2019) The potential of a technosol and tropical native trees for reclamation of copper-polluted soils. Chemosphere 220:892–899
5. Astuti SNC, Nugroho AP (2022) Toxicity of Copper (Cu) and Chromium (Cr) on the seed germination of Mung Bean (*Vigna radiata*) L. J Biotechnol Nat Sci 2(2):40–48
6. Badiaa O, Yssaad HAR, Topcuoglu B (2020) Effect of heavy metals (copper and zinc) on proline, polyphenols and flavonoids content of tomato (*Lycopersicon esculentum* Mill.). Plant Arch 20(2)
7. Bagdonas E, Vosylienė MZ (2006) A study of toxicity and genotoxicity of Copper, Zinc, and their mixture to rainbow trout (*Oncorhynchus mykiss*). Biologija 52(1)
8. Benatti RM, Yookongkaew N, Meetam M, Guo WJ, Punyasuk N, AbuQamar S, Goldsbrough P (2014) Metallothionein deficiency impacts copper accumulation and redistribution in leaves and seeds of *Arabidopsis*. New Phytol 202(3):940–951
9. Blotevogel S, Oliva P, Denaix L, Audry S, Viers J, Schreck E (2022) Stable Cu isotope ratios show changes in Cu uptake and transport mechanisms in *Vitis vinifera* due to high Cu exposure. Front Plant Sci 12:755944
10. Boutigny S, Sautron E, Finazzi G, Rivasseau C, Frelet-Barrand A, Pilon M, Rolland N, Seigneurin-Berny D (2014) HMA1 and PAA1, two chloroplast-envelope PIB-ATPases, play distinct roles in chloroplast copper homeostasis. J Exp Bot 65:1529–1540
11. Brengi SH (2024) Alleviation of copper stress in tomato plants using proline and melatonin in an open soilless system. Alex Sci Exch J 45(1):175–190

12. Chen CT, Chen TH, Lo KF, Chiu CY (2004) Effects of proline on copper transport in rice seedlings under excess copper stress. Plant Sci 166(1):103–111
13. Chen G, Li J, Han H, Du R, Wang X (2022) Physiological and molecular mechanisms of plant responses to copper stress. Int J Mol Sci 23(21):12950
14. Chen WR, Feng Y, Chao YE (2008) Genomic analysis and expression pattern of OsZIP1, OsZIP3, and OsZIP4 in two rice (*Oryza sativa*) genotypes with different zinc efficiency. Russ J Plant Physiol 55:400–409
15. Chen XF, Hua D, Zheng ZC, Zhang J, Huang WT, Chen HH, Huang ZR, Yang LT, Ye X, Chen LS (2021) Boron-mediated amelioration of copper-toxicity in sweet orange [*Citrus sinensis* (L.) Osbeck cv. Xuegan] seedlings involved reduced damage to roots and improved nutrition and water status. Ecotoxicol Environ Saf 212:112579
16. Chu CC, Lee WC, Guo WY, Pan SM, Chen LJ, Li HM, Jinn TL (2005) A copper chaperone for superoxide dismutase that confers three types of copper/zinc superoxide dismutase activity in *Arabidopsis*. Plant Physiol 139:425–436
17. Covre WP, da Silveira Pereira WV, Gonçalves DAM, Teixeira OMM, do Amarante CB, Fernandes AR, (2020) Phytoremediation potential of *Khaya ivorensis* and *Cedrela fissilis* in copper-contaminated soil. J Environ Manage 268:110733
18. Daryabeigi Zand A, Mühling KH (2022) Phytoremediation capability and copper uptake of maize (*Zea mays* L.) in copper-contaminated soils. Pollutants 2(1):53–65
19. Deng F, Yamaji N, Xia J, Ma JF (2013) A member of the heavy metal P-type ATPase OsHMA5 is involved in xylem loading of copper in rice. Plant Physiol 163:1353–1362
20. DiDonato RJ Jr, Roberts LA, Sanderson T, Eisley RB, Walker EL (2004) *Arabidopsis* yellow stripe-like2 (YSL2): a metal-regulated gene encoding a plasma membrane transporter of nicotianamine-metal complexes. Plant J 39:403–414
21. Dou C, Cui H, Zhang W, Yu W, Sheng X, Zheng X (2023) Copper and cadmium accumulation and phytorextraction potential of native and cultivated plants growing around a copper smelter. Agronomy 13(12):2874
22. El-Beltagi HS, Sofy MR, Aldaej MI, Mohamed HI (2020) Silicon alleviates copper toxicity in flax plants by up-regulating antioxidant defense and secondary metabolites and decreasing oxidative damage. Sustainability 12(11):4732
23. Fan W, Liu C, Cao B, Qin M, Long D, Xiang Z, Zhao A (2018) Genome-wide identification and characterization of four gene families putatively involved in cadmium uptake, translocation and sequestration in mulberry. Front Plant Sci 9:879
24. Farid M, Farooq MA, Fatima A, Abubakar M, Ali S, Raza N, Alhaithloul HAS, Soliman MH (2021) Copper-induced responses in different plant species. Approaches to the remediation of inorganic pollutants. Springer, Singapore, pp 259–280
25. Feil SB, Pii Y, Valentinuzzi F, Tiziani R, Mimmo T, Cesco S (2020) Copper toxicity affects phosphorus uptake mechanisms at molecular and physiological levels in *Cucumis sativus* plants. Plant Physiol Biochem 157:138–147
26. Flores-Cáceres ML, Hattab S, Hattab S, Boussetta H, Banni M, Hernández LE (2015) Specific mechanisms of tolerance to copper and cadmium are compromised by a limited concentration of glutathione in alfalfa plants. Plant Sci 233:165–173
27. Ghazaryan KA, Movsesyan HS, Minkina TM, Nevidomskaya DG, Rajput VD (2022) Phytoremediation of copper-contaminated soil by *Artemisia absinthium*: comparative effect of chelating agents. Environ Geochem Hlth 1–13
28. Giannakoula A, Therios I, Chatzissavvidis C (2021) Effect of lead and copper on photosynthetic apparatus in citrus (*Citrus aurantium* L.) plants. The role of antioxidants in oxidative damage as a response to heavy metal stress. Plants 10(1):155
29. González A, Sáez CA, Moenne A (2018) Copper-induced activation of TRPs and VDCCs triggers a calcium signature response regulating gene expression in *Ectocarpus siliculosus*. PeerJ 6
30. Grotz N, Fox T, Connolly E, Park W, Guerinot ML, Eide D (1998) Identification of a family of zinc transporter genes from *Arabidopsis* that respond to zinc deficiency. Proc Natl Acad Sci U S A 95(12):7220–7224

31. Hendgen M, Hoppe B, Döring J, Friedel M, Kauer R, Frisch M, Dahl A, Kellner H (2018) Effects of different management regimes on microbial biodiversity in vineyard soils. Sci Rep 8:9393
32. Hoppen C, Müller L, Hänsch S, Uzun B, Milić D, Meyer AJ, Weidtkamp-Peters S, Groth G (2019) Soluble and membrane-bound protein carriers mediate direct copper transport to the ethylene receptor family. Sci Rep 9:1–11
33. Hussain D, Haydon MJ, Wang Y, Wong E, Sherson SM, Young J, Camakaris J, Harper JF, Cobbett CS (2004) P-type ATPase heavy metal transporters with roles in essential zinc homeostasis in *Arabidopsis*. Plant Cell 16:1327–1339
34. Hussain Q, Ye T, Li S, Nkoh JN, Zhou Q, Shang C (2023) Genome-wide identification and expression analysis of the copper transporter (COPT/ctr) gene family in *Kandelia obovata*, a typical mangrove plant. Int J Mol Sci 24(21):15579
35. Iqbal MZ, Nayab S, Shafiq M (2018) Effects of copper on seed germination and seedling growth performance of *Lens culinaris* Medik. J Plant Dev 25:85
36. Javaid S, Ashraf K, Sultan K, Siddiqui MH, Ali HM, Chen Y, Zaman QU (2022) Risk assessment of potentially toxic metals and metalloids in soil, water and plant continuum of fragrant rice. Agronomy 12(10):2480
37. Javed SB, Pii Y, Valentinuzzi F, Tiziani R, Mimmo T, Cesco S (2020) Copper toxicity affects phosphorus uptake mechanisms at molecular and physiological levels in *Cucumis sativus* plants. Plant Physiol Biochem 157:138–147
38. Javed T, Ali MM, Shabbir R, Anwar R, Afzal I, Mauro RP (2021) Alleviation of copper-induced stress in pea (*Pisum sativum* L.) through foliar application of gibberellic acid. Biology 10(2):120
39. Jin MF, You MX, Lan QQ, Cai LY, Lin MZ (2021) Effect of copper on the photosynthesis and growth of *Eichhornia crassipes*. Plant Biol 23(5):777–784
40. Khalid M, Nan HUI, Kayani SI, Kexuan TANG (2020) Diversity and versatile functions of metallothioneins produced by plants: a review. Pedosphere 30(5):577–588
41. Ku HM, Tan CW, Su YS, Chiu CY, Chen CT, Jan FJ (2012) The effect of water deficit and excess copper on proline metabolism in *Nicotiana benthamiana*. Biol Plant 56:337–343
42. Kumar S, Yadav A, Kumar A, Verma R, Lal S, Srivastava S, Sanyal I (2021) Plant metallothioneins as regulators of environmental stress responses. Int J Plant Environ 7(01):27–38
43. Kumar V, Pandita S, Preet GSS, Sharma A, Khanna K, Kaur P, Shreeya A, Setia R (2021) Copper bioavailability, uptake, toxicity and tolerance in plants: a comprehensive review. Chemosphere 262:127810
44. Kumar V, Pandita S, Sidhu GPS, Sharma A, Khanna K, Kaur P, Bali AS, Setia R (2020) Copper bioavailability, uptake, toxicity and tolerance in plants: a comprehensive review. Chemosphere 262:127810
45. Li XY, Lin ML, Lu F, Zhou X, Xiong X, Chen LS, Huang ZR (2023) Physiological and ultrastructural responses to excessive-copper-induced toxicity in two differentially copper-tolerant citrus species. Plants 12(2):351
46. Li Y, Shi S, Zhang Y, Zhang A, Wang Z, Yang Y (2023) Copper stress-induced phytotoxicity associated with photosynthetic characteristics and lignin metabolism in wheat seedlings. Ecotoxicol Environ Saf 254:114739
47. Zouari M, Hassena AB, Trabelsi L, Rouina BB, Decou R, Labrousse P (2019) Exogenous proline-mediated abiotic stress tolerance in plants: possible mechanisms. In: Osmoprotectant-mediated abiotic stress tolerance in plants: recent advances and future perspectives, pp 99–121
48. Maharajan T, Krishna TPA, Ceasar SA, Ignacimuthu S (2023) Zinc supply influenced the growth, yield, zinc content, and expression of ZIP family transporters in sorghum. Planta 257(2):44
49. Marques DM, Veroneze Júnior V, da Silva AB, Mantovani JR, Magalhães PC, de Souza TC (2018) Copper toxicity on photosynthetic responses and root morphology of *Hymenaea courbaril* L. (Caesalpinioideae). Water Air Soil Pollut 229:1–14
50. Memon A, Meraklı N (2022) Comparative structural analysis of heavy metal ATPases in *Arabidopsis thaliana*, *Arabidopsis halleri*, *Brassica rapa*, and *Brassica juncea*. Turk J Agric Food Sci Technol 10:2988–2995

51. Meychik N, Nikolaeva Y, Kushunina M (2021) The significance of ion-exchange properties of plant root cell walls for nutrient and water uptake by plants. Plant Physiol Biochem 166:140–147
52. Mira H, Martínez-García F, Peñarrubia L (2001) Evidence for the plant-specific intercellular transport of the *Arabidopsis* copper chaperone CCH. Plant J 25:521–528
53. Mosa KA, El-Naggar M, Ramamoorthy K, Alawadhi H, Elnaggar A, Wartanian S, Hani H (2018) Copper nanoparticles induced genotoxicity, oxidative stress, and changes in superoxide dismutase (SOD) gene expression in cucumber (*Cucumis sativus*) plants. Front Plant Sci 9:872
54. Mostofa MG, Seraj ZI, Fujita M (2014) Exogenous sodium nitroprusside and glutathione alleviate copper toxicity by reducing copper uptake and oxidative damage in rice (*Oryza sativa* L.) seedlings. Protoplasma 251:1373–1386
55. Mujeeb A, Abideen Z, Aziz I, Sharif N, Hussain MI, Qureshi AS, Yang HH (2023) Phytoremediation of potentially toxic elements from contaminated saline soils using *Salvadora persica* L.: seasonal evaluation. Plants 12(3):598
56. Navarrete A, González A, Gómez M, Contreras RA, Díaz P, Lobos G, Moenne A (2019) Copper excess detoxification is mediated by a coordinated and complementary induction of glutathione, phytochelatins and metallothioneins in the green seaweed *Ulva compressa.* Plant Physiol Biochem 135:423–431
57. Noreen S, Akhter MS, Yaamin T, Arfan M (2018) The ameliorative effects of exogenously applied proline on physiological and biochemical parameters of wheat (*Triticum aestivum* L.) crop under copper stress condition. J Plant Interact 13(1):221–230
58. Nualla-Ong A, Phongdara A, Buapet P (2020) Copper and zinc differentially affect root glutathione accumulation and phytochelatin synthase gene expression of *Rhizophora mucronata* seedlings: implications for mechanisms underlying trace metal tolerance. Ecotoxicol Environ Saf 205:111175
59. Ravet K, Danford FL, Dihle A, Pittarello M, Pilon M (2011) Spatiotemporal analysis of copper homeostasis in *Populus trichocarpa* reveals an integrated molecular remodeling for a preferential allocation of copper to plastocyanin in the chloroplasts of developing leaves. Plant Physiol 157:1300–1312
60. Romero P, Gabrielli A, Sampedro R, Perea-García A, Puig S, Lafuente MT (2021) Identification and molecular characterization of the high-affinity copper transporters family in *Solanum lycopersicum.* Int J Biol Macromol 192:600–610
61. Sakouhi L, Ferjani EE (2022) Effects of excess copper on sunflower seedling growth, mineral nutrition, and cellular redox state. Euro-Mediterr J Environ Integr 7(4):583–591
62. Sameena PP, Puthur JT (2021) Differential modulation of photosynthesis and defense strategies towards copper toxicity in primary and cotyledonary leaves of *Ricinus communis* L. J Photochem Photobiol 8:100059
63. Sancenón V, Puig S, Mira H, Thiele DJ, Peñarrubia L (2003) Identification of a copper transporter family in *Arabidopsis thaliana.* Plant Mol Biol 51:577–587
64. Sanz A, Pike S, Khan MA, Carrió-Seguí À, Mendoza-Cózatl DG, Peñarrubia L, Gassmann W (2019) Copper uptake mechanism of *Arabidopsis thaliana* high-affinity COPT transporters. Protoplasma 256:161–170
65. Sarwar N, Imran M, Shaheen MR, Ishaque W, Kamran MA, Matloob A, Rehim A, Hussain S (2017) Phytoremediation strategies for soils contaminated with heavy metals: modifications and future perspectives. Chemosphere 171:710–721
66. Shabbir Z, Sardar A, Shabbir A, Abbas G, Shamshad S, Khalid S, Shahid M (2020) Copper uptake, essentiality, toxicity, detoxification and risk assessment in soil-plant environment. Chemosphere 259:127436
67. Shams M, Yildirim E, Guleray AGAR, Ercisli S, Dursun A, Ekinci M, Raziye KUL (2018) Nitric oxide alleviates copper toxicity in germinating seed and seedling growth of *Lactuca sativa* L. Not Bot Horti Agrobot Cluj-Napoca 46(1):167–172
68. Song Y, Zhang H, Chen C, Wang G, Zhuang K, Cui J, Shen Z (2014) Proteomic analysis of copper-binding proteins in excess copper-stressed rice roots by immobilized metal affinity chromatography and two-dimensional electrophoresis. Biometals 27:265–276

69. Srivastava S, Mishra S, Tripathi RD, Dwivedi S, Gupta DK (2006) Copper-induced oxidative stress and responses of antioxidants and phytochelatins in *Hydrilla verticillata* (Lf) Royle. Aquat Toxicol 80(4):405–415
70. Tasar N (2024) Genotoxic effects of copper nanoparticles on *Vicia pannonica* Crantz root meristem cells. J Nanoparticle Res 26(8):185
71. Tie SG, Tang ZJ, Zhao YM, Li W (2012) Oxidative damage and antioxidant response caused by excess copper in leaves of maize. Afr J Biotechnol 11(19):4378–4384
72. Wairich A, De Conti L, Lamb TI, Keil R, Neves LO, Brunetto G, Sperotto RA, Ricachenevsky FK (2022) Throwing copper around: how plants control uptake, distribution, and accumulation of copper. Agronomy 12(5):994
73. Wang H, Du H, Li H, Huang Y, Ding J, Liu C, Wang N, Lan H, Zhang S (2018) Identification and functional characterization of the ZmCOPT copper transporter family in maize. PLoS One 13
74. Wang Q, Wei N, Jin X, Min X, Ma Y, Liu W (2021) Molecular characterization of the COPT/Ctr-type copper transporter family under heavy metal stress in alfalfa. Int J Biol Macromol 181:644–652
75. Wang Y, Zhang P, Yang C, Guo Y, Gao P, Wang T, Lui Y, Xu L, Zhou G (2024) Responses in plant growth and root exudates of *Pistia stratiotes* under Zn and Cu stress. Plants 13(5):736
76. Waters BM, Chu HH, DiDonato RJ, Roberts LA, Eisley RB, Lahner B, Salt DE, Walker EL (2006) Mutations in *Arabidopsis* yellow stripe-like1 and yellow stripe-like3 reveal their roles in metal ion homeostasis and loading of metal ions in seeds. Plant Physiol 141:1446–1458
77. Wenli Z, Chang LI, Yajun Z, Jingguang C, Haohua HE, Guoyou YE (2020) Rice heavy metal P-type ATPase OsHMA6 is likely a copper efflux protein. Rice Sci 27(2):143–151
78. Xin J, Ma S, Xiao H, Tian R (2024) Copper accumulation and its effect on glyoxalase system, AsA-GSH cycle, and antioxidase activity in *Pontederia cordata* leaves. Water Air Soil Pollut 235(6):1–13
79. Xu E, Liu Y, Wang G, Zhan X, Li J, Zhou K, Zou Y (2024) Molecular mechanisms of plant responses to copper: From deficiency to excess. Int J Mol Sci 25(13):6993
80. Yuan M, Li X, Xiao J, Wang S (2011) Molecular and functional analyses of COPT/Ctr-type copper transporter-like gene family in rice. BMC Plant Biol 11:69
81. Yuce M, Ekinci M, Turan M, Agar G, Aydin M, Ilhan E, Yildirim E (2024) Chrysin mitigates copper stress by regulating antioxidant enzyme activity, plant nutrient, and phytohormone content in pepper. Sci Hortic 328:112887
82. Zhang C, Lu W, Yang Y, Shen Z, Ma JF, Zheng L (2018) OsYSL16 is required for preferential Cu distribution to floral organs in rice. Plant Cell Physiol 59:2039–2051
83. Zheng L, Yamaji N, Yokosho K, Ma JF (2012) YSL16 is a phloem-localized transporter of the copper-nicotianamine complex that is responsible for copper distribution in rice. Plant Cell 24:3767–3782

Zinc: Uptake, Toxicity, and Tolerance in Crop Plants and Management of Zn-Contaminated Soil

Sharfa Naaz, Ragib Husain, Swati Sachdev, and Mohammad Israil Ansari

Abstract Micronutrients play a very important role in the growth and development of crop plants as well as zinc (Zn) is one of them. However, optimum concentration of Zn is very essential for the metabolism of plants because both deficiency and excess amount of Zn may result in adverse effects on plant health and crop productivity. During this present chapter, an attempt has been made to elaborate on the mechanistic understanding of Zn absorption from the soil, with special reference to the role of transporter proteins such as ZIP (Zinc-Iron Permease) family and factors governing Zn availability in the soil. These include explorations on the effect of Zn toxicity on physiological and molecular processes that highlighted the presence of apparent visible symptoms of toxicity and the disturbance caused by the excess Zn on the metabolic processes of the plant. In addition, emphasis has been drawn to effective management practices to treat Zn-contaminated soils, which are very crucial for sustainable agriculture. Altogether, this chapter would illustrate aims at providing detailed understanding roles of Zn in agriculture and provides practical solutions for management of Zn deficiencies and toxicities, leading towards enhanced crop productivity and food security for the nation.

Keywords Zinc transporter proteins · Zinc deficiency · Contaminated soil management · Phytotoxicity · Phytoremediation

S. Naaz · R. Husain
Plant Ecology and Climate Change Science Division, CSIR—National Botanical Research Institute, Lucknow, Uttar Pradesh, India

S. Naaz · M. I. Ansari (✉)
Department of Botany, University of Lucknow, Lucknow, Uttar Pradesh, India
e-mail: ansari_mi@lkouniv.ac.in

R. Husain
Department of Botany, Bareilly College, Bareilly, Uttar Pradesh, India

S. Sachdev
Department of Liberal Education, Era University, Lucknow, Uttar Pradesh, India

A. Husen (ed.), *Enhanced Crop Production Setup Under Various Micro and Nano Contaminants*, Smart Nanomaterials Technology,
https://doi.org/10.1007/978-981-96-6099-5_7

1 Introduction

For the growth of plants, zinc (Zn) is an essential micronutrient that plays a vital role in number of physiological processes, such as growth, regulation, enzyme activation, protein synthesis, and hormone regulation [70]. Despite its role in growth and development, acquisition of Zn in excess may have adverse effect on the quality and productivity of crops [41, 89]. Thus, both deficiency as well as abundance of Zn is detrimental, affecting the nutritional quality and productivity of crops [70, 80]. Zinc is mostly found in calcareous, sandy, and highly weathered soils and soils with a high level of phosphorus [7]. The interveinal chlorosis (yellowing between the leaf veins), dwarfism, reduced size and number of leaves, and delayed maturity are the most common symptoms of Zn deficiency [69]. However, a small amount of Zn is required for the growth and development of the plants [42]. On the other hand, occurrence of excess amount of Zn in the soil is toxic to the plants and leads to number of abnormalities such as chlorosis and necrosis (death of plant tissue), reduced root and shoot growth, disruption of water and nutrient uptake and impaired reproductive development [41].

The plants uptake Zn from soil through their roots [33]. The soil physicochemical characteristics, like pH, organic matter, and soil texture, play a very crucial role in the process of absorption [105]. The availability of Zn is generally enhanced by the acidic conditions, whereas alkaline conditions reduce it [6]. In a similar manner, high levels of organic matter can bind with the Zn, making it less available to the plants [28]. The soil texture also affects the availability of Zn for the plants, for example, sandy soils may leach Zn faster than clayey soils [49]. For the absorption of Zn, plants have developed specialized transporters, such as the ZIP (Zinc-Iron Permease) family, which control the Zn uptake and maintain homeostasis [26]. For maintaining optimal plant health and development, the presence of Zn in the soil is very critical. Various enzymes such as DNA, RNA, and protein synthesizing enzymes, involved in the developmental processes, require Zn as a cofactor for catalyzing the activities [56]. Auxins which regulate cell elongation and division, root development, flowering, and fruit development in plants, require Zn for its own synthesis, thus indirectly increasing the importance and role of Zn in the developmental processes [79]. The integrity of the membrane is maintained through the Zn stabilizing membrane protein, which ensures proper nutrient transport and signal transduction [23, 51]. Zn also plays a very important protective role in preventing damage to membrane lipids caused by oxidation, preserving cellular integrity [58]. Zn is an important component of the enzyme superoxide dismutase (SOD), which detoxifies reactive oxygen species (ROS) generated under stress [65, 81]. In this way, the optimum level of Zn improves the ability of plants to cope up drought and heat stress by maintaining cellular homeostasis and reducing oxidative damage. Zn exhibits paramount role in proper functioning and growth of plants, if available in adequate amount. Whereas, availability of Zn in excess can hinder the normal physiological and biochemical activities of plants, reducing their growth and development. Due to increased industrial pollution, mining activities, and excessive application of Zn-containing fertilizers and pesticides, the unrestricted accumulation of Zn in soil has taken place [71]. This led to uptake and acquisition of Zn in plants to toxic level that eventually results in reduced plant growth, abridged photosynthetic activity, imbalanced nutrient acquisition, and production of ROS instigating oxidative stress [41]. Altogether, both deficiency and excess accumulation hinder the physiological and biochemical processes in plants. Thus, an understanding of the molecular pathways and mechanisms of zinc uptake and transport, accumulation, and homeostasis is necessary to fully realize the benefits while limiting its toxicity on plants. This chapter presents the absorption mechanisms of zinc, toxicity, and tolerance of the crop plants to it, and maintains focus on some practices that can be adopted for the management of zinc-contaminated soils to bring good plant health and high productivity.

2 Significance of Zinc for Plant Growth

Zinc, being essential only in trace amounts for proper growth and development in plants, microorganisms, and humans, plays different roles, but mainly as a cofactor for many enzymes [22]. Indeed, these biocatalysts do facilitate critical metabolic reactions in plants [29]. Some enzymes require Zn as a cofactor, and their role includes protein and carbohydrate synthesis, chlorophyll, and other important biosynthesis processes. Actually, it has been estimated that over 300 enzymes might need Zn for activity. Among these enzymes are zinc finger proteins, RNA polymerases, and DNA polymerases [15]. It is also an important component in the process of photosynthesis, where plants convert light energy into chemical energy effectively. It interferes with the activity of an enzyme called carbonic anhydrase, this enzyme assimilates CO_2 to form sugar [64]. Zn is necessary for nutrient absorption and transport in the plant, this implies that essential elements have to reach all parts of the growth structures [40]. As a transition element, Zn occurs in all six enzyme groups, which include ligases, hydrolases, lyases, isomerases, and oxidoreductases. Besides this, Zn is also involved in maintaining cell membrane integrity by regulating the movement of water and nutrients [17].

Zn also contributes to the resistance of a plant against biotic stress. This is because it is instrumental in many physiological and biochemical activities [35]. It creates stability in the cell membrane, produces antioxidants [81], and offers protection against pathogens. SOD is just one such vital protein whose synthesis is carried out by Zn. This helps minimize the oxidative stress caused by pathogen attacks [12]. Instead, Zn deficiency promotes susceptibility of the plant to biotic stress due to its weakened resistance. Zn-deficient plants present weakened cell membranes, decreased activity of antioxidant enzymes, and an imbalanced regulation of signaling pathways associated with pathogen-responsive events, which results in increased sensitivity to infections. Moreover, these plants cannot recover the pathogen-induced tissue damage, which generally results in improved disease progression [18]. Therefore, Zn plays a significant role in keeping the plant healthy and minimizing its susceptibility to biotic stress [18, 34].

Zn is critical for physiological processes in the plant, among which are hormone synthesis (e.g., tryptophan, a precursor for the IAA) and the functions of mitogen-activated protein kinases (MAPKs) [60]. MAPKs are crucial in signal transduction and are coupled with the repair process of PSII under photo-inhibition. In addition, Zn regulates the concentration of CO_2 in mesophyll cells and helps in the functioning of Rubisco, which is the key enzyme participating in CO_2 fixation in the Calvin cycle. Zn is also important for flowering, seed development, and pollen formation, affecting direct pollen viability and reproductive success in the plant [74]. Elevated levels of Zn increase more prolific fruit set and seed quality, thereby leading to better crop productivity [74]. Inadequate levels of Zn in the soil produce a variety of visual symptoms on the plant, including stunted growth, interveinal chlorosis, or yellowing between the veins, and delayed maturity. In extreme cases, necrosis may be triggered in the leaves [62]. Consequently, crop productivity and quality may be decreased due to Zn deficiency [89] (Fig. 1).

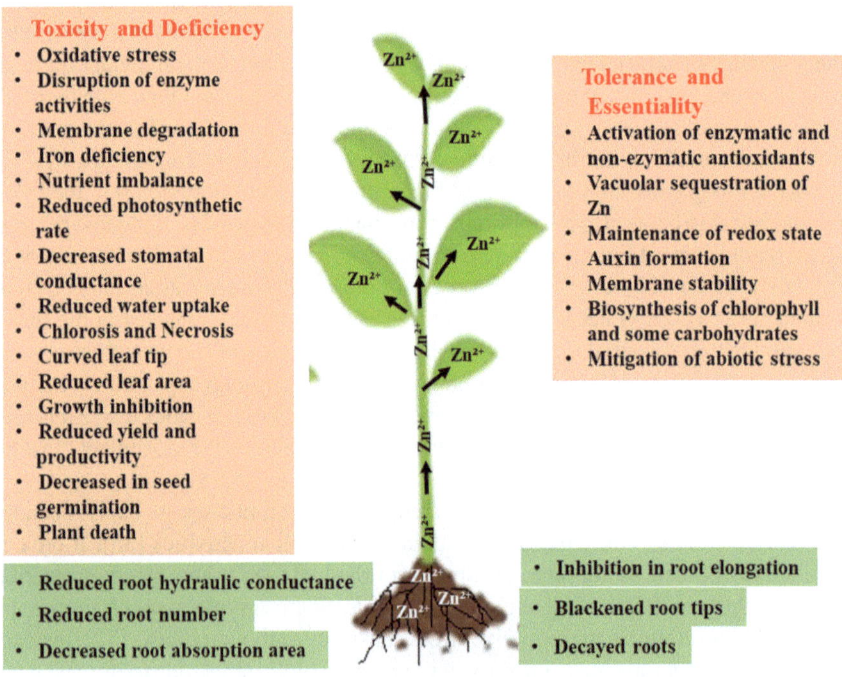

Fig. 1 Impact of Zn deficiency and toxicity, essentiality, and mechanism of tolerance in plants

3 Mechanism of Zn Uptake in Plants

Zn uptake in plants is a complex process involving multiple mechanisms for absorption, transport, and homeostasis. Based on recent studies, some key aspects of Zn uptake in plants are discussed.

3.1 Absorption of Zn by Root

Zinc requirements and concentration can vary among some plant species, including between closely related ones. Such variability is due to differences in the capacities of the different species of plants to absorb, transport, and sequester Zn. The control at the genetic level regarding uptake and distribution of zinc in roots allows survival of the cell within the root as it holds a tight balance between absorption and homeostasis. This regulation includes certain genes and transporters, which strictly regulate Zn uptake from the soil to avoid both deficiency and toxicity [19]. The proper level of Zn is vital for plant health because Zn plays a crucial role in enzymatic activities and cellular processes [96]. This would make the uptake and use of Zn by root cells adjusted by the general condition, such that a recommended amount is received for both growth and defense mechanisms, without otherwise impairing imbalances that could conceivably cause harm in such imbalances. Such balance is fundamental to basic functions such as transpiration and nutrient assimilation, in which roots normally play an important part [108].

Zn is typically absorbed as divalent Zn^{2+} ions through roots, but can also be chelated by organic acids to become absorbed by the roots as Ligand-Zn complexes. Zn in soil is primarily released from its bound forms through root acidification by H^{+} ions and by the secretion of organic acids from the roots for chelation, thereby providing a pathway through which Zn^{2+} ions are absorbed by the epidermal cells of the root. There are two general mechanisms of Zn uptake depending on the exudates formed by the root system of the plants. Organic acids, which include citric acid, malic acid, oxalic acid, and tartaric acid, are exuded directly by epidermal cells or as part of root exudates. Phytosiderophores or phytometallophores secreted into the rhizosphere by roots form stable complexes with Zn that can be absorbed by the epidermal root cells [101]. These two processes are mainly used by the graminaceous plants. Phytosiderophores such as nicotianamine and deoxymugineic acid are low molecular weight organic compounds with a high binding affinity for metals, facilitating their chelation and absorption [47]. Nicotianamine is a non-protein amino acid crucial for the transport and homeostasis of metal ions such as Zn, Fe, and Cu in plants. It is synthesized through biochemical condensation with S-adenosylmethionine (SAM). Plant roots release chemicals, like sugars, proteins, amino acids, and minerals, which favor Zn uptake.

Zn can be absorbed passively through diffusion, as influenced by the gradient of concentration between the soil solution and the plasma membrane of the root cell

(RCPM) [24]. Such availability of Zn in the soil solution is dependent but relatively slow because it is a mechanism involving passivity (Saleem). The main reason for the passive transport of Zn is the hyperpolarization of the RCPM, which is sustained by the RCPM ATPase ion system [96]. Due to the pumping of H^+ ions outside the root through the hydrolysis of ATP, RCPM becomes hyperpolarized, and soil pH gets reduced. Increasing the pH of the rhizosphere increases the solubility of Zn, which leads to its uptake through the roots [37]. Since the water molecules are neutrally charged, they can easily penetrate into the plasma membrane. However, in the case of Zn^{2+} ions, it requires a transporter protein for its transport across the membrane. The ZIP (Zinc/Iron-regulated Transporter-like Protein) family proteins help in the entry of Zn in root cells of plants. OsZIP1, OsZIP3, and OsZIP4 regulate zinc uptake as well as homeostasis in rice [38]. Other transport proteins involved in Zn transport and sequestration in the plant include Heavy Metal ATPase (HMA), Natural Resistance-Associated Macrophage Protein (NRAMP), and Metal Tolerance Protein (MTP) families [35, 52]. HMA proteins are implicated in efflux of Zn into the apoplastic space, while MTP proteins have been reported to assist in sequestrating Zn in vacuoles [33]. Hence, the entire process of Zn absorption from the soil mainly occurs in the form of Zn^{2+} ions through passive diffusion or active transport assisted by transporter proteins, ZIP and Iron-Regulated Transporters (IRT) [5]. The epidermal and cortex cells of the root itself contribute to a great extent to the uptake of Zn from the soil solution.

3.2 *Radial Transport (Within Roots)*

Zinc translocation in plants begins with its uptake by the roots and progresses through several tissues before reaching the xylem. These tissues include the epidermis, cortex, pericycle, and endodermis (Fig. 2), with the thickness of these tissues decreasing as they approach the central region of the stele. The movement of minerals and water within the plant is known as radial transport. Transporter proteins located in these tissues regulate the flow of minerals and water, with the expression of the genes encoding these proteins being tightly controlled to maintain proper transport and homeostasis. A number of factors can contribute to the induction of Zn translocation in the radial pathway, including Zn concentration in the root tissues, efficiency of translocation, and the demand of plant for Zn in the shoots, sink tissues, and at different stages of growth [67]. These factors alter the translocation of Zn and cause differences in the symplastic movement of Zn within the plant [86]. The other mechanism by which radial translocation occurs is by symplastic and apoplastic transport. Molecules move from cell-to-cell in the symplastic pathway through the plasmodesmata, the channels between the adjacent cells. However, because of the deposition of suberin lamellae, the casparian strip seals the endodermis, which thereby blocks the apoplastic route, allowing for movement, actually through spaces between cell walls. Suberin, which is highly impermeable to water and chelated ions, creates a barrier, forcing mineral ions and water molecules to enter cells before reaching the

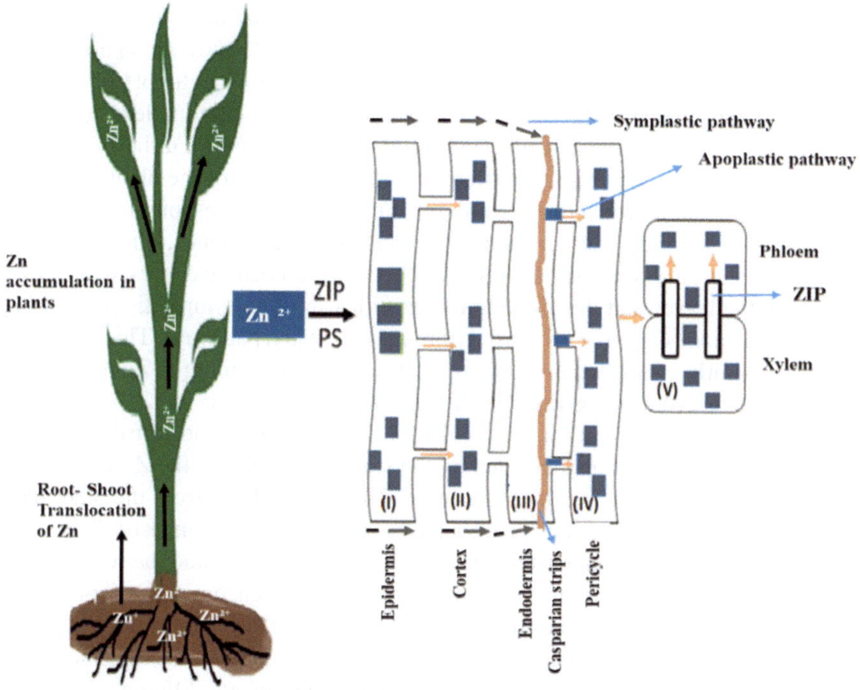

Fig. 2 Zn uptake, root-to-shoot translocation and allocation within plants

casparian strip. After crossing the membrane, they continue their journey through the symplastic pathway (Fig. 2). Both the apoplastic and the symplastic routes are responsible for contributing to the total Zn amount importation into the shoot [16]. In the case of the apoplastic pathway, Zn moves through the apoplast space of the cell walls without passing across the membranes. This route is not particularly selective compared with the symplastic route. P-type ATPases and other metal transporters regulate transport of Zn from the root surface to the central vascular tissues, or xylem, thus ensuring that all parts of the plant receive adequate Zn.

3.3 *Long-Distance Transport of Zinc*

Once Zn is penetrated into the central root tissues, it gets loaded into the xylem responsible for transporting water and nutrients from the roots to other parts of the plant. This process has been shown to be regulated by such proteins as heavy metal ATPases (HMAs) [85]. In addition, transpiration pull also facilitates upward movement of Zn loaded in xylem tissues. After crossing the casparian strip of the endodermis and pericycle, the metal ions cross into the xylem parenchyma cells through a symplastic route. The participation of specific transporters, such as the HMA family or P1B-type ATPases, in the pericycle and xylem parenchyma is essential for the xylem loading of metal ions, which is an active process [75]. The xylem parenchyma possesses plasma membrane H^+ ATPase responsible for the acidic pH of 5–6.5 of the xylem sap, which in fact favors Zn transport as a metal complex along with asparagine, histidine, organic acids, and nicotianamine [92, 96]. Furthermore, Zn is unloaded out of the xylem and transferred to the aerial parts of the plant, including stems, leaves, and reproductive organs, and into tissues like mesophyll cells, which it employs in key activities, including the activation of enzymes, photosynthesis, and the protection against oxidative stress. Generally, unloading occurs in the xylem mostly in the leaves and the growing tissues either through symplastic (cell-to-cell through plasmodesmata) or apoplastic (through cell walls) pathways. This unloading process is tightly controlled in order not to affect the right amount of Zn distribution in the plant to avoid being toxic [11]. Transporter proteins in the ZIP family are considered to be important for the regulation of Zn transport and distribution [72]. Furthermore, toxicity is also avoided by the storage of some of the Zn in vacuoles or binding to ligands such as nicotianamine. Zinc is mobilized from the older parts of the plant to the growing regions through transport in the phloem, especially when Zn is deficient. The system ensures that the developing tissues and the reproductive structures are fed. The entire process of Zn translocation encompasses root uptake, transport across the root tissues to the xylem, and long-distance movement through the xylem, followed by redistribution within the shoots and the leaves. Zinc movement is regulated in its efficiency and direction through several different transporter proteins maintaining proper zinc homeostasis.

4 Factors Influencing Zinc Uptake

The uptake of zinc would depend not only on the plant-specific characteristics like root architecture, root exudation, and mutualistic mycorrhizal relationship but also on the soil properties. Interplay between these variables dictates the efficiency of Zn uptake and its consequences for plant growth and health. Options related to optimizing Zn uptake in crops can improve agricultural productivity along with nutrient management [104].

The bioavailability and uptake of Zn in plants will depend on a complex interaction between abiotic and biotic factors. Soil pH is a major factor that determines the bioavailability of Zn. Zinc remains more soluble and is thus bioavailable in soils with a pH < 6.0, whereas it tends to form the compounds as insolubles at pH > 7.5 in alkaline soils, hence less available for plant uptake [98]. Organic matter in the soil also plays an important role by promoting the precipitation of stable organo-metallic complexes, which then affects Zn uptake positively or negatively depending on the quantity and organic matter decomposition stage [68]. Furthermore, soil texture and cation exchange capacity (CEC affect the mobility of Zn,clay-rich soils retain higher concentrations of Zn than sandy soils [27]. The presence of other ions, for example, phosphorus, acts as a barrier to Zn uptake by forming insoluble zinc-phosphate complexes or competitive inhibition at the root surface, whereby phosphorus and zinc both compete for the available sites [77, 86]. Zn uptake is not just determined by the soil and environmental conditions but also is plant-specific.

The root morphology, for instance, length, surface area, and density, is highly significant in Zn acquisition. Higher and more developed root systems enable the exploration of a larger volume of soil, hence access to a higher amount of available Zn [102]. Second, root exudates which include organic acids, amino acids, and sugars are able to alter the rhizosphere environment through the solubilization of Zn, thereby increasing its availability to the plant for uptake. Mycorrhizal associations where the symbiotic fungi extend the root's deep into soil, which significantly enhances the absorption of Zn, thus increasing availability to plants [59].

5 Zn Toxicity in Plants

Zinc toxicity in the soil is very harmful for physiology of plants and is highly toxic to the range of growing and functioning of plants [43]. The amount of excess metal could increase the potential threat to human health due to the contamination of the food chain [2]. Furthermore, the concentration of metal accumulated in plant is a function of its concentration in soil where plant is growing. An example of this includes excess accumulation of Zn in grains reported to be found because of high concentrations of Zn in a growth substrate [88]. The particular effect of high Zn levels depends upon the species of plants, their growth stages, and the bioavailability of Zn in soil, which is determined by soil pH, organic matter, clay percentage, cation exchange capacity, and soil texture, respectively [2]. Excess of Zn in soils above the threshold level reduces the soil fertility, microbial biomass, and increases macronutrient deficiency [2]. High Zn levels in the plants generate nutritive imbalance, oxidative stress, and reduce photosynthesis [82, 87]. Higher concentration of Zn may inhibit photosynthesis process through some alterations in electron transport chain, synthesis of chlorophyll, and capacity of Rubisco enzyme to convert carbon dioxide [2, 39]. At high concentrations of Zn, a study reported the damage to root cortical cells and necrosis in mesophyll tissues [78]. Zn toxicity can lessen seed germination, root elongation, total plant growth and biomass [61], crop yield and debase the quality of the produce [2, 41] (Fig. 1). For example, in Cicer arietinum, the high Zn levels can even suppress amylase activity, thus leading to the retardation of carbohydrate hydrolysis at seed germination, which further leads to sugar transport inhibition to the embryo axis [50]. Likewise, with increasing Zn concentrations, Cucumis sativus presented diminished seed germination as well as root development [13]. A high concentration of Zn, i.e., 400–1600 mM, has been shown to cause inhibitory effects on the growth of the plant after five days of exposure [94]. Further, it affects the growth rate of the plants, biomass, and the elongation and division of the roots [94]. However, in Coriandrum sativum and Medicago sativa, Zn stress fails to significantly impact seed germination but had-restrained root length, indicating that the response to Zn toxicity is concentration and species-specific [57, 103]. This is because Zn toxicity affects cellular functions at the molecular level.

Zn stress impairs the activity of a proton pump, V-ATPase, leading to cellular pH and ion transport regulation with chlorosis and necrosis in *Arabidopsis thaliana* [100]. Too much Zn also affects mitochondrial structure in plants such as tea, reducing the level of nicotinamide levels [41], therefore reducing the level of energy metabolism while stopping general plant growth entirely [106]. In addition, Zn toxicity often induces root tip blackening along with root length reduction, accompanied by altered architecture of the root leading to a reduction in water and nutrient uptake efficiency [11]. Root hair development has been documented to be impaired with less proteins, such as MRH5 (morphogenesis of root hair 5) and RHD3 (root hair defective 3), further limiting the effectiveness of the root through arrest of shoot growth in A. thaliana [36]. Furthermore, in A. thaliana Zn and Brassica napus stress affects the root growth with inhibition of primary root elongation and enhancement of more vertical growth of lateral roots; it can be some sort of adaptation to dig deep into layers of soil [63, 99]. The species-specific variability is observed in the threshold for toxicity of Zn as well as its impact on plant growth. In case of Zn-stressed plants, suppression of growth is caused by both direct and indirect effects. The direct effects of excess Zn are upon the induction of Zn toxicity, wherein elevated Zn concentrations produce ROS, which inflict oxidative cell damage on proteins, lipids, and nucleic acids. This impacts essential processes, such as photosynthesis and respiration, and retards plant growth [84]. Zn toxicity interferes with the simple mechanism of enzymatic activity by the interplay of Zn ions competing for essential cofactors in metalloproteins, thus impeding metabolism [20]. It also influences hormone regulation, particularly auxin that alters cell division and elongation, hence restrictive growth [31]. In addition, Zn stress indirectly influences uptake of some vital nutrients, such as Fe, Mn, or Mg, which leads to deficiencies that weaken plant health [93].

This nutritive imbalance along with the impaired root growth, negatively affects the water uptake in the plant, further increasing the stress condition in the plant [41]. In plants, high levels of Zn affect nitrogen metabolism and photosynthesis by affecting nitrate uptake and translocation [41]. Therefore, Zn stress results in a combination of chlorosis, necrosis, wilting, and stunted growth, thus, decreasing the productivity of the plant. These changes in Zn toxicity are not only due to the direct accumulation of Zn by tissues but also by the modifications in the uptake of nutrients and water, which significantly impair plant growth and development.

6 Management of Zn-Contaminated Soils

Overly excessive Zn contamination in the soil creates significant concerns regarding environmental health and crop productivity. To effectively mitigate the problem of Zn contamination, an integrated approach must focus on physical, chemical, and biological methods. In this regard, different ways and means for managing Zn content in soil and alleviating Zn-induced phytotoxicity have been discussed.

6.1 Soil Amendment

Inorganic and organic amendments, including lime, biochar, compost, gypsum, and organic matter, are utilized to reduce the content of heavy metals in the soil. They alter the properties such as pH and cation exchange capacity (CEC), that determine metal mobility and bioavailability in soil. For example, lime increases pH, thus reducing the Zn solubility by formation of less mobile compounds; similarly, biochar will have an adsorbing potential of metals and organic matter, improve soil structure, and enhance water retention. Amendments work by either changing the soil chemical environment through precipitation, adsorption, or complexation mechanisms. The increasing soil pH with lime reacts the Zn, with insoluble hydroxides or carbonates, resulting in a lesser amount of plant availability for uptake [35]. In contrast, biochar adsorbs Zn at its surface, thereby blocking Zn leaching and plant absorption [1]. Application of biochar in mine polluted soil was found to reduce the uptake and bioaccumulation of Zn and other heavy metals in Spinach and Cilantro [44]. Organic amendments, for example, compost add organic acids that chelate with Zn, thus reducing its free ionic form and toxicity. Soil amendments reduce the bioavailability of Zn as well as improve the health of the soil as a whole. For example, these amendments will immobilize Zn by improving soil structure as well as retaining water [9]. Another possible approach is achieved through phosphate fertilizers that have been shown to decrease mobility of Zn by forming insoluble Zn-phosphate complexes [66]. The effectiveness of these amendments would mainly depend on factors that include the level of contamination and environmental conditions.

6.2 Soil Washing

The technique of soil washing is an efficient remediation technique to remove Zn from contaminated soils. It has been used in the mobilization and removal of Zn by a range of washing agents, such as acids and chelating agents. Acidic solutions can lower the pH of the soil and dissolve zinc along with other metals, thus simplifying removal processes [90]. The other chelating agents are nitrilotriacetic acid (NTA) and ethylenediaminetetraacetic acid (EDTA), which bind strongly to Zn, increase solubility, and enhance mobility. Surfactants are SDS, oxalic acid, and citric acid. These surfactants are the most frequent agents that alter surface characteristics in soils to enhance Zn removal. For instance, [21] showed that SDS improved Zn removal through the alteration of the soil surface. In addition, [53] indicated that oxalic acid removed over 80% of Zn from polluted soils. Biodegradable and non-toxic organic acids including citric acid, are known to be good for the sustainable remediation of soils. Efficiency in soil washing is primarily a function of the washing agent and some properties of soil, including texture, pH, organic matter content, and initial Zn concentration [46]. These factors inform strategy optimization for soil washing techniques in order to obtain efficient and sustainable remediation outcomes.

6.3 Phytoremediation and Microbial Remediation

Bioremediation is a green, socially-accepted, sustainable, and economical technique for restoration of contaminated soil and water. Bioremediation process involves action of plants (phytoremediation) and microorganisms to alleviate the toxic effect of pollutants either individually or in combination. Application of bioremediation approach to address the challenge of Zn contamination has been investigated by several researchers, and some of them are outlined in Table 1. Phytoremediation is a plant-based technique [10], where the uptake, concentration, and detoxification of pollutants, particularly heavy metals, takes place in the environment [4]. Phytoextraction is used by plants to absorb metals through roots and store them in the aboveground shoots [32]. Phytostabilization is referred to as the process through which metals are immobilized within the soil by retarding their movement. It is a process, whereby metal movement is retarded in the soil, thereby, immobilizing the metals in place. [14, 87]. Phytovolatilization is considered the transmutation of metals to less toxic forms, which then become volatiles and are emitted into the surroundings [73]. Phytodegradation is the process of altering pollutants into harmless products through metabolic processes to eliminate such harmful substances from the environment [91]. Many studies have demonstrated the success of phytoremediation in reclamation of Zn-contaminated environments. *Brassica juncea* and *Salix* spp. can accumulate heavy metals Zn, Pb, and Cd from contaminated soils [8, 95]. *Arabidopsis halleri* and *Thlaspi caerulescens* hyper-accumulate Zn and have been shown to efficiently mitigate Zn contamination in soils [25, 83].

Microbial remediation for instance, microbes such as bacteria and fungi engage in the detoxification of metals. It includes biosorption—adsorption of heavy metals by the cell walls of microbes; bioaccumulation—absorption and accumulation of metals within microbial cells; biotransformation—conversion of toxic metals to less harmful forms; and biomineralization—precipitation of metals into stable minerals [30]. Production of chelating agents, biosurfactants, exopolysaccharides, enzymes, and acids mediate microbial remediation process. These approaches result in metal immobilization, reducing their bioavailability, thereby lowering their uptake and accumulation as well as toxicity in plants [2]. For instance, species of Pseudomonas and Bacillus are particularly effective in metal detoxification and can enhance metal uptake in plants. The combination of plant and microbial remediation offers a powerful strategy for the enhancement of metal uptake efficiency. In this plant–microbe duo, microbes solubilize metals that then enhance the bioavailability of metals for uptake by the plants, while the interaction of the microorganisms with the nutrient cycles of plants and the suitable environment provided by the latter boosts their growth. The synergy between the plant and microbes results in efficient removal of metals from soils while improving soil health, hence making it a likely remediation solution for mitigating environmental pollution.

Table 1 Potential of microbes and plants in bioremediation of Zn

Bioremediator (microbe/plant)	Action response	Key references
Desulfuromonas palmitatis	Removed 50% Zn from contaminated soil. Further addition of EDTA enhanced Zn removal efficiency to 80%	[97]
Bacillus sp.	Zn resistant isolates displayed zinc removal efficiency around 40, 25, and 50% at pH 5,7, and 9, respectively	[48]
Glomus fasciculatum, *Pseudomonas putida*, and *Helianthus annuus*	Integration of microbes with plant in contaminated soil significantly enhanced accumulation of Zn upto 42.56 and 29.84 mg/kg in root and shoot, respectively	[55]
Rhodobacter sphaeroides	Reduced overall bioavailable fraction of Zn to 100% in soil and increased stable fractions after bioremediation	[76]
Sporosarcina pasteurii	Urease hydrolyzing enzyme produced by bacteria-mediated precipitation of Zn by increasing soil pH. Reduced soluble exchangeable fraction of Zn by 85.90% after bioremediation, resulting in improved metal stabilization and reduced bioavailability was observed	[54]
Bacillus altitudinis MT422188	Bacterial cells mediated surface adsorption of Zn ions and facilitated removal upto 81 mg/L and 87 mg/L after 4 and 8 days, respectively. Synthesis of metallothionein under Zn stress was observed along with active role of hydroxyl, carbonyl, and amide group in management of Zn stress	[45]
Fusarium incarnatum and Rye grass	Reduced 49.35% Zn from contaminated soil in 45 days	[107]

7 Conclusion and Future Prospects

Zinc is one of the major micronutrients for plants, contributing enormously towards various biochemical and physiological processes. However, imbalance in Zn availability—either deficiency or excess- is critically deleterious to crops. Deficiency would result in stunted growth and reduce crop yield. Too much Zn, on the other hand, may cause toxicity. Zn uptake primarily occurs through the roots of the plant via transporter proteins, particularly the ZIP family, which transport across the plasma membrane. A number of biotic factors and abiotic factors have a quite profound impact on the uptake and availability of Zn in soil. When Zn is present in too low or in excess in substrate, plant growth and productivity can be highly impeded. Plants have developed various mechanisms to cope with the challenge posed by Zn deficiency as well as toxicity by regulating Zn uptake, sequestering excess Zn in vacuoles, and activating antioxidant defense systems; however, these mechanisms at times fail to cope with this challenge; therefore management of soils is important. Management of Zn availability in soils can be very efficiently done through adjustments in soil pH, addition of organic amendments, and promotion of plant–microbe interactions. Organic matter not only improves the soil texture but also regulates Zn mobility through interaction with it. While growth under Zn deficiency, colonization by mycorrhizal fungi may improve Zn uptake, especially in nutrient-poor soils, by extending the root length for Zn. Similar to this, excessive concentration of Zn can be detrimental to plant growth, and for regulating the content of Zn, physical, chemical, and/or biological methods may be considered. Of these approaches, biological approach that incorporates phytoremediation, microbial remediation, and rhizoremediation is sustainable, effective, and eco-friendly. These methods have been known to minimize Zn toxicity and regain soil health by immobilizing or mobilizing Zn in soil. Further research will need to lay a focus on an understanding of Zn homeostasis in plants, improving crop tolerance to Zn toxicity, and the development of more effective innovative strategies for the effective management of Zn-contaminated soils.

Competing Interests The authors declare that they have no competing interests.

Declaration of Generative AI and AI-Assisted Technologies in the Writing Process During the preparation of this work, the book chapter author(s) have not taken any assistance from AI, and the book chapter author(s) reviewed and edited the content as needed and take(s) full responsibility for the content of the publication.

References

1. Abdel-Fattah TM, Mahmoud ME, Ahmed SB, Huff MD, Lee JW, Kumar S (2015) Biochar from woody biomass for removing metal contaminants and carbon sequestration. J Ind Eng Chem 22:103–109
2. Abedi T, Gavanji S, Mojiri A (2022) Lead and zinc uptake and toxicity in maize and their management. Plants 11(15):1922
3. Ahmad S, Sachdev S (2023) Bioremediation: an alternative tool for restoration of urban agroecosystem contaminated with harmful xenobiotics. In: Xenobiotics in urban ecosystems: sources, distribution and health impacts. Springer, Cham, pp 291–319
4. Ahmadpour P, Ahmadpour F, Mahmud TMM, Abdu A, Soleimani M, Tayefeh FH (2012) Phytoremediation of heavy metals: a green technology. Afr J Biotechnol 11(76):14036–14043
5. Ajeesh Krishna TP, Maharajan T, Victor Roch G, Ignacimuthu S, Antony Ceasar S (2020) Structure, function, regulation and phylogenetic relationship of ZIP family transporters of plants. Front Plant Sci 11:662
6. Alloway BJ (2009) Soil factors associated with zinc deficiency in crops and humans. Environ Geochem Health 31(5):537–548
7. Alloway BJ (2008) Zinc in soils and crop nutrition. IZA and IFA. Brussels, Belgium and Paris, France, p 139
8. Anamika S, Eapen S, Fulekar MH (2009) Phytoremediation of cadmium, lead and zinc by Brassica juncea L. Czern and Coss. J Appl Biosci 13:726–736
9. Antonangelo JA, Sun X, Zhang H (2021) The roles of co-composted biochar (COMBI) in improving soil quality, crop productivity, and toxic metal amelioration. J Environ Manag 277:111443

10. Ashraf S, Ali Q, Zahir ZA, Ashraf S, Asghar HN (2019) Phytoremediation: environmentally sustainable way for reclamation of heavy metal polluted soils. Ecotoxicol Environ Saf 174:714–727
11. Balafrej H, Bogusz D, Triqui ZEA, Guedira A, Bendaou N, Smouni A, Fahr M (2020) Zinc hyperaccumulation in plants: a review. Plants 9(5):562
12. Bastakoti S (2023) Role of zinc in management of plant diseases: a review. Cogent Food & Agric 9(1):2194483
13. Biswas SK (2017) Growth and yield of cucumber (Cucumis sativus) influenced by stem pruning and foliar application of zinc. Doctoral dissertation, Department of Horticulture
14. Bolan NS, Park JH, Robinson B, Naidu R, Huh KY (2011) Phytostabilization: a green approach to contaminant containment. Adv Agron 112:145–204
15. Brito S, Lee MG, Bin BH, Lee JS (2020) Zinc and its transporters in epigenetics. Mol Cells 43(4):323–330
16. Broadley MR, White PJ, Hammond JP, Zelko I, Lux A (2007) Zinc in plants. New Phytol 173(4):677–702
17. Broadley M, Brown P, Cakmak I, Rengel Z, Zhao F (2012) Function of nutrients: micronutrients. In: Marschner's mineral nutrition of higher plants. Academic, pp 191–248
18. Cabot C, Martos S, Llugany M, Gallego B, Tolrà R, Poschenrieder C (2019) A role for zinc in plant defense against pathogens and herbivores. Front Plant Sci 10:1171
19. Cardini A, Pellegrino E, White PJ, Mazzolai B, Mascherpa MC, Ercoli L (2021) Transcriptional regulation of genes involved in zinc uptake, sequestration and redistribution following foliar zinc application to Medicago sativa. Plants 10(3):476
20. Chasapis CT, Ntoupa PSA, Spiliopoulou CA, Stefanidou ME (2020) Recent aspects of the effects of zinc on human health. Arch Toxicol 94:1443–1460
21. Chen Y, Zhang S, Xu X, Yao P, Li T, Wang G, Gong G, Li Y, Deng O (2016) Effects of surfactants on low-molecular-weight organic acids to wash soil zinc. Environ Sci Pollut Res 23:4629–4638
22. Chen J, Calderone LA, Pan L, Quist T, Pandelia ME (2023) The Fe and Zn cofactor dilemma. Biochim Biophys Acta (BBA)-Proteins Proteomics 1871(5):140931
23. Ciftci-Yilmaz S, Mittler R (2008) The zinc finger network of plants. Cell Mol Life Sci 65:1150–1160
24. Claus J, Bohmann A, Chavarría-Krauser A (2013) Zinc uptake and radial transport in roots of Arabidopsis thaliana: a modelling approach to understand accumulation. Ann Bot 112(2):369–380
25. Cosio C, Martinoia E, Keller C (2004) Hyperaccumulation of cadmium and zinc in Thlaspi caerulescens and Arabidopsis halleri at the leaf cellular level. Plant Physiol 134(2):716–725
26. Eide DJ (2005) The ZIP family of zinc transporters. Zinc finger proteins: from atomic contact to cellular function. Springer US, Boston, pp 261–264
27. Fan TT, Wang YJ, Li CB, He JZ, Gao J, Zhou DM, Friedman SP, Sparks DL (2016) Effect of organic matter on sorption of Zn on soil: elucidation by Wien effect measurements and EXAFS spectroscopy. Environ Sci Technol 50(6):2931–2937
28. Ferreira GW, Lourenzi CR, Comin JJ, Loss A, Girotto E, Ludwig MP, Brunetto G (2023) Effect of organic and mineral fertilizers applications in pasture and no-tillage system on crop yield, fractions and contaminant potential of Cu and Zn. Soil Tillage Res 225:105523
29. Freake HC, Sankavaram K (2013) Zinc: physiology, dietary sources, and requirements.
30. Gadd JM (2008) Transformation and mobilization of metals, metalloids, and radionuclides by microorganisms. In: Biophysico-chemical processes of metals and metalloids in soil environments. Wiley-Jupac series, vol 1, pp 53–96
31. Gangwar S, Singh VP, Tripathi DK, Chauhan DK, Prasad SM, Maurya JN (2014) Plant responses to metal stress: the emerging role of plant growth hormones in toxicity alleviation. In: Emerging technologies and management of crop stress tolerance. Academic, pp 215–248
32. Ghori Z, Iftikhar H, Bhatti MF, Sharma I, Kazi AG, Ahmad P (2016) Phytoextraction: the use of plants to remove heavy metals from soil. In: Plant metal interaction. Elsevier, pp 385–409

33. Gupta N, Ram H, Kumar B (2016) Mechanism of Zinc absorption in plants: uptake, transport, translocation and accumulation. Rev Environ Sci Bio/Technol 15:89–109
34. Hajiboland R (2012) Effect of micronutrient deficiencies on plants stress responses. In: Abiotic stress responses in plants: metabolism, productivity and sustainability, pp 283–329
35. Hamzah Saleem M, Usman K, Rizwan M, Al Jabri H, Alsafran M (2022) Functions and strategies for enhancing zinc availability in plants for sustainable agriculture. Front Plant Sci 13:1033092
36. Harmanjit K, Neera G (2021) Zinc toxicity in plants: a review. Planta 253(6)
37. Haruta M, Gray WM, Sussman MR (2015) Regulation of the plasma membrane proton pump (H+-ATPase) by phosphorylation. Curr Opin Plant Biol 28:68–75
38. Huang S, Sasaki A, Yamaji N, Okada H, Mitani-Ueno N, Ma JF (2020) The ZIP transporter family member OsZIP9 contributes to root zinc uptake in rice under zinc-limited conditions. Plant Physiol 183(3):1224–1234
39. Islam F, Yasmeen T, Riaz M, Arif MS, Ali S, Raza SH (2014) Proteus mirabilis alleviates zinc toxicity by preventing oxidative stress in maize (Zea mays) plants. Ecotoxicol Environ Saf 110:143–152
40. Ji C, Li J, Jiang C, Zhang L, Shi L, Xu F, Cai H (2022) Zinc and nitrogen synergistic act on root-to-shoot translocation and preferential distribution in rice. J Adv Res 35:187–198
41. Kaur H, Garg N (2021) Zinc toxicity in plants: a review. Planta 253(6):129
42. Kaur H, Kaur H, Kaur H, Srivastava S (2023) The beneficial roles of trace and ultratrace elements in plants. Plant Growth Regul 100(2):219–236
43. Kaur H, Srivastava S, Goyal N, Walia S (2024) Behavior of zinc in soils and recent advances on strategies for ameliorating zinc phyto-toxicity. Environ Exp Bot 105676
44. Khan AZ, Ding X, Khan S, Ayaz T, Fidel R, Khan MA (2020) Biochar efficacy for reducing heavy metals uptake by Cilantro (Coriandrum sativum) and spinach (Spinaccia oleracea) to minimize human health risk. Chemosphere 244:125543
45. Khan M, Ijaz M, Chotana GA, Murtaza G, Malik A, Shamim S (2022) Bacillus altitudinis MT422188: a potential agent for zinc bioremediation. Bioremediat J 26(3):228–248
46. Klik B, Kulikowska D, Gusiatin ZM, Pasieczna-Patkowska S (2020) Washing agents from sewage sludge: efficiency of Cd removal from highly contaminated soils and effect on soil organic balance. J Soils Sediments 20:284–296
47. Kluska K, Adamczyk J, Krężel A (2018) Metal binding properties, stability and reactivity of zinc fingers. Coord Chem Rev 367:18–64
48. Krishna MP, Varghese R, Babu VA, Jyothy S, Hatha, AM (2013) Bioremediation of zinc using Bacillus sp. isolated from metal-contaminated industrial zone. In: Prospects in bioscience: addressing the issues. Springer India, pp 11–18
49. Lahive E, Matzke M, Svendsen C, Spurgeon DJ, Pouran H, Zhang H, Lawlor A, Pereira MG, Lofts S (2023) Soil properties influence the toxicity and availability of Zn from ZnO nanoparticles to earthworms. Environ Pollut 319:120907
50. Larbi S, Ben Kaab S, Teixeira da Silva JA, Ben Kaab LB (2020) Cadmium and copper stresses affect germination and enzymatic activities in chickpea (Cicer arietinum L.). Agrochimica: Int J Plant Chem Soil Sci & Plant Nutr Univ Pisa 64(2)
51. Lee SR (2018) Critical role of zinc as either an antioxidant or a prooxidant in cellular systems. Oxid Med Cell Longev 2018(1):9156285
52. Liu P, An L, Ma L, Zou L, Du S, Shen Y (2024) MTP family analysis and association study reveal the role of ZmMTP11 in lead (Pb) accumulation. Plant Physiol Biochem 212:108740
53. Liu S, Tian S, Li K, Wang L, Liang T (2018) Heavy metal bioaccessibility and health risks in the contaminated soil of an abandoned, small-scale lead and zinc mine. Environ Sci Pollut Res 25:15044–15056
54. Liu P, Zhang Y, Tang Q, Shi S (2021) Bioremediation of metal-contaminated soils by microbially-induced carbonate precipitation and its effects on ecotoxicity and long-term stability. Biochem Eng J 166:107856
55. Mani D, Kumar C, Patel NK (2015) Integrated micro-biochemical approach for phytoremediation of cadmium and zinc contaminated soils. Ecotoxicol Environ Saf 111:86–95

56. Maret W (2013) Zinc and the zinc proteome. In: Metallomics and the cell, pp 479–501
57. Marichali A, Dallali S, Ouerghemmi S, Sebei H, Hosni K (2014) Germination, morpho-physiological and biochemical responses of coriander (Coriandrum sativum L.) to zinc excess. Ind Crops Prod 55:248–257
58. Marreiro DDN, Cruz KJC, Morais JBS, Beserra JB, Severo JS, De Oliveira ARS (2017) Zinc and oxidative stress: current mechanisms. Antioxidants 6(2):24
59. Marschener HJFCR (1998) Role of root growth, arbuscular mycorrhiza, and root exudates for the efficiency in nutrient acquisition. Field Crop Res 56(1–2):203–207
60. McCall KA, Huang CC, Fierke CA (2000) Function and mechanism of zinc metalloenzymes. J Nutr 130(5):1437S-1446S
61. Meng Y, Xiang C, Huo J, Shen S, Tang Y, Wu L (2023) Toxicity effects of zinc supply on growth revealed by physiological and transcriptomic evidences in sweet potato (Ipomoea batatas (L.) Lam). Sci Reports 13(1):19203
62. Mithare P (2019) Nutrient deficiencies, symptoms, stages of development and control measures in major field crops. Chief Editor Dr. Neeraj Kumar, 99
63. Molnár A, Kondak S, Benkő P, Janovszky P, Kovács K, Szőllősi R, Gondor OK, Oláh D, Gémes K, Galbács G, Janda T, Kolbert Z (2022) Limited Zn supply affects nutrient distribution, carbon metabolism and causes nitro-oxidative stress in sensitive Brassica napus. Environ Exp Bot 202:105032
64. Mondal M, Khanra S, Tiwari ON, Gayen K, Halder GN (2016) Role of carbonic anhydrase on the way to biological carbon capture through microalgae—a mini review. Environ Prog Sustain Energy 35(6):1605–1615
65. Mondola P, Damiano S, Sasso A, Santillo M (2016) The Cu, Zn superoxide dismutase: not only a dismutase enzyme. Front Physiol 7:594
66. Montalvo D, Degryse F, Da Silva RC, Baird R, McLaughlin MJ (2016) Agronomic effectiveness of zinc sources as micronutrient fertilizer. Adv Agron 139:215–267
67. Moreira A, Moraes LA, dos Reis AR (2018) The molecular genetics of zinc uptake and utilization efficiency in crop plants. In: Plant micronutrient use efficiency, pp 87–108
68. Moreno-Lora A, Delgado A (2020) Factors determining Zn availability and uptake by plants in soils developed under Mediterranean climate. Geoderma 376:114509
69. Morya R, Morya R, Mittoliya VK, Chhirouliya JK (2023) Functions and deficiency symptoms of micro nutrients. Dr. Nareshkumar B. Desai, 35
70. Mousavi SR, Galavi M, Rezaei M (2013) Zinc (Zn) importance for crop production-a review
71. Okereafor U, Makhatha M, Mekuto L, Uche-Okereafor N, Sebola T, Mavumengwana V (2020) Toxic metal implications on agricultural soils, plants, animals, aquatic life and human health. Int J Environ Res Public Health 17(7):2204
72. Pacheco DDR, Santana BCG, Pirovani CP, De Almeida AAF (2023) Zinc/iron-regulated transporter-like protein gene family in Theobroma cacao L: characteristics, evolution, function and 3D structure analysis. Front Plant Sci 14:1098401
73. Padmavathiamma PK, Li LY (2007) Phytoremediation technology: hyper-accumulation metals in plants. Water Air Soil Pollut 184:105–126
74. Pandey N, Pathak GC, Sharma CP (2006) Zinc is critically required for pollen function and fertilisation in lentil. J Trace Elem Med Biol 20(2):89–96
75. Pasricha S, Mathur V, Garg A, Lenka S, Verma K, Agarwal S (2021) Molecular mechanisms underlying heavy metal uptake, translocation and tolerance in hyperaccumulators-an analysis: heavy metal tolerance in hyperaccumulators. Environ Chall 4:100197
76. Peng W, Li X, Song J, Jiang W, Liu Y, Fan W (2018) Bioremediation of cadmium-and zinc-contaminated soil using Rhodobacter sphaeroides. Chemosphere 197:33–41
77. Prasad R, Shivay YS, Kumar D (2016) Interactions of zinc with other nutrients in soils and plants-a review. Indian J Fertil 12(5):16–26
78. Rout GR, Das P (2009) Effect of metal toxicity on plant growth and metabolism: I. Zinc. In Sustainable agriculture; Springer, Dordrecht, The Netherlands, pp 873–884
79. Rudani K, Vishal P, Kalavati P (2018) The importance of zinc in plant growth-a review. Int Res J Nat Appl Sci 5(2):38–48

80. Sachdev S, Ansari SA, Ansari MI (2023a) ROS generation in plant cells orchestrated by stress. In: Reactive oxygen species in plants: the right balance. Springer, Singapore, pp 23–43
81. Sachdev S, Ansari SA, Ansari MI (2023b) Antioxidant defensive mechanisms to regulate cellular redox homeostatic balance. In: Reactive oxygen species in plants: the right balance. Springer, Singapore, pp 143–172
82. Sachdev S, Ansari SA, Ansari MI (2023c) Photosynthetic apparatus: major Site of oxidative damage. In: Reactive oxygen species in plants: the right balance. Springer, Singapore, pp. 75–92
83. Sarret G, Saumitou-Laprade P, Bert V, Proux O, Hazemann JL, Traverse A, Marcus MA, Manceau A (2002) Forms of zinc accumulated in the hyperaccumulator Arabidopsis halleri. Plant Physiol 130(4):1815–1826
84. Sharma P, Jha AB, Dubey RS, Pessarakli M (2012) Reactive oxygen species, oxidative damage, and antioxidative defense mechanism in plants under stressful conditions. J Bot 2012(1):217037
85. Sidhu GPS (2016) Physiological, biochemical and molecular mechanisms of zinc uptake, toxicity and tolerance in plants. J Glob Biosci 5(9):4603–4633
86. Stanton C, Sanders D, Krämer U, Podar D (2022) Zinc in plants: Integrating homeostasis and biofortification. Mol Plant 15(1):65–85
87. Subba P, Mukhopadhyay M, Mahato SK, Bhutia KD, Mondal TK, Ghosh SK (2014) Zinc stress induces physiological, ultra-structural and biochemical changes in mandarin orange (C itrus reticulata Blanco) seedlings. Physiol Mol Biol Plants 20(4):461–473
88. Subbaiah LV, Prasad TNVKV, Krishna TG, Sudhakar P, Reddy BR, Pradeep T (2016) Novel effects of nanoparticulate delivery of zinc on growth, productivity, and zinc biofortification in maize (Zea mays L.). J Agric Food Chem 64(19):3778–3788
89. Suganya A, Saravanan A, Manivannan N (2020) Role of zinc nutrition for increasing zinc availability, uptake, yield, and quality of maize (Zea mays L.) grains: an overview. Commun Soil Sci Plant Anal 51(15):2001–2021
90. Sur IM, Hegyi A, Micle V, Gabor T, Lăzărescu AV (2023) Influence of the extraction solution on the removal of heavy metals from polluted soils. Materials 16(18):6189
91. Suresh B, Ravishankar GA (2004) Phytoremediation—a novel and promising approach for environmental clean-up. Crit Rev Biotechnol 24(2–3):97–124
92. Takahashi M, Terada Y, Nakai I, Nakanishi H, Yoshimura E, Mori S, Nishizawa NK (2003) Role of nicotianamine in the intracellular delivery of metals and plant reproductive development. Plant Cell 15(6):1263–1280
93. Tiwari S, Patel A, Pandey N, Raju A, Singh M, Prasad SM (2020) Deficiency of essential elements in crop plants. In: Sustainable solutions for elemental deficiency and excess in crop plants, pp 19–52
94. Tsonev T, Cebola Lidon FJ (2012) Zinc in plants-an overview. Emir J Food & Agric (EJFA) 24(4)
95. Turan M, Esringu A (2007) Phytoremediation based on canola (Brassica napus L.) and Indian mustard (Brassica juncea L.) planted on spiked soil by aliquot amount of Cd, Cu, Pb, and Zn. Plant Soil Environ 53(1):7
96. Umair Hassan M, Aamer M, Umer Chattha M, Haiying T, Shahzad B, Barbanti L, Nawaz M, Rasheed A, Afzal A, Liu Y, Guoqin H (2020) The critical role of zinc in plants facing the drought stress. Agriculture 10(9):396
97. Vaxevanidou K, Papassiopi N, Paspaliaris I (2008) Removal of heavy metals and arsenic from contaminated soils using bioremediation and chelant extraction techniques. Chemosphere 70(8):1329–1337
98. Veena M, Puthur JT (2022) Seed nutripriming with zinc is an apt tool to alleviate malnutrition. Environ Geochem Health 44(8):2355–2373
99. Wang X, de Souza MF, Li H, Qiu J, Ok YS, Meers E (2022) Biodegradation and effects of EDDS and NTA on Zn in soil solutions during phytoextraction by alfalfa in soils with three Zn levels. Chemosphere 292:133519

100. Wang J, Yang S (2021) Dose-dependent responses of Arabidopsis thaliana to zinc are mediated by auxin homeostasis and transport. Environ Exp Bot 189:104554
101. Xuan Y, Scheuermann EB, Meda AR, Jacob P, von Wirén N, Weber G (2007) CE of phytosiderophores and related metal species in plants. Electrophoresis 28(19):3507–3519
102. Xue YF, Zhang W, Liu DY, Yue SC, Cui ZL, Chen XP, Zou CQ (2014) Effects of nitrogen management on root morphology and zinc translocation from root to shoot of winter wheat in the field. Field Crop Res 161:38–45
103. Yahaghi Z, Shirvani M, Nourbakhsh F, Pueyo JJ (2019) Uptake and effects of lead and zinc on alfalfa (Medicago sativa L.) seed germination and seedling growth: Role of plant growth promoting bacteria. S Afr J Bot 124:573–582
104. Yu BG, Chen XX, Cao WQ, Liu YM, Zou CQ (2020) Responses in zinc uptake of different mycorrhizal and non-mycorrhizal crops to varied levels of phosphorus and zinc applications. Front Plant Sci 11:606472
105. Zeng F, Ali S, Zhang H, Ouyang Y, Qiu B, Wu F, Zhang G (2011) The influence of pH and organic matter content in paddy soil on heavy metal availability and their uptake by rice plants. Environ Pollut 159(1):84–91
106. Zhang Y, Wang Y, Ding Z, Wang H, Song L, Jia S, Ma D (2017) Zinc stress affects ionome and metabolome in tea plants. Plant Physiol Biochem 111:318–328
107. Zhang H, Xu Y, Kanyerere T, Wang YS, Sun M (2022) Washing reagents for remediating heavy-metal-contaminated soil: a review. Front Earth Sci 10:901570
108. Zhao YY, Cao CL, Liu YL, Wang J, Li J, Li SY, Deng Y (2020) Identification of the genetic requirements for zinc tolerance and toxicity in *Saccharomyces cerevisiae*. G3 Genes Genomes Genet 10(2):479–488

Nickel: Uptake, Toxicity, and Tolerance in Crop Plants and Management of Ni-Contaminated Soil

Nidhi Prajapati, Arati Chaudhary, Anamika Pathak, Mahenk Patel, Ankita Adesra, Anil Patani, and Dharmendra Prajapati

Abstract Nickel (Ni) is a micronutrient essential to plants at very low concentrations but toxic at high ones. The amount absorbed by crop plants from the soil through their roots depends on the pH and organic matter content, among other ions, such as those of the other metals. Toxic accumulation occurs because of chlorosis, stunted growth, and imbalanced nutrient statuses due to the fact that it is an essential micronutrient that participates in various physiological functions, including urease activation. The three ways in which plants adapt to this environmental stress are through the activation of antioxidant systems to minimize oxidative stress and interaction with organic acids; Ni is also sequestrated in the vacuoles. Breeding programs can take advantage of natural differences in Ni tolerance shown by some crops. Ni-contaminated soils must therefore be managed for safe agricultural production. It can be done through soil amendments, phytoremediation, and Ni-tolerant crop types. In fact, though time-consuming, the use of hyperaccumulator plants in phytoremediation is a sustainable way of reducing the Ni contents in soils. The addition of phosphate or lime or organic matter into the soil reduces the Ni bioavailability resulting in the uptake of lesser amounts of the element by the plants. Therefore, the present study with a focus on understanding the mechanisms of Ni uptake and tolerance in crops provides avenues for controlling Ni contamination in agricultural soils to ensure food safety and sustainable crop production.

Keywords Ni uptake · Ni toxicity · Crop plants · Ni tolerance · Ni-contaminated soil · Sustainable agriculture

---

N. Prajapati · A. Chaudhary · A. Pathak · M. Patel · A. Patani · D. Prajapati (✉)
Department of Biotechnology, Smt. S.S. Patel Nootan Science and Commerce College, Sankalchand Patel University, Visnagar, Gujarat, India
e-mail: prajapatidharmendra88@gmail.com

A. Patani
e-mail: anilpatani89@gmail.com

A. Adesra
Department of Microbiology, Smt. S.S. Patel Nootan Science and Commerce College, Sankalchand Patel University, Visnagar, Gujarat, India

A. Husen (ed.), *Enhanced Crop Production Setup Under Various Micro and Nano Contaminants*, Smart Nanomaterials Technology,
https://doi.org/10.1007/978-981-96-6099-5_8

1 Introduction

Naturally occurring metallic element Ni is found in around 0.008% of the earth's crust. It is widely available as trace element in soil, water, and air, and it can also be found in all forms of life. Some plants and microbes require Ni as an essential micronutrient [9]. There are various chemical forms in which Ni is found, including sulfides, oxides, and silicates. Although Ni plays a minor but important role in various enzymatic processes in biological systems, its concentration in the environment is of course very low but can rise due to human activity [23]. The most prevalent natural sources of Ni in the soil and water are rocks' weathering and volcanic activities. Although this can vary with geology, average Ni concentration in the Earth's crust is between 50 and 200 mg/kg [54]. Much higher concentrations exist in Ni-rich soils, which typically occur in lateritic soils or ultramafic rocks. However, volcanic eruptions, as well as forest fires, both being relatively rare events, represent other natural sources of Ni to the environment [17]. Although Ni naturally occurs in the environment, its presence largely in the atmosphere is attributed to human activities—industrial processes, more particularly. Various industries employ Ni, such as aluminum fabrication, batteries, electroplating, and stainless steel [23]. Several tons of dust and particulate matter containing Ni are released into the atmosphere from mining, smelting, and refining processes that are considered the most important sources of Ni emissions [9].

Ni compounds are discharged into aquatic environments by effluents from industries. Phosphate fertilizers, applied in large volumes in the fertilizer industry, contain trace amounts of the incorporated metal in the soils [96]. Eventually, this may cause high concentrations of Ni in crops and groundwater resulting from leaching and runoff with potentially deleterious effects on the environment and human health. Another major Ni source in the environment is through the combustion of coal, oil, and other fossil fuels in autos, power plants, and industrial processes. During the combustion process, released Ni may be deposited onto surfaces of soil and water bodies, thus leading to contamination [9]. These human activities have resulted in an increase in Ni concentration in the environment, and it normally outweighs the natural background concentration present within areas since this has mostly occurred in urbanized and industrial sites. Elevated levels of Ni pose a risk to humans, animals, and plants depending on the type of chemical of Ni used and exposure levels. To minimize Ni's impact on the environment and humans, strict regulatory scrutiny as well as control over emissions is being contemplated [23] (Fig. 1).

Ni has been increasingly realized in the agricultural sector to be a micronutrient that is both necessary and potential polluting agent. As minute levels of Ni are essential for its health, there must be controlled concentrations of this element within plant life. At correct amounts, Ni facilitates numerous metabolic processes within crops that enhance yields and health [10]. Though high levels of Ni may pose a threat to food safety and agricultural production, they might also be harmful to plant growth. Moreover, the chain of metabolism may pose a risk to human health. The key enzyme involved in the metabolism of nitrogen is urease, which contains Ni. The conversion

Fig. 1 Plant responses to environmental Ni toxicity

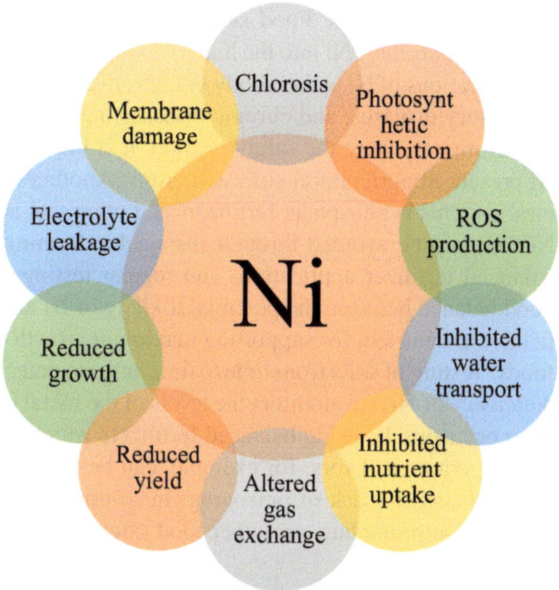

of urea to ammonia and carbon dioxide—by urease, for example—allows plants to utilize nitrogen more efficiently [2]. This is particularly important for general growth and in the synthesis of protein. When Ni is deficient in plants, this then leads to the impairment of urea's ability to be metabolized followed by a shortage in nitrogen, and then, development is stunted [83]. Growth requirements for seed germination require adequate Ni, particularly for nitrogen-fixing plants like legumes. Ni also helps in more effective root development, increasing the uptake of other essential elements like iron and zinc [8]. Ni has been reported to increase the production of defense chemicals, thereby improving resistance of plants to many diseases. Some phenolic compounds that also protect the plants against pathogens and environmental stress may at times be induced by Ni. While plant tissues, in general, can tolerate concentrations below 0.1 ppm of Ni, a deficiency in this element can cause many symptoms including poor growth, chlorosis, or leaf yellowing, and necrosis, death of tissue, particularly in crops like soybeans, wheat, and barley. Although Ni is required by plants, the amount of Ni present in the environment has to be sternly in control [63].

High concentrations of Ni can cause harmful effects on plants and result in a wide array of physiological changes. Characteristics that are present with Ni poisoning include stunted growth, chlorosis of the leaves, reduced photosynthesis, and nutrient assimilation. Ni has the tendency to accumulate within plant tissues, primarily roots and leaves. This affects crop productivity and, more importantly, its health considerably. This accumulation can lead to crops grown in soils contaminated with industrial activities or overfertilization yielding less [78]. When it is present at high levels in farming soils, Ni can become accumulated in the tissues of edible plants like grains, fruits, and vegetables. Food safety is compromised because eating contaminated crops can introduce Ni into the human food chain. Exposed people who possess high concentrations of Ni in their body systems over a long time exhibit allergic reactions, respiratory diseases, and chronic diseases including cancer and renal disease [68]. Therefore, to ensure the safety of food, it is important to monitor the quantities of Ni present in agricultural soils and crops. Another source of Ni entry into agricultural systems is phosphate fertilizers and other soil additives. The formation of soil Ni can only be avoided through sustainable farming practice use, mainly through balanced fertilizer applications and regular testing of the soil. There should be a good balance between the potential toxicity of Ni and the fact that it acts as a beneficial micronutrient for supporting maximum growth of crops and ensuring safety in foods. Testing of soils from time to time helps to detect areas with a risk of toxicity or deficiency of Ni and monitors the levels of the metal [67]. Pollution can be prevented from occurring by using balanced fertilizers which have very low concentrations of Ni. Hyperaccumulators, for example, can be planted to absorb excessive Ni in the soil reducing the risk of poisoning. In so doing farmers will increase crop output without compromising consumer food safety as levels of Ni in agricultural systems are controlled [73].

This chapter discusses the complex issues of Ni in agriculture including mechanisms for plant tolerance, toxicity, uptake, and management strategies of excess Ni contamination in soils. This chapter tries to give readers an overall understanding of the role that Ni plays in agriculture, from its biochemical impact on crops to mitigating actions needed to ensure food safety and sustainable farming practice.

2 Ni Uptake in Crop Plants

2.1 Sources and Forms of Ni in Soil

Although Ni is an essential element to plants, its contents and availability for uptake in soils alter. There are two major types of sources of Ni in the soil:

2.1.1 Naturally Occurring Ni in Soils

Ni occurs naturally because of the weathering of parent rock materials in soils. Amounts and kinds of Ni that Ni-bearing rocks contain are indicators of how much Ni is available in soils. Ni-rich soils originate mainly because of the presence of mafic and ultramafic rocks [50]. Ni is substituted within many minerals of these soils: Oxides and hydroxides; silicates, especially olivine and pyroxenes; sulfides and carbonates. Yet, the apparent bioavailability of naturally occurring Ni in the soil is influenced by the pH of the soil, the quantity of organic matter present, and other factors that affect the solubility and structure [51].

2.1.2 Contribution of Human Activities to Ni Levels

The following are the main routes in which human activity has resulted in higher Ni concentrations in various soils: (1) Industrial activities: Ni is released into the environment during mining, smelting, and refining operations. This is the primary method through which Ni accumulates in surrounding soils. Ni pollution is also caused by the burning of fossil fuels, particularly coal. (2) Agricultural inputs: Ni can be introduced into agricultural soils by fertilizers, particularly phosphate fertilizers and the application of sewage sludge as an amendment. Elevated levels of Ni may be built up over time through the repeated use of such materials. (3) Waste disposal: Municipal and industrial wastes, which include effluents and solid wastes, may contain Ni and ultimately contribute to its accumulation in soils [23]. Since Ni is an essential micronutrient, these anthropogenically introduced sources of the metal may provide a flux of Ni into the root environment of plants; this will have both beneficial and negative impacts when levels increase. Directly injected anthropogenically introduced Ni is often more readily available than soil derived Ni which can sometimes prove detrimental to crop development and food quality [19].

2.2 *Mechanisms of Ni Uptake by Plants*

Plants take in Ni into the soil largely through uptake by the roots, and there are various factors that may affect the process. The mechanisms by which Ni gets into the plant roots involve both passive and active transport mechanisms, and many physiological and environmental variables are involved in Ni absorption [2].

2.2.1 Root Absorption Processes (Passive and Active Transport)

There are two major routes by which Ni is absorbed in plants. The two ways include passive transport and active transport: (1) Passive transport: In passive transport, diffusion occurs due to a concentration gradient as Ni and water diffuse into the plant root cells. This method does not require any energy at all. Ni ions move from concentration-rich soil solution to concentration-poor regions within the root cells. When Ni concentrations in the soil solution are quite high at the beginning of Ni uptake, most of it occurs by passive transport. (2) Active transport: Ni ions have to be transported against a concentration gradient and therefore need energy—mostly in the form of ATP—for its active transport [49]. Some transport proteins in plasma membrane root cell transport Ni. Even at negligible concentrations of Ni are presented in the outer environment, the transporters facilitate Ni transport into plant cells and make sure that there is enough amount of it. Since the active transport process is strictly regulated, the uptake of Ni by plants may vary according to each plant's needs. Once inside the root cells, Ni can be streamed to other areas of the plant through the xylem in order to perform critical biological functions including nitrogen metabolism and urease activity [38].

2.2.2 Factors Influencing Ni Uptake

The following factors have been identified as affecting the Ni bioavailability both in the soil and its uptake by plants: (1) Soil pH: The soil pH is the main variable controlling the uptake of Ni. As a rule, in acidic or low pH soils, Ni has great solubility, and hence its availability for uptake by plants is higher. As the soil pH increases toward neutral or alkaline values, Ni starts forming some insoluble compounds and becomes less available to plants. Therefore, such plants growing in acidic soils become more prone to absorb higher concentrations of Ni, which in high enough concentration can be toxic [15]. (2) Organic Matter: The organic matter is one of the major variables that influence the uptake of Ni in soil. If Ni ions strongly get bound by organic matter in less accessible forms, then such complexes with Ni may form that reduce the availability of the metal to plants [36]. Alternatively, organic material may undergo decomposition and release bound Ni in organic complexes, which becomes available for uptake by plants. In addition to these factors, the association between Ni and organic materials is the other factor which contributes toward the mobility of Ni in the soil [98]. (3) Other Nutrients: Ni uptake may also be dependent on the availability of other nutrients in the soil. For example: Iron (Fe), manganese (Mn), and zinc (Zn):These cations will compete with Ni for uptake sites on the membranes of the root of plant. This interaction is strengthened if the nutrient concentrations are high compared to Ni [27]. Phosphorus (P): At high levels of phosphorus in the soil, Ni uptake may be inhibited through the formation of insoluble Ni-phosphate complexes and consequently reduced Ni availability to the plant. In addition to factors related to Ni itself, such as temperature, moisture content of the soil, and the type or variety of plants present, other environmental factors also affect the efficiency of Ni uptake from the soil. These together determine the amount of Ni absorbed by plants and all effects resulting from Ni to their growth and development [91].

2.3 Translocation and Accumulation Within Plants

After this, the absorbed Ni by the roots of the plant is translocated to different parts of the plant through the xylem. Here, Ni plays a role in several physiological roles; however, several factors like the type of plant, the environment surrounding it, and the quantity of Ni present in the soil are different from the translocation and distribution of Ni in the plant [2].

2.3.1 Movement of Ni from Roots to Shoots

After being absorbed by the cells of the root, Ni is transported by the water-conducting tissue or xylem of the plant. Some transport proteins facilitate the transfer of Ni3 + ions into the xylem vessels often. Water loss through the leaf causes the Ni ions to start out in the transpiration stream [4]. This movement transfers Ni to the above-ground parts of the plant, including its stems, leaves, flowers, and eventually its seeds. The ability of the plant to control the movement of this ion, as well as how much Ni the plant itself needs, determine how efficiently this occurs. In some plant species, it moves at a very fast rate from roots to shoots, while in others, it stays mainly in the root tissues [92].

2.3.2 Accumulation in Different Plant Tissues (Leaves, Stems, and Seeds)

Since the physical needs of the plant and the amount of Ni absorbed influence its distribution in different plant tissues, Ni tends to accumulate in specific plant tissues.

- **Leaves**: One of the primary deposition sites of Ni is in leaves. Here, the Ni transported through the xylem water accumulates in leaf tissue and can possibly also take part in enzymatic action. It may be with the urease that is necessary for the breakdown of nitrogenous compounds [83]. In some plants, referred to as Ni hyperaccumulators, Ni may at times reach very high concentrations in the leaves without being toxic [19].
- **Stems**: Compared to leaves, the general tendency for stems is that they may have lower concentrations of Ni. Nevertheless, it is believed that in cases where translocation has taken place, a part of the Ni may still be retained in the stem tissues. In this context, the Ni concentration level in the stem of a plant may be considered more or less in the middle of the range as found between roots and leaves [43].
- **Seeds**: Ni can also accumulate in seeds, but this is often a characteristic of species for which Ni is necessary for germination and viability of the seed. The Ni concentration in seeds typically is lower than that in leaves or roots but is necessary for adequate germination and seed development for the succeeding generation of plants. Ni accumulation in seeds of edible crops can be sufficiently high to pose a food safety problem [44].

2.3.3 Influence of Plant Species and Varieties on Ni Uptake and Accumulation

The ability of different plant species to absorb and accumulate Ni varies significantly, as do even variants within the same species (Fig. 2).

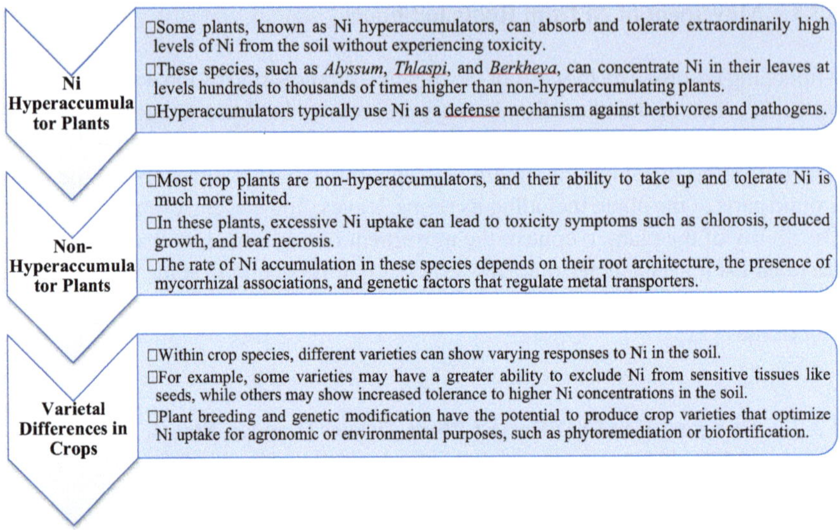

Fig. 2 Influence of plant species and varieties on Ni uptake and accumulation

3 Ni Toxicity in Crop Plants

Although Ni is an essential element, high concentrations of it may make crop plants poisonous. High doses of Ni can elicit a number of detrimental impacts on the growth, metabolism, and health of a plant if they are beyond the tolerance level. The symptoms of Ni toxicity are a combination of physiological and observable signs [2].

3.1 Symptoms and Indicators of Ni Toxicity

Beside physiological and metabolic changes, Ni toxicity may change the physical aspect of the exposed plants.

3.1.1 Observable Symptoms

In fact, the first signs of Ni toxicity in many cases are physical changes inside a plant's leaves. The most familiar visual symptoms that people have seen are: (1) Chlorosis: probably one of the typical indicators of Ni poisoning—a yellow color of plant leaves based on a depression of chlorophyll biosynthesis [65]. The older leaves are usually the first to manifest signs of chlorosis, which spread from the tips or edges toward the center. Chlorosis involves limited photosynthesis as well as limited uptake of nutrients [46]. (2) Necrosis: At a more intense level of Ni exposure, the tissue necrosis begins to show up, usually in the form of small brown or black spots on the leaves. These spots grow into each other and then coalesce, leading to extensive tissue damage and killing. Necrosis is mostly irreversible and thus regarded as one extreme effect of the plant [22]. (3) Inhibited Growth: High levels of Ni can limit the growth potential of plants, resulting in small or stunted plants. The infected plant will likely have fewer leaves or flowers, and the shoots and roots will be abnormal, having a shorter than normal stature. Sometimes the interference occurs in a specific manner causing problems in the root development which remains shorter, thicker, or deformed [63]. (4) Leaf Curling and Deformation: In Ni poisoning, the plants show abnormal growth patterns, curling, or deformation of leaves. Most these distortions are due to cellular injury as well as abnormalities in the absorption of nutrients and control of water [40].

3.1.2 Biochemical and Physiological Disruptions

There are also important physiological and biochemical changes due to the toxic effects of Ni, often before the onset of overt symptoms.

- **Inhibition of Enzymes**: Ni poisoning affects an important function of enzymes in plants. One such enzyme that catalyzes the breakdown of urea in plant tissues is urease, which Ni inhibits. This will result in the accumulation of toxic levels of urea, inhibition of protein synthesis, and complete use up of available nitrogen[83]. Moreover, many enzymes that catalyze vital metabolic functions such as respiration, photosynthesis, and antioxidant defense are hampered by Ni poisoning. Inhibition of some of these enzymes can disturb a line of metabolic disorders interfering with plant growth and yield [12].
- **Oxidative Stress**: The overproduction of reactive oxygen species (ROS) including hydroxyl radicals (OH·), superoxide radicals (O_2^-), and hydrogen peroxide (H_2O_2) could be one of the effects of an excess of Ni[64]. Through such oxidative stress, ROS can damage DNA, lipids, proteins, and cellular structures. The major effect of oxidative stress is lipid peroxidation affecting cellular function and membrane integrity [14]. Enzymatic activities of superoxide dismutase (SOD) and catalase (CAT), typical antioxidant enzymes, have been reported to increase significantly in plants subjected to Ni-induced oxidative stress as a means of recuperating the losses. On the other side, over-oxidation can breach the plant's defense system and even lead to permanent cellular damage [74].
- **Nutrient Balance Disruption**: Other vital nutrients involved include iron (Fe), zinc (Zn), magnesium (Mg), and calcium (Ca) which would be taken up and utilized less because of the excess Ni. Growth-related problems resulted from this imbalance, and poor deficiency of nutrients occurred. The inadequacy of magnesium causes photosynthesis to be less than adequate. Inadequate iron often results in interveinal chlorosis [93].

- **Impaired Photosynthesis**: Ni toxicity that damages the chloroplasts' (photosystem II)PSII and reduces the amount of chlorophyll present will prevent photosynthesis from becoming complete[65]. This consequently reduces electron transport and decreases light absorption leading to a reduced energy level in a plant. Since photosynthesis is an initial step in production, lowered plant growth and productivity will occur [53].

3.2 *Impact on Plant Growth and Development*

The exposure of Ni toxicity onto plant growth and development is vast. Exposure to large amounts of Ni is harmful to important physiological functions such as respiration and photosynthesis, disrupts the balance of nutrients, and finally reduces crop quality and production [2].

3.2.1 Effect on Photosynthesis and Respiration

Photosynthesis and respiration are two important processes for growth and energy production in plants that are adversely affected by Ni toxicity.

- **Photosynthesis**:
  High concentrations of Ni interfere with the photosynthetic machinery in many ways:
  - **Chlorophyll Degradation**: High amounts of Ni can cause lower chlorophyll content and subsequently lead to chlorosis or the inability of the plant in the production of light energy. Ni content can also decrease the concentration of green pigments on the leaves and may impede light absorption in their surfaces by blocking or eliciting the production of chlorophyll [2].
  - **Photosystem Damage**: Ni can damage PSII, thus weakening the chain of electron transport in chloroplasts. It reduces the plant's potential to transform sunlight into chemical energy by disrupting light-dependent photosynthesis processes [86].
  - **Reduced CO2 Fixation**: In addition to the inactivation of the enzyme involved in photosynthesis, the Rubisco responsible for carbon fixation during the Calvin cycle is deactivated by Ni toxicity[65]. This deters the growth and development of the plant to produce biomass by rendering it unable to alter CO_2 into organic molecules [33].
- **Respiration**: The synthesis of ATP, an energy-generating compound, depends on the action of mitochondrial respiration, and thus, Ni poisoning adversely affects its activity. Excess levels of Ni alter the mitochondrial electron transport chain that is affecting cellular respiration [70]. This may cause low amounts of energy for various crucial cellular activities such as growth, nutrient uptake, and stress response. On the other hand, Ni produces reactive oxygen species that degrade

Table 1 Disruption of essential nutrient balance

Iron (Fe)	Zinc (Zn)	Magnesium (Mg) and Calcium (Ca)
Iron insufficiency is frequently the result of excess Ni competing with iron for absorption and transport in plants [57] Interveinal chlorosis is caused by iron shortage, especially in young leaves, as iron is a crucial element of chlorophyll and is necessary for photosynthesis [46] Chlorophyll production may be further inhibited by the rivalry between iron and Ni for absorption sites, which would worsen the impact on photosynthesis [56]	Zinc is a necessary micronutrient that is involved in protein synthesis and enzyme function, although zinc intake can be inhibited by Ni [72] Processes including DNA replication, growth regulation, and defense against oxidative stress are all impacted by zinc deficiency [81] Deformities in plant growth and poor health might result from an imbalance between zinc and Ni [27]	Elevated levels of Ni have the potential to obstruct the absorption and dispersion of magnesium and calcium, two elements crucial for the structure and operation of plants [9] Calcium is necessary for signal transmission and cell wall stability, while magnesium is essential for the synthesis of chlorophyll [37] Weak cell walls, low photosynthetic efficiency, and heightened vulnerability to external stressors can result from the disruption of these nutrients [41]

mitochondria membranes and disrupt even more the production of energy. Crop plants respond to decreased photosynthesis and respiration by reducing growth because of lower available energy [21].

3.2.2 Disruption of Essential Nutrient Balance

The imbalance of vital nutrients brought about by Ni toxicity increases nutritional shortages and worsens plant stress. Interaction with other essential nutrients accounts for poor growth, development, and function by Ni [63] (Table 1).

Ni toxicity induces an imbalance in nutrition that leads to several deficiencies and reduction in the disease resistance, impaired development, and metabolic dysfunctions of the plants [45].

3.2.3 Reduction in Crop Yield and Quality

Ni toxicity affects the quantity and quality of the agricultural produce apart from directly affecting the productivity and crop quality.

- **Yield Reduction**: Crop productivity is greatly reduced as a result of stunted growth, poor nutrient balance, and reduced photosynthesis. Toxicity may also result in reduced biomass in above-ground parts such as leaves, stems, and fruits, as well as below-ground roots. In extreme cases it can also lead to early senescence, wherein the development of plants will have stopped, and a plant will die off before it attains optimal maturity [41].

- **Quality Deterioration**: The edible tissue parts, such as seeds, fruits, or leaves, may be the sites of Ni poisoning in quality. As Ni is toxic at high levels of accumulation, the livestock and finally the consumers may suffer health problems if Ni accumulates in considerable amounts within such tissues. Besides, Ni-polluted fruits and vegetables will also lose some flavor, color, and size which may not be that marketable [23].
- **Nutritional Imbalance**: Besides these effects, crop damage caused by Ni poisoning may also result in nutrient deficiencies, such as iron, zinc, or magnesium levels. Its suitability for consumption by human beings and animals is further affected with a corresponding decrease in its nutritional value. Nutrient-reduced crops could also pose regulatory problems if harvested for human consumption, mainly where the levels of Ni in them far surpass the permissible levels [5].

3.3 *Molecular Mechanisms of Ni Toxicity*

That is the reason Ni toxicity impacts at a molecular level: it implicates crucial components such as DNA, proteins, and cell membranes and throws cellular homeostasis into disarray. Other complicating factors in absorption and metabolism are interactions with other metal ions, ROS generation, and gene expression alteration [11].

3.3.1 Interaction with Other Metal Ions and Competition for Uptake

The Ni toxicity reduces the absorption of other important metal ions due to competition for transporters and other binding sites in the cells.

- **Competition for Transporters**: Ni can compete with other divalent cations metal for uptake by metal transporters; among such are iron (Fe^{2+}), zinc (Zn^{2+}), manganese (Mn^{2+}), magnesium (Mg^{2+}), and calcium (Ca^{2+}) [55]. As an example, it can be transported by the ZIP transporters responsible for the absorption of iron and zinc. At high levels, Ni may outcompete the essential metals for absorption, causing them to be deficient in the plant [76]. The outcome of this competition is: (1) Iron deficiency: This results in decreased chlorophyll synthesis and causes chlorosis. (2) Zinc deficiency: It interferes with the synthesis of proteins and the activity of enzymes. (3) Magnesium deficiency: It inhibits photosynthesis because magnesium is a constituent part of the chlorophyll molecule [84].
- **Altered Metal Homeostasis**: A cascade of nutrient deficiencies results because Ni toxicity interferes with metal homeostasis. Damage is amplified if necessary metals cannot be targeted inside the plant or not taken up at all [65]. For example, too much Ni causes iron-deficiency chlorosis, even in soils containing adequate amounts of iron. This is because there is an excess amount of Ni competing

for immobilization into necessary proteins such as ferritin or blockage of iron transport into the plant [24].

- **Sequestration and Detoxification**: Plant cells cope with the toxic effect of Ni through sequestrating it inside vacuoles or by offering it to ligands such as organic acids (citrate), amino acids (histidine), metallothioneins, and phytochelatins, in which Ni contains lower bioavailability and more cell-damaging effects due to their binding characteristics. Even though the sequestration and detoxification process is varied between different plant species, after Ni overloads a system, cellular damage begins [63].

4 Ni Tolerance in Crop Plants

Crop plants that are tolerant to Ni can tolerate high levels of Ni. Although the majority of the plants have developed their complicated physiological and biochemical protections against excessive Ni, they survive and grow well in metal-contaminated soils [36]. Some of these include activation of antioxidant defense systems, sequestration, and exclusion of Ni from sensitive tissues [97].

4.1 Physiological Mechanisms of Tolerance

4.1.1 Ni Exclusion from Sensitive Tissues

In plants, there are many different strategies employed to limit the intake and dispersal of Ni. Some metal transporter proteins at the root level control the amount of Ni that enters the root cells, limiting absorption and thus the total amount reaching delicately sensitive tissues like leaves, flowers, and seeds [2]. But some species also provide a physical barrier; for instance, the Casparian strip in the endodermis of the root prevents Ni from being transported from the soil into the xylem and, consequently, the shoot [71]. The limiting of Ni to the roots protects the above-ground parts from lethal doses. Ni-tolerant plants may restrict its entry into potentially hazardous sites such as leaves and chloroplasts, where it would interfere with light capture that is used for photosynthesis, even if Ni is absorbed by the plant [2]. Alternatively, Ni could be stored in less metabolically active parts of the plant, such as old leaves, or in specific types of cells, such as epidermal cells, thereby minimizing interference with critical functions [19].

4.1.2 Sequestration in Vacuoles or Cell Walls

Ni ions are therefore sequestered in vacuoles as part of a critical mechanism for Ni tolerance referred to as vacuolar sequestration. Once absorbed, Ni is mobilized to vacuoles, where it is sequestered from the cytoplasm and other biological processes [62]. This makes it impossible for Ni to interact with DNA, enzymes, and other important functions. Ni often forms complexes with organic acids like citrate or malate in the vacuoles and makes it less toxic [1]. In this manner, the vacuole can be termed as a safe lodging for Ni so that it cannot cause harm to other cellular components. Besides, some plants bind Ni to cellulose or pectins that are present within the cell walls [85]. This immobilization in the matrix of the cell wall would reduce the amount of free Ni, which can then translocate to the cytoplasm and cause damage. It is one process that works particularly well in roots where this Ni can be accumulated without interfering with the vital physiological functions [88].

4.1.3 Activation of Antioxidant Defense Systems

Tolerance plants activate a large number of antioxidant defense mechanisms to respond to Ni-induced oxidative stress, including the production of some key enzymes [69]. While CAT breaks hydrogen peroxide down to water and oxygen to prevent further accumulation of reactive species, SOD converts dangerous O_2^- into H_2O_2 [89]. Further, ascorbate and glutathione serve as electron donors by which several enzymes, such as ascorbate peroxidase (APX) and glutathione peroxidase (GPX), reduce hydrogen peroxide [47]. Such cooperative action reportedly aids in reducing oxidative damage and also provides enhanced plant tolerance to high Ni concentrations by scavenging the overproduced ROS that develop under Ni stress [16]. Beside enzymic defense mechanisms, plants produce non-enzymic antioxidants like glutathione, ascorbate (vitamin C), and tocopherols (vitamin E). These antioxidants scavenge free radicals and thereby provide additional protection against Ni-induced oxidative stress [69]. Apart from these, the small, cysteine-rich proteins called phytochelatins and metallothioneins are also produced by plants and bind to Ni ions to form stable complexes. These complexes enable Ni detoxification as a result of its prevention from coming into contact with key cellular components and also through being sequestered in the vacuoles or through the action of transport proteins, which remove Ni from the cell [28].

4.2 Genetics of Ni Tolerance

The identification of key genes involved in tolerance processes, metal homeostasis, and generation of defense responses would massively increase the understanding of the genetics of Ni tolerance in plants [36]. Understanding these genetic factors, therefore, is crucial for breeding crops with enhanced Ni tolerance in the soil, especially where metals are involved [26]. Other key genes contributing to tolerance are HMA, for heavy metal ATPases, and natural resistance-associated macrophage proteins (Nramp), which control the transport and sequestration of Ni as well as metal transporter genes such as those in the ZIP family controlling Ni, zinc, and iron uptake and distribution [35]. These genes regulate Ni absorption, transport, and storage in vacuoles which subsequently regulates Ni concentrations in the plants. In addition, the genes encoding antioxidant enzymes including APX, CAT, and SOD are essential to dampen the oxidative stress of Ni and hence from causing damage to the cellular machinery [29]. Phytochelatin and metallothionein also play a crucial role in detoxification of Ni through encoding genes for proteins that bind to and sequester Ni in vacuoles or form stable complexes that prevent the interaction of Ni with sensitive cellular components [77].

The tolerance-related genes are up-regulated by increased Ni concentration such that the plant can respond quickly [45]. Controlling these genes is, therefore, the key to controlling how the plant would react to Ni stress. The WRKY, bZIP, and MYB families of transcription factors mediate regulation of these genes; however, the long-term adaptation to high-Ni conditions may depend on epigenetic modifications, such as DNA methylation and histone modification [34]. Crop breeding for Ni resistance requires a multidisciplinary approach and thus demands both breeding and genetic engineering strategies. While more conventional breeding approaches—such as hybridisation or natural variant selection—will increase tolerance to Ni, MAS accelerates the process because it identifies molecular markers that are linked to genes that encode for the tolerance [3]. The focused focus of genetic manipulation is the ability selectively to overexpress crucial genes—antioxidant enzymes or metal transporters—to enhance resistance to Ni. New gene-editing tools, such as CRISPR/Cas9, allow the precise modification of the genes that control Ni tolerance, while transgenic methods move genes from Ni-tolerant species into sensitive crops [52]. The integration of these approaches should provide means toward developing Ni-tolerant crops that can live safely in Ni-polluted environments [45].

4.3 Plant–Microbe Interactions

Plant–microbe interactions are a vital factor in increasing Ni tolerance for crops. The soil microbiome, comprising mycorrhizal fungi and plant growth-promoting rhizobacteria (PGPR), is one of the most significant factors determining uptake, distribution, and health of plants concerning Ni [75]. Organic acids like citric or oxalic acid produced by microorganisms in the soil dissolve Ni through chelation of metal ions. This raises the absorption of Ni but needs to be balanced carefully in order not to create toxicity [98]. The other entity that interacts with the plant root system is the microbes that sometimes change the pH in the soil or produce compounds that will alter the solubility of Ni in the roots. Auxins and gibberellins, known for stimulating root growth and enhancing the availability of nutrients, are referred to as growth-promoting substances produced by PGPR for enhanced Ni tolerance in plants [6]. Plants become tolerant of Ni stress due to the induction of systemic resistance and activation of antioxidant defense mechanisms by PGPR [60, 61]. Some PGPR strains, like those of *Bacillus subtilis*, *Rhizobium spp.*, and *Pseudomonas fluorescens*, also facilitate plant sequestration of Ni by inducing synthesis of phytochelatins and metallothioneins, which will chelate the Ni and make it available for storage within cell walls or vacuoles [75]. Mycorrhizal fungi, in their symbiotic association, further alleviate Ni stress by improving the uptake of nutrients, especially phosphorus, that is often associated with Ni toxicity [65]. Mycorrhizae also decrease the availability of Ni through encapsulation of the metal within their structures or by transforming it into less soluble forms and, thus, protect the plant roots from damage [87].

5 Management of Ni-Contaminated Soil

Effective management of Ni-contaminated soil helps mitigate the adverse impacts of Ni pollution on agriculture and the environment. It involves assessing the degree and type of Ni contamination by various methods of testing and evaluating the soils [18].

5.1 Assessment of Ni Contamination

Accurate soil testing is required for the measurement of Ni pollution levels and assessment of the possible impact on plant health. Representative soil samples should be collected at different points and depths within the contaminated area in order to proceed. In order to ensure uniformity of the samples, they should be oven-dried, ground, and sieved prior to analysis [7]. Chemical methods are used to determine soil total Ni concentrations. Atomic Absorption Spectroscopy: In this technique, soil samples are dissolved by acids, and then Ni ions absorb the light. Inductively Coupled Plasma Mass Spectrometry: The presence of Ni ions is measured according to their mass-to-charge ratio with high accuracy [95]. Analysis of several elements, including Ni, can be performed simultaneously using ICP-OES by measuring light emitted by atoms in a plasma. X-ray fluorescence can be consulted for rapid, non-destructive examination as it measures fluorescent X-rays metals in the sample give off after being subjected to X-rays [13]. The extraction techniques measure the amount of Ni available for absorption into the plants that will help evaluate the bioavailability of Ni. Chelating agents, including EDTA, give an indication of the percent bioavailable under conditions that mimic those which prevail in plant uptake [58]. Acid extractants like HCl and ammonium acetate quantify Ni that is readily available to plants for absorption. The bioavailability of Ni in the soil is dependent on a number of factors. It is also the pH of the soil that is an important variable since acidity soils with a high titration value render Ni more soluble and hence more plant available, while low acidity soils reduce Ni availability [32].

5.2 Remediation Technologies

Various technologies are applied in the remediation of Ni-polluted soils aimed at reducing the concentration of the metal, immobilizing it, or eliminating it altogether. There is one of the highly researched techniques on phytoremediation—the use of hyperaccumulator plants. This refers to plants that can uptake Ni in reasonably large quantities without harming [79]. Some of the hyperaccumulators include *Alyssum murale*, *Alyssum bertolonii*, *Thlaspicaerulescens*, and *Berkheyacoddii*. The first three hyperaccumulators uptake Ni from the soil directly and generally stock Ni in their tissues, especially shoots and leaves. These plants can be safely harvested for the accumulated Ni to be extracted for processing and metal recovery [42]. However, since it is a green technique, phytoremediation may take some time and proper management of plant disposal and needs proper care to avoid recontamination. Ni in contaminated soils can also be sequestered from the biogeochemical cycle by chemical additions that make it inert. Lime application is raising the pH of the soil; it reduces Ni's solubility and the quantity that can be available to plants [30]. Soil washing can easily extract Ni from soil. This process involves the process of washing out contaminated soil particles using either water or chemical solutions to separate the two types of particles. The wash water must be disposed of appropriately. Electrokinetic remediation, on the other hand, can only be used for fine-particle soils. It involves forcing metal ions to electrodes with the use of an electric field where it is collected [99].

5.3 Sustainable Agricultural Practices

What are applied on polluted soils with Ni are sustainable agriculture practices that help in the management of these soils and mitigate the harmful effects on crop productivity and soil health [60, 61]. Crop selection and rotation, where crops like certain grasses, legumes, and forage crops exhibit increased Ni tolerance, is one of the most important strategies [31]. Another method would be to use hyperaccumulator crops or plants, for example, brassicas or sunflowers that stabilize Ni in soil. Crop rotation is also essential to maintain the fertility of the soil due to the reduction of soil stress and improvement of its structure as well as breaking pest and disease cycles, which are likewise most often exacerbated in contaminated environments [80]. Legumes enhance fertility and structure, while legumes assist in stabilizing or sequestering Ni. These varieties, whether selected through breeding, genetic engineering, or natural selection, possess the following attributes: higher antioxidant defense mechanisms, reduced Ni uptake, and better metal sequestration for survival in contaminated environments. Field tests in the local environment are required to prove their efficiency [59].

6 Applications

6.1 Economic and Environmental Impact

The effectiveness and sustainability of soil remediation techniques contaminated with Ni will only be understood based on their impact on the economy and environment. Such an approach will require balancing the various merits and demerits of these approaches with respect to their impact on soil health and productivity [39]. Remedy's costs, for instance, are the testing of soil, purchase of amendments, planting hyperaccumulators, and use of chemical or physical remediation technologies. Successful remediation has many benefits. These include reduction of environmental and human health risks while soil fertility, productivity, and porosity increase to provide for better crop yields and ecosystem functions [90]. Remediation also raises property value and makes environmental regulations feasible, therefore preventing fines and lawsuits. For example, phytoremediation can be expensive upfront but, in the long-term, increases the productivity of lands and quality of soils, which can benefit a given economy in terms of regained metal [48]. Remedial action is effective in the long term through adaptive management. Economically viable techniques become preferable, which may equilibrate initial costs and long-term gains as well as lead to increased land value and production to ensure a return on investment. Remediation actions enhance resilience, promote biodiversity, and restore ecological functioning to the environment [94]. Societally, it promotes public health by reducing pollution in residential and agricultural areas, thus enhancing safer living and food production conditions. Involving local communities in decision-making supports remediation efforts and ensures that the requirements of the community are matched [20].

6.2 Future Challenges and Research Needs

Research in Ni contamination of the soil and its effects on agriculture continues, and new challenges are to be encountered [60, 61]. The growing rate of new sources of pollution due to industrial processes like mining, smelting, and improper disposal of electronic waste, e-waste, raises Ni contamination issues [23]. Climatic changes exacerbate the problems in the soil with respect to plant uptake as well as availability of Ni. Drastic rains and other extreme weather conditions also influence the mobility of Niand and thus necessitate adaptative techniques for managing the soil [66]. Research into genetic variants, stress pathways, and processes that could lead to more tolerant types is expanding the understanding of how the plant responds to Ni stress. Crop tolerance must also be enhanced in polluted settings through an understanding of how Ni impacts nutrient interactions and metabolic pathways of the plant [25]. There is also research going on in regards to recent studies of soil microbiome function that is related to plant health and Ni cycling, which may improve remediation by stimulating desired microbial functions [79]. Some of the most promising applications of biochar, nanomaterials, and electrokinetic remediation are metal sequestration and decontamination [82]. Improving leaching and washing techniques contribute to increased productivity by decreasing the impairment of the environmental surroundings [31]. Monitoring and detection techniques have reached a better level by considering real-time sensors and remote sensing technologies. It can lead to an extensive tracking of plant health as well as soil contamination [59].

7 Conclusion

Soil pollution of Ni seriously hampers the health of the environment and agriculture. The crucial knowledge behind effective management is the dynamics of Ni uptake, toxicity, and tolerance. Plants absorb Ni through root, and the uptake is controlled by the pH of the soil, quantity of organic matter present, and the availability of nutrients. The effects of these accumulations are toxic, with visible chlorosis, necrosis, and stunted development shown in plants by excessive concentrations of Ni. These effects result in disturbances in nutrient uptake, reduced crop productivity and yield, and disruption of vital physiological activities such as respiration and photosynthesis. In the remediation of Ni-polluted soils, in many cases, the use of integrated approaches is a necessity, considering the amalgamation of several approaches. Among the integrated approaches is the combination of chemical additives and electrokinetic remediation, soil washing, and phytoremediation. Crop rotation and the selection of Ni-tolerant crops are examples of sustainable agriculture techniques that increase the productivity and health of the soil. It should ensure that the cleanup costs and environmental benefits are balanced in order to ensure that methods for managing the soil are effective while long-term viable. Future technological developments and continuing research will chart the future course of Ni research and farming. New problems, particularly new sources of pollution and other effects due to climate change, will require new creative solutions. Advances in biochar, genetic engineering, and monitoring in real time will improve effectiveness of remediation techniques. Multifaceted solutions for Ni contamination would include sustainable behaviors, cooperative research, and advancements in technology. Management and restoration of contaminated soils will be supported by ensuring evidence-grounded practices and policies and by interacting with stakeholders.

Competing Interests The authors declare that they have no competing interest.

Declaration of Generative AI and AI-Assisted Technologies in the Writing Process During the preparation of this work, the book chapter author(s) have not taken any assistance from AI, and the book chapter author(s) reviewed and edited the content as needed and take(s) full responsibility for the content of the publication.

References

1. Agrawal B, Czymmek KJ, Sparks DL, Bais HP (2013) Transient Influx of Ni in root mitochondria modulates organic acid and reactive oxygen species production in Ni hyperaccumulator *Alyssum murale*. J Biol Chem 288(10):7351–7362
2. Ahmad MS, Ashraf M (2011) Essential roles and hazardous effects of Ni in plants. Rev Environ Contam Toxicol 214:125–167
3. Ahmar S, Gill RA, Jung KH, Faheem A, Qasim MU, Mubeen M, Zhou W (2020) Conventional and molecular techniques from simple breeding to speed breeding in crop plants: recent advances and future outlook. Int J Mol Sci 21(7):2590
4. Alonso F, Riente P, Yus M (2011) Ni nanoparticles in hydrogen transfer reactions. Acc Chem Res 44(5):379–391
5. Amjad M, Raza H, Murtaza B, Abbas G, Imran M, Shahid M, Naeem MA, Zakir A, Iqbal MM (2019). Ni Toxicity Induced Changes in Nutrient Dynamics and Antioxidant Profiling in Two Maize (*Zea mays* L.) Hybrids. Plants (Basel, Switzerland), 9(1): 5.
6. Ansari M, Devi BM, Sarkar A, Chattopadhyay A, Satnami L, Balu P, Choudhary M, Shahid MA, Jailani AAK (2023) Microbial exudates as biostimulants: role in plant growth promotion and stress mitigation. J Xenobiot 13(4):572–603
7. Ashraf MA, Maah MJ, Yusoff I (2014) Soil contamination, risk assessment and remediation. Environ Risk Assess Soil Contam 1:3–56
8. Bai C, Reilly CC, Wood BW (2006) Ni deficiency disrupts metabolism of ureides, amino acids, and organic acids of young pecan foliage. Plant Physiol 140(2):433–443
9. Begum W, Rai S, Banerjee S, Bhattacharjee S, Mondal MH, Bhattarai A, Saha B (2022) A comprehensive review on the sources, essentiality and toxicological profile of Ni. RSC Adv 12(15):9139–9153
10. Brown PH, Welch RM, Cary EE (1987) Ni: a micronutrient essential for higher plants. Plant Physiol 85(3):801–803
11. Cameron KS, Buchner V, Tchounwou PB (2011) Exploring the molecular mechanisms of Ni-induced genotoxicity and carcinogenicity: a literature review. Rev Environ Health 26(2):81–92
12. Carter EL, Flugga N, Boer JL, Mulrooney SB, Hausinger RP (2009) Interplay of metal ions and urease. Metallomics: Integr Biometal Sci 1(3):207–221
13. Chojnacka K, Samoraj M, Tuhy Ł, Michalak I, Mironiuk M, Mikulewicz M (2018) Using XRF and ICP-OES in biosorption studies. Molecules 23(8):2076
14. Collin F (2019) Chemical basis of reactive oxygen species reactivity and involvement in neurodegenerative diseases. Int J Mol Sci 20(10):2407
15. De Macedo FG, Bresolin JD, Santos EF, Furlan F, Lopes da Silva WT, Polacco JC, Lavres J (2016) Ni availability in soil as influenced by liming and its role in soybean nitrogen metabolism. Front Plant Sci 7:1358
16. Dvořák P, Krasylenko Y, Zeiner A, Šamaj J, Takáč T (2021) Signaling toward reactive oxygen species-scavenging enzymes in plants. Front Plant Sci 11:618835
17. Economou-Eliopoulos M,Laskou M, Eliopoulos DG,Megremi I,Kalatha S, Eliopoulos GD (2021) Origin of critical metals in Fe–Ni laterites from the Balkan Peninsula: opportunities and environmental risk. Minerals 11:1009
18. El-Naggar A, Ahmed N, Mosa A, Niazi NK, Yousaf B, Sharma A, Chang SX (2021) Ni in soil and water: sources, biogeochemistry, and remediation using biochar. J Hazard Mater 419:126421

19. Fabiano CC, Tezotto T, Favarin JL, Polacco JC, Mazzafera P (2015) Essentiality of Ni in plants: a role in plant stresses. Front Plant Sci 6:754
20. Fanzo J, Bellows AL, Spiker ML, Thorne-Lyman AL, Bloem MW (2021) The importance of food systems and the environment for nutrition. Am J Clin Nutr 113(1):7–16
21. Frazier AE, Thorburn DR (2012) Biochemical analyses of the electron transport chain complexes by spectrophotometry. In: Mitochondrial disorders: biochemical and molecular analysis, pp 49–62
22. Galluzzi L, Kepp O, Krautwald S, Kroemer G, Linkermann A (2014) Molecular mechanisms of regulated necrosis. Semin Cell & Dev Biol 35:24–32
23. Genchi G, Carocci A, Lauria G, Sinicropi MS, Catalano A (2021) Ni: human health and environmental toxicology. Int J Environ Res Public Health 17(3):679
24. Ghasemi R, Ghaderian SM, Krämer U (2009) Interference of Ni with copper and iron homeostasis contributes to metal toxicity symptoms in the Ni hyperaccumulator plant Alyssum inflatum. New Phytol 184(3):566–580
25. Godoy F, Olivos-Hernández K, Stange C, Handford M (2021) Abiotic stress in crop species: improving tolerance by applying plant metabolites. Plants (Basel). 10(2):186
26. Grennan AK (2009) Identification of genes involved in metal transport in plants. Plant Physiol 149(4):1623–1624
27. Hamzah Saleem M, Usman K, Rizwan M, Al Jabri H, Alsafran M (2022) Functions and strategies for enhancing zinc availability in plants for sustainable agriculture. Front Plant Sci. 13:1033092
28. Hasan MK, Cheng Y, Kanwar MK, Chu XY, Ahammed GJ, Qi ZY (2017) Responses of plant proteins to heavy metal stress-a review. Front Plant Sci 8:1492
29. Hasanuzzaman M, Bhuyan MHMB, Anee TI, Parvin K, Nahar K, Mahmud JA, Fujita M (2019) Regulation of ascorbate-glutathione pathway in mitigating oxidative damage in plants under abiotic stress. Antioxidants (Basel). 8(9):384
30. Hu Y, Wang J, Yang Y, Li S, Wu Q, Nepovimova E, Kuca K (2024) Revolutionizing soil heavy metal remediation: cutting-edge innovations in plant disposal technology. Sci Total Environ 170577
31. Hussain MI, Khan ZI, Ahmad K, Naeem M, Ali MA, Elshikh MS, Yang HH (2024) Toxicity and bioassimilation of lead and Ni in farm ruminants fed on diversified forage crops grown on contaminated soil. Ecotoxicol Environ Saf 283:116812
32. Jakubus M, Graczyk M (2020) Availability of Ni in soil evaluated by various chemical extractants and plant accumulation. Agronomy 10:1805
33. Jensen RG (2000) Activation of Rubisco regulates photosynthesis at high temperature and CO2. Proc Natl Acad Sci USA 97(24):12937–12938
34. Kajla M, Roy A, Singh IK, Singh A (2023) Regulation of the regulators: transcription factors controlling biosynthesis of plant secondary metabolites during biotic stresses and their regulation by miRNAs. Front Plant Sci 14:1126567
35. Khan NM, Ali A, Wan Y, Zhou G (2024) Genome-wide identification of heavy-metal ATPases genes in Areca catechu: investigating their functionality under heavy metal exposure. BMC Plant Biol 24(1):484
36. Khare N, Patani A, Khare P, Singh S (2024) Recent updates on the use of smart nanostructures for food packaging applications. In: Metal and metal-oxide based nanomaterials. smart nanomaterials technology, pp 133–155
37. Kleczkowski LA, Igamberdiev AU (2021) Magnesium signaling in plants. Int J Mol Sci 22(3):1159
38. Kotyk A (2012) Cell membrane transport: principles and techniques. Springer Science & Business Media
39. Koul B, Taak P (2018) Biotechnological strategies for effective remediation of polluted soils. Springer
40. Kumar S, Wang M, Liu Y, Fahad S, Qayyum A, Jadoon SA, Chen Y, Zhu G (2022) Ni toxicity alters growth patterns and induces oxidative stress response in sweetpotato. Front Plant Sci 13:1054924

41. Kumari VV, Banerjee P, Verma VC, Sukumaran S, Chandran MAS, Gopinath KA, Venkatesh G, Yadav SK, Singh VK, Awasthi NK (2022) Plant nutrition: an effective way to alleviate abiotic stress in agricultural crops. Int J Mol Sci 23(15):8519
42. Küpper H, Lombi E, Zhao FJ, Wieshammer G, McGrath SP (2001) Cellular compartmentation of Ni in the hyperaccumulators Alyssum lesbiacum, Alyssum bertolonii and Thlaspigoesingense. J Exp Bot 52(365):2291–300
43. Leitenmaier B, Küpper H (2013) Compartmentation and complexation of metals in hyperaccumulator plants. Front Plant Sci 4:374
44. Léon V, Rabier J, Notonier R, Barthelémy R, Moreau X, Bouraïma-Madjèbi S, Viano J, Pineau R (2005) Effects of three Ni salts on germinating seeds of Grevillea exul var. rubiginosa, an endemic serpentine Proteaceae. Ann Bot 95(4):609–618
45. Leškovï A, Zvarï K M, Araya T, Giehl RFH (2020) Ni toxicity targets cell wall-related processes and pin2-mediated auxin transport to inhibit root elongation and gravitropic responses in Arabidopsis. Plant Cell Physiol 61(3):519–535
46. Li J, Cao X, Jia X, Liu L, Cao H, Qin W, Li M (2021) Iron deficiency leads to chlorosis through impacting chlorophyll synthesis and nitrogen metabolism in Areca catechu L. Front Plant Sci 12:710093
47. Li S (2023) Novel insight into functions of ascorbate peroxidase in higher plants: More than a simple antioxidant enzyme. Redox Biol 64:102789
48. Lone MI, He ZL, Stoffella PJ, Yang XE (2008) Phytoremediation of heavy metal polluted soils and water: progresses and perspectives. J Zhejiang Univ Sci B 9(3):210–220
49. Mandel LJ, Balaban RS (1981) Stoichiometry and coupling of active transport to oxidative metabolism in epithelial tissues. Am J Physiol-Renal Physiol 240(5):F357–F371
50. Massoura ST, Echevarria G, Becquer T, Ghanbaja J, Leclerc-Cessac E, Morel JL (2006) Control of Ni availability by Ni bearing minerals in natural and anthropogenic soils. Geoderma 136(1–2):28–37
51. Misra V, Mall AK, Pandey H, Srivastava S, Sharma A (2023) Advancements and prospects of CRISPR/Cas9 technologies for abiotic and biotic stresses in sugar beet. Front Genet 14:1235855
52. Muhammad I, Shalmani A, Ali M, Yang QH, Ahmad H, Li FB (2021) Mechanisms regulating the dynamics of photosynthesis under abiotic stresses. Front Plant Sci 11:615942
53. Mustafa A, Zulfiqar U, Mumtaz MZ, Radziemska M, Haider FU, Holatko J, Brtnicky M (2023) Ni (Ni) phytotoxicity and detoxification mechanisms: a review. Chemosphere 328:138574
54. Neves MI, Prajapati DH, Dekok LJ (2017) Manganese toxicity hardly affects sulfur metabolism in brassicarapa. In: Sulfur metabolism in higher plants - fundamental, environmental and agricultural aspects. Springer
55. Ning X, Lin M, Huang G, Mao J, Gao Z, Wang X (2023) Research progress on iron absorption, transport, and molecular regulation strategy in plants. Front Plant Sci 14:1190768
56. Nishida S, Aisu A, Mizuno T (2012) Induction of IRT1 by the Ni-induced iron-deficient response in Arabidopsis. Plant Signal Behav 7(3):329–331
57. Olaniran AO, Balgobind A, Pillay B (2013) Bioavailability of heavy metals in soil: impact on microbial biodegradation of organic compounds and possible improvement strategies. Int J Mol Sci 14(5):10197–228
58. Pandey AK, Zorić L, Sun T, Karanović D, Fang P, Borišev M, Wu X, Luković J, Xu P (2022) The anatomical basis of heavy metal responses in legumes and their impact on plant-rhizosphere interactions. Plants (Basel) 11(19):2554
59. Patani A, Balram D, Yadav VK, Lian KY, Patel A, Sahoo DK (2023) Harnessing the power of nutritional antioxidants against adrenal hormone imbalance-associated oxidative stress. Front Endocrinol 14:1271521
60. Patani A, Patel A, Prajapati D, Khare N, Singh S (2023) Various agriculture crop plant-based bioactive compounds and their use in nanomaterial synthesis and applications. In: Husen A (ed) Nanomaterials from agricultural and horticultural products, pp 223–241
61. Peng JS, Gong JM (2014) Vacuolar sequestration capacity and long-distance metal transport in plants. Front Plant Sci 5:19

62. Pishchik V, Mirskaya G, Chizhevskaya E, Chebotar V, Chakrabarty D (2021) Ni stress-tolerance in plant-bacterial associations. PeerJ 9:12230
63. Prajapati D, Patani A, Jain T, Patel A, Singh S (2023) ROS responsive silica nanoparticles for controlled and targeted drug delivery. In: Chawla S, Singh S, Husen A (eds) Smart nanomaterials targeting pathological hypoxia. smart nanomaterials technology, pp 327–346
64. Prajapati DH, Ausma T, De Boer J, Hawkesford MJ, De Kok LJ (2020) Nickel toxicity on Brassica rapa seedlings. J Cultiv Plants 72(9):473–478
65. Pugnaire FI, Morillo JA, Peñuelas J, Reich PB, Bardgett RD, Gaxiola A, Wardle DA, van der Putten WH (2019) Climate change effects on plant-soil feedbacks and consequences for biodiversity and functioning of terrestrial ecosystems. Sci Adv 5(11):eaaz1834
66. Rabinovich A, Di R, Lindert S, Heckman J (2024) Ni and soil fertility: review of benefits to environment and food security. Environments 11(8):177
67. Rai PK, Lee SS, Zhang M, Tsang YF, Kim KH (2019) Heavy metals in food crops: Health risks, fate, mechanisms, and management. Environ Int 125:365–385
68. Rajput VD, Harish, Singh RK, Verma KK, Sharma L, Quiroz-Figueroa FR, Meena M, Gour VS, Minkina T, Sushkova S, Mandzhieva S (2021) Recent developments in enzymatic antioxidant defence mechanism in plants with special reference to abiotic stress. Biology (Basel) 10(4):267
69. Reddam A, McLarnan S, Kupsco A (2022) Environmental chemical exposures and mitochondrial dysfunction: a review of recent literature. Curr Environ Health Reports 9(4):631–649
70. Robbins NE 2nd, Trontin C, Duan L, Dinneny JR (2014) Beyond the barrier: communication in the root through the endodermis. Plant Physiol 166(2):551–559
71. Roohani N, Hurrell R, Kelishadi R, Schulin R (2013) Zinc and its importance for human health: an integrative review. J Res Med Sci 18(2):144–157
72. Sabreena Hassan S, Bhat SA, Kumar V, Ganai BA, Ameen F (2022) Phytoremediation of heavy metals: an indispensable contrivance in green remediation technology. Plants (Basel, Switzerland) 11(9):1255
73. Sachdev S, Ansari SA, Ansari MI, Fujita M, Hasanuzzaman M (2021) Abiotic stress and reactive oxygen species: generation, signaling, and defense mechanisms. Antioxidants (Basel, Switzerland) 10(2):277
74. Saeed Q, Xiukang W, Haider FU, Kučerik J, Mumtaz MZ, Holatko J, Naseem M, Kintl A, Ejaz M, Naveed M, Brtnicky M, Mustafa A (2021) Rhizosphere bacteria in plant growth promotion, biocontrol, and bioremediation of contaminated sites: a comprehensive review of effects and mechanisms. Int J Mol Sci. 22(19):10529
75. Segawa S, Shibamoto M, Ogawa M, Miyake S, Mizumoto K, Ohishi A, Nishida K, Nagasawa K (2014) The effect of divalent metal cations on zinc uptake by mouse Zrt/Irt-like protein 1 (ZIP1). Life Sci 113(1–2):40–44
76. Seregin IV, Kozhevnikova AD (2023) Phytochelatins: sulfur-containing metal(loid)-chelating ligands in plants. Int J Mol Sci. 24(3):2430
77. Shahzad B, Tanveer M, Rehman A, Cheema SA, Fahad S, Rehman S, Sharma A (2018) Ni; whether toxic or essential for plants and environment-a review. Plant Physiol Biochem 132:641–651
78. Sharma JK, Kumar N, Singh NP, Santal AR (2023) Phytoremediation technologies and their mechanism for removal of heavy metal from contaminated soil: an approach for a sustainable environment. Front Plant Sci 14:1076876
79. Skuza L, Szućko-Kociuba I, Filip E, Bożek I (2022) Natural molecular mechanisms of plant hyperaccumulation and hypertolerance towards heavy metals. Int J Mol Sci 23(16):9335
80. Song Y, Leonard SW, Traber MG, Ho E (2009). Zinc deficiency affects DNA damage, oxidative stress, antioxidant defenses, and DNA repair in rats. J Nutr 139(9):1626–1631
81. Srivastava A, Prajapati D, Singh S, Jain T (2023) Novel strategies in radiotherapy to reduce hypoxia using nanomaterials. In: Smart nanomaterials targeting pathological hypoxia. Springer Nature Singapore, pp 115–129
82. Stuiver CEE, Zuidersma EI, Aghajanzadeh T, Prajapati DH, Parmar S, Hawkesford MJ (2016) Molybdate toxicity and sulfur and nitrogen metabolism in Chinese cabbage. Berichteaus dem Julius Kühn-Institut 185:54

83. Sun Y, Luo J, Feng P, Yang F, Liu Y, Liang J, Wang H, Zou Y, Ma F, Zhao T (2022) *MbHY5-MbYSL7* mediates chlorophyll synthesis and iron transport under iron deficiency in *Malus baccata*. Front Plant Sci 13:1035233
84. Tan X, Li K, Wang Z, Zhu K, Tan X, Cao J (2019) A review of plant vacuoles: formation, located proteins, and functions. Plants (Basel) 8(9):327
85. Tikkanen M, Mekala NR, Aro EM (2014) Photosystem II photoinhibition-repair cycle protects photosystem I from irreversible damage. Biochem Biophys Acta 1837(1):210–215
86. Vithanage M, Kumarathilaka P, Oze C, Karunatilake S, Seneviratne M, Hseu ZY, Gunarathne V, Dassanayake M, Ok YS, Rinklebe J (2019) Occurrence and cycling of trace elements in ultramafic soils and their impacts on human health: a critical review
87. Wahab A, Muhammad M, Munir A, Abdi G, Zaman W, Ayaz A, Khizar C, Reddy SPP (2023) Role of arbuscular mycorrhizal fungi in regulating growth, enhancing productivity, and potentially influencing ecosystems under abiotic and biotic stresses. Plants (Basel) 12(17):3102
88. Wang L, Hu C, Shao L (2017) The antimicrobial activity of nanoparticles: present situation and prospects for the future. Int J Nanomedicine 12:1227–1249
89. Wang Y, Branicky R, Noë A, Hekimi S (2018) Superoxide dismutases: dual roles in controlling ROS damage and regulating ROS signaling. J Cell Biol 217(6):1915–1928
90. Wang J, Delavar MA (2023) Techno-economic analysis of phytoremediation: a strategic rethinking. Sci Total Environ 165949
91. Wang X, Ai S, Liao H (2023) Deciphering interactions between phosphorus status and toxic metal exposure in plants and rhizospheres to improve crops reared on acid soil. Cells 12(3):441
92. Wang Y, Mostafa S, Zeng W, Jin B (2021) Function and mechanism of jasmonic acid in plant responses to abiotic and biotic stresses. Int J Mol Sci 22(16):8568
93. Weyh C, Krüger K, Peeling P, Castell L (2022) The role of minerals in the optimal functioning of the immune system. Nutrients 14(3):644
94. Williams BK, Brown ED (2014) Adaptive management: from more talk to real action. Environ Manag 53(2):465–479
95. Wilschefski SC, Baxter MR (2019) Inductively coupled plasma mass spectrometry: introduction to analytical aspects. Clin Biochem Rev 40(3):115–133
96. Wuana RA, Okieimen FE (2011) Heavy metals in contaminated soils: a review of sources, chemistry, risks and best available strategies for remediation. Int Sch Res Not 1:402647
97. Yu H, Li W, Liu X (2024) Physiological and molecular bases of the Ni toxicity responses in tomato. Stress Biol 4:25
98. Zheng Y, Lu J, Zhang Z, Li Y, Tan Y, Cai W, Ma C, Chen F (2022) Effect of low-molecular organic acids on the migration characteristics of Ni in reclaimed soil from the panyi mine area in China. Toxics 10(12):798
99. Zoghi P, Mafigholami R (2023) Optimisation of soil washing method for removal of petroleum hydrocarbons from contaminated soil around oil storage tanks using response surface methodology. Sci Rep 13(1):15457

Silicon: Uptake, Toxicity, and Tolerance in Crop Plants and Management of Si-Contaminated Soil

Ishita Jayesh Oza, Amit B. Verma, Nimisha Patel, Ahmad Ali, and Hina Alim

Abstract Silicon (Si) plays an important role in plant physiology, especially in improving resistance to various biotic and abiotic stresses. This present study deals with treatment options for silicon-contaminated soils, as well as examining the mechanisms of silicon uptake in plants, potential toxicity, and tolerance mechanisms. Some transporters protein, such as Lsi1 (Low silicon rice 1) and Lsi2 (Low silicon rice 2), facilitate the uptake of Si by plants, which are affected by variables such as soil pH and water availability. Si can reduce absorption problems by reducing the bioavailability and tissue retention of heavy metals such as lead and cadmium. Silicon also increases plant resistance to salt and aluminum stress by improving ionic balance and reducing the utilization of harmful ions. The health benefits and proper management of silica are important because high levels of silica in soil can be harmful. Crop management such as adding straw and soil amendments such as biochar and organic matter can help to control silica levels and reduce risk. This book chapter highlights the important role of Si in sustainable agriculture and provides suggestions for its use to improve soil health and crop production.

Keywords Silicon uptake · Silicon toxicity · Crop tolerance · Heavy metal mitigation · Biochar

1 Introduction

The eighth most abundant silicon element in the earth's crust and is rarely found in pure form. It is widely distributed in cosmic dust, planets, and interplanetary objects in space as various forms of silica or silicate. It is important to learn how plants use

I. J. Oza · A. B. Verma · A. Ali · H. Alim (✉)
Department of Life Sciences, University of Mumbai, Vidyanagari Campus, Mumbai, India
e-mail: hina.alim@lifesciences.mu.ac.in

N. Patel
Department of Life Sciences, J C Bose University of Science and Technology, YMCA, Faridabad, Haryana, India

A. Husen (ed.), *Enhanced Crop Production Setup Under Various Micro and Nano Contaminants*, Smart Nanomaterials Technology,
https://doi.org/10.1007/978-981-96-6099-5_9

silicon, the potential dangers of high silicon production, the process by which plants develop silicon tolerance, and how to manage silicon-rich soil.

1.1 Importance of Silicon in Plant Physiology

Silicon often prescribed as the supreme element and plays an important role in a plant physiology as a treatment of stress tolerance, although it is not considered an essential nutrient for all plants. Its importance in corporate nutrition has been highlighted in recent decades, particularly in the context of stress physiology. As the global demand for agriculture increases, land scarcity will also increase due to population growth, insufficient supply, and increasing land density. While poor soil management causes problems such as soil pollution disease, and soil erosion is important for soil management and security. The importance of protecting the environment includes population awareness of food and food product [1, 2, 31]. Environmental factors such as drought, temperature, and soil salinity can negatively affect plant growth and cause stress, thereby reducing agricultural productivity and quality [23, 27]. It is now known that silicon is beneficial to plants. It increases integrity and resistance to damage by forming amorphous silica in cell walls, thus increasing plant strength and hardiness. Silicon also increases drought tolerance by enabling stomata to close and reducing water loss. It also increases resistance to pests and diseases by activating protective molecules, creating a physical defense system for the plant. Silicon can also promote root growth and improve soil structure by improving nutrient uptake and utilization. In general, silicon plays an important role in increasing plant resistance to biotic and abiotic stresses and promoting the formation of cell walls, thus improving plant resilience and health [17, 21, 26].

1.2 Beneficial Effects of Silicon on Plants

Silicon has long been considered as a non-essential element for plant growth and production [5]. It has many benefits for plants: (a) It makes plants strong and vigorous—Silicon accumulates in the form of amorphous silica in the walls of plants, forming a hard layer that strengthens the structure of the plant. This improvement in integrity provides greater support for the plant and better protection against mechanical stress from weather conditions such as wind and rain. Another benefit of strong cell walls is resistance to lodging, a problem in cereal plants when stems bend or break under the weight of the grain. Silicon may also help to reduce lead toxicity in barley plants [11]. (b) Improve drought tolerance—It helps maintain water balance by disrupting the closure of stomata, the small parts of leaves that regulate water vapor exchange. Plants with high silica content can control the opening and closing of their stomata, thus reducing water quality and increasing their ability to withstand heat stress. Silicon can also help protect tissues from oxidative damage caused by stress. (c) Increased diseases and pests resistance—The defense system formed by silicon in plant walls prevents insects and other pests from feeding on the plant. Silicon can also cause plants to produce antibodies such as phytoalexins, which help to protect plants from infection. Silicon can also increase plant resistance to various diseases. (d) Improved nutrition and utilization—Silica improves the plant's ability to absorb and utilize important nutrients such as potassium and phosphorus. This is thought to be due to silica's ability to improve soil structure and nutrient availability, as well as being beneficial for root growth and development. Silicon also promotes the absorption and utilization of nutrients, (e) promotes the formation of cell walls, and protects plants from diseases—As mentioned earlier, silicon is an important part of plant walls thus help to improving the overall health and growth of plants. Silica deposits in the cell wall help to strengthen the plant's defenses mechanism, creating a barrier against diseases and pests. Si can also enhance plant's resistance to biotic and abiotic stresses by promoting the activation of defense genes and pathways. Si may also contribute to the activation of defence-related genes and pathways, thereby further strengthening the plant's resistance to biotic and abiotic stresses [8, 15, 33].

2 Silicon Uptake in Crop Plants

2.1 Mechanisms of Silicon Uptake

Silicon (Si) is recognized as a beneficial element for rice crop growth, as it can increase production by alleviating both biotic and abiotic stresses during cultivation [7]. The main form of silicon that is taken through the roots of plant is mono-silicic acid. Particularly transport proteins, such as Lsi1, an aquaporin family member that permits passive silicon migration into plant cells, and Lsi2, an efflux transporter that transfers silicon from root cells into the xylem for transportation to other plant parts, aid in this uptake. Because silicon and water are translocated through the xylem of the plant, the rate of transpiration has a significant impact on the distribution of silicon inside the plant [20] (Fig. 1).

Several key factors influence the uptake of silicon by plants, starting with the availability of silicon in the soil. The concentration of Si, primarily present as silicic acid, determines its uptake; soils rich in organic matter and clay typically exhibit higher Si availability, while sandy soils with minimal organic matter often have limited Si, especially in regions lacking Si-rich parent materials. Soil pH and temperature further impact Si availability, as silicic acid is more soluble in acidic conditions (pH 5–6) compared to alkaline soils (pH 7–8), and higher temperatures can enhance its solubility. Water availability is also crucial, as adequate moisture facilitates the transport of silicic acid to plant roots; however, drought conditions can hinder Si uptake due to reduced silicic acid availability and restricted root growth, while excessive waterlogging can limit oxygen diffusion to roots, creating unfavorable conditions for Si absorption. Additionally, plant species and genotypes play a significant role,

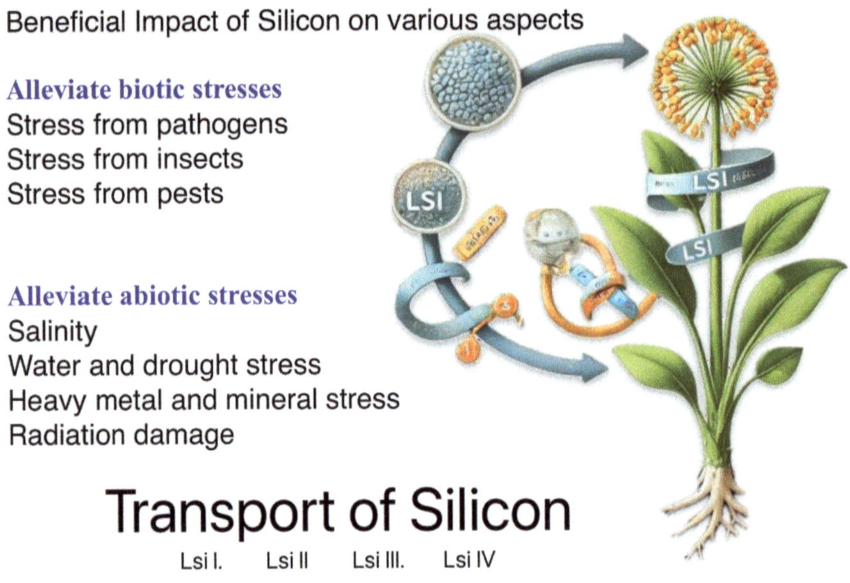

Fig. 1 Uptake of silicon with the help of LSI

with certain "Si accumulators" like rice, wheat, barley, and sugarcane demonstrating enhanced Si uptake capabilities. Genetic variation within species can further influence Si absorption based on root morphology and physiological traits. Lastly, while Si is not commonly applied as a fertilizer, the use of specific fertilizers can indirectly affect Si uptake; fertilizers that acidify the soil can increase silicic acid solubility, and those promoting root development can enhance the root system's surface area for better Si absorption [8, 15, 33].

2.2 *Silicon Accumulators*

Silicon accumulators vary significantly in their silicon content across different plant tissues, primarily due to chemical transformations during storage, deposition, and transport, as well as differences in uptake and transport mechanisms [8, 16]. Based on silicon uptake relative to water, plants are classified into three categories: active silicon accumulators, such as rice, wheat, and barley, which take up more silicon than water, passive accumulators, like oats, which have similar uptake rates; and rejective accumulators with lower uptake rates [13, 30]. The uptake mechanism involves specific trans-membrane proteins known as silicon transporters, which are passive and do not require energy for the transport of silicic acid across the plasma membrane. In addition, the use of silica-accumulating plants in phytoremediation provides a way to control silica levels by removing excess silica from contaminated soil. Plants known to have high silicon content include rice *(Oryza sativa)*. These plants store large amounts of silica in their leaves and stems to improve integrity and protect against pests and diseases. Wheat (*Triticum aestivum*) which provides comparable benefits [19], barley (*Hordeum vulgare*, which effectively accumulates silicon; and sugarcane (*Saccharum officinarum*), which is known for its increased stem strength and improved resistance to lodging. On the other hand, cucumber (*Cucumis sativa)* is classified as a silicon-storing plant [19]. This means that it can absorb and retain silicon, which can lead to more infection and stress. In addition, plant genotypes (which may differ in their ability to store silica), soils with higher silicate mineral concentrations (improve health), and agricultural practices that affect total silica (such as cultivation, fertilization, and watering) are important factors affecting silicon accumulation [10, 14, 28].

3 Toxicity

High concentrations of silica in soil can be toxic to some plants. Silica poisoning can cause symptoms such as stunted growth, leaf necrosis, and root damage. Detoxification mechanisms in plants have developed mechanisms to detoxify excess silicon. This process involves the production of silicon materials such as silicates and silica polymers, which may be less harmful than pure silica [17].

4 Role of Silicon in Mitigating Toxicity

4.1 Mitigation of Heavy Metal Toxicity

The ability of silicon to mitigate the toxic effects of heavy metals such as lead (Pb), cadmium (Cd), and arsenic (As) in plants. Silicon can form complexes with these metals, reducing their bioavailability or accumulating in tissues, making them unhealthy. Application of silicon in vegetable leaves can protect plants from oxidative stress and other damage associated with heavy metal toxicity by reducing the concentration and uptake of aluminum and cadmium [22]. Salt stress, particularly from sodium chloride (NaCl), is a major abiotic stress that negatively impacts plant growth by causing ionic imbalance and osmotic stress. Silicon mitigates these effects by enhancing plant's ability to maintain higher potassium (K+) levels while reducing sodium (Na+) accumulation, thus improving plant's ionic balance. Some studies on barley (*Hordeum vulgare*) have shown that silicon supplementation can increase photosynthesis efficiency and potassium utilization, reduce sodium toxicity, and promote growth in saltwater [4]. Silicon is not just an inert substance found in plants, it is but also important for plants to recover from stress, especially heavy metal toxicity. Understanding the physical and biochemical structure of silicon can help improve environmental planning and agriculture [12].

4.2 Aluminum Toxicity and Silicon Interaction

Aluminum ($Al+^3$) is more soluble in acidic soils and is toxic to plants, inhibiting root growth and impairing nutrient uptake. To reduce these effects, silicon and aluminum act on the soil and tissues. In a study on white clover (*Trifolium repens*), silicon ointment reduced aluminum absorption and increased the acidity of organic acids such as citric acid, which chelated aluminum ions and inhibited absorption [32] (Fig. 2).

5 Mechanisms of Silicon Tolerance in Crop Plants

Silicon plays a multifaceted role in plant physiology and biochemistry, particularly in the following: It promotes various adaptations to frustration. One of the main benefits of silicon is that it protects against increased photosynthesis by increasing chlorophyll content. It protects photosynthetic organisms from damage caused by oxidative stress and stabilizes plant cell walls, increasing light absorption and energy conversion. Si also affects nutrient absorption, promotes root development, and balances nutrients in the soil. It also improves stress by reducing oxidative stress, improves water use, and rehabilitates stressed tissue roots. High concentrations of silicon can be toxic to plants, leading to several negative effects. These includes stunted growth, impaired chlorophyll production, can cause chlorosis and necrosis of leaves, nutrient imbalance and reduced crop yield due to high silicon content. Symptoms of silica toxicity include excess silica in leaves and stems, abnormal leaf growth (such as curling or curling), slow growth, and reduced yields. Factors that cause silica toxicity include high silica concentrations. Sensitivity varies with soil or water sources, nutrient imbalances, extreme pH values that affect silica solubility, plant species, and

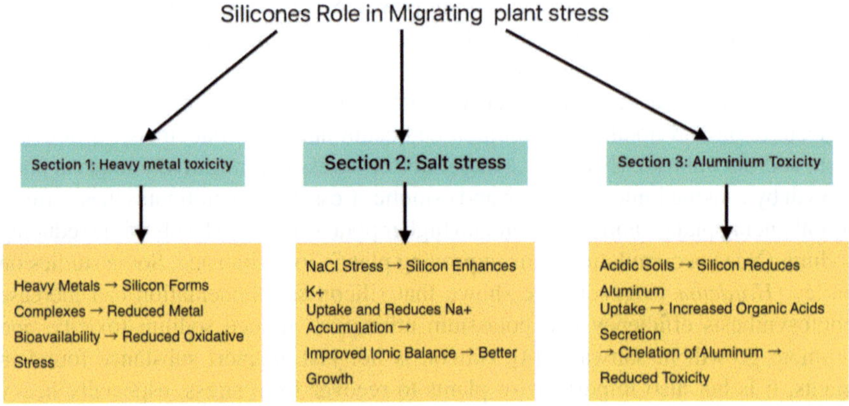

Fig. 2 Silicon's role in mitigating plant stress

genotypes. Some crops, especially rice, wheat, corn, soybeans, etc., are particularly affected by silicon toxicity due to their excessive silicon content. It will adversely affect growth and yield, especially under certain soil conditions. Si is important for plant growth and development. However, it is very important to manage it. Good silicon content to avoid toxicity. By understanding the factors that affect silica uptake and based on the toxicity and response of plants, growers can use the appropriate management method. Maximize the benefits of Si while minimizing its negative effects [8, 18, 28].

5.1 Silicon-Mediated Mechanisms of Metal Detoxification

Numerous explanations have been put out for how Si-derived benefits aid in metal detoxification. Toxic metal immobilization in the soil (at the soil level), co-precipitation of metals, metal ion chelation, compartmentation, structural changes in plant tissues, and changes in molecular responses are among the commonly acknowledged mechanisms [6].

6 Silicon's Role in the Activation of Antioxidant Enzymes

The role of silicon has been found in the activity of major antioxidant enzymes, including superoxide dismutase (SOD), catalase (CAT), and peroxidase (POD). The enzyme SOD helps to convert superoxide radicals to oxygen and hydrogen peroxide. They usually contain metal cofactor such as copper, zinc, manganese, or iron. SOD images are usually spherical or nearly spherical with metal cofactors located in the center of the structure.

Catalase decomposes hydrogen peroxide into water and oxygen. It usually contains a heme group, essential for its enzymatic activity. Catalase is often illustrated as a tetramer (composed of four subunits) with a central heme group and typically appears as a roughly spherical or donut-like shape, and peroxidases catalyze the reduction of hydrogen peroxide and other peroxides using various electron donors. They frequently contain heme groups as well. Images of peroxidases often resemble those of catalase but may vary depending on the specific type. They may present a more elongated or irregular shape compared to catalase which lessen oxidative stress brought on by environmental factors such as heavy metals, salt, and drought. Silicon deposits in plant cell walls, enhancing their structural integrity and resistance to physical harm and disease attack. Particular instances include the way in which silicon priming, even in the face of stress, significantly increased grain yield and growth in maize (*Zea mays*) when water was scarce [29] (Table 1).

Table 1 Role of silicon in activating antioxidant enzymes

Antioxidant enzyme	Function	Metal cofactor	Silicon's role
Superoxide Dismutase (SOD)	Converts superoxide radicals into hydrogen peroxide and oxygen	Copper, zinc, manganese, or iron	Enhances SOD activity, potentially by stabilizing the enzyme structure or providing metal cofactors
Catalase (CAT)	Decomposes hydrogen peroxide into water and oxygen	Heme group	Increases CAT activity, possibly by improving the stability of the enzyme or the heme group
Peroxidase (POD)	Catalyzes the reduction of hydrogen peroxide and other peroxides	Heme group	Enhances POD activity, potentially by stabilizing the enzyme structure or the heme group

7 Genetic and Molecular Mechanisms

Genetic factors play an important role in the uptake and transport of silicon in plants, mainly through specific proteins called silicon transporters. These proteins are encoded by genes, and their expression and activity levels vary among plants and cultivars are affected by environmental conditions such as soil silica availability, pH, and plant growth stage. The ability of plant species to store silicon and their susceptibility to silicon toxicity depend on a variety of genetic variations. This variation results from changes in the expression and activity of silicon transporters and other genes, suggesting an opportunity to develop plants with improved silicon tolerance and utilization. The primary method of silicon uptake is passive diffusion of silicon dioxide from roots. The rate of absorption depends on soil pH, silica concentration, and the activity of silica transporters [24]. Once absorbed, transport silicon reaches the roots and is ejected from cortical cells and xylem as silica [15, 25]. In the flower, silicon polymerizes to form silica, which provides structural support and protection against environmental stress. Species, developmental stage, and environmental factors influence the extent of silicon accumulation in different tissues. Some plants can accumulate silica while others cannot. Considering all these, it is important to understand the genetic and molecular factors for the use of silicon in agriculture.

8 Management of Silicon-Contaminated Soils

The amount of SiO_2 available to plants is 105–120 mg/kg during dormancy, and bioavailable silicon (Si) in the soil is generally 50–250 mg/kg. The effectiveness of silicon is affected by many factors such as soil type, pH, moisture, temperature, and the presence of ions during the absorption–desorption process. In particular, the level of silicon contained will vary among soil types, and some soil types will have silicon deficiency [12]. It is known that the use of silicon can improve the silicon cycle and reduce various environmental stress factors by providing simple silicon products to the soil [10]. Silicon is essential for the regulation of nutrition in monocots and dicots and helps plants resist biotic and abiotic stresses [12]. Plants with high silicon content in their roots or stems have strong resistance to stresses such as drought, low temperature, salinity, and heavy metals, thus reducing their susceptibility to pests. In addition to providing significant economic and environmental benefits such as improving soil health by increasing crop yields, removing metals and reducing salinity, the use of silicon fertilizers in agriculture has been shown to be beneficial as an addition to the fertilization process, increasing yields. However, more research is needed to understand the use of silica fertilizers in science [12]. However, silicon has many uses. Biochar gained much attention due to its unique characteristics such as large specific surface area, stable structure, high cation exchange capacity, and carbon content. Furthermore, the use of silicon-accumulating plants in phytoremediation provides a viable way to remove excess silicon from contaminated soils, thereby efficiently controlling silicon levels [9] to silica-contaminated soil to improve soil structure and moisture retention [3], and using advanced technologies such as nanomaterial or controlled composting. The Si delivery mission is one way to manage these risks. In addition, the use of silica-accumulating plants in phytoremediation provides a way to control silica content by removing excess silica from contaminated soil (Fig. 3).

9 Conclusion and Future Prospects

It is an essential component of contemporary agricultural practices due to its potential to improve plant resilience and productivity, particularly in reducing the negative effects of many biotic and abiotic stresses. By understanding the processes of silicon absorption, its potential toxicity, and management techniques for silicon-contaminated soils, farmers and researchers can efficiently utilize the benefits of this beneficial element. By enhancing soil health, increasing crop yields, and reducing reliance on chemical inputs, incorporating silicon into agricultural practices can ultimately support a more resilient and sustainable agricultural system. Future studies should focus on investigating how silicon interacts with other nutrients, creating more effective ways to manage silicon, and looking at potential uses of silicon in different agricultural systems. By further exploring the complexities of silicon's function in plant physiology, we can realize its full potential and help create a more productive and sustainable agricultural sector.

Fig. 3 Biochar an organic compound

Competing Interests The authors declare that they have no competing interest.

Declaration of Generative AI and AI-Assisted Technologies in the Writing Process During the preparation of this work, the book chapter author(s) have not taken any assistance from AI, and the book chapter author(s) reviewed and edited the content as needed and take(s) full responsibility for the content of the publication.

References

1. Abdul Rahmam NSN, Abdul Hamid NW and Nadarajah K (2021) Effects of antibiotic stress on soil micro biome. Int J Mol Sci 22(16). https://doi.org/10.3390/ijms22169036
2. Agullar PA, Valdes G, Araneda N, Valdebenito ER, Hansen F, Nuti M (2023) Microbial community in the composting process and its positive impact on the soil micro biota in sustainable agriculture. Agronomy 13:542. https://doi.org/10.3390/agronomy13020542
3. Ahmad R, Hadi F, Jan A, Ditta A (2022) Straw incorporation in contaminated soil enhances drought tolerance but simultaneously increases the accumulation of heavy metals in rice 14(17). https://doi.org/10.3390/su141710578
4. Akhter MS, Noreen S, Saleem N, Saeed M, Ahmad S, Khan TM, Saeed M, Mahmood S (2021) Silicon can alleviate the toxic effect of NaCl stress by improving K+ and Si uptake,

photosynthetic efficiency with reduced Na+ toxicity in barley (*Hordeum Vulgare).* SILICON 14(1):1241-252. https://doi.org/10.1007/s12633-021-01270-6

5. Bat EO, Szegő A, Gyöngyik M, Mir MI, Papp I (2021) Effects of silicon in plants with particular reference to horticultural crops. Int J Hortic Sci 27:95–105. https://doi.org/10.31421/ijhs/27/2021/9096
6. Bhat JA, Shivaraj SM, Singh P, Navadagi DB, Tripathi DK, Dash PK, Solanke AU, Sonah H, Deshmukh R (2019) Role of silicon in mitigation of heavy metal stresses in crop plants. Plants (Basel, Switzerland) 8(3):71. https://doi.org/10.3390/plants8030071
7. Chaiwong N and Prom-u-thai C (2022) Significant roles of silicon for improving crop productivity and factors affecting silicon uptake and accumulation in rice: a review. Agronomy 12(3). https://doi.org/10.1007/s42729-022-00787-y
8. Epstein E (1994) The anomaly of silicon in plant biology. Proc Natl Acad Sci 91(1):11–17. https://doi.org/10.1073/pnas.91.1.11
9. Ghulam M, Zeeshan A, Mohammad V, Iftikhar A, Muhammad U, Rashid I, Usman Z, Muhammad R, Salman M, Abd U, Muhammad A, Muhammad HR, Allah D, Akash T (2024) Recent trends and economic significance of modified/functionalized biochars for remediation of environmental pollutant, Article number: 217. https://doi.org/10.1038/s41598-023-50623-1
10. Guntzer F (2015) Silicon and plant stress tolerance. J Bot 1–13. https://doi.org/10.1155/2015/815896
11. Hammond KE, Evans DE, Hodson MJ (1995) Aluminium/silicon interactions in barley (*Hordeum vulgare* L.) seedlings. Plant Soil 173:89–95. https://doi.org/10.1007/BF00155521
12. Keqi Z, Yuan Y, Lihua Z, Jiachao Z, Yaoyu Z, Hongli H, Shuang L, Lin L (2022) Silicon-based additive on heavy metal remediation in soils: toxicological effects, remediation techniques, and perspectives. Environ Res 205. https://www.sciencedirect.com/science/article/pii/S0013935121015450
13. Liang Y, Si J, Römheld V (2005) Silicon uptake and transport is an active process in *Cucumis sativus.* New Phytol 167(3):797–804. https://doi.org/10.1111/j.1469-8137.2005.01463.x. PMID: 16101916
14. Liang Y, Nikolic M, Bélanger R, Gong H, Song A (2015) Silicon uptake and transport in plants: physiological and molecular aspects. In: Silicon in agriculture. Springer, Dordrecht. https://doi.org/10.1007/978-94-017-9978-2_4
15. Ma JF, Yamaji N (2008) Silicon uptake and accumulation in higher plants. Trends Plant Sci 11(8):392–397. https://doi.org/10.1016/j.tplants.2006.06.007. Science Direct
16. Ma JF, Miyakeb Y, Takahashi E (2001) Si as a beneficial element for crop plants. In: Datnoff LE, Snyder GH, Korndörfer GH (eds) Si in agriculture, vol 8. Elsevier, Amsterdam, pp 17–39
17. Ma JF, Yamaji N (2008) Functions and transport of silicon in plants. Cell Mol Life Sci 65(19):3049–3057. https://doi.org/10.1007/s00018-008-7580-x
18. Marschner H (2012) Marschner's mineral nutrition of higher plants. Academic, pp 651–653. ISBN: 9780123849052.
19. Miroslav N, Nina N, Yongchao L, Ernest AK, Volker R (2007) Germanium-68 as an adequate tracer for silicon transport in plants. Characterization of silicon uptake in different crop species. Plant Physiol 143(1):495–503. https://doi.org/10.1104/pp.106.090845
20. Mitani N, Ma JF (2005) Uptake system of silicon in different plant species. J Exp Bot 56(414):1255–1261. https://doi.org/10.1093/jxb/eri121
21. Murtaza G, Ahmed Z, Usman M, Deng G, Chen S, Alwahibi MS, Rizwana H, Iqbal J, Ahmad S, Iqbal R (2025) Protective effects of multiple-chemical engineered biochar on hormonal signalling, antioxidant pathways and secondary metabolites in lavender exposed to chromium and fluoride toxicity. J Crop Health. https://www.google.com/search?q=//doi.org/10.1007/s10343-025-01134-8&authuser=1
22. Ngugi M, Gitari H, Muui C, Gweyi-Onyango J (2022) Growth tolerance, concentration, and uptake of heavy metals as ameliorated by silicon application in vegetables. Int J Phytoremediation 24(6):495–504. https://doi.org/10.1080/15226514.2022.2045251
23. Onyekachi OG, Boniface OO, Gemlack NF, Nicholas N (2019) The effect of climate change on abiotic plant stress: a review. Intec Open. https://doi.org/10.5772/intechopen.82681

24. Raven JA (2001) Si transport at the cell and tissue level. In: Studies in plant science, vol 8, pp 41–55. https://doi.org/10.1016/S0928-3420(01)80007-0
25. Reynolds OL, Keeping MG, Meyer JH (2009) Si-augmented resistance of plants to herbivorous insects: a review. Ann Appl Biol 155:171–186. https://doi.org/10.1111/j.1744-7348.2009.00348.x
26. Rivero M, Cárdenas S, Benavides S, González M, Morales M, González A, Rincón S (2022) Silicon supplementation as a promising approach to induce thermotolerance in plants: current understanding and future perspectives. J Soil Sci Plant Nutr 23(2):1–22. https://doi.org/10.1007/s42729-022-00913-9&authuser=1
27. Santisree P, Jalli LC, Bhatnagar-Mathur P, Sharma KK (2020) Emerging roles of salicylic acid and jasmonates in plant abiotic stress responses. https://doi.org/10.1002/9781119552154.ch17
28. Singh VK (2017) Silicon fertilization in crop plants: a review. J Plant Nutr 40(10):1431–1446. https://doi.org/10.1080/01904167.2017.1287513
29. Sirisuntornlak N, Ullah H, Sonjaroon W, Arirob W, Anusontpornperm S, Datta A (2021) Effect of seed priming with silicon on growth, yield and nutrient uptake of maize under water-deficit stress. J Plant Nutr 44(6):794–806. https://doi.org/10.1080/01904167.2021.1884707
30. Takahashi E, Ma JF, Miyake Y (1990) The possibility of Si as an essential element for higher plants. Comments Agric Food Chem 2:99–122
31. Vincze EB, Becze A, Laslo E, Mara G (2024) Beneficial soil microbiones and their potential role in plant growth and soil fertility. Agriculture 14:152. https://doi.org/10.3390/agriculture1401015
32. Yang W, Feng H, Zhou J, Jia T, Tang T, Zhang H, Peng Y (2024) Exogenous silicon induces aluminium tolerance in white clover (*Trifolium repens*) by reducing aluminium uptake and enhancing organic acid secretion. Peer J 12:e17472. https://doi.org/10.7717/peerj.17472
33. Zhao C, McGrath SP (2002) The role of silicon in plant nutrition and crop productivity. Adv Agron 77:137–180

Nanofertilizers in Improving Soil Quality and Crop Plant Growth

Shahnoush Nayeri and Maryam Zakavi

Abstract Nanotechnology in agriculture and the global economy may pose health risks. In this work, we discuss the impact of nanofertilizers (NFs) on soil quality and crop growth. We analyze recent developments in the form of case studies and show if NFs are viable to contribute toward sustainable agriculture, improve soil quality, and enhance crop productivity. The NFs were found to supplement the nutritional values of the soil, upgrade the ecosystem, and increase the quantity of microbes in the soil. This work highlights the potential of nanofertilizers as a sustainable alternative to conventional fertilizers, emphasizing their ability to enhance nutrient delivery, improve soil quality, and promote crop growth while minimizing environmental impact. It underscores the need for further research and regulatory frameworks to ensure these innovative agricultural practices' safe and effective application.

Keywords Nanoparticle · Nanofertilizer · Plant growth · Soil quality · Sustainable agriculture

1 Introduction

With the growing demand for food production, novel agricultural practices have emerged [7]. The application of conventional chemical fertilizers, while very effective in raising crop yields, has adverse environmental consequences with respect to soil degradation, water pollution, and reduction in biodiversity [28, 39]. Nanofertilizers are another branch of nanotechnology applied in response. Nanofertilizers are engineered materials for efficient and effective nutrient delivery to plants, enabling

S. Nayeri (✉)
Department of Biology, Pennsylvania State University, University Park, PA, USA
e-mail: smn5869@psu.ac.ir

M. Zakavi
Department of Cellular and Molecular Biology, Faculty of Life Sciences and Biotechnology, Shahid Beheshti University, Tehran, Iran
e-mail: m_zakavi@sbu.ac.ir

A. Husen (ed.), *Enhanced Crop Production Setup Under Various Micro and Nano Contaminants*, Smart Nanomaterials Technology,
https://doi.org/10.1007/978-981-96-6099-5_10

the plants to increase their intake of nutrients while reducing harm to the environment. Evidence that these fertilizers are made of nanoparticles is provided by [8, 37], and [91], which have the potential to be designed in such a way as to realize controlled release for a very long period, helping them continue being available to crops [8, 37, 91]. This also allows for better quality soil because of reduced accumulation of harmful chemicals. It also improves qualitative and quantitative aspects of crop growth [8, 27].

Because of the high-surface area-to-volume ratio of the nanoparticles, they will interact more with the roots than anything else for better nutrient absorption. Among the major advantages of nano-fertilizers, probably gaining greater prominence would relate to improving the quality of the soils. Traditional fertilizers tend to create imbalances within the soils because of the build-up of excess nutrients. On the other hand, nano-fertilizers are designed such that their nutrient release is controlled in nature; as such, the possibility of nutrient leaching and runoff is reduced. This helps not only in maintaining soil fertility but also in reducing the environmental impact of agricultural practices. Besides, nano-fertilizers may improve soil structure and microbial activity, which might contribute to a more resilient and productive agricultural ecosystem [76]. Nano-fertilizers have also proved promising for the growth and development of crops. Plant growth, yield, and product quality, in return, are likely to be increased accordingly with the increased availability of nutrients through such fertilizers [6, 57]. Some research studies also pointed out the fact that nano-fertilizer increases the intake of essential vital nutrients required by the plants for their development, such as nitrogen, phosphorus, and potassium [59]. It has also been credited with enhanced pesticide and disease resistance, which contributes to the gain in the performance of crops [2, 36, 73].

It is considered that with respect to the challenges that have to be overcome for their proper dissemination and acceptance, application of nano-fertilizers in agriculture is still at the nascent stage. In this respect, urgent needs exist for comprehensive studies concerning long-term impacts of nano-fertilizers on the state of health of the soil and environmental condition, and in relation to the aforementioned, the determination of cost-effective methods of production. In addition, the regulatory frameworks have to be devised on safe and responsible use of nanotechnology in agriculture [48].

This chapter discusses the impact of nano-fertilizers on soil quality and crop growth. We analyze recent developments in the form of case studies and show if nano-fertilizers are viable to contribute toward sustainable agriculture, improve soil health, and enhance crop productivity. This work presents insight into how nano-fertilizers will transform existing agricultural practices and give a clue to responding to increasing demands for sustainability in food production.

2 Green Synthesis of Nanomaterials as Nanofertilizers

Common techniques in nanoparticle production are top-down and bottom-up methods [31, 30, 84]. The top-down approach is the breaking down of the bulk material into nano-units while the bottom-up way is building up the nanoparticles from atoms, molecules, and smaller monomers [43]. These methods include chemical processes, physical processes, green synthesis, and biological processes (Fig. 1). All the methods have their own advantages coupled with their challenges, which range from environmental conditions to expensive equipment. Green synthesis methods are more reliable, cost-effective, and easier to characterize. They bring along reduced toxicity levels and have led to exploration into microorganisms, plant extracts, graphene nanosheets, and enzymes as alternatives. This trend in the development of cost-effective and eco-friendly methods for synthesizing nanoparticles is driven by the need to improve nanoparticle production and its many applications [26]. Green synthesis is a cost-effective, nontoxic, and sustainable method for producing nanoparticles (NPs) using natural resources like plant extracts and microorganisms [25]. This process involves combining noble metal salt mixtures with biomaterials such as proteins, alkaloids, flavonoids, reducing sugars, and polyphenols serving as reducing agents [31–34, 49, 72].

2.1 *Microbial Synthesized NFs*

Microorganisms can be used as nanofactories for the quick production of NPs compared to the traditional physicochemical processes [20]. Nanomaterials from microbes are derived through intracellular and extracellular biosynthesis methods. Intracellular biosynthesis is the deposition of positively charged metal ions by elec trostatic interactions, and the latter converts metal ions to their respective NPs by the nitrate-reductase synthesis method. Microorganisms have become one of the most favored entities for biogenic–biosynthesis metallic nanoparticles (MNPs) synthesis due to their rapid growth rate, easy cultivation, and ability to grow under ambient conditions [29, 74]. Through various studies, it has been shown that MNPs are synthesized both intracellularly and extracellularly by different microorganisms, viz., bacteria, fungi, yeasts, and microalgae [13, 40, 44, 47, 79, 80]. Bacteria have been used very widely as nanofactories in the production of several metal nanoparticles because they inhabit almost every niche of the earth's environment and can thrive in extreme conditions [40, 92–96]. Microorganisms such as bacteria, actinomycetes, fungi, yeasts, and microalgae, with reductase enzymes, can accumulate and detoxify heavy metals, producing silver and gold nano particles (NPs). However, microbially produced metal nanoparticles have some limitations, including laborious procedures and downstream processing [26, 43, 49].

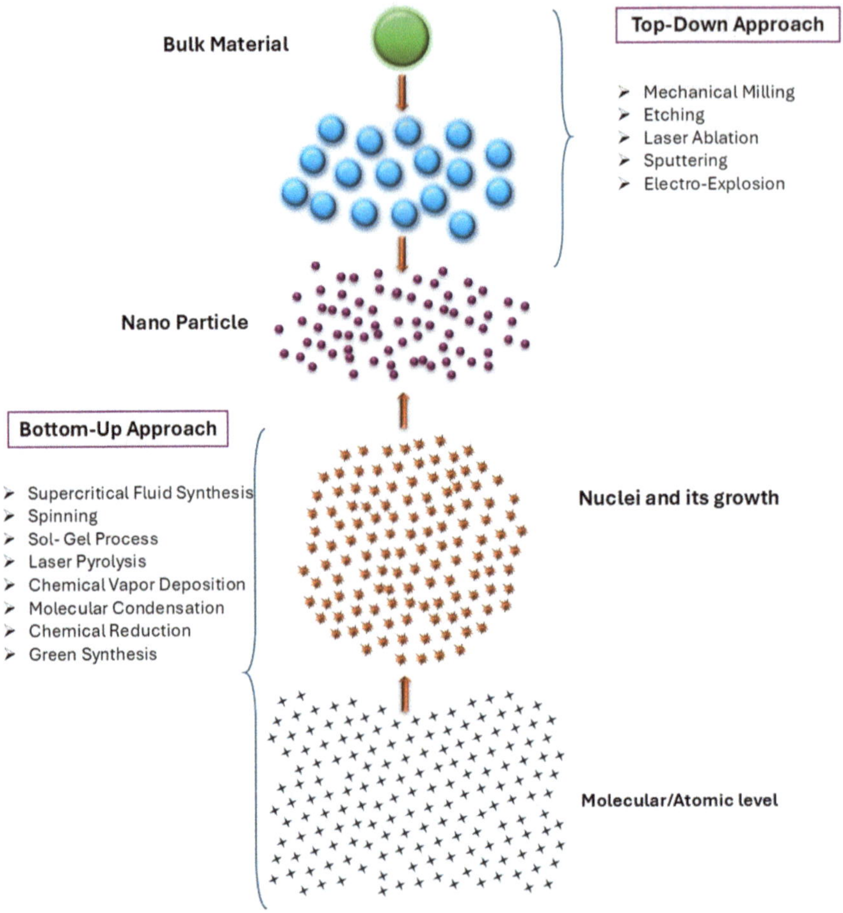

Fig. 1 The top-down and bottom-up approaches to efficient synthesis of nanomaterials via chemical, mechanical, biological, and green synthesis methods

2.2 *Plant-Mediated Synthesis*

The use of biomolecular reducing agents in plant extracts ensures the synthesis of metallic nanoparticles by the extracts, which is responsible for improved growth and performance in the plants [1]. Nanoparticles form after the cleansing, drying, boiling, and filtration of the plant matter because phytochemicals are deposited electron donors such as terpenoids, flavones, quinones, ketones, and aldehydes [1, 98]. The plant extracts contain many reducing agents that act on the capping agents, stabilizing the process of metal nanoparticle formation [1].

It has recently been reported that synthesized Fe_2O_3 NPs can promote barley seedling growth up to 42%, stimulate the root biomass by 42%, and also enhance the iron transfer from the root to the shoot [69]. The adsorption of Fe_2O_3 NPs attached to soil particles could also reduce the loss of nutrients and improve the fertilizers' economics [70].

Ahmed [77] designed a highly effective green chemistry approach for the synthesis of Zn, Mn, and Fe nano-oxides, which further improved the plant growth, performance, and quality of the squash plant [77]. Manpreet [82] published a paper based on the green synthesis of ZnO NPs, using an aqueous extract of *Azadirachta indica* leaves. The obtained ZnO NPs were applied on seed germination, which highly increased seed germination compared with control seeds [82].

The *Caotropis* plant extract can be applied to the eco-friendly synthesis of ZnO NPs, which results in significant growth acceleration [16]. *Coriandrum sativum* leaf extract due to eco-friendly synthesis of ZnO NPs has been investigated for their potential in biofertilizer application [87].

Prakashraj et al. [61] applied *Citrus medica* L. in green synthesis of nanoparticles to improve nitrogen content that is essential for good growth and development of the applied plants. The extract acts as a reducing and capping agent for the synthesis of TiO_2 NPs and particularly provides an effective role in enhancing the seed germination of *Capsicum annuum* and its diverse growth parameters [61]. Murgueitio-Herrera et al. [56] studied the potential of ZnO MnO-NPs and FeO ZnO NPs nanofertilizer nanoparticles in promoting the growth of Andean lupine and cabbage crops. It could thus be concluded from the above that the use of green-produced zinc NPs highly improved pea plant productivity and growth [56]. In another study associated with Zn NPs along with titania-treated Zn P, seed germination, and growth characteristics of amaranth, moringa leaf extract was used for its synthesis. The height of the plant and the fresh weight were higher in Zn NPs treated plants, and it was observed that nutrient recovery, yield, and productivity were also improved in that study [65]. In this way, studies have emphasized the scope of bio-compatible nano-agriculture on the plant growth parameters related to germination and growth.

3 Types of NFs Application

Foliar application, seed nanopriming, and soil treatment can be used to apply nanofertilizers. Nanofertilizers are sprayed directly on plant leaves as a foliar application and taken up by the plants very surly, thus benefitting from rapid nutrient absorption. Nanofertilizer solution NP acts as a seed nanopriming agent, which is to be applied on treated seeds before sowing aimed at enhancing the rate of germination and nutrient uptake. Stirring nano-fertilizers with soil helps in the prevention of nutrient loss (soil treatment). This relies on the quality of soil, how many nutrients are available in the land, and how climatic conditions affect nutrient uptake [90]. Knowledge regarding these factors can help increase the yield of food crops, minimize environmental contamination, and engage in ecologically friendly agricultural practices.

3.1 Foliar Applications

Nanofertilizers, upon being dusted over the foliage, are absorbed by plants through their stomata, stigma, hydathodes, and bark texture and move through symplastic or apoplastic pathways to different organs and tissues [43]. Liquid fertilizers can be absorbed through the leaves because foliar spray method is used to apply it. Ensuring optimal and precise transfer of nutrients: It employs nanofertilizer-based delivery. Nanofertilizers, fungicides, herbicides and preservatives are applied by foliar application of nanoparticles (NP), which shows exciting future perspectives. As far as leaf uptake is concerned, the mechanisms of absorption are governed by a number of factors that determine its pathway through stomata or endocytosis and efficient direct absorption but with restrictions faced in terms cell walls holdup before following into shoots. Foliar spray advantage is immediate response, better nutrient utilization and least leaching and runoff [90].

3.2 Seed Nanopriming

Seed priming is a presowing treatment that will cause physiological changes in the seeds, which will then result in quicker germination and improved growth of the plant. Half the application of nanofertilizers by immersion of seeds in them is one way of getting extraordinary results by reducing the application of fertilizers. Growth promotion by nanofertilizers occurs due to their penetration into the pores of seeds and stimulation of plant hormones. They also suppress reactive oxygen species and modulate the hormones controlling development in plants. Seed priming also leads to the induction of genes that are related to plant vigor, resulting in high resistance, especially under stressed conditions [90].

3.3 Soil Application

The current of nanofertilizers can penetrate rhizoder-mis lateral root junctions, but the former parts become impermeable as a result of the suberin, making the foliar application a better method than the soil application [43]. Nanofertilizers that were put into the soil by classical methods could affect the roots of plants and also the soil particles and alter the actions of both. They regulate the amounts of nutritive elements that are released and enhance the growth and productivity of plants. However, the method is inevitably associated with unpredictable long-term effects, higher expenses, and regulatory issues [90].

4 Classification of Nanofertilizer

Nanofertilizers are classified by their components (carbon nanotubes, polymers, or metals) or by action. These characteristics increase nutrient efficiency, control over release rate and place, enhanced plant growth stimulation as well as water and natural resource loss mitigation which makes them an important asset for sustainable agriculture. Controlled-release nanofertilizers are a new group of fertilizers, in which nutrient delivery is maintained to minimum losses and that resulting in increased nutrient use efficiency while minimizing environmental pollution. These in-carrier material encapsulated nanoparticles provide higher nutrient utilization efficiency, less environmental pollution, and increased crop productivity for which still more research is needed to feed the global population [90].

4.1 Carbon-Based NFs

It is so because carbon plays an essential role in plant growth, especially via biochar. Biochar surfaces coated with carbon nanostructures pin down micronutrients and provide water retention that increases soil moisture crucial for plant growth. Start by using carbon-based nanofertilizers like carbon nanotubes (CNTs) and graphene that may improve nutrient delivery, and absorption, as well as enhance the growth of your plants. Utilizing their high aspect ratio and chemical properties, these nanofertilizers can enhance crop productivity [90].

4.2 Chitosan-Based NFs

Chitosan nanofertilization exerts key effects on seed germination and plant growth by generating biomolecules related to plant developmental and protective processes against stressors [21]. Recent studies have demonstrated that nanofertilizers may enhance plant growth parameters of wheat, maize and rice seedlings, as well as coffee plants [15, 57]. Nanocomposite films in tropical peatland have been found to demonstrate a positive trend in growth, displaying synergistic effects on fungi and mycorrhiza. In greenhouse conditions, nanofertilizers enhance nutrient uptake, photosynthesis, and overall development of the plants, hence, they have the potential for improved fertilizer use efficiency in promoting better growth of coffee plants [10].

4.3 Clay-Based NFs

Nanoparticles made of clay have expansive surface areas and nanolayer reactivity, making them ideal for controlled-release fertilizer formulations. Organoclays demonstrate higher specific surface area, hydrogel capacity, and charge due to even smaller particle sizes than nanoparticles [90].

4.4 Layer Double Hydroxides (LDH)-Based NFs

Layered double hydroxides (LDH) are two-dimensional compounds with intercalated anionic materials, providing controlled release of anions. Their overall nature can be technologically structured through modifications in composition and synthetic approaches, thus affecting anion exchange and thermal stability. In recent studies, the hydrotalcite-like structure developed from LDH was used for phosphate fertilization and intercalated anionic herbicides to reveal their potential in obtaining slow-release formulations for acid herbicides [12].

4.5 Nanocapsule-Based NFs

Nanocapsules are small, encapsulated substances composed of organic or inorganic elements that carry and release fertilizers [23]. They slowly release the nutrient content with time, hence giving efficient utilization of applied fertilizer through controlled accurate application and maintaining nutrient stability. It is shown that with nanofertilizers containing nanocapsules, crop yields can be enhanced in such a way that photosynthetic rates of maize plants and nitrogen release were increased under cobalt stress [5, 23, 59, 90].

4.6 Nanogel-Based NFs

Nanogels are porous nitrogen, phosphorus, and potassium-bearing materials applied in fertilizers, drug delivery, and water purification. They absorb and store water, a process that confers protection against erosion. Nanogels are biodegradable and nontoxic, thus eco-friendly. In one of the studies, a type of composite fertilizer containing calcium phosphate nanogel combined with dipotassium hydrogen phosphate was found to have improved Abelmoschus esculentus germination rates, enzyme activities, and fruit weights [21, 90].

4.7 Polyurethane-Based NFs

Nanofertilizers based on polyurethane are a new generation of fertilizers utilizing polyurethane NPs for the delivery of essential nutrients into the soil. These newly designed nanofertilizers, which are fabricated by using ring-opening metathesis polymerization, showed higher efficiency and longer time of release, concomitant release of various types of nutritional elements, enhanced water-holding capacity and reduced the erosion propensity of the soil. Conventional slow-release fertilizers utilized some polymers like polyurethane for coating, but some nutrient elements could not be combined effectively with polyurethane. One such innovation in this regard would be the coating of the fertilizer with wheat straw-derived polyurethane and its liquefaction using the solvents. This biomass-based polyurethane is more economically feasible compared to synthetic polymers [53].

4.8 Starch-Based NFs

Starch-based nanofertilizers are environmentally benign, readily soluble nanocrystals applied in agriculture. In this regard, one study developed a polymeric formulation of polyvinyl alcohol starch for the gradual release of Cu–Zn micronutrient-carrying carbon nanofibers. The nanofertilizer increased plant height and demonstrated ROS-scavenging activity with a lower metal release profile in comparison with CNFs [54].

4.9 Zeolite-Based NFs

Zeolites are microporous aluminosilicate minerals that can, as soil amendments, absorb and retain water, nutrients, and organic compounds. Zeolite-based nanofertilizers could enter into the soil and get into the roots of crops to store the nutrients for crop uptake. They can be tailored to release the required combination of crop nutrients and thus reduce fertilization costs. Zeolite-based nanofertilizers have the characteristics of controlled release balanced with biodegradability and ease of modification, thus holding great promise to support sustainable agriculture [75].

5 Agro-Applications of Green Synthesized Nanoparticles

Macronutrient-based nanofertilizers have shown significant improvement in crop yield growth and productivity, obtaining an 18–29% higher nutrient use efficiency as against traditional chemical fertilizers like nitrogen, phosphorus, and potassium

[64] (Table 1). Nanocarriers such as zeolites, clay, and chitosan can release N in a controlled manner to enhance their absorption by plants [63, 66, 75]. The nano-zeolites and their blends have been widely used to develop N-NFs that outperform ordinary urea regarding maize growth, productivity, and nutrient absorption. Urea-loaded zeolite can act as a controlled-release fertilizer and increase nitrogen reserves in yield and soil hydration. It also increases soil hydration and plant-available nitrogen [75]. The hydroxyapatite nano-fertilizers have been shown to increase P usage efficiency and reduce phosphorus losses. According to the studies, hydroxyapatite enhances soybean growth rates and seed yields compared to standard Phosphorous fertilizers [45, 50, 60]. Some studies have shown potassium nanofertilizers to increase crop responsiveness to plants compared to traditional fertilizers (Table 1). They improve basil leaf area, chlorophyll content, biomass output, and the resistance of crops to infestation by insects. K-NF also enhances photosynthesis, development, and productivity of plants. They also enhance nutrient uptake in peanut crops [11, 97]. Calcium phosphate nanoparticles can regulate nitrogen usage in grapevines by adding urea [86]. Seed germination and maize growth were enhanced by treatment with Mg hydroxide NPs at 500 ppm [78]. Foliar application of magnesium-nanoparticles and iron-NPs enhanced cowpea development [19]. Sulfur nanoparticles are a viable alternative to sulfur fertilizers and enable the release of nitrogen and phosphorus in a slow manner. Further, S-NFs enhance the sulfur metabolism, and water uptake, and reduce dehydrating effects in sunflower seedlings. When added to NPK fertilizer and changed into S-NFs, it enhances the reproductive development of maize crops [38].

Micronutrients play an important role in the physiological activities of plants; therefore, key elements for plant metabolism include Mn, B, Zn, Fe, Cu, Cl, Ni, and Mo. Researchers have tested NPs as a potential Zn supplement in plants and have discovered that with the addition of this constituent at specific concentrations, significant improvements in the growth of plants are evident [24, 85] (Table 1). The effects of ZnO NPs on different plant species are complex and hence context-dependent, thus requiring more exploration to be understood in terms of application in agriculture [89]. The CuNPs-mediated fertilizers are effective in the rapid release of nutrients, and the agricultural industry is fast embracing them not only as pesticides but also as fertilizers. Seedling treatment with 0.25 mg/L of Copper NPs increased waterweed photosynthesis by 35%. Though CuNPs at 200 mg/L or higher showed risks for mung bean, wheat, and yellow squash, micro-CuO, bulk CuO, and $CuSO_4$ raised the allicin content in leaves of scallions by 56–187% and 42–90%, respectively [38]. Photocatalytic chemicals, such as titanium dioxide, are capable of handling environmental issues of agriculture. The TiO_2 photocatalytic NPs exhibited superior performance compared with the unbalanced metallic and absorbent C-based NMs in promoting root elongation and increasing barley kernel amino acids, Ca, Mg, and Zn at low dosages [68]. Normally, the application of nanoparticles in plant growth and nutrition is quite important for the maintenance of the health and productivity of plants. Boron-based nanofertilizers complexed borate with humic acid to provide foliar spray concentrated delivery of boron to crops. Low doses of zinc and boron nanofertilizers increased pomegranate fruit yield and quality [90]. Nickel nanofertilizers of various formulations increase the content of chlorophyll in wheat seedlings

Table 1 Nanoparticles applications in crop productivity [26, 43]

NPs	Concentration (μgmL^{-1})	Benefits	Crop
Ag NP	20, 40 and 60 μgmL^{-1}	20 and 40 μgmL^{-1}increased morphological criteria and metabolites while, 60 μgmL^{-1}decreased plant growth	wheat
Ag NPs	20, 40 and 60 μgmL^{-1}	20 and 40 μgmL^{-1}boosted morphological criteria, while 60 μgmL^{-1}reduced plant growth	Fenugreek
Ag NPs	25, 50, 75, and100 μgmL^{-1}	25, 50 and 75 μgmL^{-1} AgNPs boosted morphological parameters, while 100 μgmL^{-1}reduced plant growth	Wheat
FeS_2 NPs	80–100 μgmL^{-1}	Boosted germination and productivity	Sesamum
Fe NPs	0.002–2 mM	Boosted plant growth and improving the chloroplast number and grana stacking	Crops
α- and γ-Fe2O3 NPs	100, 200, and 400 μgmL^{-1}	Ascorbic acid boosted in fruits	Watermelon
CuO NPs	200 mg kg^{-1}	Increased photosynthesis and biomass of spinach	Spinach
$Mg(OH)_2$ NPs	50–1000 μgmL^{-1}	The concentration of 500 μgmL^{-1} exhibited highest seed germination and corn growth	Maize
CeO2 NPs	50 and 250 μgmL^{-1}	250 μgmL^{-1} boosted the fruit dry weight and lycopene content	Tomato
ZnO NPs	10 μgmL^{-1}	Boosted arabian coffee growth, biomass and photosynthesis	Coffee
ZnO NPs	20 μgmL^{-1}	Boosted grain yield and biomass	Wheat
TiO2 NPs	0, 50, 100 and 200 μgmL^{-1}	Improved antioxidant enzyme activities and essential oil content	Moldavian balm

(continued)

Table 1 (continued)

NPs	Concentration (μgmL^{-1})	Benefits	Crop
Multi-walled carbon	0–200 μgmL^{-1}	At concentration 70–90 μgmL^{-1} increased grains production	Wheat
MgO NPs	50, 100 and 150 μgmL^{-1}	Boosted seed germination % and index, vigor seedling index and morphological parameters	Maize
Mn_3O_4 NPs, ZnO NPs, and Fe2O3 NPs	20 μgmL^{-1}	Increased the growth and yield	Pumpkin
CeO_2 NPs and nanofertilizers (N, P, K, Zn, Fe)	–	Improved the uptake of NPK nanofertilizer and increased head weigh of cabbage	Cabbage
N NF	100 ppm	Boosted yield	Wheat, sesame, pearl millet and mustard
Zn NF	20 ppm		
Chitosan-silicon	0.04–0.12%, w/v	Increased yield and antioxidant activity	Maize
Ag NPs	1, 5, 10, 50, 100 mg kg^{-1}	Boosted Phragmites australis biomass, and antioxidant enzyme activities	Phragmites australis
CuO NPs	0, 25, 50, 75, 100, 150 mg 100 mL^{-1}	25 and 30 mg 100 mL^{-1} increased the seed germination efficiency	*Brassica nigra*
ZnO NPs	0.21 μM	Increased cultivated tobacco growth and antioxidant activity	Tabacco
TiO_2 NPs and SiO_2 NPs	20 and 30 μgmL^{-1}	Boosted growth and antioxidant activity	Rice
Ag NPs	0.0, 2.5, 5, 10, and 25 mg L^{-1}	Boosted tomato seed germination rate, germination speed index and secondary metabolites	Tomato
ZnO NPs	0, 50 or 100 mg L^{-1}	Boosted growth and photosynthetic pigments and oil content	Canola

(continued)

Table 1 (continued)

NPs	Concentration (μgmL^{-1})	Benefits	Crop
CuO NPs	0, 0.5, 1, 2, 4 and 6 mg mL^{-1}	Enhanced vigor, germination, and seedling growth of wheat and phytochemical compounds	Wheat
Zn NF (N > 1.2%; P_2O_5 > 0.001%; K_2O > 0.0001%)	100ppm NF to 1000 parts of water 300 gha	Better nutrient absorption by plant cells. Improvement in protein content	Rice Rice
Nano Si	300 g ha	Adds structural color to plants and improves the plant growth and yield	Rice
Nano chelated silicon	2 ml/L	Increased chlorophyll content	Rice
Nano complete	2 g/L	Leaves	Rice
Fertilizer super micro plus SMP nanofertilizer	(N 5%, P 4%, K 2%, Mg 1%, Ca 1.5%, Fe 4%, Cu 1%, Zn 5%, Mn 2%, Mo 0.04% and B 0.06%)	Easily absorbance by the epidermis of leaves translocated NFs to stems which enhanced the uptake of active	Wheat
Nano chelated super fertilizer	(N 6%, P 3%, K 17%, Mg3%, Ca 1%, S 6%, Fe	Molecules	Wheat
(NCSF) Nano NPK + Nano-K	4%, Zn 4%, Cu 0.5%, Mn 2%, B 0.5%, Mo 0.1%) @ 1 kg ha^{1} Nano NPK (liquid) sprays at 20, 30 & 45 DAS @ 3 ml/L of water + 2 Nano-K (L) sprays	Improves in seed germination rate	Wheat
Chitosan poly-methacrylic acid (CS-PMAA) nanoparticles (NPK)	At 110 and 125 DAS @ 4ml/L of water Nano-100 (NPK 500:60:400 ppm loaded on CS-PMAA)	Shorter life cycle	Wheat
Nano Mn-oxide (30 nm)	–	The proportion of 23.5% Higher shoot and grain Mn content was recorded with the spray of Nano-Mn	Wheat
Nano Fe/SiO_2	25 mg/kg	–	Barley

(continued)

Table 1 (continued)

NPs	Concentration (μgmL^{-1})	Benefits	Crop
Nano-Zn and Nano-B	Foliar spray of 2000 ppm of NFs at nine-leaf stage, stem elongation, heading	Response NFs also depends on different cultivars	Maize
Nano Fe/SiO_2	25 mg/kg	Root dry weight increased gradually	Maize
Nano-zeourea		Quality parameters and nutrient uptake was increased	Maize
Nano-silica (SiO_2-NPs) Nano-K fertilizer	10 gkg ^{1} 1.33 g MOP in 1.11 L of nano polymer/ 15 kg soil	Controlled release of K and supply of K for extended time as per the requirement of crop through nano-K have resulted in gaining the grain yield	Maize Maize
Complete NF Nano NPK (slow release fertilizer)	2000 ppm during the 9-leaf stage, stem elongation, heading SRF @ NPK: 223:285:83 kg ha^{1}	Slow release fertilizers are more eco-friendly, easy to synchronize and have greater benefits	Maize Maize
Nano N fertilizer	1.38-g urea in 1.15-L nanopolymer	Nanofertilizer (1.38-g urea in 1.15-L nanopolymer) increased the seed yield of the crop	Sorghum
Nano-ZnO (30 nm)	500 ppm	Reduction of doses of fertilizers, consumption of fertilizers, environmental hazards and increased NUE	Sorghum
Nano NPK	(8% N, 5% P, 3% K, 10% micronutrients, 5% amino acids, and 5% seaweed extract) @ 5 L ha^{-1}	Substituting 50% recommended NPK through Nano NPK not only fulfills the requirements but also helps in increasing the yield by reducing the weed population	Sorghum

(continued)

Table 1 (continued)

NPs	Concentration (μgmL^{-1})	Benefits	Crop
Nano-Zn	16 L ha^{-1}	Nano-Zn enhanced plant dry matter (12.5%), grain yield (37.7%), and plant zinc concentration (10.4%) over the control	Pearl millet
$NP_{2.5}$: 2.5times reduction of RDP through ZnO NPs 200 ppm	–	Improved plant height, fresh and dry weight of the crop seedlings in the greenhouse	Pearl millet

[22, 90]. Nanoparticles of TiO_2 also showed extended protection from pests and diseases due to their antimicrobial properties [93]. They efficiently decreased the soil salinity of broad bean crops, improved growth, and increased the vigor and germination potential of canola seeds. In addition to these functions, the germination of wheat seeds was promoted in the presence of high concentrations of TiO_2 NPs while low concentrations had inhibitory effects [9, 90, 91].

Green silver NPs, because of their increased surface area, depicted important antimicrobial effects. It is possible for the nanoparticles to penetrate into bacteria and enhance the growth and development of plants. In this respect, it has been reported that silver nanoparticles accumulate in *Escherichia coli* cells, leading to cell damage and cell death. As a result, it enhances the efficiency of use of nutrients, consequently indicating their potential in scientific and agricultural applications [10].

Nutrient-controlled delivery with copper-based nano-biofertilizers is another approach for improving crop growth and increasing yield by enhancing uptake efficiency. The nanoparticles containing copper may be potential players in the physiological process of plants, photosynthesis, and pathogen defense; it also increases the stress tolerance of plants and antioxidant enzyme activity. However, further research is needed for a full understanding of long-term effects to optimize the application of copper-based nano biofertilizers in an agroecosystem [38]. The effect of copper nanoparticles on maize crops includes better growth and reduced leaf curling and wilting. Stressed plants show increased pigments, increased number of seeds, enhanced yield, and improved water retention capacity [10, 38].

Fe is one of the essential elements participating in metabolic processes such as respiration, photosynthesis, and the production of chlorophyll [46]. Green synthesis of Fe NPs involves plant extracts and microorganisms to reduce the contamination risks. Fe NPs of a stable nature have been formed using *Azadirachta Indica* leaf extract. Fe NPs are unique nanofertilizers because of their magnetic nature and size on a nanoscale. They can increase the rate of plant growth and make them more stress-tolerant. Fe_2O_3 NPs increased the root and stem length, biomass, height, antioxidant enzyme, and phytohormone contents of peanut plants. Iron is an essential element required for metabolic functions such as respiration, photosynthesis, and the production of chlorophyll [4, 10, 14].

ZnO NFs increase crop yields while decreasing soil leaching, thus enabling plants to cope with stress [3, 28]. ZnO NFs were found to enhance the growth and productivity of crops such as sesame, barley, and pearl millet [42]. In a recent experiment, 20 mg/L ZnO NPs exhibited a considerable increase in the biochemical content and growth indices of *Citrus aurantium* fruit [62].

6 Nano-Fertilizers Versus Conventional Chemical Fertilizers

Nanofertilizers offer enhanced efficiency and reduced environmental impact compared to traditional fertilizers. They deliver vital nutrients directly to plants reducing fertilizer requirements and improves plant uptake. Additionally, they improve soil properties, which reduces fertilizer usage and environmental impacts. This enhances plant growth and yield while reducing nutrient leaching. Implementing nanofertilizers can enhance fertilizer efficiency while minimizing environmental impacs [59, 76, 90, 91]. Here, we presented some advantages of NFs applications in soil and crop growth and development.

6.1 Greater Surface Area

Nanofertilizers with smaller particle sizes improve plant nutrient uptake. They enhance metabolic functions [3]. This leads to increased photosynthate production. Their heightened surface area also enhances nutrient utilization efficiency [8]. Additionally, it increases reactivity with other substances [90].

6.2 High Solubility

Nanofertilizers, due to their reduced size and increased surface area, have higher solubility than conventional fertilizers. Examples include nano-hydroxyapatite and nano-zinc oxide [90, 91].

6.3 Encapsulation of Fertilizers Within NPs

Encapsulating fertilizers in nanoparticles (NPs) improves nutrient absorption in crops. Methods include polymer-based encapsulation. Also inorganic shell formation and layer-by-layer assembly. Zeolite-based nanofertilizers which use zeolites as carriers, enhance nutrient availability and reduce nutrient loss. They also fix soil nutrients like NO3 and NH4. These methods help improve crop development and soil health [41, 76, 90].

6.4 Easy Penetration and Controlled Release of Fertilizers

Nanofertilizers improve fertilizer penetration into plant tissues, increasing the uptake of nutrients. With their ultra-high-surface area-to-volume ratio, nano-sized particles can diffuse across the plant cell wall and membrane. Functionalization with targeting molecules—molecules that attach to specific plant-based targets—ensures that they are selectively taken up by plant tissues, thus the precision delivery of important nutrients to plants, reducing the usage of fertilizers, minimizing nutrient leaching, and mitigating environmental pollution for improved crop yield and agricultural sustainability [35, 36].

6.5 High Nutrient Absorption Efficiency

Nanofertilizers improve the efficiency of applied fertilizers and enhance the quantity of soil nutrients taken up in crop yield, hence minimizing excess use and leaching loss [35]. ZnO NPs increase zinc uptake, root length, and shoot dry weight in wheat plants while iron oxide NPs increase iron uptake and chlorophyll content, resulting in higher yields [57, 62, 82, 85].

6.6 Effective Duration of Nutrient Release

The nutrient release time in nanofertilizers affects agricultural productivity and sustainability [3]. They provide continuous supplies of nutrients, hence reducing the need for frequent applications that normally result in environmental pollution. Studies show that chitosan-coated nanofertilizers are capable of releasing nutrients for up to 45 days, and those coated with polymers could last up to 60 days. Other factors influencing this duration include nanoparticle size and environment [66].

6.7 Improved Microbial Activity

Nanobiofertilizers can be used to enhance natural and plant-available sources of nutrients in secondary and micronutrients by providing balanced nutrient uptake, improved plant growth, and soil fertility [5, 21, 35]. New improved structure, function, and vegetation attributes were later engineered or invented. For example, gold nanoparticles act with rhizobacteria [51], on the other hand, chitosan-based nanoparticles allow the nitrogen-fixing bacteria to enter the plant's body for greater availability of nitrogen and, in turn, better yield [67]. Whereas the only disadvantage is that such biofertilizers have limited shelf life, which on the other hand can be tackled with nanomaterials. Desiccation, heat, and UV damage resilience of biofertilizers can be enhanced with nanoformulation [90].

6.8 Improved Soil Activity

Soil activity is a crucial aspect of soil health. It is primarily driven by soil microorganisms like bacteria and fungi. Nanofertilizers enhance soil activity by delivering nutrients efficiently to the root zone. They stimulate soil microorganisms and promote nutrient cycling. They amplify nitrogen mineralization rate and enhance soil bacteria and fungi activity [28, 41, 43, 48].

6.9 Improved Soil Water-Holding Capacity

Nanofertilizers enriched with humic acid and clay, improve soil structure. They increase water-holding capacity and promote organic matter. Additionally, they create a better microorganism environment. They reduce water loss and improve crop yields. They enhance soil health by forming larger aggregates [90]. For example, a study has shown that a combination of conventional fertilizers and nano-fertilizers can very significantly improve soil mineral nitrogen, dehydrogenase activity, and soil microbial biomass carbon in crops such as maize, wheat, pearl millet, and mustard [88]. Zinc, key to photosynthesis, antioxidant defense, and disease resistance, can be optimized by nano-fertilizers applied in conjunction with conventional fertilizers [24, 56, 65, 85]. Dapkekar et al. [17] and Srivastav et al. [83] maintained at least 75% of the recommended nitrogen with two sprays of nano-urea or nano-urea and nano-Zn to prevent nutrient depletion and thus maintained the ecological balance of plant growth and soil health [88]. It, therefore, can be an effective approach toward sustainable agriculture practices.

6.10 Ecofriendly Nature

Nanofertilizers are environmentally friendly and efficient versions of traditional fertilizers. Their slow release allows better control over nutrient balance in the soil, reducing damage to the environment [27]. Thus, this efficiency saves farmers money by reducing the frequency of applications and labor costs. They will become even more efficient with the advancement in nanotechnology.

6.11 Low Production Cost

Nanofertilizers reduce production costs, enhance nutrient-use efficiency due to controlled release and targeted delivery, reduce wastage, and therefore are cost-effective; they have lower labor intensity and enhanced absorption rates in plants, with a longer retention time in the soil [90].

6.12 Fulfills the Goal of Precision Farming

Precision farming is a methodology that brings out the extent of the input used in crop growth, including fertilizers, pesticides, and water, with the aid of advanced technology [81]. If nanotechnology had been discovered, nanofertilizers would be quite plausible, giving accurate application as compared to the conventional product, cost reduction, less effect on the environment, precise crop nutrition management, and supply of essential substances [38, 76, 88].

6.13 Improves Plant Stress Tolerance

Plants respond to environmental stresses by genetic, biochemical, and physiological alterations. Nanoscale particles enhance events associated with plant growth and development by crossing the cell walls, supplying the plant with nutrients, and facilitating endurance against stress factors like drought or high and low temperatures. They also supply critical nutrients and defensive chemicals like antioxidants, enhancing health status under water shortages and hot summer periods [5, 9, 97].

6.14 Stimulates Plant Growth

The NPs enhance plant growth by improving access to nutrients, through efficient uptake, and with the improvement in soil structures. The other way that these NPs contribute to balancing is by the soil pH level, its absorption of nutrients, and minimizing the amount of fertilizer runoff. Nanofertilizers coat the soil surface with a thin protective layer so that available nutrients can be delivered to the root zone with the minimum amount of loss, promoting the growth of the plant [4, 59, 73].

7 The NFs Effects on Soil Quality and Its Microflora Diversity

NFs are considered to have a significant impact on the soil microflora, hence influencing ecosystem performance. The NFs were found to supplement nutritional values of the soil, upgrade the ecosystem, and enhance the quantity of microbes in the soil. NFs generate humic acid, increasing fertility in the soil, which allows better growth of soil microbes. NFs release nutrients slowly, providing continuous support to plant growth while helping conserve biodiversity useful to microorganisms [55]. However, a high concentration of titanium oxide can disrupt bacterial symbiosis and inhibit clover growth [52]. Metallic silver NFs increase the abundance of proteobacteria, planctomycetes, and acidobacteria, this promotes metabolic activity and decomposes soil pollutants. Higher concentrations of metallic silver NFs decrease the abundance of acidobacteria, actinomycetes, and cyanobacteria but increase humic and fulvic acids. Natural environment risks and impacts associated with NFs are still challenging to quantify and understand. NFs have been found to be significantly affecting the microflora of the soil and ecosystem performance [43, 58, 71].

8 Environmental Safety and Human Health Concerns

Nanofertilizers offer promises of better fertilizer application, less chemical pollution, and higher agricultural productivity; however, concerns regarding their toxicity and safety also prevail. It is evidenced from the literature that both ZnO NPs and $ZnSO_4$ show an enhancement of wheat germination and growth, but ZnO NPs are more effective than $ZnSO_4$. Their effectiveness factors are correlated among plant species, nanoparticle size, concentration, composition, and different chemical properties. Further, the hydrophilic nature of the particles allows ease of translocation in the plant through plant tissues; also, the small size of the particles enhances the surface-area-to-mass ratio for nutrient uptake. Regulatory concerns include an understanding of nanoparticle size, characterization, and distinguishing toxic effects. Nanotechnology is now set to revamp agricultural applications, as both governing and researchers are working to develop a working product on NFs concerning those issues [43, 48, 92]. NFs are eco-friendly and can be regulated using biosensors for monitoring soil nutrient status, crop growth stage, and environmental conditions [18, 43].

Nanotechnology in agriculture and the global economy may pose health risks. In order to protect the environment and human health, it is very important to identify an effective dose of NFs. The NFs can cause acute harmful effects due to protein denaturation, generation of reactive oxygen species, and mitochondrial dysfunction. Exposure to NFs for a longer time can result in organ growth or malfunction. TiO_2 NPs, as a food additive, induce inflammation and oxidative stress, resulting in organ growth or malfunction. Their use in individuals is still arguable until certain issues about fate, transport, and toxicity are raised. Long-term studies are required for the assessment of the impacts nanomaterials have on the reproductive system of the animal and hormone secretion [18]. Nanofertilizers pose a serious threat to ecosystem health through nontarget organisms, including insects, fish, and birds. This can result in negative effects on reproduction, growth, and development in nontarget organisms. Further research on these impacts should be conducted in relation to beneficial microorganisms like nitrogen-fixing bacteria and mycorrhizal fungi [43, 91].

9 Conclusion and Future Remarks

In summary, the use of nanofertilizers presents a promising solution to improve soil quality, enhance crop growth, and promote sustainable agriculture by providing controlled nutrient release, reducing environmental impact, and increasing crop productivity. While challenges remain in terms of long-term impacts and regulatory frameworks, further research and development in this area offer potential benefits for the future of food production.

Competing Interests The authors declare that they have no competing interest.

Declaration of Generative AI and AI-Assisted Technologies in the Writing Process During the preparation of this work, the book chapter author(s) have not taken any assistance from AI, and the book chapter author(s) reviewed and edited the content as needed and take(s) full responsibility for the content of the publication.

References

1. Adeyemi JO, Oriola AO, Onwudiwe DC, Oyedeji AO (2022) Plant extracts mediated metal-based nanoparticles: synthesis and biological applications. Biomolecules 12(5):627
2. Adisa IO, Pullagurala VLR, Peralta-Videa JR, Dimkpa CO, Elmer WH, Gardea-Torresdey JL, White JC (2019) Recent advances in nano-enabled fertilizers and pesticides: a critical review of mechanisms of action. Environ Sci Nano 6(7):2002–2030

3. Akhtar N, Ilyas N, Meraj TA, Pour-Aboughadareh A, Sayyed R, Mashwani Z-U-R, Poczai P (2022) Improvement of plant responses by nanobiofertilizer: a step towards sustainable agriculture. Nanomaterials 12(6):965
4. Amin H, Minkina T, Sushkova S, Mandzhieva S (2021) Role of metal-based nanomaterials in plant growth. In: International symposium of soil science and plant nutrition
5. Amjad S, Serajuddin M (2021) Applications of nanobiotechnology to mitigate mineral nutrients deficiency stress in crop plants. In: Nanobiotechnology: mitigation of abiotic stress in plants, pp 437–452
6. Asadu CO, Ezema CA, Ekwueme BN, Onu CE, Onoh IM, Adejoh T, Ezeorba TPC, Ogbonna CC, Otuh PI, Okoye JO (2024) Enhanced efficiency fertilizers: overview of production methods, materials used, nutrients release mechanisms, benefits and considerations. Environ Pollut Manag
7. Ashraf SA, Siddiqui AJ, Abd Elmoneim OE, Khan MI, Patel M, Alreshidi M, Moin A, Singh R, Snoussi M, Adnan M (2021) Innovations in nanoscience for the sustainable development of food and agriculture with implications on health and environment. Sci Total Environ 768:144990
8. Avila-Quezada GD, Ingle AP, Golińska P, Rai M (2022) Strategic applications of nano-fertilizers for sustainable agriculture: benefits and bottlenecks. Nanotechnol Rev 11(1):2123–2140
9. Badshah I, Mustafa N, Khan R, Z-u-R M, Raja NI, Almutairi MH, Aleya L, Sayed AA, Zaman S, Sawati L (2023) Biogenic titanium dioxide nanoparticles ameliorate the effect of salinity stress in wheat crop. Agronomy 13(2):352
10. Bairwa P, Kumar N, Devra V, Abd-Elsalam KA (2023) Nano-biofertilizers synthesis and applications in agroecosystems. Agrochemicals 2(1):118–134
11. Basavegowda N, Baek K-H (2021) Current and future perspectives on the use of nanofertilizers for sustainable agriculture: the case of phosphorus nanofertilizer. 3 Biotech 11(7):357
12. Benício LPF, Silva RA, Lopes JA, Eulálio D, dos Santos RMM, de Aquino LAD, Vergütz L, Novais RF, da Costa LM, Pinto FG (2015) Layered double hydroxides: nanomaterials for applications in agriculture. Rev Bras Ciênc Solo 39(1):1–13
13. Bhardwaj AK, Naraian R (2021) Cyanobacteria as biochemical energy source for the synthesis of inorganic nanoparticles, mechanism and potential applications: a review. 3 Biotech 11(10):445
14. Bolade OP, Williams AB, Benson NU (2020) Green synthesis of iron-based nanomaterials for environmental remediation: a review. Environ Nanotechnol Monit & Manag 13:100279
15. Channab B-E, Idrissi AE, Ammar A, Dardari O, Marrane SE, El Gharrak A, Akil A, Essemlali Y, Zahouily M (2024) Recent advances in nano-fertilizers: synthesis, crop yield impact, and economic analysis. Nanoscale
16. Chaudhuri SK, Malodia L (2017) Biosynthesis of zinc oxide nanoparticles using leaf extract of calotropis gigantea: characterization and its evaluation on tree seedling growth in nursery stage. Appl Nanosci 7(8):501–512
17. Dapkekar A, Deshpande P, Oak MD, Paknikar KM, Rajwade JM (2018) Zinc use efficiency is enhanced in wheat through nanofertilization. Sci Rep 8(1):6832
18. Dehghanian Z, Lajayer BA, Atigh ZBQ, Nayeri S, Ahmadabadi M, Taghipour L, Senapathi V, Astatkie T, Price G (2023) Micro (nano) plastics uptake, toxicity and detoxification in plants: challenges and prospects. Ecotoxicol Environ Saf 268:115676
19. El-Qurashi MA, Al-Yahya FA, Al-Hazmi AS, Saleh AA (2023) Efficacy of biologically synthesized nanoparticles on suppression plant-parasitic nematodes: a review. Egypt J Agronematol 22(2):41–60
20. Eweis AA, El-Raheem HA, Ahmad MS, Hozzein WN, Mahmoud R (2024) Green fabrication of nanomaterials using microorganisms as nano-factories. J Clust Sci 1–28
21. Faizan M, Singh A, Eren A, Sultan H, Sharma M, Djalovic I, Trivan G (2024) Small molecule, big impacts: nano-nutrients for sustainable agriculture and food security. J Plant Physiol 154305
22. Fatima F, Hashim A, Anees S (2021) Efficacy of nanoparticles as nanofertilizer production: a review. Environ Sci Pollut Res 28(2):1292–1303

23. Hamed R, Jodeh S, Alkowni R (2024) Nano bio fertilizer capsules for sustainable agriculture. Sci Rep 14(1):13646
24. Hamzah Saleem M, Usman K, Rizwan M, Al Jabri H, Alsafran M (2022) Functions and strategies for enhancing zinc availability in plants for sustainable agriculture. Front Plant Sci 13:1033092
25. Hano C, Abbasi BH (2021) Plant-based green synthesis of nanoparticles: production, characterization and applications, vol 12. MDPI
26. Haris M, Hussain T, Mohamed HI, Khan A, Ansari MS, Tauseef A, Khan AA, Akhtar N (2023) Nanotechnology–a new frontier of nano-farming in agricultural and food production and its development. Sci Total Environ 857:159639
27. Haydar MS, Ghosh D, Roy S (2024) Slow and controlled release nanofertilizers as an efficient tool for sustainable agriculture: recent understanding and concerns. Plant Nano Biol 100058
28. Hossain ME, Shahrukh S, Hossain SA (2022) Chemical fertilizers and pesticides: impacts on soil degradation, groundwater, and human health in Bangladesh. In: Environmental degradation: challenges and strategies for mitigation. Springer, pp 63–92
29. Hulkoti NI, Taranath T (2014) Biosynthesis of nanoparticles using microbes—a review. Colloids Surf B 121:474–483
30. Husen A, Siddiqi KS (2014) Phytosynthesis of nanoparticles: concept, controversy and application. Nanoscale Res Lett 9(229):1–24. https://doi.org/10.1186/1556-276X-9-229
31. Husen A (2020) Introduction and techniques in nanomaterials formulation. In: Husen A, Jawaid M (eds) Nanomaterials for agriculture and forestry applications. Elsevier Inc., USA, pp 1–14. https://doi.org/10.1016/B978-0-12-817852-2.00001-9
32. Husen A (2017) Gold nanoparticles from plant system: synthesis, characterization and their application. In: Ghorbanpourn M, Manika K, Varma A (eds) Nanoscience and plant–soil systems, vol 48. Springer International Publishing AG, Cham, pp 455–479. https://doi.org/10.1007/978-3-319-46835-8_17
33. Husen A (2019) Natural product-based fabrication of zinc oxide nanoparticles and their application. In: Husen A, Iqbal M (eds) N*anomaterials* and plant potential. Springer International Publishing AG, Cham, pp 193–291 https://doi.org/10.1007/978-3-030-05569-1_7
34. Husen A (2023) Secondary metabolites based green synthesis of nanomaterials and their applications. Springer Nature Singapore Pte Ltd., Singapore. https://doi.org/10.1007/978-981-99-0927-8
35. Iqbal M, Umar S, Mahmooduzzafar F (2019) Nano-fertilization to enhance nutrient use efficiency and productivity of crop plants. In: Nanomaterials and plant potential, pp 473–505
36. Iqbal MA (2019) Nano-fertilizers for sustainable crop production under changing climate: a global perspective. Sustain Crop Prod 8:1–13
37. Jakhar AM, Aziz I, Kaleri AR, Hasnain M, Haider G, Ma J, Abideen Z (2022) Nano-fertilizers: a sustainable technology for improving crop nutrition and food security. NanoImpact 27:100411
38. Jha A, Pathania D, Damathia B, Raizada P, Rustagi S, Singh P, Rani GM, Chaudhary V (2023) Panorama of biogenic nano-fertilizers: a road to sustainable agriculture. Environ Res 235:116456
39. Jote CA (2023) The impacts of using inorganic chemical fertilizers on the environment and human health. Organ Med Chem Intl J 13(3):555864
40. Khan SS, Kour D, Kaur T, Sharma A, Kumar S, Kumari S, Ramniwas S, Singh S, Negi R, Sharma B (2024) Microbial nanotechnology for precision nanobiosynthesis: innovations, current opportunities and future perspectives for industrial sustainability. Curr Microbiol 81(8):251
41. Khan MR, Rizvi TF (2017) Application of nanofertilizer and nanopesticides for improvements in crop production and protection. In: Nanoscience and plant–soil systems, pp 405–427
42. Kumar R, Dhiman M, Sharma L, Dadhich A, Kaushik P, Sharma MM (2022) Nanofertilizers: the targeted nutrient supplier and enhance nutrients uptake by pearl millets (Pennisetum glaucum). Biocatal Agric Biotechnol 45:102524
43. Kumar N, Samota SR, Venkatesh K, Tripathi S (2023) Global trends in use of nano-fertilizers for crop production: advantages and constraints–a review. Soil Tillage Res 228:105645

44. Kumari KA, Mangatayaru KG, Reddy GB (2023) Fungal and yeast-mediated biosynthesis of metal nanoparticles: characterization and bio applications. In: Fungal cell factories for sustainable nanomaterials productions and agricultural applications. Elsevier, pp 309–336
45. Li C, Rippner DA, Manavalan LP, Parikh SJ (2021) Evaluation of Bacillus seed coatings on soybean phosphorus uptake in an oxisol fertilized with 32 P-labeled hydroxyapatite. Plant Soil 464:273–287
46. Li P, Wang A, Du W, Mao L, Wei Z, Wang S, Yuan H, Ji R, Zhao L (2020) Insight into the interaction between Fe-based nanomaterials and maize (Zea mays) plants at metabolic level. Sci Total Environ 738:139795
47. Madkour LH (2018) Biogenic–biosynthesis metallic nanoparticles (MNPs) for pharmacological, biomedical and environmental nanobiotechnological applications. Chron Pharm Sci J 2(1):384–444
48. Mahesha K, Singh N, Amarshettiwar S, Singh G, Gulaiya S, Das H, Kumar J (2023) Entering a new agricultural era through the impact of nano-fertilizers on crop development: a review. Int J Plant & Soil Sci 35(20):94–102
49. Malik AQ, Mir TuG, Kumar D, Mir IA, Rashid A, Ayoub M, Shukla S (2023) A review on the green synthesis of nanoparticles, their biological applications, and photocatalytic efficiency against environmental toxins. Environ Sci Pollut Res 30(27):69796–69823
50. McKnight MM, Qu Z, Copeland JK, Guttman DS, Walker VK (2020) A practical assessment of nano-phosphate on soybean (Glycine max) growth and microbiome establishment. Sci Rep 10(1):9151
51. Mishra VK, Kumar A (2009) Impact of metal nanoparticles on the plant growth promoting rhizobacteria. Dig J Nanomater Biostruct 4(3):587–592
52. Moll J, Gogos A, Bucheli TD, Widmer F, van der Heijden MG (2016) Effect of nanoparticles on red clover and its symbiotic microorganisms. J Nanobiotechnol 14:1–8
53. Moradi S, Babapoor A, Ghanbarlou S, Kalashgarani MY, Salahshoori I, Seyfaee A (2024) Toward a new generation of fertilizers with the approach of controlled-release fertilizers: a review. J Coat Technol Res 21(1):31–54
54. Motamedi E, Safari M, Salimi M (2023) Improvement of tomato yield and quality using slow release NPK fertilizers prepared by carnauba wax emulsion, starch-based latex and hydrogel nanocomposite combination. Sci Rep 13(1):11118
55. Mubeen B, Hasnain A, Wang J, Zheng H, Naqvi SAH, Prasad R, Rehman AU, Sohail MA, Hassan MZ, Farhan M (2023) Current progress and open challenges for combined toxic effects of manufactured nano-sized objects (MNO's) on soil biota and microbial community. Coatings 13(1):212
56. Murgueitio-Herrera E, Falconí CE, Cumbal L, Gómez J, Yanchatipán K, Tapia A, Martínez K, Sinde-Gonzalez I, Toulkeridis T (2022) Synthesis of iron, zinc, and manganese nanofertilizers, using Andean blueberry extract, and their effect in the growth of cabbage and lupin plants. Nanomaterials 12(11):1921
57. Nayeri S, Dolatyari M, Mouladoost N, Nayeri S, Zarghami A, Mirtagioglu H, Rostami A (2023) Ag/ZnO core–shell NPs boost photosynthesis and growth rate in wheat seedlings under simulated full sun spectrum. Sci Rep 13(1):14385
58. Nayeri S, Tohidfar M (2017) Comparative assessment of biosafety aspects between nanotechnology and GMOs. J Biosaf 9(4):33–44
59. Ninama J, Sachan K, Yadav B, Satapathy SN, Kumar J, Singh BV (2023) Effects and consequences of nano fertilizer application on plant growth and developments: a review. Int J Environ Clim Change 13(10):2288–2298
60. Pilotto L, Zuluaga MYA, Scalera F, Piccirillo C, Marchiol L, Civilini M, Pii Y, Cesco S, Fellet G (2024) Sustainable crop fertilization by combining biogenic nano-hydroxyapatite and P solubilizing bacteria: observations on barley. Plant Nano Biology 9:100091
61. Prakashraj R, Vijayakumar S, Punitha V, Vidhya E, Nilavukkarasi M, Praseetha P (2021) Fabricated TiO 2 nanofertilizers for foliar assimilation to enhance yield and disease resistance in Capsicum annuum L. J Plant Growth Regul 1–8

62. Punitha V, Vijayakumar S, Alsalhi MS, Devanesan S, Nilavukkarasi M, Vidhya E, Kumar SP, Kim W (2023) Biofabricated ZnO nanoparticles as vital components for agriculture revolutionization–a green approach. Appl Nanosci 13(9):5821–5827
63. Rakhimol K, Thomas S, Kalarikkal N, Jayachandran K (2021) Nanotechnology in controlled-release fertilizers. In: Controlled release fertilizers for sustainable agriculture. Elsevier, pp 169–181
64. Reshma Anjum M, Maheswari J, Anusha K, Sravya B, Narasimha G, Abd-Elsalam KA (2023) Maximizing crop yield with macro and micro nano enhanced fertilizers. In: Nanofertilizers for sustainable agroecosystems: recent advances and future trends. Springer, pp 1–33
65. Reshma Z, Meenal K (2022) Foliar application of biosynthesised zinc nanoparticles as a strategy for ferti-fortification by improving yield, zinc content and zinc use efficiency in amaranth. Heliyon 8(10)
66. Riseh RS, Vazvani MG, Kennedy JF (2023) The application of chitosan as a carrier for fertilizer: a review. Int J Biol Macromol 126483
67. Rkhaila A, Chtouki T, Erguig H, El Haloui N, Ounine K (2021) Chemical proprieties of biopolymers (chitin/chitosan) and their synergic effects with endophytic Bacillus species: unlimited applications in agriculture. Molecules 26(4):1117
68. Rodríguez-González V, Terashima C, Fujishima A (2019) Applications of photocatalytic titanium dioxide-based nanomaterials in sustainable agriculture. J Photochem Photobiol C 40:49–67
69. Rostamizadeh E, Iranbakhsh A, Majd A, Arbabian S, Mehregan I (2020) Green synthesis of Fe 2 O 3 nanoparticles using fruit extract of Cornus mas L. and its growth-promoting roles in Barley. J Nanostructure Chem 10:125–130
70. Rui M, Ma C, Hao Y, Guo J, Rui Y, Tang X, Zhao Q, Fan X, Zhang Z, Hou T (2016) Iron oxide nanoparticles as a potential iron fertilizer for peanut (Arachis hypogaea). Front Plant Sci 7:815
71. Sambangi P, Gopalakrishnan S, Pebam M, Rengan AK (2022) Nano-biofertilizers on soil health, chemistry, and microbial community: benefits and risks. Proc Indian Natl Sci Acad 88(3):357–368
72. Samuel MS, Ravikumar M, John JA, Selvarajan E, Patel H, Chander PS, Soundarya J, Vuppala S, Balaji R, Chandrasekar N (2022) A review on green synthesis of nanoparticles and their diverse biomedical and environmental applications. Catalysts 12(5):459
73. Sarkar N, Chaudhary S, Kaushik M (2021) Nano-fertilizers and nano-pesticides as promoters of plant growth in agriculture. plant-microbes-engineered nano-particles (PM-ENPs) nexus in agro-ecosystems: understanding the interaction of plant, microbes and engineered nano-particles (ENPS), pp 153–163
74. Shah AH, Rather MA (2021) Intracellular and extracellular microbial enzymes and their role in nanoparticle synthesis. In: Microbial nanotechnology: green synthesis and applications, pp 41–59
75. Sharma V, Javed B, Byrne H, Curtin J, Tian F (2022) Zeolites as carriers of nano-fertilizers: from structures and principles to prospects and challenges. Appl Nano 3(3):163–186
76. Sharma S, Singh S, Bahuguna A, Yadav B, Barthwal A, Nandan R, Khatana R, Pandey A, Thakur R, Singh H (2022) Nanotechnology: an efficient tool in plant nutrition management. In: Ecosystem services: types, management and benefits, pp 165–188
77. Shebl A, Hassan A, Salama DM, Abd El-Aziz M, Abd Elwahed MS (2019) Green synthesis of nanofertilizers and their application as a foliar for Cucurbita pepo L. J Nanomater 1:3476347
78. Shinde S, Paralikar P, Ingle AP, Rai M (2020) Promotion of seed germination and seedling growth of Zea mays by magnesium hydroxide nanoparticles synthesized by the filtrate from Aspergillus niger. Arab J Chem 13(1):3172–3182
79. Siddiqi KS, Husen A (2016) Fabrication of metal and metal oxide nanoparticles by algae and their toxic effects. Nanoscale Res Lett 11(363):1–11
80. Siddiqi KS, Husen A (2016) Fabrication of metal nanoparticles from fungi and metal salts: scope and application. Nanoscale Res Lett 11(98):1–15
81. Singh AK (2010) Precision farming. Water Technology Centre, IARI, New Delhi

82. Singh M, Singh J, Sharma D, Kaur B, Rawat M Plant leaves mediated synthesis of semiconductor ZnO nanoparticles and its application for seed germination. In: AIP Conference Proceedings, 2018. vol 1. AIP Publishing
83. Srivastav A, Ganjewala D, Singhal RK, Rajput VD, Minkina T, Voloshina M, et al (2021) Effect of ZnO nanoparticles on growth and biochemical responses of wheat and maize. Plan Theor 10:2556 https://doi.org/10.3390/plants10122556
84. Subramanian KS, Manikandan A, Thirunavukkarasu M, Rahale CS (2015) Nano-fertilizers for balanced crop nutrition. In: Nanotechnologies in food and agriculture, pp 69–80
85. Thounaojam TC, Meetei TT, Devi YB, Panda SK, Upadhyaya H (2021) Zinc oxide nanoparticles (ZnO-NPs): a promising nanoparticle in renovating plant science. Acta Physiol Plant 43:1–21
86. Torres-Díaz LL, Pérez-Álvarez EP, Parra-Torrejón B, Marín-San Román S, de Sáenz de Urturi I, Ramírez-Rodríguez GB, Murillo-Peña R, González-Lázaro M, Delgado-López JM, Garde-Cerdán T (2024) Effects of foliar application of methyl jasmonate and/or urea, conventional or via nanoparticles, on grape volatile composition. J Sci Food Agric
87. Ukidave VV, Ingale LT (2022) Green synthesis of zinc oxide nanoparticles from coriandrum sativum and their use as fertilizer on Bengal gram, Turkish gram, and green gram plant growth. Int J Agron 1:8310038
88. Upadhyay PK, Singh VK, Rajanna G, Dwivedi BS, Dey A, Singh RK, Rathore SS, Shekhawat K, Babu S, Singh T (2023) Unveiling the combined effect of nano fertilizers and conventional fertilizers on crop productivity, profitability, and soil well-being. Front Sustain Food Syst 7:1260178
89. Vignesh A, Amal TC, Sivalingama R, Selvakumar S, Vasanth K (2024) Unraveling the impact of nanopollution on plant metabolism and ecosystem dynamics. Plant Physiol Biochem 108598
90. Yadav A, Yadav K, Abd-Elsalam KA (2023) Nanofertilizers: types, delivery and advantages in agricultural sustainability. Agrochemicals 2(2):296–336
91. Yadav RK, Singh N, Singh A, Yadav V, Bano C, Khare S, Niharika (2020) Expanding the horizons of nanotechnology in agriculture: recent advances, challenges and future perspectives. Vegetos 33:203–221
92. Yaseen R, Ahmed AIS, Omer AM, Agha MKM, Emam TM (2020) Nano-fertilizers: bio-fabrication, application and biosafety. Nov Res Microbiol J 4(4):884–900
93. Younis AB, Haddad Y, Kosaristanova L, Smerkova K (2023) Titanium dioxide nanoparticles: recent progress in antimicrobial applications. Wiley Interdiscip Rev: Nanomed Nanobiotechnol 15(3):e1860
94. Zakavi M, Askari H, Shahrooei M (2022) Bacterial diversity changes in response to an altitudinal gradient in arid and semi-arid regions and their effects on crops growth. Front Microbiol 13:984925
95. Zakavi M, Askari H, Shahrooei M (2022) Characterization of bacterial diversity between two coastal regions with heterogeneous soil texture. Sci Rep 12(1):18901
96. Zakavi M, Askari H, Shahrooei M (2022) Maize growth response to different bacillus strains isolated from a salt-marshland area under salinity stress. BMC Plant Biol 22(1):367
97. Zhang H, Liu S, Ren T, Niu M, Liu X, Liu C, Wang H, Yin W, Xia X (2023) Crucial abiotic stress regulatory network of NF-Y transcription factor in plants. Int J Mol Sci 24(5):4426
98. Zuhrotun A, Oktaviani DJ, Hasanah AN (2023) Biosynthesis of gold and silver nanoparticles using phytochemical compounds. Molecules 28(7):3240

Exploring Genetic and Molecular Methods to Induce Crop Plant Tolerance Under Various Contamination

Haripriya Kavidayal, Aman Dabral, Manmohan J. Dobriyal, Biplov C. Sarkar, and Indra Singh

Abstract The intensification of farming practices, urban expansion, and land use changes has led to increased pollution levels, and pose threat to global agricultural productivity and food security. However, with the advent of genetic and molecular approaches, the crop plant's resilience to contaminants can be improved. Formerly, for remediation of toxic pollutant from the soil, phytoremediation, and chemical amendment-based approaches were used due to cost-effectiveness, however, these methods have several limitations. Importantly biotechnology has transformed modern-day agricultural practices, crop breeding, and improvement. These biotechnological techniques involved innovations, such as genome editing, marker-assisted selection, and genomics-assisted breeding. Notably, transgenic techniques involve inserting and deleting specific gene even from distinct species used for genetic modification of plants to acquire new traits. Moreover, technologies, such as genome editing and DNA markers, also revolutionize plant breeding and can help improve crop productivity. This chapter highlights the potential application of genic, molecular approaches for improved crop stress resilience and productivity.

H. Kavidayal
Genetics and Tree Improvement Division, ICFRE-Forest Research Institute, P.O.I.P.E., Dehradun 248 195, Uttarakhand, India

A. Dabral (✉)
Department of Forest Biology and Tree Improvement, College of Forestry, Ranichauri, VCSG Uttarakhand University of Horticulture and Forestry, Pauri Garhwal, Uttarakhand, India
e-mail: amandabral93@gmail.com

M. J. Dobriyal
Department of Silviculture and Agroforestry, College of Horticulture and Forestry, Rani Lakshmi Bai Central Agricultural University Jhansi, Jhansi, Uttar Pradesh, India

B. C. Sarkar
Faculty of Agricultural Sciences, GLA University, Mathura, Uttar Pradesh, India

I. Singh
Department of Silviculture and Agroforestry, College of Forestry, Ranichauri, VCSG Uttarakhand University of Horticulture and Forestry, Pauri Garhwal, Uttarakhand, India

A. Husen (ed.), *Enhanced Crop Production Setup Under Various Micro and Nano Contaminants*, Smart Nanomaterials Technology,
https://doi.org/10.1007/978-981-96-6099-5_11

Keywords Agroecosystem · Productivity · Gene regulation · Omics · Gene editing

1 Introduction

Globally, in recent time, we have witnessed the rapid expansion of contemporary intensive farming practices, urbanization, changes in land use system, rapid population expansion, and industrialization. These are the main causing factors leading to contamination of land, water, and air due to contaminants, such as pesticides, insecticide, various pollutant including nano/micropollutants, particulate matter, salinity, and so on [53, 54, 56, 53, 54]. These emerging pollutants pose a significant danger to global agricultural productivity and food security. These contaminants are toxic, infiltrating the food chain and impeding agricultural productivity [108]. These challenges can be mitigated using genetic and molecular approaches. Additionally, using these approaches, agricultural crop resilience against various contaminants can be improved. Moreover, these approaches could be an effective tool to achieve sustainable agriculture production that enhance global food security and nutrition. Previously, phytoremediation, soil amendments, and chemical treatments were the conventional methods used for reclamation of contaminated sites. However, these methods have some limitation such as less efficacy, high cost, lack long-term sustainability, and most importantly in case of severe contamination, these method prove inadequate and fail to tackle the essential requirement of augmenting the plant's intrinsic stress tolerance. However, recent advancement in genetic engineering techniques has transformed plant breeding and improvement, resulting in creation of genetically improved better-adapted agriculture crop varieties that are tolerant to contamination and climate changes. Genetic engineering techniques includes genomics-assisted breeding, next-generation sequencing methods, omics technology, gene expression modulation, genome editing, etc. [91]. Despite the potential of these methods, their acceptance is hindered by various limitations, including widespread acceptance, regulatory constraints, and field trials. This chapter is covering genetic and molecular approach potential for developing agricultural crops that can tolerate contaminated sites.

2 Stress Adaptation and Impact of Various Stress on Agricultural Crop

Agricultural crops are generally subjected to a wide range of stresses, which results in a reduction in the productivity of crops. These stresses are broadly categorized into two types, viz., abiotic and biotic stress [46]. These stresses are the most critical obstacle to the overall growth and development of crop plants and key limiting causes affecting crop production yield as well as quality [79]. Notably, for stress adaptation in agricultural crops, many regulatory processes, such as physiological, biochemical, genetic, and molecular mechanisms, involved these mechanisms usually depend on altered gene expression [55, 57, 148]. Stress adaptation is vital for agricultural crops as there is need to develop climate-resilient crops to cope with changing climatic conditions around the globe, therefore, understanding these mechanisms is essential [35]. Moreover, agriculture crop productivity is mainly influenced by several biotic and abiotic stresses, necessitating adaptation to these stresses, the plant develops efficient strategies to adapt and defend against these stresses [33, 43, 51, 52, 77]. Importantly, crops are subjected to many stresses like abiotic, i.e., drought, salinity, temperature extremes, and biotic such as pathogens and herbivores. Moreover, these stresses have a substantially impact crop growth and production. Plants adapt to biotic and abiotic stresses by developing strategies at morphological, anatomical, biochemical, and molecular levels (Table 1). These strategies include epigenetic memories, molecule crosstalk, reactive oxygen species signaling, hormone accumulation, redox status changes, R-gene resistance, and systemic acquired resistance. New "omics" technologies help researchers identify genetics behind plant stress response and adaptation [68].

Comprehending and manipulating these mechanisms via genetic and molecular techniques might substantially improve the stress resilience of crops and be used in developing and promoting highly stress-tolerant agricultural crops [148]. In addition, researchers are investigating genetic and molecular approaches to improve crop stress tolerance. Conventional breeding, genetic engineering, CRISPR/Cas9 genome editing, marker-aided selection (MAS), and omics technologies are the some significant approaches [130]. Importantly, CRISPR/Cas9 genome editing allows precise editing of plant genomes [113], MAS reflects specific genetic markers associated with desirable qualities [112], though omics technologies offer new ways to understand stress responses [62]. In contrarily, traditional breeding procedures are time-consuming and have limited genetic variety.

3 Types and Sources of Contamination that Affect Agricultural Crop Growth and Yield

Recently, explosion of the global population, climate change, coronavirus pandemic (COVID-19), significantly created an imbalance in all sectors globally, including agricultural crop production and possess a serious threat to food security [123]. Such factors leads to use excessive fertilizers and inorganic supplements to increase agricultural productivity. Furthermore, the potential threats to agriculture as that can accumulate in crops, and affect the growth, development, and product quality are, emerging contamination (ECs) in the form of nano- and microplastics, per- and poly-fluoroalkyl matters or perfluorinated chemicals (PFAS), endocrine-disrupting chemicals (EDCs), antibiotics, hormones and pain relievers [96]. The main sources

Table 1 Adaptation strategies of plant during various stress

S. No.	Class	Adaptation strategies	Examples	Agricultural crops	Key references
1	Morphological adaptation	Leaf shedding, leaf size reduced, more branching to conserve water, small roots and lateral roots for better water uptake, Sclerophylly	Traits, such as Xerometric in plants growing in dry areas, drought-tolerant plant	*Sorghum bicolor*, *Triticum aestivum*	[2, 6]
2	Anatomical adaptation	Thick cuticle, thick-walled epidermal cells to reduce water loss, aerenchyma formation in roots for oxygen transport under waterlogged conditions, increased stomatal size in response to drought	Sclerophyllous leaves in drought-tolerant species	Rice, maize, wheat	[61]
3	Biochemical adaptation	Accumulation of compatible solutes (proline, glycine betaine) for osmotic adjustment under drought stress, synthesis of oxidants (SOD, CAT) to combat ROS under oxidative stress	TaTip2;2 in B2, an abscisic acid	Wheat	

(continued)

Table 1 (continued)

S. No.	Class	Adaptation strategies	Examples	Agricultural crops	Key references
4	Molecular Adaptation	Activation of transcription factors (MYB, ERF, NAC) under biotic/abiotic stress, epigenetic changes for stress memory (e.g., priming for faster response in repeat stress events)	*AtMYB15* for cold stress tolerance in *Arabidopsis*, Defense priming in plants	*Arabidopsis* Spp.	[143]
5	Defense mechanism	Hypersensitive response (cell death around infection site), Systemic acquired resistance (SAR) to enhance immunity throughout the plant, development of physical barriers such as thickened cell walls and cuticle to block pathogen entry	-	Tomato	[1]

of these contaminants include agricultural practices, industrial activities, pharmaceuticals, and urban runoffs. Additionally, excessive use of fertilizers, contaminated irrigation water, soil amendments for a long time resulting in heavy metal, lead, cadmium, arsenic accumulation in agricultural soil, reduce fertility, simultaneously, decreases agricultural crop and productivity [109]. In the context of molecular and genetic research, it is crucial to comprehend the various types and sources of contamination. This understanding is crucial for developing crop varieties that are able of withstanding and lessen the effect of these pollutants. Broadly, types and sources of contamination are categorized as shown below (Fig. 1).

Fig. 1 Types and sources of contamination in agriculture

3.1 *Chemical Contaminants*

Chemical contaminants in agricultural practices are among the most widespread and harmful, originating from both agricultural and nonagricultural sources.

3.1.1 Fertilizers

In the last few decades' crop production increased to satisfy the basic needs of the increasing human population. This rise in agriculture production increases the demand for fertilizer consumption. It was estimated that total fertilizer requirement increased by 1.6% per annum globally, for fulfilling demands for essential nutrients for instance nitrogen, phosphorus, and potassium. These key nutrients are essential for growth and development of crops, so they form a large part of the fertilizer industry. Importantly, nitrogenous fertilizers, such as ammonium nitrate and ammonium phosphate, nitrate, and nitrous oxide contribute to nitrate pollution. While, phosphatic fertilizers contain cadmium (Cd), arsenic (As), radium (Ra), etc. These contaminants can accumulate in soil and plants leading to toxicity in crops. Potassium fertilizers, mainly, potassium chloride (KCl), are potential sources of accumulation of chloride ions.

3.1.2 Heavy Metals

Agricultural systems are affected by various pollutants, including heavy metals (cadmium (Cd), lead (Pb), mercury (Hg), and arsenic (As)). Sources of these heavy metals include fertilizers, pesticides, and sewage sludge. Organic and synthetic fertilizers contribute to heavy metal accumulation in soils and plants. As, Cu, Zn, and Cd have high accumulation potential. Inorganic fertilizers, such as phosphate fertilizers, liming materials, and biofertilizers, contribute to heavy metal release in agricultural soils and reach animals and humans [5].

3.1.3 Pesticides

Pesticides are nowadays essential in agricultural production as farmers are utilizing them to manage weeds and insects, and significant enhancements in agricultural yields. However, poor utilization and excessive application may result in contamination. Some persistent organochlorine pollutants like DDT and its derivatives can endure in the soil for years, impacting microbial activity and plant vitality. Moreover, organophosphates are harmful to several organisms and can infiltrate water sources via runoff, adversely affecting aquatic ecosystems and agriculture. Although carbamates decompose more rapidly than organochlorines, excessive application may still result in soil and water contamination.

Other than the above said contaminant sources industrial contaminants also entered in agricultural systems through atmospheric deposition or tainted irrigation water, causing soil acidification and reducing crop nutritional availability.

3.2 Physical Contaminants

These nonorganic materials, such as microplastics and particulate matter from industrial or urban sources, can impede plant growth in soil or water. Since these contaminants are not easily degradable, they can persist for a longer time in water and soil. These contaminants can have long-term impacts on soil structure and water quality. The increasing use of plastics in agriculture, such as mulch, greenhouse films, and irrigation pipes, has led to the accumulation of plastic trash in agricultural soils. The utilization of plastic has significantly escalated. Soil serves as a major reservoir for these contaminants, which can infiltrate the soil environment through multiple pathways such as sewage sludge, compost, contaminated irrigation water, plastic mulching, atmospheric deposition, and littering. These contaminants can disrupt soil structure, impede water infiltration, interfere with root growth, and reduce crop productivity. Dust, soot, and other pollutants can deposit as particulate matter onto crops and soil surfaces, thereby reducing sunlight, photosynthesis, and crop yields [60].

3.3 Biological Contaminants

Biological contamination results from the presence of bio-organisms including bacteria, algae, viruses, and protozoa. Pathogenic bacteria in plants lead to significant yield reductions in various economically important crops, causing economic, and social adversity. Human activities, such as monoculture agriculture and international trade, exacerbate the dissemination of plant infections and the generation of novel diseases. The origin of such infectious entities is contaminated soil, water, or plant germplasm [129]. Moreover, invasive plant species also negatively impact agriculture ecosystems, causing significant crop production reduction due to their widespread distribution in various ecosystems. Invasive plant species not only compete with crops for water, nutrients, and light, but also alter the soil composition, thereby affecting nutrient cycling and soil fertility [10]. Other contaminants are soil-borne parasitic nematodes and soil-born pests that cause severe deterioration of crops health by feeding on plant roots, reducing their ability to take up water and nutrients. These factors are also leading to vast monetary losses in the agriculture and livestock industry globally [85].

3.4 Nano-Based Contamination

The scientific community is working tirelessly to increase the crop yield to meet global population needs. It is projected that the population will escalated to 10.0 billion by 2050. Equally, due to climate change, scientists will face more challenges in maintaining or increasing agricultural yield and food security. Besides traditional fertilizer applications, a technology called nanoparticles (NPs) is an emerging technology in agriculture. This technology has wide scope due to its advantages such as enhancing energy efficacy, cost, and comparatively higher productivity to traditional approaches [90]. Notably, nanotechnology used in agriculture as nano-fertilizers, nanopesticides, nanosensor and nano-remediation technology. Notably, nanotechnology also has toxic effects and risk to plant and human due to its unique properties of nanoscale materials.

3.4.1 Nanofertilizers

Nanofertilizers (NFs) technology in agriculture can transform modern agriculture system, increase yields with minimizing environmental impact as these are a promising solution for increasing agricultural efficiency and productivity. Nanofertilizers mainly classified into five categories, action based, controlled-release, targeted delivery, plant growth-promoting, and control water and nutrient losses. These are new generation fertilizers can easily absorbs by organism through nose and mouth, leading to potential toxicity and neurological damage. The excessive and improper use of nanofertilizers can decrease the crop growth. It releases in the environment and pollute soil, water, and air, lowers soil ecosystems, and decreasing fertility. Nanofertilizers entered into aquatic ecosystems and possess serious threats to aquatic organisms, through bioaccumulation and biomagnification. Furthermore, NFs also possesses serious risk to nontargeted organism, for example, insects, fish, and birds, as it disrupt reproduction, growth, and development, potentially resulting in population decreases [136].

3.4.2 Nanopesticides

Nanopesticides, like conventional pesticides, can enter the environment through accidental or deliberate emissions. Characterizing their biological safety is challenging due to their complexity. Glyphosate's toxicity is a concern for manufacturers, workers, and food safety. Nanoparticles can be contaminated with solvents, posing risks to animals and humans. RNA directed against insect target genes provides pest control, but gaps such as delivery mechanism, low environmental stability, toxicity, and long-term pest resistance have not been addressed. Atrazine, an environmental persistence insecticide, is used extensively in corn and sugarcane crops, but its adverse effects on soil and aquatic ecosystems are significant. Imazethapyr, a hazardous agrochemical ingredient, is considered irritant to the eyes and respiratory system. In vitro tests show that fipronil negatively affects mouse embryo development and vertebrate reproduction [25]. Additional research is necessary to comprehend the long-term environmental and health impacts, optimize safe nanomaterial designs, and implement risk assessment frameworks for sustainable nanotechnology in agriculture.

4 Genetic Approaches for Tolerance Against Contamination

Contamination, including heavy metals, pesticides, and other pollutants, poses significant threats to agricultural productivity and environmental health due to their harmful effects on the natural environment. Genetic approaches can enhance tolerance in crops, and these genetic tolerance mechanisms involve the modification of specific genes to allow organisms, particularly crops, to thrive and survive in the contaminated environments. Numerous processes, including improved detoxification, altered metabolism, and higher absorption or sequestration of harmful substances, can lead to tolerance. Scientists and practitioners are increasingly focusing on these mechanisms to develop crops that can maintain productivity in contaminated soils. Understanding the genetic concept of tolerance allows facilitates common approaches such as traditional breeding, genetic engineering, marker-assisted selection, editing tools and techniques, among others, to achieve these modifications (Fig. 2).

Fig. 2 Genetic approaches for tolerance against contamination in agroecosystems

(a) **Breeding approaches**: Researchers have employed the conventional breeding method to improve crop productivity through phenotypic selection, utilizing wild germplasm, and measuring and developing hybrid varieties that are resistant to stressors, diseases, and contaminants. However, limitations include uncertainty of gene combinations among crosses, transfer of both desirable and undesirable traits, and time/cost constraints [3]. Hence, the breeders have moved on to advance breeding techniques, such as molecular breeding. According to most studies, As and Cd contaminants affect some crops, such as rice, wheat, barley, maize etc. Since rice cultivated in flooded conditions, it is a severely affected staple food crop as compared to other nonflooded crops [89]. Notably, genes, such as *OsNRAMP5, OsZIP5* and *OsZIP9,* and *OsNRAMP1,* etc., are present in the rice that helps to enhance tolerance and reduce pollutants accumulation especially Cd and As [140]. Therefore, breeders have employed the genomic approach, called molecular breeding, to detect the quantitative trait loci (QTLs) with specific heavy metals (HM) accumulation, translocation, and tolerance mechanism in rice aiming to reduce HM pollutants [58]. Moreover, arsenic toxicity in rice two QTLs for arsenic contents were found in the root and shoot on chromosome 8 and chromosome 2, 5, 6, and 9, respectively. The backcross population derived from a cross between WTR1 (indica) and Hao-an-nong (japonica) was hydroponically cultivated and subjected to As (10 ppm) and genotyped by using 704 single nucleotide polymorphism (SNP) markers. The WTR1 showed higher tolerance for chlorophyll content and biomass accumulation while Hao-an-nong showed more tolerance for root length and plant height under arsenic stress. Furthermore, hydroponic is also one of the best screening methods to estimate the tolerance level in crops [93]. Additionally, 16 hydroponic rice cultivars and four rice genotypes, viz., Xiushui 123, Xiushui 134, Jiahe 991, and Xianghu 301 exhibited improvements in root length, dehydrogenase activity, and root dry weight under copper stress, and the selected genotypes were utilized for local agriculture [88]. Similarly, the rice variety BRRI-29 from the Bangladesh Rice Research Institute (BRRI), traditional rice varieties, viz., Beto and Soria Sali of Assam, India have been found tolerance against arsenic, and aluminum toxicity under phosphorous deficiency, respectively [75, 118]. This modernized approach of breeding not only ensures crop tolerance against contaminants but also contributes to reduce the health risks in animals and humans.

Limitations:

Like conventional breeding techniques, the QTL mapping has also several limitations.

1. The QTL mapping is less accurate.
2. Required large and diverse population. It is less effective in small populations and involves many assumptions.
3. As distance between QTL and markers increases, the chance of detecting QTL decreases.
4. Single experiments can detect only QTLs with a relatively large effect on the trait of interest.

(b) **Marker-assisted selection (MAS)**: A promising technique for boosting produce and safeguarding crops from biotic and abiotic stress is marker-assisted selection. Unlike conventional breeding, MAS facilitates the effective pyramiding of desirable traits within a single genotype, achieving this more precision and fewer selection cycles, hence minimizing unintentional losses, thereby reducing time and resources while enhancing the crop yield [12]. The MAS is widely used for drought, water lodging, and cold tolerance in crops including rice, tomato, wheat, maize, and pigeon pea. Some crop varieties, such as Barley—SloopSA, Tango [50] (Barr et al. 2000), maize varieties—Vivek QPM 9, Sunrise (Hash et al. 2006) [12], rice varieties—Cadet and Jacinto, Pusa 1460, Samba Mahsuri [42, 48, 115]; wheat varieties—Mace, Westmore, UC1113 and Lassik ([21, 44, 144]; and Dubcovsky et al. 2008); and tomato variety—AB2 [74] are successful crop varieties through marker-assisted breeding. Breeding programs use markers like KASP (kompetitive allele-specific PCR) to accelerate efforts to improve drought tolerance in crops. Eltaher et al. [37] found that TaDreb-B1, a KASP marker, has the potential and consistency to enhance drought tolerance in wheat. However, besides being a good modern technique, it has some limitations to work in the field of genetics.

Limitations:

1. MAS is dependent on QTL mapping and as the number of QTL alleles increases it shows limited effectiveness.
2. It has limited marker availability which can lead to limited selection options.
3. When recombination occurs between markers and gene of interest leading to false-positive results.
4. It is challenging to transfer traits from wild to elite varieties.
5. It can be more expensive than conventional techniques.

(c) **Transgenic approaches**: Transgenic plants are organisms that have undergone genetic modification to acquire a new trait—such as resistance to contamination—that they do not naturally possess. These plants are synthesized by artificial induction of different genes from other resistant plants or bacteria [103]. The transgenic approach is being harnessed to enhance plants' ability to detoxify and tolerance against stresses and environmental contaminants. Transgenic plants with particular bacterial genes transformed mercury and selenium into less toxic varient. Besides these transgenic methods, using endophytes, which are microbes that live inside plants, also helped plants handle chemicals that normally hurt them better and got rid of more pollutants [32]. *Agrobacterium tumefaciens*-mediated plant transformation is the predominant approach for plants transformation by transferring T-DNA from the *A. tumefefaciens* to genome of host plant [121]. The transgenic alfalfa (*Medicago sativa* L.), an *Agrobacterium*-mediated transgenic plant, generated by expressing ATP sulfurylase gene from *Arabidopsis*, human CYP2E1 and Glutathione S-transferase (GST) genes from hypocotyl segments to enhance their accumulation efficiency against mixed contamination in soil including heavy metals and organic pollutants [67, 147]. Recently, a new genotype, *Brassica napus* cv. BARI Sarisha-8 has been recognized as a highly resistant *B. nupus* genotype to Cd and lead (Pb) stress through genetic engineering utilize the *Arabidopsis thaliana* ABC transporter associated with mitochondrion 3 (*AtATM3*) gene [124]. Researchers are developing genetically engineered plants as effective tools to better understand phytoremediation, which paves the way for more effective treatment of heavy metal contaminations [137]. Moreover, transgenic technologies employ RNA interference (RNAi), microRNA (miRNA)- or artificial microRNA (amiRNA) for gene silencing techniques to enhance biotic and abiotic stresses in agricultural crop [17, 19, 65, 127]. This phenomenon plays an important role in plants' physiological and development processes and is preferred for studying gene functions of the plants [17, 63]. The *Arabidopsis* plant was utilized for RNAi by target *ACR2* gene to enhance arsenic hyperaccumulation for phytoremediation purposes (https://www.isaaa.org/resources/publications/pocketk/34/default.asp). These kinds of studies accelerate the transgenic process for crop improvement under stressful conditions and open a way to address further efficient protocols to treat heavy metal contamination.

Limitations:

Genetic engineering presents numerous disadvantages. Certain people contend that it is an inadequate method for both the environment and humanity. The transfer of genes across plants can result in cross-contamination and heightened weediness in the environment. It may be a contributing factor to health hazards in humans. Consequently, obtaining regulatory clearances for GM foods is quite tough to guarantee food safety. The process is costly, and not all nations, particularly impoverished countries in Africa, can purchase GM crop seeds. Due to ethical concerns, individuals refrain from consuming genetically modified foods. Likewise, the *Agrobacterium*-mediated transformation necessitated stringent regulatory authorization. The transformation efficiency is low, transgene integration is sporadic and potentially unexpected, and the entire process often requires several months to complete.

5 DNA Marker Applications for Tolerance Against Contamination

Breeding programs can use DNA markers, also known as molecular markers, to track inheritance by identifying DNA variation sites through DNA fragments or genes linked to specific genomic regions. It can enhance a plant's genetic structure in many ways, such as to find the parents' genes, measure genetic variation, find and develop agronomic trait-marker linkage, marker polymorphism, determining dominant or co-dominant inheritance, genetic mutations, and variation [8, 142]. The list of DNA markers is huge and is classified based on their nature such as PCR and non-PCR or hybridization, locus specific or non-specific, dominantly and co-dominantly inherited, as well as the degree of marker polymorphism and repeatability efficiency. Molecular markers provide all inheritance details related to allelic position including heterozygosity, maternal homozygosity, and paternal homozygosity, for each progeny or lineage within the population which may be readily verified and documented. Therefore, researchers have been utilizing these marker techniques in crop plants to identify genetic markers linked with pollutant tolerance and evaluate the effect of pollution on plant genetic diversity as well [83]. Crop plants are constantly negatively exposed to unfavorable environmental conditions that affect their growth, development, or productivity. The frequent rise of agricultural contamination poses a significant threat to food security, environment, and human health. Continuous contact with pollutants, viz., heavy metals (HMs), pathogens, and fertilizers, can harm or damage DNA (deletion, insertion, and point mutation) producing genotoxicity, which can lead to numerous negative health effects that may increase the risk of cancer and genetic abnormalities [49]. The change in DNA can be analyzed with DNA markers such as amplified fragment length polymorphism (AFLP), random amplified polymorphic DNA (RAPD), expressed sequence tags (ESTs), simple sequence repeats (SSRs), and single nucleotide polymorphism (SNP).

Crop species including rice, wheat, and maize are majorly affected by the stresses and heavy metals. In rice, heavy metal transporter genes, viz., *OsHMA3, OsNRAMP5, OsZIP5* and *OsZIP9,* and *OsNRAMP1* are associated with tolerance and reduce Cd and As accumulation [140]. Similarly, metal detoxification transporters *TaHMA2* and *TaABCC2/3/4* [116] and P-type ATPase heavy metal transporter, *ZmHMA3* protein [73] in wheat and maize, respectively, are linked with Cd, Pb, and Zn tolerance. Miftahudin et al. [87] conducted a study on *Secale cereal* L. in which they found three AFLP markers (AMAL1, AMAL4 and AMAL5), by analyzing recombinant inbred line (RIL) population, to be tightly linked to aluminum-tolerant Alt3 gene on chromosome 4R. This concluded that these markers can identify the genetic determinants of aluminum tolerance in Rye. Besides a co-dominating marker, it has some disadvantages including less reproducibility, complexity, and time-consuming due to which it began to be used less. Hence, in technological advancement, researchers used other DNA markers which has more reproducibility, less time constraints, and efficient. SSR markers are one of them which are highly polymorphic, abundant, and highly reproducible. Due to its efficient nature, researchers utilize SSRs to map QTLs associated with stresses and heavy metals tolerance. Barakat et al. [13] identified three genes viz., *Xgwm132*, *Xgwm617*, and *Xgwm577*, by SSR analysis for heat tolerance in wheat crops. Likewise, SSR markers in rice, *RM3586* (chromosome 3), and *RM3735* (chromosome 4) had a significant association with heat tolerance exhibiting 3 and 17% total genetic variation, respectively [13, 40, 146]. In addition, [49] also implemented AFLPs and SSRs to assess the genotoxic impact and genetic stability on three polluted plant species (lettuce, mint, and rocca) under polluted conditions and found significant for the other polluted plant species. Similarly, [100] studied on *Miscanthus sinensis* and developed EST-SSRs linked to Cr tolerance. The four genes associated with ABC transporters, two high-affinity sulfate transporters, and wall-associated receptor kinases were identified significant under Cr stress. Many studies have been found SNPs role in tolerance against heavy metals such chromium (Cr), chlorine (Cl) and ammonium [7, 78]. SNPs are genetic polymorphisms when variation occurs in a single position in a DNA sequence among the individual. The SNPs associated with important trait could assist in molecular breeding for tolerance to Cr toxicity [7]. Similarly, in *L. vannamei* 12 SNPs were identified as being linked with ammonia tolerance [78].

The SNP markers associated with ammonia tolerance offer significant potential for to genetic selection to reduce mortality and enhance production. Numerous SNPs were recognized that either enhance or reduce the associated risk of chlorine (Cl). These SNPs compromise markers adjacent to genes associated with mast cell activation pathways. Researchers discovered that four substances (bisphenol A, valproic acid, aflatoxin B, and benzo(a)pyrene) significantly modified the gene expressions of the detected SNPs. Today, researchers also use association mapping to study abiotic stresses in crop plants, as it evaluates a greater number of alleles in a larger population than linkage analysis does. However, traditional methods tend to limit the identification of rare alleles in plant populations, prompting researchers to adopt SNP chips for both association and linkage mapping. Studies showed that SNP chips yield high-quality data, precise genotyping data, and extensive genetic information. Due to the abundance and variation, SNPs are often used for linkage analysis compared to SSRs. Therefore, these advanced molecular markers are used in identifying functional genes and genetic difference [139]. In this view, [47] discovered 27 SNPs associated with genetic variations associated to drought tolerance in maize lines by detecting functional genetic differences.

In conclusion, DNA markers play a crucial role in understanding tolerance against various contaminations. These methods are very cost-effective, efficient, and reliable opening a new way to identify the specific genetic traits associated with resistance and/or tolerance to environmental pollutants. This approach not only helps to identify the genes and markers but also contributed in bioremediation, conservation, and resilience against the pollutants, ultimately promoting environmental sustainability and biodiversity conservation.

6 Advanced Genome Editing Tools

Nowadays, people widely use genome editing technology, which uses engineered endonucleases, in addition to molecular marker technology. In genome editing technology, an organism's genome is altered by knocking a desired gene for targeted traits, while in transgenic approaches, foreign genes are incorporated into the host genome [94]. This approach is most promising for elucidating plant genomes to improve productivity, stresses tolerance, and pollutantion resistance. Such techniques are zinc finger nucleases (ZFNs) [23] transcription activator-like effector nucleases (TALENs) [72, 80], and clustered regularly interspaced short palindromic repeat (CRISPR)-associated Cas9 [28] Importantly, ZFNs and TALENs are effective tools for generating genome modification, widely accepted tools due to its effectiveness, accessibility, cost-effectiveness, and ease of engineering [117]. Whereas the CRISPER/Cas9 is more advanced and promising technology compared to ZFNs and TALENs.

Nevertheless, one of the promising gene editing techniques is ZFNs, this technique are assemblages of DNA recognition modules and the cleavage domain of the *Fok*I restriction endonuclease combined with zinc finger proteins [41, 136, 138]. These techniques includes deletion, insertion inversions, translocations, and point mutations and can be used for range of genetic modification and introduced into a targeted genome [23]. Importantly, in eukaryotes, the zinc finger motif Cys2-His2 is popular DNA-binding domains. The ZFN-mediated gene knockouts were first successfully executed in tobacco (*ALS* gene) for the development of herbicide-resistant plant [59, 128] followed by soyabean [29], *Arabidopsis thaliana* (for tolerance to abiotic stress and seed development) [71, 104, 145], and maize (for herbicide resistance) [122].

However, there are several limitations to ZFN-mediated technology including off-target binding and DNA sequence cleavage, complexity, time-consuming, and high expense of constructing protein domains for each specific genome locus [24, 94]. Hence, the search for a novel method for genome editing continued and new tools such as TALENs and CRISPR/Cas9 were developed for genome modification [99].

Transcription activator-like effector nucleases (TALENs) are engineered restriction enzymes that target and cleave the specific DNA sequences. Importantly, TALEs (transcription activators like effectors) are produced naturally by *Xanthomonas* spp a bacterium that secrete proteins delivered to plant cell in response to increase susceptibility against to pathogen [99]. Generally, TALENs are typically employed to induce mutations through homologous recombination (HR) and nonhomologous end-joining (NHEJ). Importantly, the TALENs has been utilized for genome editing in range of plants, including Arabidopsis, Nicotiana, Brachypodium, barley, potatoes, tomatoes, sugarcane, flax, rapeseed, soybean, rice, maize, and wheat [81, 84]. Notably, rice was the initial crop in which TALENs were utilized for improvement [72]. Currently, researchers are exploring TALENs which have emerged as a novel approach for genome editing in various species. TALENs have also been utilized in the model plant *A. thaliana* to target five Arabidopsis genes such as ADH1, TT4, MAPKKK1, DSK2B, and NATA2 to enhance the somatic mutagenesis frequencies [125]. In potato tuber, due to the high temperatures, the reducing sugars react with amino acids, producing a high amount of acrylamide, which is a carcinogen. Therefore, to regulate the accumulation of reducing sugar, TALENs and RNAi technology were employed to knockout the vacuolar invertase gene (*VInv*) in commercial potato variety, Ranger Russet [27]. However, studies are ongoing on TALENs in tolerance against contamination such as heavy metals. Due to its limited role in improving crop tolerance to contaminants, the CRISPR/Cas9 system is found as an efficient, simple, and flexible technique for genome editing [132, 133].

Clustered regulatory interspaced short palindromic repeats (CRISPR) associated with protein 9 (Cas9) technologies is a robust RNA-guided genome editing tool. It was initially found from *Streptococcus pyogenes* as a component of the adaptive immune response against invading nucleic acids [117]. This genomic editing technique facilitates precise and targeted modifications without the incorporation of foreign DNA, proving effective in plants [114]. In modern times, it has become a transformative instrument for redesigning the genomes of numerous crop plants to withstand several stresses [15] along with gene editing in plant growth-promoting rhizobacteria (PGPR) can enhance the production of bioactive compounds that facilitate tolerance to pollutants, accumulation, and detoxification of hazardous pollutants [76]. The CRISPR-mediated precision in genetic engineering is widespread. In this view, the advanced molecular biology facilitated by CRISPR/Cas9 tremendously impacted and enabled a strong, dynamic, and effective change that can reach sustainable development and ecological restoration through phytoremediation [97, 126].

CRISPR-mediated genome editing is very particular and dependent on the availability of plants' genome sequences. Some phytoremediators, such as *Thlaspi caerulescens, Arabidopsis halleri, Hirschfeldia incana, Pteris vittata,* and *Brassica juncea,* have been sequenced partially or completely which can accumulate heavy metals such as Cd, Zn, Ni, Pb [11, 14, 22, 82]. For instance, the CRISPR-mediated regulation of the metal transporter gene *OsNRAMP5* led to reduced Cd-accumulation in indica rice without compromising yield [126]. Recently, *oxp1* CRISPR-mediated mutants in *Arabidopsis* shown enhanced resistance to Cd. Similarly, in rice several knockout mutants, viz., *OsLCT1* and *OsNramp5* (reduce Cd level), *OsARM1* (tolerance to As), *OsHAK1* (controls the uptake and translocation of Cesium) were generated by CRISPR [4, 34, 120, 134]. This advanced technique is an immense breakthrough in the world for crop productivity, pollution tolerance, and detoxification of hazardous pollutants while also mitigating the health risks in animals and humans. However, it also raises ethical, safety, and regulatory challenges that require careful consideration.

7 Omics Approach

Contaminants, at low concentrations, act as cofactors to stimulate the molecular, physiological, and metabolic pathways of plants, thereby aiding in their thorough growth and development. It is a very challenging task to decipher the contaminants and their physiological, biochemical, and molecular responses at the cellular level [30, 107]. Modern biotechnological tools and techniques have enabled a more in-depth probe to understand interaction between metal and plant [66] comprehensively. Omics offers a superior and more holistic perspective on the physiological, biochemical, and molecular processes of stressed cells' [66]. In recent times, numerous omics-based methodologies have been advocated as viable tools for plant systems. Omics techniques, including genomics, transcriptomics, proteomics, metabolomics, ionomics, and phenomics, are emerging approaches in crop science. This has been made possible solely by new advancement in next-generation sequencing (NGS) technologies. Omics approaches have been applied to several key crops including rice (*Oryza sativa* L.), wheat (*Triticum aestivum* L.), tomato (*Solanum lycopersicum*), barley (*Hordeum vulgare* L.), soybean (*Glycine max*), maize (*Zea mays* L.), millet (*Setaria italica* L.), *Medicago truncatula* cotton (*Gossypium hirsutum* L.) etc.

7.1 Genomics

Genomics is the study of all of an organism's genes (genome) primarily aims on structure, organization, functional assessment, mapping, and modification of genomes [95]. Genomics has perceived remarkable advances in recent time and playing key role in the crop improvement by facilitating the sequence of a large part of gene, making the whole genome sequencing (WGS) relatively easier, cost-effective, reliable, and faster as compared to other traditional approaches. Notably, several genomic methods include structural genomics, functional genomics epigenomics, mutagenomics, and pangenomics, allowing crop breeders to identify the desired traits present in plant genome that is linked to tolerance [141]. A high-quality genome sequencing of introgressed lines of *Solanum pennellii* and *S. lycopersicum* able to find a candidate genes linked stress tolerance [20]. Significantly, genome sequencing is essential in the genomics and transcriptomics research of range of plants, highlighting the fact that individual genome content is more limited than previously believed. Furthermore, metagenomics approaches thoroughly investigate the microbial communities' dynamic ecosystem [31, 45]. Microbes including bacteria and fungi produce enzymes that help break down, remove, and detoxify harmful environmental pollutant. One of the key benefit of metagenomocs in enhancing the tolerance of induced crops against contamination by providing an understanding the complex interaction between plants soil microbes and environmental contamination [119].

In silico studies on environmental pollutants help us understand how microbes break down pollutants and find new genes that are responsible for microbial remediation [92, 110].

7.2 Transcriptomics

The regulation of gene expression is essential for plants to develop and tolerate to heavy metal stress. Transcriptomics is an investigation of the entirely of an organism's RNA transcripts generated by its genome within a cell or tissue. Observing the changes in gene expression within plants could give important information. This approach mainly aims at structure, classification, function, and expression of gene transcription [98, 111]. Importantly, transcriptomics reveals the expression at the whole genome level under stress, it makes easier to find new genes that help crop plants withstand against stress [131]. The effect of cadmium was observed on mung beans where cadmium remarkably inhibits the overall growth, biomass, and have disrupting effect on cell wall cell division and DNA replication and repair. The transcriptomic analysis reveals that 25,688 genes were differentially expressed (DEGs), which means that the body's response to cadmium is complex. It was found that increase in genes in response to detoxification, antioxidant defense mechanism, and stress response [70]. In addition, Cd-induced differentially abundant protiens (DAPs) assist plants in resisting and sequestering cadmium and other heavy metals [101].

7.3 Proteomics

Proteomics is the comprehensive analysis of entire protein sets expressed by an organism's genome. After genomics, research in proteomics is the significant advancement in the exploration of biological systems [135]. Moreover, proteomics offers a better understanding about structure and function of the organism compare to genomics. However, proteomics is much more complex than genomics, as they constantly changing among cells due to biochemical interactions with the genome and environmental influences [18]. Cell splicing, protein modification (glycosylation, phosphorylation), and protein degradation are other factors that make proteomics more complex. Notably, proteomics approaches helpful in unraveling about response of different cell to stress. These studies have found possible links between the amount and changes in proteins and how these factors affect a plant's ability to handle stress [18]. For instance, stress response of *Lactuca sativa* and *Mirabilis jalapa* to environmental contamination was assessed using proteomics [26, 69].

7.4 Metabolomics

It deals with the identification and quantification of all low molecular weight metabolites within a given organism during the development stage [9]. Plants contain thousands of metabolites, and most of them are still unknown [9, 102]. The metabolic level amplifies small changes in gene and protein expression because metabolites are smaller than genes and proteins [39, 64]. Metabolomics is an emerging omics methodologies widely used for crop improvement and heavy metal stress tolerance. Many studies [16, 36, 38, 86] have identified metabolomics as an effective omics tool for investigating the metabolic pathways in crop plants (Soybean, Pea, Alfalfa, and Maize) exposed to harsh climate conditions and abiotic stresses. These stresses affect the plants' metabolism (physiological and morphological processes) by damaging the membrane and antioxidant system in plants, ultimately reducing productivity and enhancing risks. Plant cells have a self-defense mechanism against foreign molecules by modifying the gene expression pattern, which is essential for understanding the plants' defensive strategies under stress conditions. Nonetheless, plant metabolomics comprehensive study and meticulous focus on data mining, annotation, assessment, processing, and evaluation. Overall, metabolomics has huge potential in crop plant research, which contributes to improving crop yield and quality, biomarker discovery, and stress tolerance.

8 Conclusion and Future Perspective

Agriculture plays a vital role worldwide which serve as the backbone of economies and societies. However, rise in environmental pollutants, including heavy metals, pathogens, fertilizers, and stressors, poses a significant threat to agricultural productivity worldwide. In the era of growing human populations, the demand for food certainly rises which requiring the development of resilient crop varieties that can survive and thrive in the adverse environmental conditions. Therefore, to enhance the crop productivity and tolerance to various contaminants, genetic and molecular approaches are one of the powerful tools which are aggressively used nowadays. The biotechnological technologies, including genetic engineering, MAS, and genome editing techniques like CRISPR/Cas9, have provided novel insights into genetic basis of environmental stresses and/or pollutants tolerance. The transgenic plants develop to express stress-resilient genes for the detoxification from harmful substances in soils and water. For example, some genes that encode for metallothioneins and associated with osmotic regulation have been shown to improve heavy metal tolerance and drought resistance in crops respectively. Moreover, to identify the candidate genes, RNAi and gene expression profiling methods are used in enhancing tolerance to both biotic and abiotic stresses. Despite these modifications, still the several challenges remain. The advanced approaches, such as GMOs, are required strict regulatory frameworks. Some people have believed that these genetically modified foods are not good for the environment and human health which is a contentious issue and needed a transparent communication and awareness regarding the benefits and safety of biotechnology in agriculture.

In the evolving era, technologies are also evolved time to time. The advanced technologies such omics, gene editing, and epigenetics offer an unprecedented opportunity to understand the plants responses to environmental stresses without introducing the foreign DNA. By integrating these methods, a standard model can be constructed to identify the key regulatory genes, plant stress responses, discover novel pathways associated with tolerance, and precise genetic modifications (CRISPR/Cas9 and CRISPR/Cas12). Similarly, the epigenetic modification is also taking place in understanding plant responses against various environmental pollutants. Future studies should explore the epigenetic landscape of stressed crops in order to understand how environmental factors influence gene expressions through DNA methylation and histone modification mechanism. Breeders, by utilizing the epigenetic knowledge, can develop tolerant crop plants exhibiting stable tolerance traits in a changing environment.

In conclusion, the integration of genetic and molecular approaches not only a scientific approach but also a societal imperative. As these challenges are global issue, thus by promoting national and international collaborations, public engagements, knowledge sharing, and education awareness can accelerate the development of innovative solutions, access to advanced technologies to adopt the sustainable agricultural practices, help misconceptions about biotechnology integration in agriculture and people believe and acceptance.

References

1. Abdallah NA, Shah D, Abbas D, Madkour M (2010) Stable integration and expression of a plant defensin in tomato confers resistance to *Fusarium* wilt. GM crops 1(5):344–350
2. Abreha KB, Enyew M, Carlsson AS, Vetukuri RR, Feyissa T, Motlhaodi T, Geleta M (2022) Sorghum in dryland: morphological, physiological, and molecular responses of sorghum under drought stress. Planta 255:1–23
3. Ahmar S, Gill RA, Jung KH, Faheem A, Qasim MU, Mubeen M, Zhou W (2020) Conventional and molecular techniques from simple breeding to speed breeding in crop plants: recent advances and future outlook. Int J Mol Sci 21(7):2590

4. Ainley WM, Sastry-Dent L, Welter ME, Murray MG, Zeitler B, Amora R, Corbin DR, Miles RR, Arnold NL, Strange TL, Simpson MA (2013) Trait stacking via targeted genome editing. Plant Biotechnol J 11(9):1126–1134
5. Alengebawy A, Abdelkhalek ST, Qureshi SR, Wang MQ (2021) Heavy metals and pesticides toxicity in agricultural soil and plants: Ecological risks and human health implications. Toxics 9(3):42
6. Ali Z, Merrium S, Habib-ur-Rahman M, Hakeem S, Saddique MAB, Sher MA (2022) Wetting mechanism and morphological adaptation; leaf rolling enhancing atmospheric water acquisition in wheat crop—a review. Environ Sci Pollut Res 29(21):30967–30985
7. Almas F, Hassan A, Bibi A, Ali M, Lateef S, Mahmood T, Rasheed A, Quraishi UM (2018) Identification of genome-wide single-nucleotide polymorphisms (SNPs) associated with tolerance to chromium toxicity in spring wheat (*Triticum aestivum* L.). Plant Soil 422:371–384. https://doi.org/10.1007/s11104-017-3436-1
8. Amiteye S (2021). Basic concepts and methodologies of DNA marker systems in plant molecular breeding. Heliyon 7(10)
9. Arbona V, Manzi M, de Ollas C, Gómez-Cadenas A (2013) Metabolomics as a tool to investigate abiotic stress tolerance in plants. Int J Mol Sci 14(3):4885–4911
10. Arun MN, Kumar RM, Sreedevi B, Padmavathi G, Revathi P, Pathak N, Venkatanna B (2022) The rising threat of invasive alien plant species in agriculture In: Resource management in agroecosystems. IntechOpen
11. Augu F, Fahr M, Moulin P, El Mzibri M, Smouni A, Filali-Maltouf A, Béna G, Doumas P (2016) Transcriptome changes in *Hirschfeldia incana* in response to lead exposure. Front Plant Sci 6:1231
12. Babu R, Nair SK, Prasanna BM, Gupta HS (2004) Integrating marker-assisted selection in crop breeding–prospects and challenges. Curr Sci 607–619
13. Barakat MN, Al-Doss AA, Elshafei AA, Moustafa KA (2011) Identification of new microsatellite marker linked to the grainfilling rate as indicator for heat tolerance genes in F2 wheat population. Aust J Crop Sci 5(2):104–110
14. Basharat Z, Novo LA, Yasmin A (2018) Genome editing weds CRISPR: what is in it for phytoremediation? Plants 7(3):51
15. Belykh ES, Maystrenko TA, Velegzhaninov IO (2019) Recent trends in enhancing the resistance of cultivated plants to heavy metal stress by transgenesis and transcriptional programming. Mol Biotechnol 61(10):725–741
16. Bernardi J, Stagnati L, Lucini L, Rocchetti G, Lanubile A, Cortellini C, De Poli G, Busconi M, Marocco A (2018) Phenolic profile and susceptibility to *Fusarium* infection of pigmented maize cultivars. Front Plant Sci 9:1189
17. Bharathi JK, Anandan R, Benjamin LK, Muneer S, Prakash MAS (2023) Recent trends and advances of RNA interference (RNAi) to improve agricultural crops and enhance their resilience to biotic and abiotic stresses. Plant Physiol Biochem 194:600–618
18. Bhatnagar D, Rajasekaran K, Payne GARY, Brown R, Yu J, Cleveland T (2008) The'omics' tools: genomics, proteomics, metabolomics and their potential for solving the aflatoxin contamination problem. World Mycotoxin J 1(1):3–12
19. Bogdanove AJ (2014) Principles and applications of TAL effectors for plant physiology and metabolism. Curr Opin Plant Biol 19:99–104. https://doi.org/10.1016/j.pbi.2014.05.007
20. Bolger A, Scossa F, Bolger ME, Lanz C, Maumus F, Tohge T, Quesneville H, Alseekh S, Sørensen I, Lichtenstein G, Fich EA (2014) The genome of the stress-tolerant wild tomato species *Solanum pennellii*. Nat Genet 46(9):1034–1038
21. Brevis J, Dubcovsky J (2008) Effect of the Gpc-B1 region from Triticum turgidum ssp. dicoccoides on grain yield and thousand grain weight
22. Briskine RV, Paape T, Shimizu-Inatsugi R, Nishiyama T, Akama S, Sese J, Shimizu KK (2017) Genome assembly and annotation of *Arabidopsis halleri*, a model for heavy metal hyperaccumulation and evolutionary ecology. Mol Ecol Resour 17(5):1025–1036
23. Carroll D (2011) Genome engineering with zinc-finger nucleases. Genetics 188(4):773–782

24. Carroll D (2014) Genome engineering with targetable nucleases. Annu Rev Biochem 83(1):409–439
25. Chaud M, Souto EB, Zielinska A, Severino P, Batain F, Oliveira-Junior J, Alves T (2021) Nanopesticides in agriculture: Benefits and challenge in agricultural productivity, toxicological risks to human health and environment. Toxic 9(6):131
26. Chen J, Zou W, Meng L, Fan X, Xu G, Ye G (2019) Advances in the uptake and transport mechanisms and QTLs mapping of cadmium in rice. Int J Mol Sci 20(14):3417
27. Clasen BM, Stoddard TJ, Luo S, Demorest ZL, Li J, Cedrone F, Tibebu R, Davison S, Ray EE, Daulhac A, Coffman A (2016) Improving cold storage and processing traits in potato through targeted gene knockout. Plant Biotechnol J 14(1):169–176
28. Cong L, Ran FA, Cox D, Lin S, Barretto R, Habib N, Hsu PD, Wu X, Jiang W, Marraffini LA, Zhang F (2013) Multiplex genome engineering using CRISPR/Cas systems. Science 339(6121):819–823
29. Curtin SJ, Anderson JE, Starker, CG, Baltes NJ, Mani D, Voytas DF, Stupar RM (2013) Targeted mutagenesis for functional analysis of gene duplication in legumes. In: Legume genomics: methods and protocols, pp 25–42
30. Dalvi AA, Bhalerao SA (2013) Response of plants towards heavy metal toxicity: an overview of avoidance, tolerance and uptake mechanism. Ann Plant Sci 2(9):362–368
31. De R, Mukhopadhyay AK, Dutta S (2020) Metagenomic analysis of gut microbiome and resistome of diarrheal fecal samples from Kolkata, India, reveals the core and variable microbiota including signatures of microbial dark matter. Gut Pathog 12:1–48. https://doi.org/10.1186/s13099-020-00371-8
32. Doty SL (2008) Enhancing phytoremediation through the use of transgenics and endophytes. New Phytol 179(2):318–333
33. Du B, Haensch R, Alfarraj S, Rennenberg H (2024) Strategies of plants to overcome abiotic and biotic stresses. Biol Rev
34. Du J, Yin N, Xie T, Zheng Y, Xia N, Shang J, Chen F, Zhang H, Yu J, Liu F (2018) Quantitative assessment of HR and NHEJ activities via CRISPR/Cas9-induced oligodeoxynucleotide-mediated DSB repair. DNA Repair 70:67–71
35. Eckardt NA, Ainsworth EA, Bahuguna RN, Broadley MR, Busch W, Carpita NC, Zhang X (2023) Climate change challenges, plant science solutions. Plant Cell 35(1):24–66
36. Ellis N, Hattori C, Cheema J, Donarski J, Charlton A, Dickinson M, Venditti G, Kaló P, Szabó Z, Kiss GB, Domoney C (2018) NMR metabolomics defining genetic variation in pea seed metabolites. Front Plant Sci 9:1022
37. Eltaher S, Hashem M, Ahmed AA, Baenziger PS, Börner A, Sallam A (2023) Effectiveness of TaDreb-B1 and 1-FEH w3 KASP markers in spring and winter wheat populations for marker-assisted selection to improve drought tolerance. Int J Mol Sci 24(10):8986
38. Feng Z, Ding C, Li W, Wang D, Cui D (2020) Applications of metabolomics in the research of soybean plant under abiotic stress. Food Chem 310:125914
39. Feng Z, Ji S, Ping J, Cui D (2021) Recent advances in metabolomics for studying heavy metal stress in plants. TrAC Trends Anal Chem 143:116402
40. Foolad MR (2005) Breeding for abiotic stress tolerances in tomato. In: Ashraf M, Harris PJC (eds) Abiotic stresses: plant resistance through breeding and molecular approaches. The Haworth Press Inc., New York, pp 613–684
41. Gaj T, Gersbach CA, Barbas CF (2013) ZFN, TALEN, and CRISPR/Cas-based methods for genome engineering. Trends Biotechnol 31(7):397–405
42. Gopalakrishnan S, Sharma RK, Anand Rajkumar K, Joseph M, Singh VP, Singh AK, Bhat KV, Singh NK, Mohapatra T (2008) Integrating marker assisted background analysis with foreground selection for identification of superior bacterial blight resistant recombinants in Basmati rice. Plant Breed 127(2):131–139
43. Gowda RSR, Kaur M, Chanchal, Kaushal B, Kaur H, Kumar V, Sharma R, Singh T, Choudhary A, Mehta S, Husen A (2024) Behavior, sources, uptake, interaction, and nutrient use efficiency in plant system under changing environment In: Husen A (ed) Essential minerals in plant-soil systems. Elsevier Inc., Cambridge, pp 97–127. https://doi.org/10.1016/B978-0-443-16082-0.00017-5

44. Graybosch RA, Peterson CJ, Baenziger PS, Baltensperger DD, Nelson LA, Jin Y, Kolmer J, Seabourn B, French R, Hein G, Martin TJ (2009) Registration of 'Mace'hard red winter wheat. J Plant Regist 3(1):51–56
45. Grossart HP, Massana R, McMahon KD, Walsh DA (2020) Linking metagenomics to aquatic microbial ecology and biogeochemical cycles. Limnol Oceanogr 65:S2–S20. https://doi.org/10.1002/lno.11382
46. Gull A, Lone AA, Wani NUI (2019) Biotic and abiotic stresses in plants. In: Abiotic and biotic stress in plants, vol 7, pp 1–9
47. Hao Z, Li X, Xie C, Weng J, Li M, Zhang D, Liang X, Liu L, Liu S, Zhang S (2011) Identification of functional genetic variations underlying drought tolerance in maize using SNP markers. J Integr Plant Biol 53(8):641–652
48. Hardin B (2000) Rice breeding gets marker assists. Agric Res 48(12):11–11
49. Hassan AH (2024) Applications of molecular genetic to detect DNA damage caused by some environmental pollutants in some plants. Egypt J Desert Res 74(1):1–16
50. Hayes PM, Corey AE, Mundt C, Toojinda T, Vivar H (2003) Registration of' Tango'barley. Crop Sci 43(2):729
51. Husen A (2021) Harsh environment and plant resilience. In: Molecular and functional aspects. Springer Nature Switzerland AG, Cham. https://doi.org/10.1007/978-3-030-65912-7
52. Husen A (2021) Plant performance under environmental stress. In: Hormones, biostimulants and sustainable plant growth management. Springer Nature Switzerland AG, Cham. https://doi.org/10.1007/978-3-030-78521-5
53. Husen A (2022) Environmental pollution and medicinal plants. CRC, Boca Raton, pp 33487–2742. https://doi.org/10.1201/9781003178866
54. Husen A (2023) Plants and their interaction to environmental pollution. In: Damage detection, adaptation, tolerance, physiological and molecular responses. Elsevier Inc., Cambridge. https://doi.org/10.1016/C2020-0-03319-0
55. Husen A, Ahmad A (2023) Genomics, transcriptomics, proteomics and metabolomics of crop plants. Elsevier Inc., Cambridge. https://doi.org/10.1016/C2021-0-02329-4
56. Husen A, Iqbal M (2023) Medicinal plants - their response to abiotic stress. Springer Nature Singapore Pte Ltd., Singapore. https://doi.org/10.1007/978-981-19-5611-9
57. Husen A, Zhang W (2023) Hormonal cross-talk, plant defense and development. Elsevier Inc., Cambridge. https://doi.org/10.1016/C2021-0-02821-2
58. Jahan N, Javed MA, Khan A, Tabassum B, Parveen S, Muzaffar N (2022) Multiple interval mapping of QTLs and epistasis for iron toxicity tolerance in segregating population of Indica rice *Notulae Botanicae*. Horti Agrobotanici Cluj-Napoca 50(4):12773–12773
59. Jia H, Wang N (2014) Targeted genome editing of sweet orange using Cas9/sgRNA. PLoS ONE 9(4):pe93806
60. Jia L, Liu L, Zhang Y, Fu W, Liu X, Wang Q, Huang L (2023) Microplastic stress in plants: effects on plant growth and their remediations. Front Plant Sci 14:1226484
61. Jia W, Ma M, Chen J, Wu S (2021) Plant morphological, physiological and anatomical adaption to flooding stress and the underlying molecular mechanisms. Int J Mol Sci 22(3):1088
62. Joshi S, Patil S, Shaikh A, Jamla M, Kumar V (2023) Modern omics toolbox for producing combined and multifactorial abiotic stress tolerant plants. Plant Stress 11:100301
63. Kaur B, Sandhu KS, Kamal R, Kaur K, Singh J, Röder MS, Muqaddasi QH (2021) Omics for the improvement of abiotic, biotic, and agronomic traits in major cereal crops: applications, challenges, and prospects. Plants 10(10):1989
64. Khan N, Ali S, Shahid MA, Kharabian-Masouleh A (2017) Advances in detection of stress tolerance in plants through metabolomics approaches. Plant Omics 10(3):153–163
65. Koch A, Kogel KH (2014) New wind in the sails: improving the agronomic value of crop plants through RNA i-mediated gene silencing. Plant Biotechnol J 12(7):821–831. https://doi.org/10.1111/pbi.12226
66. Kumar S, Dubey RS, Tripathi RD, Chakrabarty D, Trivedi PK (2015) Omics and biotechnology of arsenic stress and detoxification in plants: current updates and prospective. Environ Int 74:221–230

67. Kumar V, AlMomin S, Al-Shatti A, Al-Aqeel H, Al-Salameen F, Shajan AB, Nair SM (2019) Enhancement of heavy metal tolerance and accumulation efficiency by expressing *Arabidopsis* ATP sulfurylase gene in alfalfa. Int J Phytorem 21(11):1112–1121
68. Lamalakshmi Devi E, Kumar S, Basanta Singh T, Sharma SK, Beemrote A, Devi CP, Wani SH (2017) Adaptation strategies and defence mechanisms of plants during environmental stress. In: Medicinal plants and environmental challenges, pp 359–413
69. Leitão I, Leclercq CC, Ribeiro DM, Renaut J, Almeida AM, Martins LL, Mourato MP (2021) Stress response of lettuce (*Lactuca sativa*) to environmental contamination with selected pharmaceuticals: a proteomic study. J Proteomics 245:104291
70. Leng Y, Li Y, Wen Y, Zhao H, Wang Q, Li SW (2020) Transcriptome analysis provides molecular evidences for growth and adaptation of plant roots in cadimium-contaminated environments. Ecotoxicol Environ Saf 204:111098. https://doi.org/10.1016/jecoenv2020111098
71. Li JF, Zhang D, Sheen J (2014) Cas9-based genome editing in *Arabidopsis* and tobacco In Methods in enzymology, vol 546. Academic, pp 459–472
72. Li T, Liu B, Spalding MH, Weeks DP, Yang B (2012) High-efficiency TALEN-based gene editing produces disease-resistant rice. Nat Biotechnol 30(5):390–392
73. Liao C, Li Y, Wu X, Wu W, Zhang Y, Zhan P, Meng X, Hu G, Yang S, Lin H (2023) ZmHMA3, a member of the heavy-metal-transporting ATPase family, regulates cd and zn tolerance in maize. Int J Mol Sci 24(17):13496
74. Lippman ZB, Semel Y, Zamir D (2007) An integrated view of quantitative trait variation using tomato interspecific introgression lines. Curr Opin Genet Dev 17(6):545–552
75. Liu XS, Feng SJ, Zhang BQ, Wang MQ, Cao HW, Rono JK, Chen X, Yang ZM (2019) *OsZIP1* functions as a metal efflux transporter limiting excess zinc, copper and cadmium accumulation in rice. BMC Plant Biol 19:1–16
76. Lovley DR (2003) Cleaning up with genomics: applying molecular biology to bioremediation. Nat Rev Microbiol 1:35–44
77. Loyal A, Pahuja SK, Vart D, Chander S, Kumar V, Ankush, Rani N, Sharma P, Husen A (2024) Exploring the role of beneficial elements in enhancing the resistance of plants to biotic stress. In: Husen A (ed) Essential minerals in plant-soil systems. Elsevier Inc., Cambridge, pp 327–346. https://doi.org/10.1016/B978-0-443-16082-0.00019-9
78. Lu X, Kong J, Meng X, Cao B, Luo K, Dai P, Luan S (2018) Identification of SNP markers associated with tolerance to ammonia toxicity by selective genotyping from de novo assembled transcriptome in *Litopenaeus vannamei*. Fish Shellfish Immunol 73:158–166
79. Mahajan M, Kuiry R, Pal PK (2020) Understanding the consequence of environmental stress for accumulation of secondary metabolites in medicinal and aromatic plants. J Appl Res Med Arom Plants 18:100255
80. Mahfouz MM, Li L (2011) TALE nucleases and next generation GM crops. GM Crop 2(2):99–103
81. Malzahn A, Lowder L, Qi Y (2017) Plant genome editing with TALEN and CRISPR. Cell Biosci 7:1–18
82. Mandáková T, Singh V, Krämer U, Lysak MA (2015) Genome structure of the heavy metal hyperaccumulator *Noccaea caerulescens* and its stability on metalliferous and nonmetalliferous soils. Plant Physiol 169(1):674–689
83. Mantiquilla JA, Elumba ME, Shiao MS, Chiang YC (2021) Screening of polymorphic microsatellite markers from the whole-genome sequences of nipa (*Nypa fruticans Wurmb*). Plant Genet Resour 19(4):316–319
84. Martínez-Fortún J, Phillips DW, Jones HD (2017) Potential impact of genome editing in world agriculture. Emerg Top Life Sci 1(2):117–133
85. Mendoza-de Gives P (2022) Soil-Borne nematodes: Impact in agriculture and livestock and sustainable strategies of prevention and control with special reference to the use of nematode natural enemies. Pathogens 11(6):640
86. Michaletti A, Naghavi MR, Toorchi M, Zolla L, Rinalducci S (2018) Metabolomics and proteomics reveal drought-stress responses of leaf tissues from spring-wheat. Sci Rep 8(1):5710

87. Miftahudin Scoles GJ, Gustafson JP (2002) AFLP markers tightly linked to the aluminum-tolerance gene Alt3 in rye (*Secale cereale* L). Theor Appl Genet 104:626–631. https://doi.org/10.1007/s00122-001-0782-3
88. Mishra N, Srivastava AP, Esmaeili N, Hu W, Shen G (2018) Overexpression of the rice gene OsSIZ1 in *Arabidopsis improves* drought-, heat-, and salt-tolerance simultaneously. PLoS ONE 13(8):pe2201716
89. Mitra A, Chatterjee S, Moogouei R, Gupta DK (2017) Arsenic accumulation in rice and probable mitigation approaches a review. Agronomy 7(4):67
90. Moulick D, Majumdar A, Choudhury A, Das A, Chowardhara B, Pattnaik BK, Dash GK, Murmu K, Bhutia KL, Upadhyay MK, Yadav P (2024) Emerging concern of nano-pollution in agro-ecosystem: flip side of nanotechnology. Plant Physiol Biochem 3:108704
91. Munaweera TIK, Jayawardana NU, Rajaratnam R, Dissanayake N (2022) Modern plant biotechnology as a strategy in addressing climate change and attaining food security. Agric Food Secur 11(1):1–28
92. Muñoz-García A, Arbeli Z, Boyacá-Vásquez V, Vanegas J (2022) Metagenomic and genomic characterization of heavy metal tolerance and resistance genes in the rhizosphere microbiome of *Avicennia germinans* in a semi-arid mangrove forest in the tropics. Mar Pollut Bull 184:114204
93. Murugaiyan V, Ali J, Mahender A, Aslam UM, Jewel ZA, Pang Y, Marfori-Nazarea CM, Wu LB, Frei M, Li Z (2019) Mapping of genomic regions associated with arsenic toxicity stress in a backcross breeding populations of rice (*Oryza sativa* L). Rice 12:1–14
94. Mushtaq M, Bhat JA, Mir ZA, Sakina A, Ali S, Singh AK, Tyagi A, Salgotra RK, Dar AA, Bhat R (2018) CRISPR/Cas approach: a new way of looking at plant-abiotic interactions. J Plant Physiol 224:156–162
95. Muthamilarasan M, Singh NK, Prasad M (2019) Multi-omics approaches for strategic improvement of stress tolerance in underutilized crop species: a climate change perspective. Adv Genet 103:1–38. https://doi.org/10.1016/bsadgen201901001
96. Naeem M, Gill R, Gill SS, Singh K, Sofo A, Tuteja N (2023) Emerging contaminants and their effect on agricultural crops. Front Plant Sci 14:1296252
97. Naz M, Benavides-Mendoza A, Tariq M, Zhou J, Wang J, Qi S, Dai Z, Du D (2022) CRISPR/Cas9 technology as an innovative approach to enhancing the phytoremediation: concepts and implications. J Environ Manage 323:116296
98. Nejat N, Ramalingam A, Mantri N (2018) Advances in transcriptomics of plants. In: Plant genetics and molecular biology, pp 161–185
99. Nemudryi AA, Valetdinova KR, Medvedev SP, Zakian SÁ (2014) TALEN and CRISPR/Cas genome editing systems: tools of discovery. Acta Naturae 6(3):19–40
100. Nie G, Liu A, Ghanizadeh H, Wang Y, Tang M, He J, Feng G, Huang L, Zhang X (2024) Natural variation in chromium accumulation and the development of related EST-SSR molecular markers in *Miscanthus sinensis*. Agronomy 14(7):1458
101. Niu L, Li C, Wang W, Zhang J, Scali M, Li W, Liu H, Tai F, Hu X, Wu X (2023) *Cadmium tolerance* and hyperaccumulation in plants–a proteomic perspective of phytoremediation. Ecotoxicol Environ Saf 256:114882
102. Noman MU, Azhar S (2023) Metabolomics, a potential way to improve abiotic stresses tolerance in cereal crops
103. Noor I, Sohail H, Sun J, Nawaz MA, Li G, Hasanuzzaman M, Liu J (2022) Heavy metal and metalloid toxicity in horticultural plants: tolerance mechanism and remediation strategies. Chemosphere 303:135196
104. Qi Y, Li X, Zhang Y, Starker CG, Baltes NJ, Zhang F, Sander JD, Reyon D, Joung JK, Voytas DF (2013) Targeted deletion and inversion of tandemly arrayed genes in *Arabidopsis thaliana* using zinc finger nucleases *G3:* Genes. Genomes Genet 3(10):1707–1715
105. Rahman S, Iqbal M, Husen A (2023) Medicinal plants and abiotic stress: an overview. In: Husen A, Iqbal M (eds) Medicinal plants - their response to abiotic stress. Springer Nature Singapore Pte Ltd., Singapore, pp. 1–34. https://doi.org/10.1007/978-981-19-5611-9_1

106. Rahman S, Mehta S, Husen A (2023) Plants and their unexpected response to environmental pollution: an overview. In: Husen A (ed) Plants and their interaction to environmental pollution (Damage detection, adaptation, tolerance, physiological and molecular responses). Elsevier Inc., Cambridge, pp 1–23. https://doi.org/10.1016/B978-0-323-99978-6.00004-2
107. Rai KK, Pandey N, Meena RP, Rai SP (2021) Biotechnological strategies for enhancing heavy metal tolerance in neglected and underutilized legume crops: a comprehensive review. Ecotoxicol Environ Saf 208:111750
108. Rai PK, Kim KH, Lee SS, Lee JH (2020) Molecular mechanisms in phytoremediation of environmental contaminants and prospects of engineered transgenic plants/microbes. Sci Total Environ 705:135858
109. Rai PK, Lee SS, Zhang M, Tsang YF, Kim KH (2019) Heavy metals in food crops: health risks, fate, mechanisms, and management. Environ Int 125:365–385
110. Rajeev AC, Sahu N, Arvind K, Deori M, Grace T, Dev SA, Yadav VP, Ghosh I (2021) Exploring prevalence of potential pathogens and fecal indicators in geographically distinct river systems through comparative metagenomics. Environ Pollut 282:117003
111. Raza A, Tabassum J, Zahid Z, Charagh S, Bashir S, Barmukh R, Khan RSA, Barbosa F Jr, Zhang C, Chen H, Zhuang W (2022) Advances in "omics" approaches for improving toxic metals/metalloids tolerance in plants. Front Plant Sci 12:794373
112. Ribaut JM, Hoisington D (1998) Marker-assisted selection: new tools and strategies. Trends Plant Sci 3(6):236–239
113. Sami A, Xue Z, Tazein S, Arshad A, He Zhu Z, Ping Chen Y, Jin Zhou K (2021) CRISPR–Cas9-based genetic engineering for crop improvement under drought stress. Bioengineered 12(1):5814–5829
114. Sattelmacher B (2001) The apoplast and its significance for plant mineral nutrition. New Phytol 149:167–192
115. Septiningsih EM, Pamplona AM, Sanchez DL, Neeraja CN, Vergara GV, Heuer S, Ismail AM, Mackill DJ (2009) Development of submergence-tolerant rice cultivars: the Sub1 locus and beyond. Ann Bot 103(2):151–160
116. Shafiq S, Zeb Q, Ali A, Sajjad Y, Nazir R, Widemann E, Liu L (2019) Lead, cadmium and zinc phytotoxicity alter DNA methylation levels to confer heavy metal tolerance in wheat. Int J Mol Sci 20(19):4676. https://doi.org/10.3390/ijms20194676
117. Shan Q, Wang Y, Li J, Zhang Y, Chen K, Liang Z, Zhang K, Liu J, Xi JJ, Qiu JL, Gao C (2013) Targeted genome modification of crop plants using a CRISPR-Cas system. Nat Biotechnol 31(8):686–688
118. Shandilya ZM, Tanti B (2019) Hydroponic screening of traditional rice varieties in Assam, India to estimate their potential resistance to Al toxicity under P deficiency. Acta Agrobot 72(4). https://doi.org/10.5586/aa.1793
119. Sharma P, Kumar S, Pandey A (2021) Bioremediated techniques for remediation of metal pollutants using metagenomics approaches: a review. J Environ Chem Eng 9(4):105684
120. Shen L, Hua Y, Fu Y, Li J, Liu Q, Jiao X, Xin G, Wang J, Wang X, Yan C, Wang K (2017) Rapid generation of genetic diversity by multiplex CRISPR/Cas9 genome editing in rice. Sci China Life Sci 60:506–515
121. Sheng J, Citovsky V (1996) *Agrobacterium*-plant cell DNA transport: have virulence proteins, will travel. Plant Cell 8(10):1699
122. Shukla VK, Doyon Y, Miller JC, DeKelver RC, Moehle EA, Worden SE, Mitchell JC, Arnold NL, Gopalan S, Meng X, Choi VM (2009) Precise genome modification in the crop species Zea mays using zinc-finger nucleases. Nature 459(7245):437–441
123. Sridhar A, Balakrishnan A, Jacob MM, Sillanpää M, Dayanandan N (2023) Global impact of COVID-19 on agriculture: role of sustainable agriculture and digital farming. Environ Sci Pollut Res 30(15):42509–42525
124. Sultana S, Chakma K, Bhuiyan MSU (2024) Development of transgenic *Brassica napus* plants with AtATM3 gene to enhance cadmium and lead tolerance. J Hortic Sci 19(1)
125. Surabhi GK, Badajena B, Sahoo SK (2019) Genome editing and abiotic stress tolerance in crop plants. In: Recent approaches in omics for plant resilience to climate change, pp 35–56. https://doi.org/10.1007/978-3-030-21687-0_2

126. Tang L, Mao B, Li Y, Lv Q, Zhang L, Chen C, He H, Wang W, Zeng X, Shao Y, Pan Y (2017) Knockout of *OsNramp5* using the CRISPR/Cas9 system produces low Cd-accumulating indica rice without compromising yield. Sci Rep 7(1):14438
127. Tiwari M, Sharma D, Trivedi PK (2014) Artificial microRNA mediated gene silencing in plants: progress and perspectives. Plant Mol Biol 86:1–18. https://doi.org/10.1007/s11103-014-0224-7
128. Townsend JA, Wright DA, Winfrey RJ, Fu F, Maeder ML, Joung JK, Voytas DF (2009) High-frequency modification of plant genes using engineered zinc-finger nucleases. Nature 459(7245):442–445
129. Venbrux M, Crauwels S, Rediers H (2023) Current and emerging trends in techniques for plant pathogen detection. Front Plant Sci 14:1120968
130. Villalobos-López MA, Arroyo-Becerra A, Quintero-Jiménez A, Iturriaga G (2022) Biotechnological advances to improve abiotic stress tolerance in crops. Int J Mol Sci 23(19):12053
131. Wang D, Lu X, Chen X, Wang S, Wang J, Guo L, Yin Z, Chen Q, Ye W (2020) Temporal salt stress-induced transcriptome alterations and regulatory mechanisms revealed by PacBio long-reads RNA sequencing in *Gossypium hirsutum*. BMC Genomics 21:1–15
132. Wang FZ, Chen MX, Yu LJ, Xie LJ, Yuan LB, Qi H, Xiao M, Guo W, Chen Z, Yi K, Zhang J (2017) *OsARM1*, an R2R3 MYB transcription factor, is involved in regulation of the response to arsenic stress in rice. Front Plant Sci 8:1868
133. Wang L, Chen L, Li R, Zhao R, Yang M, Sheng J, Shen L (2017) Reduced drought tolerance by CRISPR/Cas9-mediated SlMAPK3 mutagenesis in tomato plants. J Agric Food Chem 65(39):8674–8682
134. Wang Y, Cheng X, Shan Q, Zhang Y, Liu J, Gao C, Qiu JL (2014) Simultaneous editing of three homoeoalleles in hexaploid bread wheat confers heritable resistance to powdery mildew. Nat Biotechnol 32(9):947–951
135. Wilkins MR, Williams KL, Appel RD, Hochstrasser DF (eds) (1997) Proteome research: new frontiers in functional genomics. Springer Science & Business Media
136. Yadav A, Yadav K, Abd-Elsalam KA (2023) Nanofertilizers: types, delivery and advantages in agricultural sustainability. Agrochemicals 2(2):296–336
137. Yadav M, Singh M, Nanda K, Singh SP (2022) Genetically engineered plants for phytoremediation of heavy metals In: Current developments in biotechnology and bioengineering. Elsevier, pp 223–239
138. Yadav RK, Tripathi MK, Tiwari S, Tripathi N, Asati R, Chauhan S, Tiwari PN, Payasi DK (2023) Genome editing and improvement of abiotic stress tolerance in crop plants. Life 13(7):1456
139. Yan J, Kandianis CB, Harjes CE, Bai L, Kim EH, Yang X, Skinner DJ, Fu Z, Mitchell S, Li Q, Fernandez MGS (2010) Rare genetic variation at *Zea mays* crtRB1 increases β-carotene in maize grain. Nat Genet 42(4):322–327
140. Yang CH, Zhang Y, Huang CF (2019) Reduction in cadmium accumulation in japonica rice grains by CRISPR/Cas9-mediated editing of *OsNRAMP5*. J Integr Agric 18(3):688–697. https://doi.org/10.1016/S2095-3119(18)61904-5
141. Yang Y, Saand MA, Huang L, Abdelaal WB, Zhang J, Wu Y, Li J, Sirohi MH, Wang F (2021) Applications of multi-omics technologies for crop improvement. Front Plant Sci 12:563953. https://doi.org/10.3389/fpls2021563953
142. Younis A, Ramzan F, Ramzan Y, Zulfiqar F, Ahsan M, Lim KB (2020) Molecular markers improve abiotic stress tolerance in crops: a review. Plants 9(10):1374
143. Yu C, Zhou F, Wang R, Ran Z, Tan W, Jiang L, Duan L (2022) B2, an abscisic acid mimic, improves salinity tolerance in winter wheat seedlings via improving activity of antioxidant enzymes. Front Plant Sci 13:916287
144. Yu LX, Abate Z, Anderson JA, Bansal UK, Bariana HS, Bhavani S, Dubcovsky J, Lagudah ES, Liu S, Sambasivam PK, Singh RP (2009) Developing and optimizing markers for stem rust resistance in wheat. Proc Oral Pap 17:117
145. Zhang F, Voytas DF (2011) Targeted mutagenesis in *Arabidopsis* using zinc-finger nucleases. In: Plant chromosome engineering: methods and protocols, pp 167–177

146. Zhang GL, Chen LY, Xiao GY, Xiao YH, Chen XB, Zhang ST (2009) Bulked segregant analysis to detect QTL related to heat tolerance in rice (*Oryza sativa* L) using SSR markers. Agric Sci China 8:482–487. https://doi.org/10.1016/S1671-2927(08)60235-7
147. Zhang Y, Liu J (2011) Transgenic alfalfa plants co-expressing glutathione S-transferase (GST) and human CYP2E1 show enhanced resistance to mixed contaminates of heavy metals and organic pollutants. J Hazard Mater 189(1–2):357–362
148. Zheng H, Dang Y, Diao X, Sui N (2023) Molecular mechanisms of stress resistance in sorghum: Implications for crop improvement strategies. J Integr Agric

Biological Tools for Improving Crop Plant Growth Under Contaminants

Bhoomika Varshney, Reena Vishvakarma, and Archana Vimal

Abstract Crop production and development under micro and nanocontamination have become a very considerate subject for farmers and researchers. These contaminants can negatively affect plant growth, soil health inclusive agricultural productivity. For the production of crops, scientists use several chemical and biological methods. The biological approach has proven more successful as compared to chemical means. Biological systems are more effective, useful, and eco-friendly. There are several biological methods or tools such as phytoremediation, plant growth-promoting bacteria, genetic engineering, and nanobiotechnology which are significant tools for crop improvement under these critical contaminants. Crop development and their productivity are affected by both environmental and genetic factors. Numerous microbes such as *Acetobacter, Bacillus, Burkholderia, and Pseudomonas* genera are naturally helpful for the growth and development of plants under unfavorable conditions. Additionally, there are many genetic transformations, nanochemicals, and nanoparticles that have arrived as a promising agent for the growth and development of crops. Recently, nanomaterials have been considered the best alternative to insecticides and pesticides. Biological tools are very beneficial to obtain a newly cultivated crop of desired traits such as high quality and quantity and easily grow in biotic and abiotic stress conditions.

Keywords Nanocontaminants · Heavy metals · Nanofertilizers · Microplastics · Nanopesticides · Phytoremediation

1 Introduction

According to predictions from the United Nations Food and Agricultural Organization (FAO), farmers have to cultivate 70% more food by 2050 to satisfy the nutritional demands of the world's forecast 9 billion population [16, 28, 74]. A significant shift

B. Varshney · R. Vishvakarma · A. Vimal (✉)
Department of Bioengineering, Integral University, Lucknow, UP, India
e-mail: vimal.archana@gmail.com

A. Husen (ed.), *Enhanced Crop Production Setup Under Various Micro and Nano Contaminants*, Smart Nanomaterials Technology,
https://doi.org/10.1007/978-981-96-6099-5_12

is a spike in the popularity of food derived from plants rather than dairy and meat to balance the world's greenhouse gas budget. However, there is also rising concern worldwide about the impact of synthetic plant protection agents on human and animal health and pollution caused by inorganic fertilizer residues [9].

To fulfill this requirement of increasing the world's population, chemical and synthetic fertilizers play a most important role. There is a broad range of chemical fertilizers used to meet the huge crop production demand [18].

These chemical fertilizers have detrimental effects on humans and the environment. At the same time, it causes water, soil, and land pollution. They collect salt, make plants wilt, and contaminate water bodies. Improper use of these chemicals leads to nutrient imbalance and acidified soil that causes the reduction of crop yields [76]. The persistent use of inorganic fertilizers also causes soil and crop abnormalities, reduced organic matter, soil acidity, and soil erosion. The pesticides used in the agriculture field may also have carcinogenic effects on long-term usage that damage fish and wildlife, poison ground and surface waters, and increase agriculture's dependency on fossil fuels [15]. Additionally, there are some micro and nano contaminants such as microplastics which are a hundred times more hazardous compared to these chemical fertilizers [5].

Nowadays, micro and nano-plastic contamination becomes a generic problem because these contaminants are life-threatening. They can act as carriers for pathogens. It provides shelter for bacteria and viruses due to their more surface area-to-volume ratio and hydrophobic nature. These contaminants can easily enter from soil to the human body via the food chain. In humans, these contaminants cause many ailments like cancer, organ damage, reproductive problems, etc. [79]. These contaminants are generated from diverse plastic polymers with distinct chemical compositions and characteristics, such as polyethylene, polystyrene, polyurethane, polypropylene, polyvinyl chloride, fibers, and polyethylene terephthalate [21]. Microplastic hotspots exist in some agricultural regions in America, Europe, and Australia. These regions generate annual microplastic loadings of approximately 44,000–300,000 metric tons, 63,000–430,000 tons, and 2800–19,000 tons, respectively [35]. There are no standard methods for detecting microplastics in soil. The analytical procedures used to detect microplastics in water and sediments are often applied to soil. Microplastics are firstly visually detected by an optical microscope. Thereafter, Raman spectroscopy and Micro-Fourier transformed infrared (m-FT-IR) are used as confirmatory tests [17, 82].

Plastic trash can break down into tiny fragments that are termed "plastic debris". This plastic debris critically affects the terrestrial and marine environments. This small plastic debris is called "microplastics (MPs)" if its size ranges from 0.1 μm to 5 mm. These are termed "nano-plastics" if their size lies in the 0.001–0.1 μm range [24, 85].

It was assessed that annually approximately 125–850 tons per million inhabitants of MPs are disposed to European agricultural soils. European and North American farmlands contribute approximately 63,000–430,000 and 44,000–300,000 tons of MPs annually, respectively [67]. Nanoplastics are more harmful than MPs. This is because they easily penetrate biological membranes and pose a high risk to living creatures [55].

These micro and nano contaminants affect adversely the whole agriculture and associated sectors. They accumulate in the soil and interact with organic pollutants, affecting the growth of plants and chlorophyll content, ultimately the photosynthetic cycle that decreases the yield of plants [68, 90]. These contaminants remain in the soil for hundreds of years and affect the physiochemical productivity and pH. The decline in earthworm population in the soil and groundwater pollution is also reported [14].

The contamination of soil with heavy metals (HMs) has become a major concern for the entire ecosystem due to rapid urbanization, industrialization, and population growth. At present, approximately 1.2 billion hectares of land are polluted with heavy metals (HMs) worldwide [81]. These HMs' pollutants are generated from different human activities like mining, sludge discharge, metal smelting, sludge discharge mining, metal smelting, extensive use of herbicides, pesticides and fertilizers, vehicle emissions, sewage irrigation, and coal combustion. The discharge of non-biodegradable HMs in soil puts both the environment's safety as well as human health in grave endanger. These HMs are extremely toxic, persistent, bioavailable, prone to buildup, difficult to decompose and alter the physicochemical properties of soil. The HMs like cadmium (Cd), lead (Pb), chromium (Cr), mercury (Hg), etc., enter the food chain and cause numerous health issues such as mental disorders, cancerous tumors, coronary artery disease, and problems with the skin and bones [26, 36, 57, 62, 72, 73, 92]. They also affect plant growth and survival if persistently present in the soil [2].

Radionuclides and HMs contamination migrate in tandem in soil and water which have adverse effects on biological systems. These radionuclide contaminants are harmful due to their chemical composition, which involves long half-lives, high solubility, substantial transferability, and quick uptake by living things. Radionuclides contamination originates from natural processes, nuclear facilities, nuclear weapons testing, nuclear waste management, and disasters [56, 84].

Soil contamination is explained as the harmful substances present in the soil that have an adverse impact on plant, animal, human health, and the ecosystems [6]. Here are some common types of soil contamination mentioned in Table 1.

To fulfill the increasing demand for food, declining yield of crops, and emerging health issues, alternative approaches and solutions to these problems are much needed. In recent years, significant improvements and advancements in technology in the sector of agriculture practices address these rising problems for sustainable production and food security [83]. There are a variety of micro and nanotools (collectively called biological tools) that can easily overcome this problem and have lots of applications. These biological tools can enhance the productivity of crops [71, 74, 86, 87]. Expanded uses of nano and biosensors in the agriculture sector will be helpful for precision farming, monitoring soil conditions, crop growth, crop diseases, use and penetration of environmental and agrochemicals pollution. It has substantially improved the human control of soil and plant health, safety assurance and quality control imparting much to sustainable agriculture and environmental management [19, 29–33, 44, 46, 52, 53, 60, 77, 78]. Biological tools such as phytoextraction,

Table 1 Different types of soil contaminants

Types	Sources	Effect	Examples
Heavy metals	Industrial activities, mining, improper disposal of electronic waste, use of contaminated fertilizers and pesticides	Can lead to toxicity in plants and animals, potentially entering the food chain and causing health issues in humans	Lead, mercury, cadmium, arsenic, chromium
Pesticides and herbicides	Agricultural activities, use of chemical pesticides and herbicides	Can harm non-target organisms, reduce biodiversity, and contaminate water supplies	DDT, glyphosate, atrazine
Industrial chemicals	Industrial processes, chemical manufacturing, waste disposal	Persistent in the environment, bioaccumulative, and can cause cancer and other health problems	Polychlorinated biphenyls (PCBs), dioxins, polycyclic aromatic hydrocarbons (PAHs)
Petroleum hydrocarbons	Oil spills, leaking underground storage tanks, improper disposal of oil products	Toxic to plants and animals, can lead to groundwater contamination, and pose health risks to humans	Benzene, toluene, ethylbenzene, xylene (BTEX)
Solvents	Industrial cleaning processes, dry cleaning, improper disposal of chemical waste	Toxic to humans and wildlife, can contaminate soil and groundwater	Trichloroethylene (TCE), perchloroethylene (PCE)
Biological contaminants	Sewage sludge, animal waste, contaminated water used for irrigation	Can cause diseases in plants, animals, and humans, and contaminate crops	Bacteria, viruses, parasites
Radioactive contaminants	Nuclear power plants, improper disposal of radioactive waste, mining activities	Can cause radiation poisoning, cancer, and environmental damage	Uranium, radon, cesium
Nutrients and fertilizers	Excessive use of fertilizers in agriculture	Can lead to nutrient imbalances in soil, water pollution (eutrophication), and harm aquatic ecosystems	Nitrogen, phosphorus
Microplastics	Breakdown of larger plastic debris, agricultural films, sewage sludge	Persistent in the environment, can affect soil structure and organisms	Small plastic particles
Salts	Irrigation with saline water, and salt application on roads	Soil degradation, reduced agricultural productivity, harm to plant life	Sodium chloride, calcium chloride, zinc sulfate

phytodegradation, phytostabilization, rhizome-filtration, biofertilization, phyto stimulation, nanotechnology, and stress alleviation are the advanced technologies that support the research and development in agricultural fields. Thus, biological tools can easily reduce uncertainty and help in the effective management of agricultural production. In this chapter, we are discussing the applications of biological tools in the agriculture sector that could ensure the environment's and agriculture's sustainability [23, 34, 63].

2 Biological Tool for Improving Crop

In recent years, biological tools applied for mitigating soil contamination have gained considerable attention due to their eco-friendliness and sustainability. Biological tools used for improving soil contamination involve the utilization of living organisms to detoxify and restore contaminated soil. These methods are often environmentally friendly, cost-efficient, and sustainable compared to traditional mechanical or chemical remediation techniques [4]. Here we discuss some biological tools for improving crop production:

2.1 Bioremediation

Bioremediation is a biological process that utilizes living creatures, mainly microorganisms like bacteria, yeast, molds, plants, or metabolites produced by them. For example, enzymes are used to break down or detoxify hazardous organic contaminants that are polluting the soil, water, and other environments. This method leverages the natural metabolic processes of these organisms to decompose pollutants into less detrimental or inert substances. There are several advantages of bioremediation such as high efficiency combined with low operating costs, reduction of chemical and biological sludge, remarkable proficiency, independence from additional nutritional requirements, metal specificity, and the possibility of metal recovery [47].

Nanobiotechnology provides a comprehension solution to reduce soil contamination and improving the crop qualities. Recently, we have integrated bioremediation along with nanobiotechnology to enhance the efficiency of the removal of environmental contamination. This approach utilizes nanoscale materials to improve the efficacy and productivity of bioremediation processes. With the help of nanobioremediation, we easily control or manage the contaminated materials at the molecular and atomic levels. These are the tiny particles that have a nanometer size range (1–100 nm) that imparts them unique physical, chemical, and mechanical properties such as high surface area and reactivity, surface charge, hydrophobicity, and hydrophilicity. Therefore, these particles have significant potential in treating various contaminants like petroleum hydrocarbons, heavy metals, pesticides, microplastics, etc. [20]. Commonly used nanoparticles include metal oxides, zero-valent metals, and carbon-based materials. The nano-bioremediation generates fast and effective results. This technique shows less damaging effects on microorganisms, so it easily enhances the microbial activity against contamination and reduces the time consumption in removing harmful contaminates [61]. Nano-bioremediation is an efficient method for the removal of heavy metals, organic and inorganic contaminates, surface and groundwater treatment, removal of radioactive wastes, hydrocarbons, and pesticides and results in improved plant development and soil health.

Nanoparticles can increase the bioavailability of contaminants by altering their physical and chemical forms, making them more accessible to microorganisms for degradation. There are certain nanoparticles, like nano zero-valent iron (nZVI), have catalytic properties that enhance the breakdown of contaminants through chemical reactions like reduction and oxidation. Nanoparticles can also stimulate microbial growth and metabolism by providing essential nutrients or creating favorable environmental conditions, leading to more efficient degradation of pollutants. Some nanoparticles can adsorb and sequester contaminants, reducing their mobility and availability in the environment [40].

2.1.1 Advantages of Bioremediation

Nano-bioremediation compared to traditional methods can significantly enhance the rate and degree of contaminant degradation. This method can be used to different types of pollutants present in any water bodies, land, soil, or sediments. The practice of nanoparticles-based bioremediation minimizes environmental pollution by limiting the use of toxic chemicals and high energy processes [7, 42].

According to recent studies, nanotechnology applied in the bioremediation of emerging contaminants like nano zero-valent iron (nZVI) is used for heavy metal remediation, carbon-based nanomaterials are implied for organic pollutant degradation, and metal nanoparticles are used for enhanced microbial degradation of petroleum hydrocarbons [22]. The summary of bioremediation is explained in Fig. 1.

2.2 Phytoremediation

Phytoremediation is an environment-friendly and economic method that uses green plants to eliminate micro and nano contamination from soil, water, and land. This process employs usage of various plant species to extract, sequester, and detoxify pollutants through natural biological processes. Phytoremediation applies to various contaminants, such as organic compounds, heavy metals, and radionuclides [88]. In situ and ex situ phytoremediation methods are used for the treatment of agriculture contamination, soil, and water characterization, mentation nutrition sustainability, and feasibility. Phytoremediation is an effective and sustainable approach to treating contaminated environments [70].

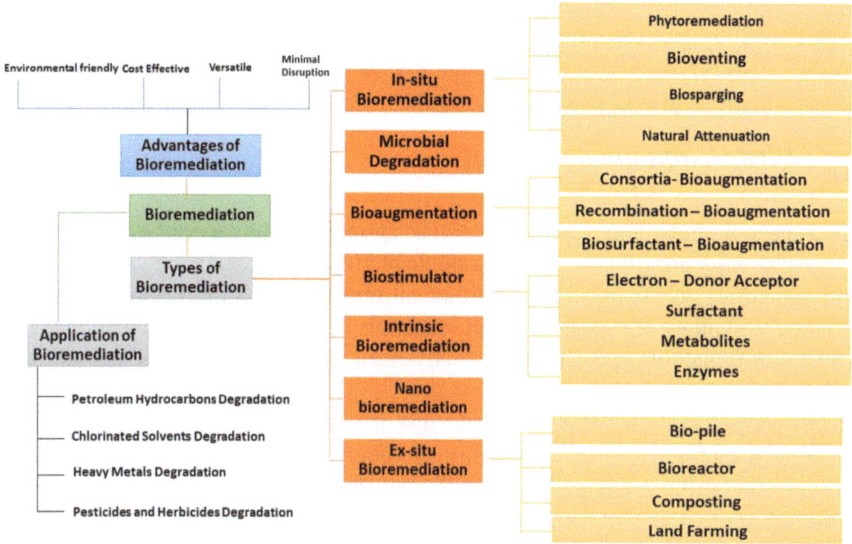

Fig.1 Summary of bioremediation

2.2.1 Types of Phytoremediation

a. **Phytoextraction**: The method plants use to absorb contaminants like heavy metals from their surrounding primarily using their roots and translocating them to the above-ground parts (e.g. shoots and leaves). *Thlaspi caerulescens* (Alpine pennycress) and *Brassica juncea* (Indian mustard) are the example of hyperaccumulator plants that are used to remove heavy metals like lead, zinc, and cadmium [38].
b. **Phytostabilization**: This method involves the use of plants with extensive root systems to immobilize contaminants in the soil and prevent their migration to groundwater or air. These plants also prevent soil erosion and neutralize heavy metals, decrease their bioavailability, and restrict their mobility.
c. **Phytodegradation (Phytotransformation)**: The metabolic process by which plants degrade or transform organic pollutants, such as pesticides and herbicides, into less toxic compounds is called phytodegradation or phtotransformation. Poplar trees and willow trees can degrade trichloroethylene (TCE) and other organic pollutants are the example of this process.
d. **Phytovolatilization**: This method is described as uptake of volatile contaminants by plants and their transpiration via leaves, releasing them into the atmosphere in a less harmful form. Some plant species can volatilize mercury and selenium, converting them into gaseous forms that are released through the pores of leaves.
e. **Rhizofiltration**: The method involves plant roots to absorb, concentrate, and precipitate contaminants from their aqueous surrounding (e.g., groundwater or

wastewater). Water hyacinths and sunflowers are the most common example that is used to minimize radionuclides and heavy metals present in water bodies [54].

2.3 *Phyto-Nanobiotechnology*

Phyto-nanobiotechnology can be defined as an interdisciplinary subject that brings nanotechnology and plant science at a single platform. It is applied in various sectors like agriculture, medicines, and environment remediation. In this approach, plants are used to synthesize and modify nanomaterials for their diverse applications. This helps in sustainability and efficiency of advanced technologies used in various areas [37].

Phytoremediation is enhanced by the use of nanoparticles (NPs) synthesized by plants as it increases the bioavailability and degradation of different contaminants present in the soil or water. NPs have ability to adsorb the heavy metals present in the surrounding environment and purify the soil and water. They also catalyze the degradation of other organic pollutants like hydrocarbons and pesticides. It breakdowns them into lesser toxic compounds. These NPs also have applicability in designing different sensors that can detect and quantify environment pollutant that helps in the real-time monitoring of the environments contaminants.

Apart from the bioremediation, phyto-nanobiotechnology also have other various other applications in agriculture science like application of nanofertilizers and nanopesticides. They are better in terms of delivery and efficiency in comparison with normal fertilizers and pesticides. They promote plant growth by increasing nutrient absorption, increasing photosynthesis, enhanced resistance against pathogens resulting in better crop yield. Nanopesticides causes less environmental pollution and has low environmental impact. NPs are also helpful in delivering antimicrobial agents, increase systemic resistance of plants, and provide them protection against various microbial attacks.

2.3.1 Advantages of Phyto-Nanobiotechnology

a. **Sustainability**: The synthesis of nanoparticle using plants employs no harsh chemicals or high-energy methods that makes the process eco-friendly.
b. **Economic**: These methods are comparatively cost effective than other conventional methods and suitable for large-scale production. It requires less labor, energy, and equipment.
c. **Environmentally Friendly**: It is greener approach that sticks to natural methods to control pollutants without the use of harmful and toxic chemical agents.
d. **Biocompatibility**: Nanomaterials synthesized via this approach are generally bio-compatible and best suited for medicinal and agriculture applications.
e. **Versatility**: This approach is used in various sectors like pharmaceutical, agriculture, environmental remediation, and other industrial applications.

f. **Soil Preservation**: It helps in maintaining soil fertility and structure that reduces the soil disposal and excavation.
g. **Aesthetic Benefits**: It also beautifies the pollutant areas by giving it scenic visual and attracts wildlife for habitat.

2.3.2 Limitations of Phyto-Nanobiotechnology

a. **Regulatory Issues**: The real-world applications of nanomaterials need approval from the respective regulatory agencies that may be complicate or delay the overall process.
b. **Scalability**: The large-scale production of nanoparticles through plants is challenging in terms of process optimization and limits its commercial application.
c. **Toxicity and Safety**: There is need of well-established safety evaluation and regulation to protect human, animals, and environment from any possible threat. It requires design and establishment of safety strategies.

2.4 Use of Plant Growth-Promoting Rhizobacteria (PGPR)

The group of naturally occurring soil bacteria that colonize inside the roots of plants and promote their growth by opting different strategies is called plant growth-promoting rhizobacteria (PGPR). It promotes plant growth by root development, enhancing nutrient uptake, producing phytohormones. Nitrogen is a limiting nutrient for plant that is required for amino acid synthesis and in turn protein synthesis. PGPR helps in nitrogen fixation in the rhizosphere. It also solubilizes the phosphate present in soil to increase their availability to the plant. At the same times, they provide stress tolerance (by secreting anti-oxidant enzymes) and immunity against plant pathogens. It is a way of sustainable agriculture practice that is beneficial for both plant and soil. By opting this green method usage of chemical pesticides and fertilizers can be minimized that will result in better environment and human health. It helps in bioremediation by removing pollutants contaminants from the soils without secreting any toxic substance. It promotes degradation of pesticides and helps in sequestration of toxic heavy metal species that improves soil health and nutritional value [13].

2.4.1 Mechanisms of PGPR

Plants/crops are dependent on chemical fertilizers to meet their nitrogen requirement for the vital functions. However, these chemical fertilizers can be substituted by PGPR as they can fix atmospheric nitrogen and make them available to the plants by converting them into simpler form. *Rhizobium*, *Azospirillum*, and *Bradyrhizobium* species are some of the commonly known nitrogen fixers for leguminous plants. The *Pseudomonas*, *Bacillus*, and *Enterobacter* species are renowned for their phosphate solubilizing ability. These bacteria convert insolube phosphate into soluble

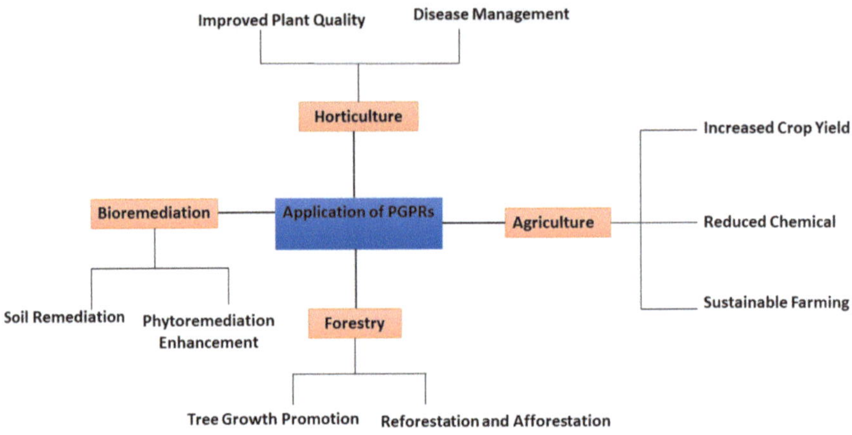

Fig. 2 Application of PGPRs as biological tools

for form, and thus, it is easily absorbed by the plants. The PGPR like *Bacillus* and *Pseudomonas* can solubilize both potassium and zinc to promote plant growth. The various applications of PGPRs in agriculture as an biological tool are explained in Fig. 2.

The combined use of PGPR (a biological tool) and nanotechnology brings down the benefit of both the modern technologies at a single platform. This combination is proved beneficial for plants in terms of better nutrition, productivity, disease resistance, stress tolerance, and soil health. This association helps in environmental remediation and promotes sustainable agriculture practice in an innovative manner [27].

2.4.2 Nanoparticle-Enhanced PGPR Functionality

The NPs and PGPR have synergistic effect on plant growth [41]. NPs promote PGPR efficiency by supporting their existence, colonization, and nutrient exchange with plant roots. PGPR also produces some bioactive compound that are plant growth promoters and provide disease resistance. NPs act as a carrier to deliver those bioactive compound to their designated targets. This association is better in terms of nutrient solubilization and uptake as compared to their individual effect.

PGPR can be used to synthesize nanoparticles in a green and eco-friendly manner, like plant-based synthesis. The bioactive compounds generated by this class of bacteria have tendency to reduce metal ions into their corresponding metallic nanoparticles. The bioactive compound also stabilizes the synthesized particles. This approach is cost effective as well as sustainable that promotes its usage. The NPs produced by this method have unique properties and functionalities.

The combination of PGPR and nanoparticles can create a synergistic effect that promotes plant growth and nutrient uptake. It also enhances stress tolerance capability. They can act as carriers for PGPR, improving their stability and efficacy. This leads to better crop yield and quality, enhanced resilience to abiotic and biotic stresses, low dependency on chemical fertilizers and pesticides, etc.

2.5 *Biochar and Soil Amendments*

Soil amendments are those substances (organic waste materials biochar, compost, and cocopeat) that are used as soil supplements to improve their physical, chemical, and biological properties. They improve soil structure, water retention, fertility, and microbial activity that results in healthier and more productive soils. Biochar and soil amendments are gaining attention as sustainable strategies to improve soil health, enhance crop productivity, and mitigate environmental issues. These practices can increase soil fertility, sequester carbon, improve water retention, and reduce greenhouse gas emissions [12].

Biochar is a stable, solid, and carbon-rich substance with high surface area and high porosity obtained by pyrolysis of organic material such as agriculture waste, wood chips, and manure. Nowadays, it is released in the soil to enhance the soil properties, plant growth, beneficial plant-microbes synergy for plant growth and survival under artificial climate conditions, nutrient content, and protection for the plant. Biochar has ability to promote plant growth because of its better physical and chemical properties like pH, electrical conductivity, particle density, surface area, pore size distribution, cation exchange capacity. It also increases total and plant-available concentrations of carbon, potassium, nitrogen, and phosphorus and contains certain minor nutrients [8].

The advantages of biochar are soil fertility enhancement, improved cation exchange capacity (CEC), water retention and management, carbon sequestration, soil pH regulation, soil microbial activity, etc. Biochar has the ability to adsorb and hold the nutrients, minimize leaching, and enhance nutrient availability to the plants. It also elevates nutrient holding capacity of soil and assists plant to exchange nutrients from the soil through its root. Therefore, it is used in the agriculture practices to get more crop yield by increasing soil fertility and water retention capacity of the soil. Biochar also modifies the soil texture and porosity to hold more water. This helps the plant to survive in the drought condition. It is also useful for acidic soil to increase their pH that helps in increasing nutrient availability. It is also provides shelter or habitat to the various useful bacteria present in the soil that act symbiotically and enhance nutrient cycling and plant growth.

It also helps in controlling plant diseases by regulating soil borne microbes. It assists in pest management enhancing soil condition and microbial activity. It increases the efficiency of phytoremediation by eliminating heavy metals and organic pollutants from the soil through adsorption over its surface and in turn decreasing their bioavailability and toxicity. It also assists in dealing climate change by sequestering carbon for several years because of its stable carbon structure. This process lowers down the emission of greenhouse gases that helps in dealing with climate change.

2.6 Genetic Engineering and Biotechnology

The utilization of biotechnology and genetic engineering tools and techniques has significantly impacted the agriculture sector. The scientists employ various recombinant DNA technology and genetic engineering methods to manipulate plants genome to express any desirable trait in a particular plant. These traits can be high nutritional value, better yield, pest resistance, stress tolerance, etc. BT cotton (insect resistant) and golden rice (rich in carotene precursor of vitamin A) are the two most successful examples of genetic engineering. These methods improvise agriculture practice by resulting better productivity in a sustainable manner [1]. **There are various** genetic engineering techniques and their mechanisms **for enhancing plant growth with suitable examples which are described in Table 2.**

One of the advance and effective biological technique is genome editing. This technique helps in altering plant DNA to give high yield and provide resistant against pathogens and pests, higher tolerance to adverse environmental conditions like drought, floods, soil or water salinity, pH change, etc. [69]. The few examples of molecular level genome editing techniques are TALENs, ZFNs, and CRISPR-Cas9. These techniques use different enzymes as molecular scissors to modify or manipulate the DNA of targeted host (plant or crop) to introduce desirable characteristics according to the need [11]. The benefits of these techniques with suitable examples and their action mechanism are explained in Table 3.

2.7 Nanotechnology-Based Solutions

Nanotechnology is emerging as a transformative field in agriculture and plant biotechnology offering advanced solutions to enhance plant growth, improve crop yields, and address various challenges related to soil health, pest management, and environmental sustainability. By manipulating materials at the nanoscale (1–100 nm), scientists can create nanomaterials with unique properties that can be applied in agriculture to boost productivity and efficiency. Nanomaterials' unique properties such as small size, large surface area-to-volume ratio, and unique physical and chemical properties that facilitate the slow release of fertilizers, site-specific nutrient distribution, enhancement of water uptake under salinity stress activation of plant defense mechanisms against biotic stress, tailored reactivity, and improved efficiency in phytoremediation by aiding the removal of pollutants from the environment [66]. The function of different nanoparticles is explained in Table 4.

Table 2 Advanced genetic engineering methods for improving plant growth

Technique	Mechanism	Applications
CRISPR-Cas9 gene editing	This precise gene-editing tool allows for the targeted modification of specific genes to introduce desirable traits or remove unwanted ones	Yield improvement: CRISPR can be used to knock out genes that limit yield or to enhance traits like drought tolerance and nutrient efficiency
Marker-assisted selection (MAS)	MAS uses molecular markers linked to desirable traits to select plants that carry these traits during breeding, speeding up the development of improved varieties	Disease resistance: MAS is used to select for disease-resistant genes, ensuring the development of robust and resilient crops
RNA interference (RNAi)	Pest resistance: RNAi is used to silence genes in pests that are essential for their survival, thereby protecting crops from damage	RNAi technology can be used to silence specific genes in plants, reducing the expression of traits that are detrimental to growth or health
Plant tissue culture	This technique involves growing plant cells, tissues, or organs in a controlled environment, allowing for the rapid propagation of genetically identical plants	Clonal propagation: tissue culture is widely used for the mass production of plants with desirable traits, ensuring uniformity and quality
Transgenic plants	Transgenic plants are those that have been genetically modified to contain genes from other species, providing them with new traits	Herbicide resistance: genes conferring resistance to specific herbicides, like glyphosate, are introduced into crops, allowing farmers to control weeds more effectively without harming the crops

2.7.1 Nanofertilizers

Nanofertilizers are an emerging class of fertilizers that incorporate nanotechnology to enhance nutrient delivery and nutrient utilization efficiency in agricultural practices than conventional fertilizers, which often suffer from significant nutrient losses due to leaching, volatilization, and poor absorption. Nanofertilizers are designed in a manner to deliver nutrients more precisely and efficiently to plants. This results in improved crop yields, reduced environmental pollution, and promotes sustainable farming practices. It has many advantages that are variable solubility, targeted activity, controlled release mechanisms, effectual and consistent performance, and minimum eco-toxicity. They have simple and safe delivery and disposal methods [86, 87].

Nanofertilizers are composed of nanoparticles whose size ranges from 1 to 100 nm. The better penetration of nanofertilizers into plant tissues is possible because of their small size which helps in efficient nutrient uptake. Nanofertilizers can be engineered to release nutrients in a controlled and sustained manner, matching the nutrient requirements of plants throughout the different phases of their growth cycle. The

Table 3 Advantages of genetic engineering in improving plant growth

Advantages	Mechanism	Examples
Increased nutrient use efficiency	Genes responsible for efficient nutrient uptake and utilization can be introduced into plants, allowing them to grow better in nutrient-poor soils	Nitrogen use efficiency: introducing genes that enable plants to utilize nitrogen more efficiently reduces the need for chemical fertilizers and improves growth in low-nitrogen soils
Enhanced photosynthesis	By modifying genes involved in photosynthesis, plants can convert sunlight into energy more efficiently, leading to faster growth and higher yields	C4 pathway engineering: introducing C4 photosynthesis traits into C3 plants like rice can increase photosynthetic efficiency, leading to improved growth rates
Drought and salinity tolerance	Genetic modifications can enhance a plant's ability to tolerate abiotic stresses like drought and high salinity, enabling growth in harsh environments	Drought-resistant crops: genes from naturally drought-tolerant plants can be introduced into crops, allowing them to survive and grow in arid regions
Pest and disease resistance	Introducing genes that confer resistance to pests and diseases can reduce the need for chemical pesticides and enhance plant health	Bt crops: *Bacillus thuringiensis* (Bt) genes are introduced into crops like cotton and corn, providing resistance to specific insect pests
Improved root architecture	Modifying genes that control root development can result in plants with deeper or more extensive root systems, enhancing nutrient and water uptake	Enhanced root growth: genes that promote root branching and depth can be engineered into crops to improve growth in nutrient-deficient or drought-prone soils
Biofortification	Genetic engineering can be used to increase the nutritional content of crops, improving human health and growth, especially in nutrient-deficient regions	Golden rice: rice has been genetically modified to produce beta-carotene, a precursor of vitamin A, addressing vitamin A deficiencies in some populations

increased surface area of nanoparticles improves the bioavailability of nutrients, ensuring that plants can absorb and utilize them more effectively. Some nanofertilizers are designed to deliver nutrients directly to the targeted parts of the plant (leaves, shoots, or roots), optimizing nutrient use and reducing waste.

These nanofertilizers consist of nanoparticles of essential nutrients such as nitrogen, potassium, phosphorus, zinc, and iron. For example, zinc is an essential micronutrient for plants, and zinc oxide nanoparticles can be used to correct zinc deficiencies in crops, leading to improved growth and yield. These are traditional fertilizers coated with nanoscale materials to control the release of nutrients like polymer-coated nanofertilizers. These fertilizers have a nanoscale polymer coating

Table 4 Function of different nanoparticles

Nanoparticles	Function	References
Carbon-based nanoparticles	Seed germination and plant growth Penetrate plant cell walls and membranes, and transport chemicals within cells	[91]
Silver-based nanoparticles	Pesticide resistance Fungal resistance Antimicrobial activity against various phytopathogens Antimicrobial activity against various phytopathogens Maintains water balance Enhances the salt-stress tolerance of crops	[39, 51]
Silicon-based nanoparticles	Resistance to biotic and abiotic stresses Enhances seed germination-related parameters Enhances photosynthetic efficiency Increases biomass and reducing oxidative stress and Cd uptake, and improving grain productivity and quality	[64, 80]
Copper-based nanoparticles	Antimicrobial capabilities Used in fungicides, insecticides, and nanofertilizers Plant disease resistance by promoting defense enzyme production	[25]
Magnesium-based nanoparticles	Enhances chlorophyll content and enzyme activity Influences the growth, chlorophyll content, and gene/miRNA expression in ornamental pineapples	[3]
Zinc-based nanoparticles	Promotes plant growth and resilience against abiotic stresses such as salinity, Cd stress, and drought Improves rice germination rates under salinity stress	[44, 46]
Iron-based nanoparticles	Improves plant metabolic processes, including photosynthesis, respiration, DNA synthesis, and the synthesis of photosynthetic pigments Reduces oxidative stress in plants grown under drought Used as nanofertilizers	[89]

that controls the release of nutrients over time, reducing the need for frequent applications. There are certain nanocarriers that transport the immobilized nutrients to the targeted plant cell in an effective way. For example, nitrogen can be immobilized inside the silica nanocarriers for its controlled and sustained release to meet the plant nutritional requirement.

Nanofertilizers are more effective than conventional fertilizers as they increase the nutrient uptake by the plant due to their higher penetration power. This leads to higher crop productivity and reduction in nutrient wastage. This helps farmers to gain financial benefit and food security. They decrease the dependency on chemical fertilizers and help in sustainable agricultural practice. It also decreases the environmental pollution by substituting the chemical fertilizers responsible for greenhouse gas emission and eutrophication and helps in nature preservation.

However, there are some limitations related with the usage of nanofertilizers that need to be addressed. Nanofertilizers may cause toxicity to the plant, microorganism present in soil, and the human beings exposed to it. Researchers are working throughout the world to assess the long-term impact of these fertilizers on human health and the environment. Its high cost also limits its use by the farmers due to budgetary concern. All these problems can be addressed by preparing proper regulatory framework to ensure safety, optimization of production process to reduce the cost and enhance the quality.

2.7.2 Nanopesticides

Nanopesticides are agrochemicals that are nanoscale structure used to eliminate plant pathogens, pests or weeds. The small size and high penetration power of nanomaterials give better results than chemical based pesticides in a sustainable manner. They are more specific than conventional pesticides as they can be tailored according to target requirement. They are highly efficient therefore required in less quantity and causes less harmful for non-target organisms. Nanopesticides also use nanomaterials as a carrier of active ingredients. They allow the controlled and sustainable release of active compound. This reduces the number of repeated usage and increases the efficiency. Target specific and sustainable release of pesticides is possible with help of silica nanoparticles with lower frequency and higher efficiency.

2.7.3 Nanosensors

Nanosensor is the analytical devices or sensors that use nanomaterials to detect and quantify the changes in the surroundings. It also has usefulness and application in agriculture sector. It is used to measure different micro and macro nutrient present in soil, soil moisture, plant diseases, heavy metal or any other pollutant, specific microorganism or pest, etc. By using nanosensor real-time data about soil and plant health can be observed and this will assist farmers in precision farming. The use of this technique helps agriculture practitioners get better productivity by choosing suitable methods of farming, irrigation, pest, and disease management according to the observed result using the nanosensors. The timely intervention to mitigate any nutrient deprivation or pathogen infection is possible due to early detection. This helps in crop protection and increasing the yield. For example, ethylene gas produced by plants during fruit ripening can be detected with the help of carbon nanotube-based nanosensors. This helps farmer to harvest the crop at appropriate time and helps in post harvest management.

2.7.4 Nanocarriers for Gene Delivery

To insert desired gene in a particular plant or cell to express a desired character in the plant carriers or vehicles are required. The nanoscale vehicles used to deliver gene or genetic material are known as nanocarriers. They assist in delivering a variety of biomolecules like RNA, DNA, plasmids, or proteins into targeted plant cell and facilitate the genetic modifications in an efficient way. The nanocarriers have smaller size that enables them to easily penetrate the cell membrane. Therefore, it can easily efficiently transfer the genetic material as compared to conventional methods like *Agrobacterium.* It is also non-invasive unlike gene gun method causing less damage to plant cell. An example is use of gold nanoparticle to deliver DNA or RNA into the plant cell to make them more tolerant to pest or develop resistance against drought.

2.7.5 Nanocatalysts

The catalyst that exists in nanoscale form and enhances the efficiency of chemical reactions is called nanocatalyst. In agriculture, they are used to promote microbial growth, degrade organic pollutants, cofactors of various enzymes, oxygen scavengers.

3 Challenges with the Biological Tool

The present era demands for more practical implementation of biological tools like PGPR, bioremediation, phytoremediation, genome editing, etc., for sustainable agriculture practices and better health of the ecosystem and the environment. These methods offer many advantages and lead to advancements in agriculture practices with sustainability. However, there are certain limitations of these biological tools that are discussed in this section [45, 59].

3.1 Limited Efficacy and Specificity

The biological tools are very specific, and their efficacy is highly dependent on type of contaminants present in the soil or water or environmental condition. For example, a biological tool designed for particular pollutants will not broadly used for all types of contaminants. The microorganisms used for degrading lipid or oil pollutants in soil may not be able to remove complex organic pollutants or heavy metal contaminants [48].

3.2 Slow Process

Another limitation associated with most of the biological tools is their time-consuming nature. The chemical and mechanical methods generally used in agriculture practices are faster as compared to biological methods like bioremediation or phytoremediation. This can be a drawback when immediate results are required. Cleaning up a polluted site using bioremediation might take several months or even years, whereas chemical remediation might achieve similar results in less time [47].

3.3 Environmental Dependence

The surrounding condition such as temperature, humidity, moisture, acidity or alkalinity of soil, and presence of other microorganisms has a huge impact on the performance of biological tools. Due to the dependency and variation in the environmental condition, inconsistency is observed when the biological tools are applied in the field [48]. This can be understood with the example of phytoremediation. The performance of this biological tool declines in arid regions as plant growth is restricted in the shortage of water. This in turn decreases the plant capacity to uptake and detoxify micro or nanocontaminants present in the soil [43].

3.4 Potential for Unintended Consequences

Introducing non-native microorganisms or genetically modified organisms (GMOs) for bioremediation or soil improvement can disrupt local ecosystems, leading to unintended ecological consequences. Introducing a non-native plant species for phytoremediation purposes might result in that plant becoming invasive, out-competing local vegetation and altering the habitat [65].

3.5 Scale-Up Challenges

Scaling up laboratory or small-scale successes to field or commercial levels can be challenging. The reason behind this is that there is difference in lab condition actual field condition. The lab condition is generally controlled and shows little variation, whereas unexpected fluctuations are observed in field that make it difficult to replicate the lab results during practical implementation in the field on the larger scale. In many cases, it was observed that certain microbial strain designed in the lab through genetic engineering showed better performance at small or piolet scale, but their efficiency is declined after large-scale application [10].

3.6 Economic Viability

There are variety of biological tools available these days in the market. However, farmers particularly small-scale farmers of developing countries still follow traditional methods of farming. The reason behind this is economic viability of conventional methods over biological tools. The advanced biotechnology is very expensive that decrease their practical usage. For example, the use of biochar without economic incentives or support is difficult because of its high cost of production [50].

3.7 Regulatory and Public Acceptance

The use of biological tools is often limited due to ethical concern and approval from concerned regulatory authority. Sometimes public is also not open to accept new technologies due to misconception and safety concerns associated with it. For example, GMOs are not widely accepted. Another example is genetically modified rhizobacteria that stimulates plant growth and have to go through series of regulatory approval and public acceptance. This is because of the potential risk associated with them that may impact human health as well as the environment [58].

3.8 Knowledge and Skill Requirements

The real-world implementation of biological tools needs knowledge and skill that limits its practical application. All practitioners might not able to get proper training or assistance for the usage of advanced tools due to remote locations or limited resources. For example, the application of bioremediation needs knowledge of microbiology, environmental science, and soil chemistry [49].

4 Conclusion and Future Perspective

The modern biological tools provide sustainable way of modern agriculture with higher efficiency and less pollution. However, to increase their potency, safety and reliability, there is need to overcome the limitations associated with them. Globally scientists are working to improvise these shortcomings and apply these techniques in a better way to boost soil health, eliminate contaminants to get better productivity. Researchers are looking for better microbial strains to be uses as biological tools like PGPR and mycorrhizal fungi to increase crop production, enhance soil quality and health, making plant stress resistant.

The techniques like phytoremediation helps farmers in multiple ways like removing contaminants, increasing soil fertility and crop yield. Phytofortification is also an emerging and attractive biological tool to improve nutritional composition of the food crops. This will help to address the problem of malnutrition. The addition of nanotechnology and biotechnology to design biological tools will upgrade and strengthen their usefulness. The example includes nanofertilizers, nanosensors, nanopesticides, etc., that promotes sustainable and efficient crop production. Nanocarriers are also productive tools to deliver desired genetic material or nutrient directly to the target sites in an efficient and sustainable manner. Development of nanosensors that monitor plant's health and soil's condition in real-time, enabling precise application of biological treatments. Developing blends of biochar with compost, microorganisms, or other organic materials to create synergistic effects on soil fertility and plants growth. Engineering crops to produce compounds that attract beneficial microbes or suppress pathogens, effectively enhancing the plant's biological toolkit.

References

1. Abdul Aziz M, Brini F, Rouached H, Masmoudi K (2022) Genetically engineered crops for sustainably enhanced food production systems. Front Plant Sci 13:1027828
2. Abouian Jahromi M, Jamshidi-Zanjani A, Khodadadi Darban A (2020) Heavy metal pollution and human health risk assessment for exposure to surface soil of mining area: a comprehensive study. Environ Earth Sci 79:1–18
3. Ali S, Ulhassan Z, Shahbaz H, Kaleem Z, Yousaf MA, Ali S, Zhou W (2024) Magnesium oxide nanoparticles as novel sustainable approach in enhancing crop tolerance to abiotic and biotic stresses. Environ Sci Nano
4. Aliyari Rad S, Nobaharan K, Pashapoor N, Pandey J, Dehghanian Z, Senapathi V, Asgari Lajayer B (2023) Nano-microbial remediation of polluted soil: a brief insight. Sustainability 15(1):876
5. Allouzi MMA, Tang DYY, Chew KW, Rinklebe J, Bolan N, Allouzi SMA, Show PL (2021) Micro (nano) plastic pollution: the ecological influence on soil-plant system and human health. Sci Total Environ 788:147815
6. Ashraf MA, Maah MJ, Yusoff I (2014) Soil contamination, risk assessment and remediation. Environ Risk Assess Soil Contam 1:3–56

7. Azubuike CC, Chikere CB, Okpokwasili GC (2016) Bioremediation techniques-classification based on site of application: principles, advantages, limitations and prospects. World J Microbiol Biotechnol 32:180
8. Bamdad H, Papari S, Lazarovits G, Berruti F (2022) Soil amendments for sustainable agriculture: microbial organic fertilizers. Soil Use Manage 38(1):94–120
9. Baweja P, Kumar S, Kumar G (2020) Fertilizers and pesticides: their impact on soil health and environment. Soil Health 265–285
10. Bellani CF, Ajeian J, Duffy L, Miotto M, Groenewegen L, Connon CJ (2020) Scale-up technologies for the manufacture of adherent cells. Front Nutr 7:575146
11. Bhattacharjee M, Meshram S, Dayma J, Pandey N, Abdallah N, Hamwieh A, Acharjee S (2024) Genetic engineering: a powerful tool for crop improvement. In: Frontier technologies for crop improvement. Springer Nature Singapore, Singapore, pp 223–258
12. Bhattacharyya PN, Sandilya SP, Sarma B, Pandey AK, Dutta J, Mahanta K, Borgohain DJ (2024) Biochar as soil amendment in climate-smart agriculture: opportunities, future prospects, and challenges. J Soil Sci Plant Nutr 24(1):135–158
13. Bigatton ED, Castillejo MA, Ayoub I, Baldessari JJ, Bruno M, Archilla MV, Haro RJ (2024) Plant growth promoting rhizobacteria (PGPR): impact on peanut flowering, seed physical quality, and yield determination (Arachis hypogaea L.). Ind Crops Prod 219:119024
14. Bodor A, Feigl G, Kolossa B, Mészáros E, Laczi K, Kovács E, Rákhely G (2024) Soils in distress: The impacts and ecological risks of (micro) plastic pollution in the terrestrial environment. Ecotoxicol Environ Saf 269:115807
15. Boudh S, Singh JS (2019) Pesticide contamination: environmental problems and remediation strategies. In: Emerging and eco-friendly approaches for waste management, pp 245–269
16. Bruinsma J (2017) World agriculture: towards 2015/2030: an FAO study. Routledge
17. Chakraborty I, Banik S, Biswas R, Yamamoto T, Noothalapati H, Mazumder N (2023) Raman spectroscopy for microplastic detection in water sources: a systematic review. Int J Environ Sci Technol 20(9):10435–10448
18. Chakraborty S, Mandal M, Chakraborty AP, Majumdar S (2022) Zinc solubilizing rhizobacteria as soil health engineer managing zinc deficiency in plants. In: Rhizosphere Engineering. Academic, pp 215–238
19. Choudhary A, Sharma S, Kaur H, Sharma N, Gadewar MM, Mehta S, Husen A (2024) Plant system, abiotic stress resilience, reactive oxygen species, and coordination of engineered nanomaterials: a review. S Afr J Bot 171:45–59. https://doi.org/10.1016/j.sajb.2024.05.053
20. Dhanapal AR, Thiruvengadam M, Vairavanathan J, Venkidasamy B, Easwaran M, Ghorbanpour M (2024) Nanotechnology approaches for the remediation of agricultural polluted soils. ACS Omega 9(12):13522–13533
21. Fred-Ahmadu OH, Bhagwat G, Oluyoye I, Benson NU, Ayejuyo OO, Palanisami T (2020) Interaction of chemical contaminants with microplastics: Principles and perspectives. Sci Total Environ 706:135978
22. Galdames Iglesias A, Ruiz Rubio L, Orueta Azcargorta M, Sánchez Arzalluz M, Vilas Vilela JL (2020) Zero-valent iron nanoparticles for soil and groundwater remediation
23. Gavrilescu M (2022) Enhancing phytoremediation of soils polluted with heavy metals. Curr Opin Biotechnol 74:21–31
24. Gigault J, Ter Halle A, Baudrimont M, Pascal PY, Gauffre F, Phi TL, Reynaud S (2018) Current opinion: what is a nanoplastic? Environ Pollut 235:1030–1034
25. Hafeez A, Razzaq A, Mahmood T, Jhanzab HM (2015) Potential of copper nanoparticles to increase growth and yield of wheat. J Nanosci Adv Technol 1(1):6–11
26. Han FX, Banin A, Kingery WL, Triplett GB, Zhou LX, Zheng SJ, Ding WX (2003) New approach to studies of heavy metal redistribution in soil. Adv Environ Res 8(1):113–120
27. Hasan A, Tabassum B, Hashim M, Khan N (2024) Role of plant growth promoting rhizobacteria (PGPR) as a plant growth enhancer for sustainable agriculture: a review. Bacteria 3(2):59–75
28. Hemathilake DMKS, Gunathilake DMCC (2022) Agricultural productivity and food supply to meet increased demands. In: Future foods. Academic, pp 539–553

29. Husen A (2022) Engineered nanomaterials for sustainable agricultural production, soil improvement and stress management. Elsevier Inc., Cambridge. https://doi.org/10.1016/C2021-0-00054-7
30. Husen A (2023) Nanomaterials and nanocomposites exposures to plants (response, interaction, phytotoxicity and defense mechanisms). Springer Nature Singapore Pte Ltd., Singapore. https://doi.org/10.1007/978-981-99-2419-6
31. Husen A (2025) Plant response to silver nanoparticles. Springer Nature Singapore Pte Ltd., Singapore
32. Husen A (2025) Plant response to gold nanoparticles. Springer Nature Singapore Pte Ltd., Singapore
33. Husen A (2025) Emerging carbon nanomaterials for sustainable agricultural practices (synthesis, plant growth, performance, production and protection). Springer Nature Singapore Pte Ltd., Singapore
34. Hussain A, Rehman F, Rafeeq H, Waqas M, Asghar A, Afsheen N, Iqbal HM (2022) In-situ, ex-situ, and nano-remediation strategies to treat polluted soil, water, and air–a review. Chemosphere 289:133252
35. Iqbal S, Xu J, Allen SD, Khan S, Nadir S, Arif MS, Yasmeen T (2020) Unraveling consequences of soil micro-and nano-plastic pollution on soil-plant system: Implications for nitrogen (N) cycling and soil microbial activity. Chemosphere 260:127578
36. Jean-Philippe SR, Labbé N, Franklin JA, Johnson A (2012) Detection of mercury and other metals in mercury contaminated soils using mid-infrared spectroscopy. Proc Int Acad Ecol Environ Sci 2(3):139
37. Jiang M, Song Y, Kanwar MK, Ahammed GJ, Shao S, Zhou J (2021) Phytonanotechnology applications in modern agriculture. J Nanobiotechnol 19:1–20
38. Kafle A, Timilsina A, Gautam A, Adhikari K, Bhattarai A, Aryal N (2022) Phytoremediation: mechanisms, plant selection and enhancement by natural and synthetic agents. Environ Adv 8:100203
39. Kale SK, Parishwad GV, Patil AS (2021) Emerging agriculture applications of silver nanoparticles. ES Food Agrofor 3:17–22
40. Kapahi M, Sachdeva S (2019) Bioremediation options for heavy metal pollution. J Health Pollut 9(24):191203
41. Kapoor D, Sharma P, Sharma MMM, Yadav S, Husen A (2024) Exploring soil microbiota and their role in plant growth, stress tolerance, disease control and nutrient immobilizer. Biocatal Agric Biotechnol 61:103358. https://doi.org/10.1016/j.bcab.2024.103358
42. Kensa VM (2011) Bioremediation-an overview. J Ind Pollut Control 27(2):161–168
43. Khilji SA, Waseem M, Tariq S, Jabeen S, Jamal A, Alomrani SO, Riaz A (2024) Microbe assisted phytoremediation of heavy metal contaminated soil by using African marigold (Tagetes erecta L.). Plant Stress 11:100369
44. Kumar A, Choudhary A, Kaur H, Mehta S, Husen A (2021) Metal-based nanoparticles, sensors and their multifaceted application in food packaging. J Nanobiotechnol 19(256):1–25. https://doi.org/10.1186/s12951-021-00996-0
45. Kumar A, Naroju SP, Langthasa M, Behera I, Tyagi S, Sharma N, Gautam RK (2023) Bioremediation potential of green wastes and plant growth promoting rhizobacteria and its enhancement by their combination: a review. Environ Adv 12:100379
46. Kumar A, Singh IK, Mishra R, Singh A, Ramawat N, Singh A (2021) The role of zinc oxide nanoparticles in plants: a critical appraisal. In: Nanomaterial biointeractions at the cellular, organismal and system levels, pp 249–267
47. Kuppan N, Padman M, Mahadeva M, Srinivasan S, Devarajan R (2024) A comprehensive review of sustainable bioremediation techniques: eco friendly solutions for waste and pollution management. Waste Manag Bull
48. Lahlali R, Ezrari S, Radouane N, Kenfaoui J, Esmaeel Q, El Hamss H, Barka EA (2022) Biological control of plant pathogens: a global perspective. Microorganisms 10(3):596
49. Lee H, Sam K, Coulon F, De Gisi S, Notarnicola M, Labianca C (2023) Recent developments and prospects of sustainable remediation treatments for major contaminants in soil: a review. Sci Total Environ 168769

50. Lokko Y, Heijde M, Schebesta K, Scholtès P, Van Montagu M, Giacca M (2018) Biotechnology and the bioeconomy—towards inclusive and sustainable industrial development. New Biotechnol 40:5–10
51. Mehmood A (2018) Brief overview of the application of silver nanoparticles to improve growth of crop plants. IET Nanobiotechnol 12(6):701–705
52. Miguel-Rojas C, Pérez-de-Luque A (2023) Nanobiosensors and nanoformulations in agriculture: new advances and challenges for sustainable agriculture. Emerg Top Life Sci 7(2):229–238
53. Mondal R, Dam P, Chakraborty J, Paret ML, Katı A, Altuntas S, Sarkar R, Ghorai S, Mandal AK, Husen A (2022) Potential of nanobiosensor in sustainable agriculture: the state-of-art. Heliyon 8(12):e12207. https://doi.org/10.2139/ssrn.4134873
54. Nedjimi B (2021) Phytoremediation: a sustainable environmental technology for heavy metals decontamination. SN Appl Sci 3(3):286
55. Ng EL, Lwanga EH, Eldridge SM, Johnston P, Hu HW, Geissen V, Chen D (2018) An overview of microplastic and nanoplastic pollution in agroecosystems. Sci Total Environ 627:1377–1388
56. Papp Z, Dezső Z, Daroczy S (2002) Significant radioactive contamination of soil around a coal-fired thermal power plant. J Environ Radioact 59(2):191–205
57. Prajapati SK, Meravi N (2014) Heavy metal speciation of soil and calotropis procera from thermal power plant basin. Proc Int Acad Ecol Environ Sci 3(1):59–75
58. Prakash D, Verma S, Bhatia R, Tiwary BN (2011) Risks and precautions of genetically modified organisms. Int Sch Res Not 1:369573
59. Qin H, Wang Z, Sha W, Song S, Qin F, Zhang W (2024) Role of plant-growth-promoting rhizobacteria in plant machinery for soil heavy metal detoxification. Microorganisms 12(4):700
60. Rajpal VR, Dhingra Y, Khungar L, Mehta S, Minkina T, Rajput VD, Husen A (2024) Metal and metal-oxide nanoparticles in nanosensing as well as managing biotic stress in plant system. Curr Res Biotechnol 7:100219. https://doi.org/10.1016/j.crbiot.2024.100219
61. Rajput VD, Minkina T, Upadhyay SK, Kumari A, Ranjan A, Mandzhieva S, Verma KK (2022) Nanotechnology in the restoration of polluted soil. Nanomaterials 12(5):769
62. Raju KV, Somashekar RK, Prakash KL (2013) Spatio-temporal variation of heavy metals in Cauvery River area. Proc Int Acad Ecol Environ Sci 4(2):68–71
63. Ramezani M, Rad FA, Ghahari S, Ghahari S, Ramezani M (2021) Nano-bioremediation application for environment contamination by microorganism. Microb Rejuvenation Pollut Environ 2:349–378
64. Rastogi A, Tripathi DK, Yadav S, Chauhan DK, Živčák M, Ghorbanpour M, Brestic M (2019) Application of silicon nanoparticles in agriculture. 3 Biotech 9:1–11
65. Rebello S, Nathan VK, Sindhu R, Binod P, Awasthi MK, Pandey A (2021) Bioengineered microbes for soil health restoration: present status and future. Bioengineered 12(2):12839–12853
66. Ridhi R, Saini GSS, Tripathi SK (2024) Nanotechnology as a sustainable solution for proliferating agriculture sector. Mater Sci Eng B 304:117383
67. Rillig MC, Ziersch L, Hempel S (2017) Microplastic transport in soil by earthworms. Sci Rep 7(1):1362
68. Russo M, Oliva M, Hussain MI, Muscolo A (2023) The hidden impacts of micro/nanoplastics on soil, crop and human health. J Agric Food Res 100870
69. Saini H, Thakur R, Gill R, Tyagi K, Goswami M (2023) CRISPR/Cas9-gene editing approaches in plant breeding. GM Crops Food 14(1):1–17
70. Salt DE, Smith RD, Raskin I (1998) Phytoremediation. Annu Rev Plant Biol 49(1):643–668
71. Saraswat P, Singh S, Prasad M, Misra R, Rajput VD, Ranjan R (2023) Applications of bio-based nanomaterials in environment and agriculture: a review on recent progresses. Hybrid Adv 100097
72. Sayadi MH, Rezaei MR (2014) Impact of land use on the distribution of toxic metals in surface soils in Birjand. Proc Int Acad Ecol Environ Sci 4(2):62–67
73. Sayadi MH, Sayyed MRG (2011) Comparative assessment of baseline concentration of the heavy metals in the soils of Tehran (Iran) with the comprisable reference data. Environ Earth Sci 63:1179–1188

74. Searchinger T, Hanson C, Ranganathan J, Lipinski B, Waite R, Winterbottom R, Ari TB (2014) Creating a sustainable food future. World resources report 2013–14: interim findings
75. Shang Y, Hasan MK, Ahammed GJ, Li M, Yin H, Zhou J (2019) Applications of nanotechnology in plant growth and crop protection: a review. Molecules 24(14):2558
76. Sharma A, Chetani R (2017) A review on the effect of organic and chemical fertilizers on plants. Int J Res Appl Sci Eng Technol 5:677–680
77. Sharma P, Pandey V, Sharma MMM, Patra A, Singh B, Mehta S, Husen A (2021) A review on biosensors and nanosensors application in agroecosystems. Nanoscale Res Lett 16(136):1–24. https://doi.org/10.1186/s11671-021-03593-0
78. Siddiqi KS, Husen A (2022) Plant response to silver nanoparticles: a critical review. Crit Rev Biotechnol 42(7):773–990. https://doi.org/10.1080/07388551.2021.1975091
79. Silva GC, Galleguillos Madrid FM, Hernández D, Pincheira G, Peralta AK, Urrestarazu Gavilán M, Fuentes-Peñailillo F (2021) Microplastics and their effect in horticultural crops: food safety and plant stress. Agronomy 11(8):1528
80. Singh A, Tiwari S, Pandey J, Lata C, Singh IK (2021) Role of nanoparticles in crop improvement and abiotic stress management. J Biotechnol 337:57–70
81. Su C (2014) A review on heavy metal contamination in the soil worldwide: situation, impact and remediation techniques. Environ Skept Crit 3(2):24
82. Šunta U, Trebše P, Kralj MB (2021) Simply applicable method for microplastics determination in environmental samples. Molecules 26(7):1840
83. Talaviya T, Shah D, Patel N, Yagnik H, Shah M (2020) Implementation of artificial intelligence in agriculture for optimisation of irrigation and application of pesticides and herbicides. Artif Intell Agric 4:58–73
84. Urushadze TF, Manakhov DV (2017) Radioactive contamination of the soils of Georgia. Ann Agrar Sci 15(3):375–379
85. Waller CL, Griffiths HJ, Waluda CM, Thorpe SE, Loaiza I, Moreno B, Hughes KA (2017) Microplastics in the Antarctic marine system: an emerging area of research. Sci Total Environ 598:220–227
86. Yadav A, Yadav K, Abd-Elsalam KA (2023) Nanofertilizers: types, delivery and advantages in agricultural sustainability. Agrochemicals 2(2):296–333
87. Yadav N, Garg VK, Chhillar AK, Rana JS (2023) Recent advances in nanotechnology for the improvement of conventional agricultural systems: a review. Plant Nano Biol 4:100032
88. Yan A, Wang Y, Tan SN, Mohd Yusof ML, Ghosh S, Chen Z (2020) Phytoremediation: a promising approach for revegetation of heavy metal-polluted land. Front Plant Sci 11:359
89. Yousaf N, Sardar MF, Ishfaq M, Yu B, Zhong Y, Zaman F, Zou C (2024) Insights in to iron-based nanoparticles (hematite and magnetite) improving the maize growth (Zea mays L.) and iron nutrition with low environmental impacts. Chemosphere 362:142781
90. Zantis LJ, Borchi C, Vijver MG, Peijnenburg W, Di Lonardo S, Bosker T (2023) Nano-and microplastics commonly cause adverse impacts on plants at environmentally relevant levels: a systematic review. Sci Total Environ 867:161211
91. Zaytseva O, Neumann G (2016) Carbon nanomaterials: production, impact on plant development, agricultural and environmental applications. Chem Biol Technol Agric 3:1–26
92. Zojaji F, Hassani AH, Sayadi MH (2014) Bioaccumulation of chromium by Zea mays in wastewater-irrigated soil: an experimental study. Proc Int Acad Ecol Environ Sci 4(2):62–67

Understanding the Impacts of Herbicides, Insecticides, and Fungicide Stress for Crop Plant Growth and Production

Amit B. Verma, Ishita J. Oza, A. Shah Rehana Anjum, Firdaus M. Quraishi, and Hina Alim

Abstract Herbicides, insecticides, and fungicides play a crucial role in managing weeds, pests, and diseases in modern agriculture. However, their use may have detrimental effects on the growth and development of crop plants and their ecosystem. The main objective of this chapter is to examine how herbicide, insecticide, and fungicide stress impact the growth and production of crop plant. It has been indicated that the use of these contaminants has a negative impact on crop productivity, nutrient uptake, hormone regulation, cell membrane integrity, and the ability to withstand drought conditions. Additionally, the impact of pesticide stress on crop yield performance and resistance to environmental stress is also discussed. Pesticides are harmful and resistant substances, and around 90% of the food one consume is exposed to herbicides. Having a thorough knowledge of herbicide, insecticide, and fungicide stress is crucial for creating sustainable agricultural practices that reduce pesticide usage while guaranteeing food safety. Strategies for reducing pesticide use encompass integrated pest management the cultivation of resistant crop varieties, conservation agriculture practices promoting beneficial insects and pollinators through habitat creation and diversification, and the development of biopesticides/organic pest control methods. Pesticides can lead to stress, oxidative stress, water stress, and disruption of the soil micro biome, nutrient deficiencies, resistance development, increased disease susceptibility, reduced seed quality, altered gene expression, soil residual effects, and some other negative impacts.

Keywords Herbicides · Insecticides · Fungicide · IPM · Crop plant

---

A. B. Verma · I. J. Oza · H. Alim (✉)
Department of Life Science, University of Mumbai, Mumbai, Maharashtra, India
e-mail: hina.alim@lifesciences.mu.ac.in

A. Shah Rehana Anjum
Department of Botany, Rizvi College of Arts, Science and Commerce, Mumbai, Maharashtra, India

F. M. Quraishi
Department of Biotechnology, Elphinstone College, Mumbai, Maharashtra, India

A. Husen (ed.), *Enhanced Crop Production Setup Under Various Micro and Nano Contaminants*, Smart Nanomaterials Technology,
https://doi.org/10.1007/978-981-96-6099-5_13

1 Introduction

Herbicides, insecticides, and fungicides are essential in modern agriculture as they help to control weeds, pests, and diseases [25]. Although these chemicals have significantly improved crop protection, they can also have unintended consequences on crops, such as stunted growth and decreased yield [45]. Studies have shown that these chemical stresses can result in reduced crop yields, impaired nutrient absorption, and decreased resistance to drought [34]. Moreover, the widespread utilization of these substances has raised concerns regarding their impact on the environment, potential health risks to humans, and the development of pesticide-resistant pests [39]. Pesticides have become indispensable in contemporary farming practices, serving as a crucial tool for safeguarding crops against pests and diseases. These chemicals are specifically used to control pests and protect agricultural and industrial products. They encompass a wide range of substances, including insecticides, herbicides, fungicides, and plant growth regulators, as shown in Fig. 1. Herbicides are employed to eradicate unwanted plants, while fungicides are utilized to inhibit the development of molds and mildew [4]. The use of pesticides is essential for meeting the increasing food requirements and safeguarding crops from pest-related damage, ensuring food safety.

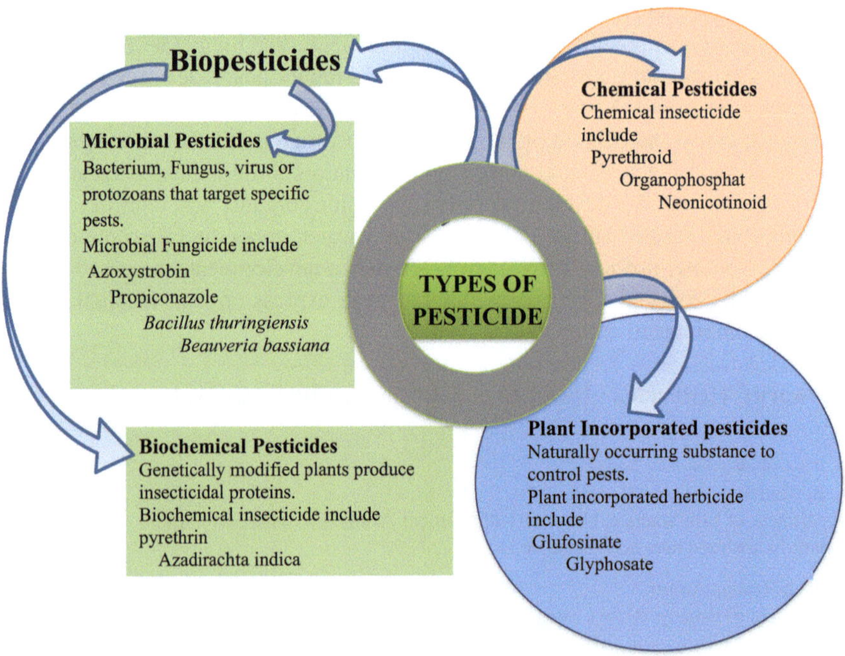

Fig. 1 Types of pesticide (chemical pesticide and biopesticide—microbial, biochemical, and plant-incorporated pesticide)

Despite the benefits of using herbicides, insecticides, fungicides, and nematicides, there are also some drawbacks to their extensive use. Numerous of these chemicals can be detrimental to non-target organisms, leading to their accumulation in the environment and causing disruptions in ecological balance [4]. Effective crop protection and agricultural productivity are essential for food safety, which relies on the careful use of pesticides [10]. It is estimated that only about 0.1% of pesticides effectively target pests, with the remainder contaminating ecosystems and negatively impacting air, water, and soil. According to the Food and Agriculture Organization, potential global food losses due to pests are about 55%, with 35% occurring before harvest and 20% after harvest [9].

Despite their importance in boosting agricultural output, pesticides are toxic and persistent. Nearly 90% of food products come into contact with various herbicides used in crops [40]. Residues of herbicides in food primarily result from direct application to plants rather than soil contamination [12]. Adopting best agricultural management practices can significantly reduce the risks of pesticide contamination in the food chain. Nevertheless, many pesticides affect more than just target pests, they also impact non-target plants and animals. They play a crucial role in crop protection and disease management, often being applied preventively. The extensive use of these chemicals can have harmful effects on plants, particularly those that are water-soluble and stable, as well as on natural flora, fauna, and aquatic life. Herbicide stress can lead to diminished growth and yield, altered plant physiology, and increased vulnerability to diseases [6]. Insecticide stress may cause acute toxicity, chronic effects, and hormonal disruption and also harm to beneficial organisms. Fungicide stress can result in phytotoxicity, reduced growth and yield, and impaired photosynthesis.

Pesticide classify in bio pesticide or chemical pesticides. Biopesticides are made from animals, plants, or microbes that can kill, harm, or repel insects. They are less toxic, biodegrade faster, and are more specific to the insect [38]. Biopesticide use is increasing because of increased environmental consciousness and global demand for organically grown food [29]. Instead of completely eliminating a particular pest, biopesticides usually try to keep a pest population under control. Biopesticides are classified into three types based on the active component used: microbial, biochemical, and plant-incorporated protectants [22]. Microbial pesticides are pest management products composed of naturally occurring or genetically modified algae, fungi, bacteria, viruses, and protozoans [20]. Biochemical pesticides/traditional chemical pesticides have a very close link [49]. It differs from traditional pesticides due to its non-toxic mechanism of action on target species and natural occurrence. Plant-incorporated protectants pesticides are those that plants directly biosynthesize from genetic material. These plant-incorporated tissues have been genetically engineered to create pesticides. Both the plant and the genetic material implanted produce PIPs. Cry proteins were the first PIPs used to kill insects in genetically modified (GM) crops with *Bacillus thuringiensis* transgenes [24].

Understanding how herbicide, insecticide, and fungicide stress affect crop plants is essential for creating sustainable agricultural practices that minimize damage and enhance crop productivity [16]. Identifying potential risks and exploring alternatives techniques are critical to this effort. Some of the strategies used to reduce dependence on pesticides, herbicide, fungicides, and insecticides include:

(a) Use of Integrated Pest Management (IPM) techniques, a combination of physical, cultural, biological, and chemical control methods [16].
(b) Use resistant crops or participate in breeding programs to increase disease resistance.
(c) By adopting conservation agriculture practices one can improve soil health and supports biodiversity [32].

By adopting above discus these strategies one can help reduce the environmental impact of pesticide use and create more sustainable food production.

2 Global Pesticide Use

The rise in pesticide usage worldwide is closely linked to the expanding human population and the demand for greater agricultural productivity [52]. The increase in pesticide usage has resulted in substantial harm to the environment. A comprehensive analysis of pesticide contamination on a global scale emphasizes the critical need to safeguard human well-being and preserve endangered species [39]. In the last five decades, there has been a notable change in the way pesticides are used worldwide. The usage of herbicides has significantly risen, whereas the application of insecticides and fungicides has decreased proportionally. China has emerged as the top global producer and exporter of herbicides. In future, advancements in pesticides are anticipated to prioritize high effectiveness and biological reactivity, with the goal of minimizing overall usage and environmental harm. These new pesticides are expected to be less harmful to the environment and more eco-friendly. Biopesticides provide numerous benefits, such as efficient pest management without harming non-target organisms, minimal residue on crops, and the ability to break down naturally. The specific impacts of this chemical may include stunted growth, yellowing of leaves (chlorosis), plant death, and contamination of the environment. The effects of herbicide stress on crops can vary based on factors such as the specific herbicide used, the quality applied, the timing of the application, as well as the particular crop species and its development stage. They are highly specific to target pests, which can lead to increased productivity and greater sustainability. Additionally, biopesticides can be improved through modern biotechnology and fermentation techniques, enhancing their effectiveness and quality. They also tend to generate lower levels of pest resistance compared to traditional pesticides.

The global pesticide market was valued at approximately $60 billion in 2021, with expectations for growth. In 2021, the Food and Agriculture Organization (FAO) reported that global pesticide consumption was about 4.2 million metric tons. Total global pesticide uses 4.4 million tons in 2022 [15]. In Asian, countries like China and India have high levels of pesticide use. In India, the total pesticide consumption was 52,466 metric tons [17], insecticides accounting for 34,441 metric tons (65.6%

Table 1 Consumption and effect (on crop plant) of pesticides in India (data adopted from Ministry of agriculture and farmers welfare 2020–2021)

Category of pesticide	Metric tons (Kg)	Percentage use	Effects on crop plants	Key references
Insecticides A. Organochlorin B. Organophosphates C. Pyrethroid D. Neonicotinoid	34,441 1,421 14,419 6,419 12,182	65.6% 2.7% 27.5% 12.2% 23.2%	Harmful to beneficial insects like bees and butterflies and other pollinators Contaminate water sources and soil Potential human health risks like cancer and neurological issues	[1] [36]
Herbicides A. Glyphosate B. Atrazine C. 2,4-D	12,351 4,419 2,351 1,581	23.5% 8.4% 4.5% 3.0%	It can lead to herbicide resistance weeds Contaminate water sources and soil Hormonal imbalances in human	[34] [19]
Fungicides A. Copper-based B. Sulfur based C. Azole	4,964 1,419 1,245 1,300	9.5% 2.7% 2.4% 2.5%	Fungicide resistant in fungal strains It can harm beneficial fungi, contaminant soil, and water Skin irritation and respiratory issues in human	[41] [14]
Other pesticides	710	1.4%	Vary depending on the specific pesticide, can include environment and health risk	[43] [51]

of total), herbicides 12,351 metric tons (23.5% of total), fungicides 4,964 metric tons (9.5% of total), and other pesticides 710 metric tons (1.4%) (Table 1).

3 Herbicides

Herbicides are chemicals designed to manage or eliminate unwanted plants, such as weeds and invasive species. The initial use of herbicides, specifically 2,4-D, occurred during war time by the USA. Herbicides can be classified based on their chemical composition, application method, site of action, or timing [48]. Herbicides are mainly categorized as systemic herbicides. These herbicides are extensively translocated in the plant through its vascular system along with water, nutrients, and other materials from site of absorption to sites of action. This herbicide moves through the system to the roots, stems, or leaves, where they exert their effects and contact herbicides. These herbicides kill only the portion of plant tissue that is in contact. These are not translocated through the plant. Plants destroy through direct contact and do not depend on internal movement of plant [48].

Types of Herbicides

Based on specificity, herbicides are categorized as selective herbicides—target-specific plant species while leaving crops unaffected, making them commonly used in agriculture to control weeds without harming the desired crops [48]. Selectivity can result from variations in translocation, differences in absorption rates, and distinct physical (morphological) or physiological characteristics among plant species. Non-selective herbicides—Have a 144 broad action and affect all plants they come into contact with, typically used to clear 145 all vegetation from that area. They are used to clear industrial sites, waste ground, railways, and railway embankments. Some of the non-selective herbicides are paraquat, glufosinate, glyphosate.

Based on application timing, the herbicides are categorized as pre-plant herbicides— these are non-selective herbicides applied to soil before planting and get mechanically incorporated into the soil. The objective for incorporation is to prevent dissipation through photodecomposition or volatility. The herbicides kill weeds as they grow through the herbicide-treated zone [48], pre-emergence herbicides—these are applied before the weed seedlings emerge through the soil surface. Herbicides do not prevent weeds from germinating, but they kill weeds as they grow through the herbicide-treated zone by affecting the cell division in the emerging seedling. Dithiopyr and pendimethalin are pre-emergence herbicides, post-emergence herbicides; these herbicides are applied after weed seedlings, have emerged through the soil surface, and generally require multiple applications for adequate control [48]. They can be foliar or root absorbed, selective or non-selective, contact or systemic. Liquid formulations of herbicides are more effective than granular formulations.

Herbicides can be classified as soil-applied herbicide based on their method of application. These are applied directly to the soil, where they are absorbed by the roots or shoots of emerging weed seedlings and are used as pre-plant or pre-emergence treatment—and foliar-applied herbicides- applied to the foliage of weeds above the soil, where they are absorbed by the exposed leaf tissues [48]. These are generally post-emergence herbicides and can either be translocated (systemic) throughout the plant or remain at specific site (contact). Certain herbicides, such as glyphosate, 2,4-D and dicamba are applied directly to the foliage.

Herbicides can affect crop plants in several ways, including through drift or overspray, soil residual activity, contamination of spray tanks, and volatilization. These issues can lead to herbicide resistance, crop damage, or reduced yields [33]. The impact of herbicide stress on crops can differ depending on factors like the type of herbicide employed, the amount applied, the timing of application, and the specific crop species and its growth stage.

3.1 *Effects of Herbicide Stress on Crop Plants*

Herbicides can have several negative effects on crop plants, significantly impacting their growth and yield. A major impact is the disruption of the electron transport chain within the chloroplast [42]. Moreover, glyphosate interferes with the production of important aromatic amino acids by obstructing the shikimic acid pathway, which is crucial for protein synthesis and overall plant growth. Herbicide disrupts cell division by interfering with microtubule formation, which is a vital process for growth [42]. These chemicals can interfere with plant hormones, leading to abnormal growth patterns like excessive stretching or unusual leaf shape, which ultimately reduces crop yield. The use of herbicides not only affects the health of plants but also has detrimental effects on soil health. These chemicals can harm beneficial soil microbes, which are essential for nutrient availability and can result in nutrient deficiencies in crops [3]. Additionally, the toxicity of herbicides can result in significant harm or even death of crop plants, especially when applied in large quantities or during critical growth periods [18]. The effects of herbicide, as depicted in Fig. 2, are commonly observed.

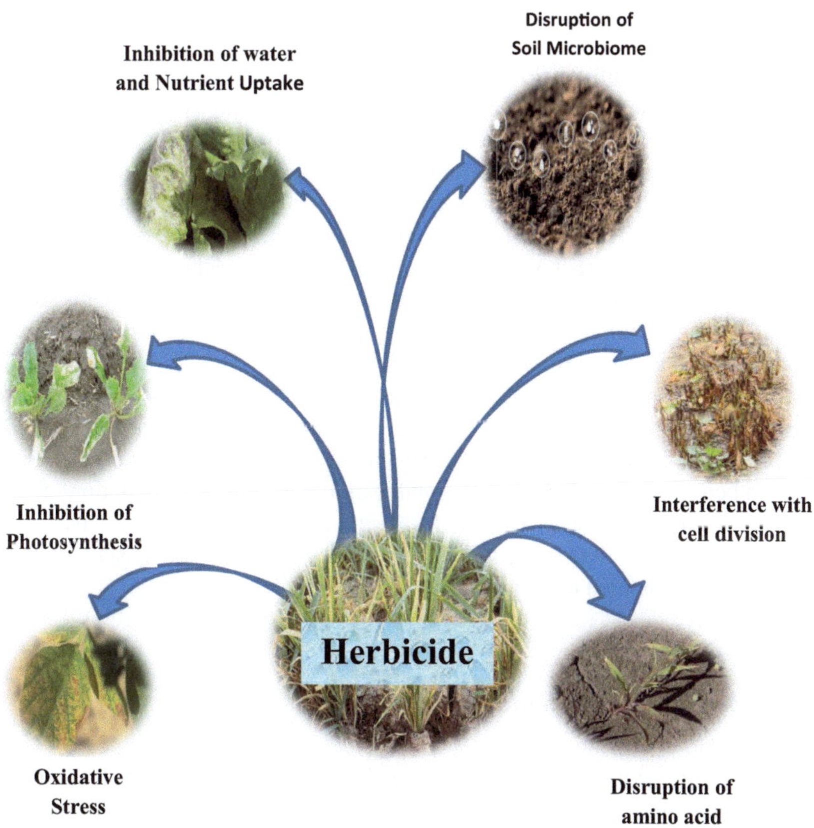

Fig. 2 Impacts of herbicide stress for crop plants

3.2 Crop Plant Affected by Usage Herbicides

Glyphosate and atrazine herbicide, frequently employed in sugarcane farming, have been linked to the development of herbicide-resistant weeds and documented cases of water pollution [2]. In soybean fields, the use of glyphosate and the unintentional movement of dicamba herbicide can result in negative consequences such as stunted growth, leaf curl, and decreased crop yields [25]. Similarly, the application of bispyribac-sodium glufosinate and atrazine in the cultivation of corn, rice, and canola plants can lead to stunted growth, yellowing of leaves, and a reduction in overall crop yield. These above findings show the significant impact of herbicide use in agricultural productivity and environmental health.

4 Insecticides

The insecticides are substances used to control or kill insects, specially that cause crop damage or disease. The insecticides are classified into systemic insecticides (antibiotics)—they are absorbed by the plant and distributed to the tissues against the pest. These type of insecticide interfere with the nervous systems of insects, disrupt their hormonal balance, inhibit enzyme functions and negatively impact their reproductive processes [30]. Contact insecticides—these insecticides kill insects by direct contact with the insect without being absorbed by the plant. Biological insecticides (biopesticides)—these insecticides are derived from natural organisms such as bacteria, viruses, or fungi, and organic insecticides—are made from natural organic compounds such as pyrethrin or rotenone [30].

Depending upon many factors, such as the type of insecticide, dose and duration of exposure, and the specific crop variety, the effect of insecticide can vary. Insecticides can affect crops through a number of mechanisms, including indirect effects, direct effects on beneficial insects, interaction with soil organisms, and disruption of plant hormone systems [30]. Some insecticides, such as neonicotinoids, can have effects on plant body because they are absorbed and distributed to the plant tissues, and they can also have unintended effects.

4.1 Impacts of Insecticide Stress on the Crop Plant

Insecticides with acute toxicity present considerable danger to crop plants during their vulnerable growth stages, which can cause damage or death. Long lasting use of insecticides can cause long-term effects such as direct effects, change in growth and plant physiology, and reduced yield. In addition, insecticides can affect the regulation of phytohormones that affect plant growth and metabolism [27]. Alternatively, the inhibition of photosynthesis caused by insecticides may result in a decrease in

Table 2 Different crop plant affected due to various class of insecticides

Insecticide	Class of insecticide	Crop plant	Effect on plant	References
Neonicotinoid	A. Imidacloprid B. Thiamethoxam C. Nitenpyram	Corn, soyabeans Cotton, rice	Stunted growth, leaf curl, and reduce yield with inhibition of seed germination	[21]
Biological insecticide	A. Bacillus thuringiensis B. Beauveria bassiana C. Metarhizium anisopliae D. Spinosad	Cotton, beans, peas, wheat and tomatoes	Leaf curl and stunted growth	[46] [21]
Organic insecticide	A. Azadirachta indica B. pyrethrin, rotenone	Potatoes, eggplant, beans, and peas	Stunted growth with curl leaf	[21]

protein content, obstruction of stomatal conductance, and an impact on the rate of photosynthesis [11]. Therefore, the consequences of insecticide use are not limited to pest control alone, but also affect the overall health and productivity of crops (Table 2).

5 Fungicides

Fungicides are chemicals used to manage or eliminate fungal diseases in plants. They come in various types such as systemic fungicides—these are the only fungicides that are absorbed into xylem and phloem and move up and down in plants. These fungicides inhibit the growth of fungi on and within treated plant surfaces, as well as in plant parts located both above and below these treated areas including phosphonates [37]. Contact fungicides— these fungicides act only on plant surfaces. These act by killing fungal spores upon direct contact and not absorbed by leaves, stems, or roots and cannot inhibit fungal development inside plants. Here are a few examples such as dithiocarbamates, nitriles, aromatic hydrocarbons, peroxides, phenylpyrroles, and cyanoimidazoles [37], preventative fungicides— applied before fungal infections occur, these fungicides help prevent the onset of fungal growth and curative fungicides— these are used after a fungal infection has been established to control and manage the spread of the disease. The effects of fungicide stress on crop plants can differ based on several factors, such as the type of fungicide used, its application rate and timing, as well as the specific crop species and its growth stage. For instance, some fungicides, such as azoles, may have systemic effects by being absorbed and distributed throughout the plant's tissues, which can also impact on non-target organisms [41].

5.1 Potential Impacts of Fungicide Stress on Crop Plants

Phytotoxicity caused by fungicides can lead to direct damage in crop plants, manifesting through symptoms such as chlorosis, necrosis, and leaf curl, which ultimately reduce crop growth, biomass, and yield by disrupting plant metabolism and nutrient uptake [44]. The application of fungicides often alters plant physiology, significantly impacting hormone regulation, consequently modifying plant development and metabolic processes. This impairment in photosynthesis can result in a reduction of photo assimilate production, further decreasing both growth and yield of the crops. For instance, contact fungicides like copper have been shown to inhibit photosynthesis by damaging chloroplasts, which affects photosystem II activity and disrupts chlorophyll biosynthesis [35].

5.2 Crop Plant Affected Due to Fungicides

The application of difenoconazole at a rate of 500 mg kg^{-1} dry weight of soil has been shown to decrease overall soil microbial activity, which negatively impacts the activity of soil enzymes, as evidenced by the respiratory quotient [31]. Additionally, some commonly used fungicides, such as chlorothalonil and dinitrophenyl, have been found to disrupt bacteria-dependent processes involved in nitrification and denitrification [26]. Furthermore, several older fungicides including sulfur, pyrazophos, and chinomethionate, effective against powdery mildew in fruits, are not recommended due to their toxicity to various beneficial organisms, such as phytoseiidae and coccinellidae.

6 Most Common Potential Impacts of Pesticides on Crop Plant

Pesticides, while important for pest control, can harm crops through a variety of mechanisms. They can cause stress and oxidative stress leading to crop damage and stunted growth. Pesticides can also affect plant water balance, making plants more susceptible to water stress and lower water temperature. Altering the soil micro biome is another factor that affects nutrition, overall soil health and plant growth. The uptake and utilization of nutrients can also be affected, leading to nutrient deficiencies and increased plant vigor. In addition, excessive use of pesticides can lead to the development of bacterial resistance, which reduces their effectiveness over time [5]. Stress from pesticides makes plants more susceptible to disease because they are less able to defend themselves. In addition, the effects of pesticides can affect seed quality, leading to reduced disease, viability, and poor seed development [4]. Pesticides may also alter gene expression in crops, influencing their development and metabolic processes. The ecological balance is disrupted as beneficial organisms, including pollinators and beneficial insects, are harmed by pesticide application [47]. Lastly, the residual effects of pesticides can persist in the soil affecting subsequent crops and the overall health of the soil ecosystem.

7 Pesticide Management Strategies

Pesticide resistance management strategies have been planned that are free from risk and successful in many situations, for example, monitoring pest populations in the field before applying pesticides, altering pesticides with different modes of operation, limiting many applications over space and time, generating refuges, and avoiding gratuitous persistence. The four basic elements of IPM are natural control, sampling economic levels, insect biology, and ecology. The first element of IPM relates to the fullest utilization of naturally occurring suppressive factors, including any practice by man which will make the total ecosystem less favorable for growth of the insect pest population. The second element is that of using sound economic threshold levels (ETLs) as the basis for applying control measures, especially chemical measures [8]. Establishing and using dynamic ETL's provide a basis for delaying the use of insecticides. This permits the maximum utilization of other control methods, such as the use of beneficial insects. The use of economic threshold levels implies adequate sampling of all harmful and beneficial insects in the agro ecosystem and particularly in any one crop at a given time. The levels found through sampling must then be measured against the economic level established for the crop, the beneficial insects, and the probable population trend of the pest species. The fourth element, insect biology and ecology, is essential to the fullest utilization of the other three elements.

It is difficult to understand natural management without a detailed knowledge of the biology and ecology of present in each plant species. This knowledge is also necessary to determine the role of each species in the system and to determine the damage caused by each type of pest [8]. By applying integrated pest management (IPM) strategies, one can reduce pesticide use and promote sustainable agriculture practices, such as crop rotation, organic farming, and many resilient agriculture. To protect ecosystem, five types of management alone can be used for IPM programs. These include chemical control, physical and mechanical control, biological control, cultivation, and management control [8].

For effective of pest management, combination of strategies designed to eliminate pests or creates an unfavorable environment for pest survival and reproduction. Environmental control is physical control or control that aims to eliminate pests or create unsuitable habitats for pests by method such as mechanical protection and thermal pest management [8]. Cultivation (planting different crops in specific area) is another essential practice that helps to maintain soil fertility, reduce pests and diseases, and increase crop yields by planting [50]. In addition, crops have developed genetic resistance to pests, thereby improving soil health and reducing dependence on chemicals, for example, the use of resistant varieties [8]. Cultural management focuses on controlling pests using natural enemies or parasite, while cultural management reduces pest problems by making change in farming practices such as pruning and sanitation [13]. Chemical management, such as the selection of pesticides and herbicides as a last resort, is based on environmental influences [13]. Pesticide rotation helps in delaying the development of resistance by changing the type of work, while precision agriculture reduces waste from methods that use pesticides only when necessary [23]. Organic agriculture refers to the need to avoid the use of synthetic pesticides through natural products and crop management, while biotechnology focuses on the development of genetically modified crops with anti-inflammatory properties [23]. Tracking and predicting the incidence of diseases and weather patterns can enable proactive measures to ready for potential outbreaks, while effective pest control initiatives can create lasting strategies that encompass various sustainable pest management and farming techniques [7] (Fig. 3).

Fig. 3 Different Steps in integrated pest management (IPM)

8 Conclusion and Future Prospects

For controlling pest pesticide application can be an effective method for controlling pests, but its indiscriminate use arrests risks to both target and non-target crops. It is essential to carefully consider the potential side effects of pesticides, taking into account their application methods and the specific agricultural systems in which they are used. The research should be focused on comprehending the consequences and longevity of pesticides in crops, as well as their influence on soil microbial communities and nitrogen metabolism. Promoting the creation of affordable biopesticides as safer substitutes is crucial, alongside the necessity for more effective techniques to eliminate pesticide residues from food grains. Moreover, additional research is required to identify the mechanisms and kinases involved in the defense of cells against pesticide toxicity. The application of herbicides, insecticides, and fungicides has a substantial effect on crop growth and production, playing a crucial role in controlling pests and diseases. This, in turn, results in increased crop yields and improved food security. These chemicals present significant challenges, such as the emergence of resistance in pests, potential harm to beneficial organisms, risks to human health, and environmental impact. The degradation of soil health and biodiversity can have lasting impacts on the sustainability of agricultural practices. To tackle these challenges, integrated pest management strategies are crucial as they promote the use of a combination of biological, cultural, and chemical methods to control pests in an environmentally responsible way. This method not only enhances crop productivity but also protects the environment and promotes the well-being of both humans and ecosystems. In order to investigate the connection between secondary plant metabolites and enzymatic diversity, research is required to explore their potential applications in pest management bioremediation and the production of fine chemicals.

Competing Interests The authors declare that they have no competing interest.

Declaration of Generative AI and AI-Assisted Technologies in the Writing Process During the preparation of this work, the book chapter author(s) have not taken any assistance from AI, and the book chapter author(s) reviewed and edited the content as needed and take(s) full responsibility for the content of the publication.

References

1. Abdullah NZ, Ishaka A, Shamsuddin N, Mohd-Rus R, Mohamed AH (2011) Chronic organophosphate pesticide exposure and coronary artery disease. Finding a bridge: IIUM Research. http://irep.iium.edu.my/id/eprint/13579
2. Aekrathok P, Songsri P, Jongrungklang N, Gonkhamdee S (2021) Efficacy of post-emergence herbicides against important weeds of sugarcane in North-East Thailand. Agronomy 2021(11):429. https://doi.org/10.3390/agronomy11030429

3. Ahemad M, Khan MS, Zaidi A, Wani PA (2009) Remediation of herbicides contaminated soil using microbes. Microb Sustain Agric 261(5):1–84. https://doi.org/10.1007/978-3-642-01979-1_2
4. Aktar W, Sengupta D, Chowdhury A (2009) Impact of pesticides use in agriculture & their benefits and hazards. J Environ Sci Health Part B 44(3):1–26. https://doi.org/10.2478/v10102-009-0001-7
5. Andrea MM, Peres TB, Luchini LC, Pettinelli A (2000) Impact of long-term pesticide applications on some soil biological parameters. J Environ Sci Health Part B 35(3):297–307. https://doi.org/10.1080/03601230009373271
6. Baweja P, Kumar S, Kumar G (2020) Fertilizers and pesticides: their impact on soil health and environment. In: Soil health, pp 265–285. https://doi.org/10.1007/978-3-030-44364-1_1
7. Begum A, Alam SN, Jalal-Uddin M (2017) Management of pesticides: purposes, uses, and concerns. In: Pesticide residue in foods, pp 53–86. https://doi.org/10.1007/978-3-319-52683-6_4
8. Belen MR, Romeo SR (2005) The concept and components of integrated pest management. Philipp J Entomol 19(1):1–15. https://www.fao.org/4/x5048e/x5048E07.htm
9. Bilal AK, Muhammad AN, Humaira N, Muhammad MA, Ghulam HA, Muhammad N, Muhammad A, Muhammad A, Muhammad MJ, Rizwan M, Muhammad I, Muhammad AA (2023) Pesticides: impacts on agriculture productivity, environment, and management strategies (Chap. 5). In: Emerging contaminants and plants, emerging contaminants and associated treatment technologies, pp 109–134. https://doi.org/10.1007/978-3-031-22269-6_5
10. Bolognesi C (2003). Genotoxicity of pesticides: a review of human biomonitoring studies. Mutat Res/Rev Mutat Res 543:251–272. https://doi.org/10.1016/S1383-5742(03)00015-2
11. Bondareva L, Fedorova N (2021) Pesticides: behavior in agricultural soil and plants. Molecules 26(17):5370. https://doi.org/10.3390/molecules26175370
12. Businelli A, Vischetti C, Coletti A (1992) Validation of Koc approach for modeling the fate of some herbicides in Italian soli. Fres Anal Chem 1:583–588. https://www.cabidigitallibrary.org/doi/full/10.5555/19931975450
13. Damalas CA, Eleftherohorinos IG (2011) Pesticide exposure, safety issues, and risk assessment indicators. Int J Environ Res Public Health 8(5):1402–1419. https://doi.org/10.3390/ijerph8051402
14. Fungicide resistant action committee (FARC) (2020) Report on fungicide resistant management. https://doi.org/10.11111/ppa.13467
15. Food and Agriculture organization (FAO) of the United Nations (2022) Global pesticide use and consumption
16. Gill HK, Garg H (2014) Pesticide: environmental impacts and management strategies. Pesticides-toxic effects. https://doi.org/10.5772/57399
17. Government of India, Ministry of agriculture and farmers welfare (2021) Report on pesticide consumption in India
18. Hasanuzzaman M, Mohsin SM, Bhuyan MHMB, Bhuiyan TF, Anee TI, Masud AAC, Nahar S (2020) Phytotoxicity, environmental and health hazards of herbicides: challenges and ways forward. In: Agrochemicals detection, treatment and remediation. Butterworth-Heinemann, Oxford, pp 55–99. https://doi.org/10.1016/B978-0-08-103017-2.00003-9
19. Heap I (2014) Herbicide-resistant weeds. In: Herbicides: chemistry, efficacy, toxicology and environmental impacts, pp 301–316. https://doi.org/10.1002/ps.3696
20. Ignatieva M, Haase D, Dushkova D, Haase A (2020) Lawns in cities: from a globalised urban green space phenomenon to sustainable nature-based solutions. Land 9(3):73. https://doi.org/10.3390/land9030073
21. Insecticide Resistance Action Committee (IRAC) (2020) Report on insecticide resistance action committee mode of action and classification, pp 1–30
22. Islam MS, Morshed A (2013) Study on homemade bio-pesticides and organic pest management in organic farming. Int J Eng Sci 2(7):18–25. Islam: Study on homemade bio-pesticides and organic... Google Scholar

23. Jorge F-C, Seth JW, Mike L, Lorraine M (2014) Genetically Engineered Crops in the United States, Department of Agriculture, Economic Research Service. https://doi.org/10.2139/ssrn.2503388
24. Kumar J, Ramlal A, Mallick D, Mishra V (2021) An overview of some biopesticides and their importance in plant protection for commercial acceptance. Plants 10(6):1185. https://doi.org/10.3390/plants10061185
25. Kumar V, Singh S, Kumar M, Gill JPK, Datta S, Singh S, Dhaka V, Kapoor D, Wani AB, Dhanjal DS, Harikumar SL, Singh J (2020) Herbicide glyphosate: toxicity and microbial degradation. J Environ Sci Health 17(20):7519. https://doi.org/10.3390/ijerph17207519
26. Lang M, Cai Z (2009) Effects of chlorothalonil and carbendazim on nitrification and denitrification in soils. J Environ Sci 21:458–467. https://doi.org/10.1016/s1001-0742(08)62292-5
27. Lopez-Perea JJ, Diaz RM, Gomez-Maqueo A (2019) Insecticide stress in crops: a review of the effects on plant growth and development. J Environ Sci Health 92(2):347–358
28. Mahmood I, Imadi SR, Shazadi K, Gul A, Hakeem KR (2016) Effects of pesticides on environment. In: Plant, soil and microbes, pp 253–269. https://doi.org/10.1007/978-3-319-27455-3_13
29. Manna MC, Rahman MM, Naidu R, Bari AF, Singh A, Thakur J, Ghosh A, Patra AK, Chaudhari S, Subbarao A (2021) Organic farming: a prospect for food, environment and livelihood security in Indian agriculture. Adv Agron 170:101–153. Manna: Organic farming: a prospect for food, environment... Google Scholar
30. Matsumura F (1985) Classification of insecticides. In: Toxicology of insecticides. Springer, Boston. https://doi.org/10.1007/978-1-4613-2491-1_3
31. Munoz-Leoz B, Garbisu C, Charcosset JY, Sanchez-Perez JM, Antigüedad I, Ruiz-Romera E (2013) Non-target effects of three formulated pesticides on microbially mediated processes in a clay-loam soil. Sci Total Environ 449:345–354. https://doi.org/10.1016/j.scitotenv.2013.01.079
32. Oirdi ME, Yaseen M, Farwa U, Raza MA, Farhan M, Sandhu ZA, Ali F, Aatif M, Hena SK, Nahvi I (2024) Crops and people: the dangers and potential benefits of pesticides. Cogent Food Agric 10(1):2334096. https://doi.org/10.1080/23311932.2024.2334096
33. Owen MDK (2008) Weed species shifts in glyphosate-resistant crops. Pest Manag Sci 64(4):319–325. https://doi.org/10.1002/ps.1539
34. Patel JK, Vyas RB, Patel HK (2022) Pesticide consumption in India: a review. J Environ Sci Health B 57:1–12
35. Petit AN, Fontaine F, Vatsa P et al (2012) Fungicide impacts on photosynthesis in crop plants. Photosynth Res 111:315–326. https://doi.org/10.1007/s11120-012-9719-8
36. Potts SG, Biesmeijer JC, Kremen C, Neumann P, Schweiger O, Kunin WE (2010) Global pollinator declines: trends, impacts and drivers. Trends Ecol Evol 25(6):345–354. https://doi.org/10.1016/j.tree.2010.01.007
37. Rachid R (2010) Introduction and toxicology of fungicides (Chap. 18). Intech open, pp 363–382. https://doi.org/10.13140/RG.2.1.2099.9125
38. Rawat N S, Lathwal SS, Panchbhai GJ, Pal RK, Jha AK, Jamra MS (2020) Physical, microbial and biochemical composition of bio-pesticides based on cow urine and dung with medicinal plants. J Pharmacogn Phytochem 9(5S):754–758. Rawat: Physical, microbial and biochemical composition... Google Scholar
39. Reddy AA, Reddy M, Mathur V (2024) Pesticide use, regulation, and policies in indian agriculture 16(17):7839. https://doi.org/10.3390/su16177839
40. Rosa M, Gonazalez R, Raquel RO, Beatriz CG, Jesus SG (2008) Occurrence, fungicide and insecticide residues in trade samples pf leafy vegetables. Food Chem 107:1342–1347. https://doi.org/10.1016/j.foodchem.2007.09.045
41. Sharma S, Singh BP, Kumar S, Sharma SS, Kaushik S (2022) Fungicide use and resistance in India. J Fungic Resist 10(1):1–10
42. Singh M, Singh N, Singh S (2020) Impact of herbicide use on soil microorganisms. In: Plant responses to soil pollution. Springer, Cham, pp 179–194. https://doi.org/10.1007/978-981-15-4964-9_11

43. Singh R, Kumar V, Singh RK, Kumar P, Singh RB (2022) Environmental and health impacts of pesticide use in India. J Environ Sci Health C 40:1–15
44. Singh R, Singh RB, Kumar P, Singh RK (2018) Fungicide stress affects seed quality and seedling establishment in wheat. J Agric Sci Technol 18(3):557–568
45. Singh R, Singh RB, Kumar P, Singh RK, Tiwari RK (2019) Resistant varieties: a review. J Agric Sci Technol 19(3):547–556
46. Sparks TC, Nauen R (2015) IRAC: Mode of action classification and insecticide resistance management. Pestic Biochem Physiol 121:122–128. https://doi.org/10.1016/j.pestbp.2014.11.014
47. Talat P, Sumira J, Sumira M, Tasneem F, Zahid HS (2016) Selective effect of pesticides on plant-a review. Crit Rev Food Sci Nutr 56(1):160–179. https://doi.org/10.1080/10408398.2013.787969
48. Vats S (2015) Herbicides: history, classification and genetic manipulation of plants for herbicide resistance. Springer science, pp 153–192. https://doi.org/10.1007/978-3-319-09132-7_3
49. Verma N, Bhardwaj A (2015) Biosensor technology for pesticides—a review. Appl Biochem Biotechnol 175(6):3093–3119. https://doi.org/10.1007/s12010-015-1489-2
50. Wang Y, Liu X, Wu X, Yang Y, Zhou X (2017) Nitenpyram, a neonicotinoid insecticide, causes stunted growth and reduced yields in rice plants. J Agric Food Chem 65(2):453–459
51. World Health Organization (WHO) (2019) The impact of pesticides on health and environment of crop plant report
52. Zhang W, Jiang F, Ou J (2011) Effects of herbicides on growth and yield of crops: a review. J Agric Sci Technol 11(4):537–548

Understanding Physiological and Biochemical Pathways Responsible for Stress Tolerance Mechanisms in Crop Plants

Rabeea Zafar and Muhammad Arshad

Abstract This chapter endeavors to enhance crop resistance to adverse factors like drought, salinity, and extreme temperatures through the identification of important physiological and biochemical mechanisms involved in stress responses. Researchers significantly highlighted morphological, cellular, and metabolic changes to gain knowledge about how these happenings contribute to improved stress resistance. These include approaches such as breeding and genetic engineering for integrating the newly discovered knowledge into practical solutions to enhance crop resilience. Future research directions include the approach toward climate change, environmental stress, and their implications on agricultural productivity. The focus is on interdisciplinary collaboration and proposed areas of study include advanced genetic techniques, functional genomics, and sustainable agriculture practices. This chapter attempts to bring strategic priorities into perspective with a better understanding of physiological and biochemical pathways governing stress tolerance toward realizing food security in a changing climate.

Keywords Stress tolerance · Physiological pathway · Biochemical pathway · Food security · Climate change · Environmental stress

List of Abbreviations

CTY	Cytoplasm
DAG	Diacylglycerol

R. Zafar (✉)
Department of Environmental Science, Faculty of Sciences, Allama Iqbal Open University, Islamabad, Pakistan
e-mail: rabeea.zafar@aiou.edu.pk

R. Zafar · M. Arshad
Institute of Environmental Science and Engineering (IESE), School of Civil and Environmental Engineering (SCEE), National University of Sciences and Technology (NUST), Islamabad, Pakistan

A. Husen (ed.), *Enhanced Crop Production Setup Under Various Micro and Nano Contaminants*, Smart Nanomaterials Technology,
https://doi.org/10.1007/978-981-96-6099-5_14

DAGK	Diacylglycerol kinase
ER	Endoplasmic reticulum
IP3	Inositol 1,4,5-trisphosphate
IP6	Inositol hexakisphosphate
IP3/IP6-R	Putative IP3 or IP6 receptors
IPKs	Inositol phosphate kinases
NUC	Nucleus
PA	Phosphatidic acid
PI-PLC	Phosphoinositide-specific phospholipase C
PKs	Protein kinases
PKCs	Putative protein kinase Cs
PM	Plasma membrane
PTases	Inositol polyphosphate phosphatases
PtdIns(4,5)P_2	Phosphatidylinositol 4,5-bisphosphate
TFs	Transcription factors
VAC	Vacuole

1 Introduction

Stress tolerance is defined as the biological processes that occur for the plant cells to withstand persistent stressors, such as the production of compatible solutes and protective proteins that counteract the harmful effects of osmotic and ionic pressures in plants exposed to conditions of drought and salinity. There are different causes of stress and multiple factors are involved in the stress tolerance mechanism in plants (Fig. 1).

Crop yilelds have high importance in addressing food security. Environmental stresses affect the crop yields to above 50% of annual productivity, thus raising serious concern of food security [1]. Throughout the years, different approaches, ranging from traditional plant breeding, genomic-assisted breeding, mutagenesis, and genetic engineering, have been used to introduce stress tolerance traits into different crop species [2]. Plant breeding has achieved partial success in developing crop varieties that resist abiotic stresses, but the progress of plant breeding has not been advanced owing to complex genetics and limited genetic variability within the germplasm. Transgenic crops have been subjected to significant controversy solely because of the environmental danger posed by the potential for transgene outcrossing of the risks associated, even though such genetic engineering may possess prospective power to transfer genes from any organism. Consequently, the cultivation of genetically modified crops is prohibited in some countries, such as Europe [3]. This way, the discovery of CRISPR technology is one way of producing gene-edited crops transgene-free, just like mutagenized crops, which have less strict regulation than GMOs [4, 5]. Therefore, genome-edited plants that have no transgenes will most likely be cultivated in agriculture without any kind of constraint.

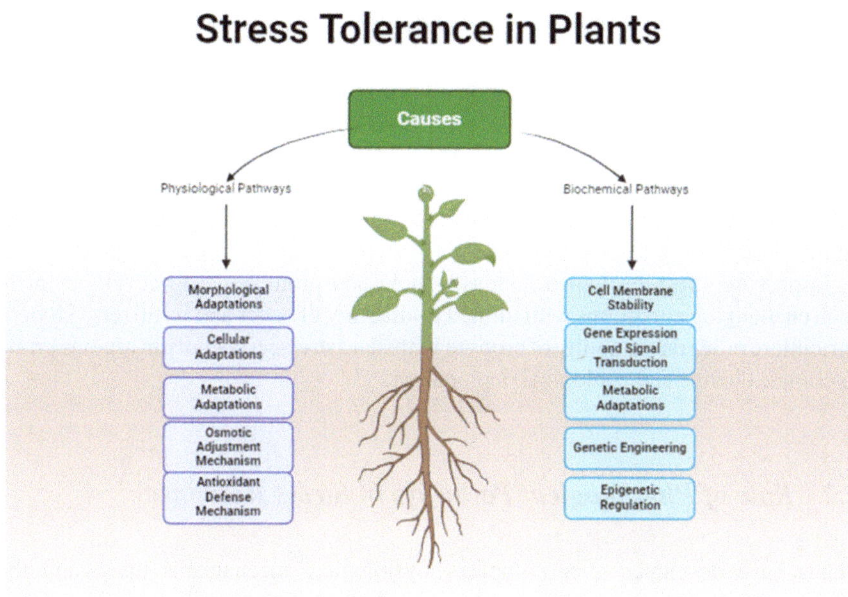

Fig. 1 Causes of stress tolerance in plants

The study of plant physiology with regard to stress tolerance has been mainly directed toward understanding the fundamental mechanisms through which plants adapt to stressful conditions. Researchers attempt to make crop plants more resistant to some of the stress factors like drought, salinity, and increased or decreased temperatures by identifying the key physiological mechanisms related to stress tolerance. This chapter focuses on biologically relevant processes that mediate stress resistance and assesses the role of morphological, cellular, and metabolic adjustments that enhance stress resistance.

2 Physiological Pathway in Crop Plants

2.1 Importance of Stress Tolerance in Crop Plants

Stress tolerance in plants is critical to global food security, especially under changing climatic conditions. Agricultural crops are threatened by different environmental stresses that impact their growth productivity and survival. Generally, these stresses can be summarized into two categories: abiotic and biotic stress [6, 7]. Abiotic stress refers to environmental factors such as drought, salinity, extreme temperature fluctuations, and nutrient deficiencies that may cause damage to growth and reproductive success. Biotic stresses, however, are derived from living organisms and include pests, diseases, and pathogens, causing massive losses in global crop yields [6].

Drought is one of the most common abiotic stresses and may reduce agricultural productivity by up to 50% in some regions. Salinity threatens nearly 20% of irrigated farmland around the world [8] and leads to significant losses in crop yield. Crop tolerance to stress is another very important trait that plant breeders and scientists are trying to enhance through conventional breeding and biotechnology-based approaches.

Improving stress resilience increases yield and promotes crop development in such challenging conditions with limited availability of water and fertilizers. Modern agriculture relies on the ability of crops to withstand stress, especially in regions prone to climate change and ecological degradation.

2.2 Role of Physiological Pathways in Stress Response

Plants have developed very complex physiological mechanisms that stand the stress and remain in homeostasis. These mechanisms are morphological, cellular, metabolic, and hormonal that enable plants to survive and grow under stressful conditions (Fig. 2). Osmotic regulation is a basic adaptation mechanism in which plants accumulate osmotically active solutes, such as proline, glycine betaine, and sugars, adjusting the osmotic potential to maintain intracellular water during drought or salinity stress [9, 10].

Another crucial mechanism of protection against oxidative stress from ROS in plants is the antioxidant defense system. Plants experience high levels of ROS while under stress, which may damage cellular components such as proteins, lipids, and DNA [11, 12]. Plants respond by generating the expressions of antioxidant enzymes, including superoxide dismutase (SOD), catalase, and peroxidases, and also some non-enzymatic antioxidants like glutathione and ascorbate.

2.3 Factors Involved in Physiological Pathways of Stress

Plant responses to stress are highly complex and involve a range of physiological processes. The ability of plants to perceive stressors and initiate adaptive responses depends largely on the following factors:

a. The synthesis of appropriate solutes, including proline, glycine betaine, and sugars, represents a crucial adjustment that takes place during osmotic regulation. These solutes facilitate the management of osmotic pressure within plant cells, thereby allowing the retention of water in conditions of drought or salinity.
b. Plants can prevent wilting and maintain metabolic processes by keeping the water potential in their cells.

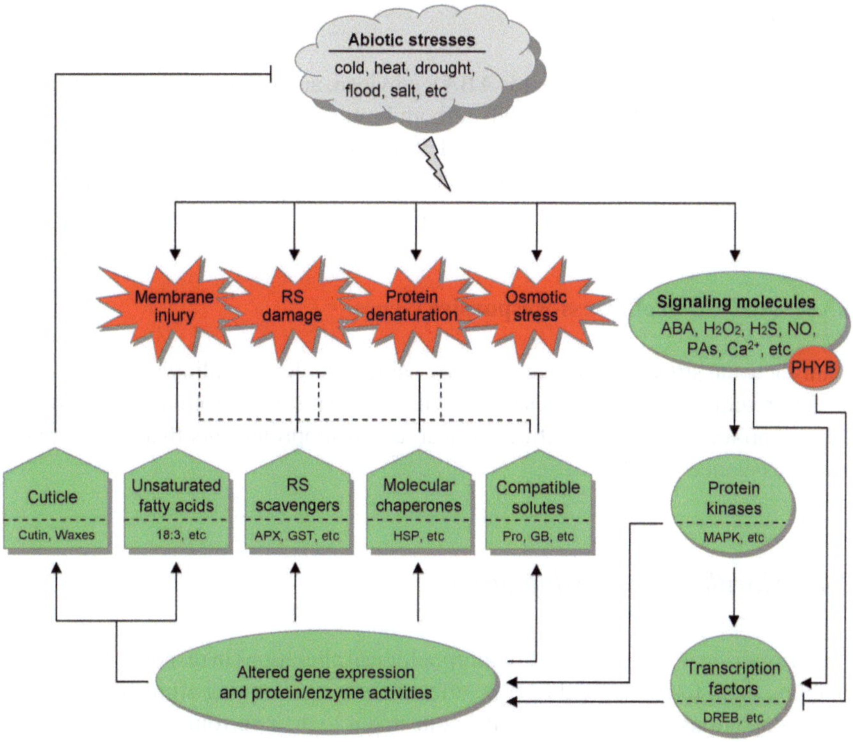

Fig. 2 Physiological pathways of stress tolerance in plants

2.3.1 Ion Homeostasis

For plants, it is fundamental to manage the uptake and distribution of ions in salt-stressed environments to maintain homeostasis. The management of ions can be achieved through sequestration of excess sodium into vacuoles through Na+/H+ antiporters in plant cells [13–15]. It counteracts the harmful effects induced by the sodium taken while maintaining the equilibrium of other important ions, such as potassium and calcium necessary for proper cellular functions.

2.3.2 ROS Scavenging

Plants increase reactive oxygen species (ROS) under stressful environmental conditions, which is a dangerous byproduct of metabolism [11, 12]. Plants respond with antioxidant defenses so as to limit ROS damage that includes enzymes such as superoxide dismutase (SOD) and catalase. This activation is vital for the preservation of cellular components as well as the maintenance of physiological functions during stress periods.

2.3.3 Hormones

Hormones including abscisic acid (ABA), ethylene, and jasmonic acid play a crucial role in governing a plant's reactions to both biotic and abiotic factors [16, 17]. For example, ABA facilitates the closure of stomata to reduce water loss in situations of drought. While ethylene is involved in controlling responses to challenges arising from mechanical stress or pathogen invasion.

2.3.4 Activation of Stress-Responsive Genes

Plants activate stress-responsive genes by using complex signaling pathways. Some of the proteins encoded by these genes include heat shock proteins, which protect cellular proteins against the effect of heat, and osmoprotectants that help maintain osmotic balance [18].

2.4 Morphological Adaptations

Plants experience a diverse array of morphological alterations in response to stressors, enabling them to endure and operate effectively in challenging climatic conditions. Such adaptations facilitate the plants' adjustment to their surrounding environment.

2.4.1 Modifications to the Leaves to Reduce Transpiration

A large percentage of plant species change the leaf morphology in ways to minimize water loss through transpiration. Some of the common changes include smaller leaves, thicker cuticles, and stomata recessed. Most plants that survive conditions characterized by arid environments display some of these changes. The cuticle is the waxy layer that covers the leaf. In well-acclimated arid environments, the cuticle tends to be thicker than in less adapted species or environments. Therefore, the amount of water evaporating has decreased [19]. Another vital adjustment is stomata control. This occurs when plants either decrease the number of stomata or develop underground stomata, which restricts the exchange of water vapor with the atmosphere.

For instance, members of the family *Cactaceae* occasionally exhibit leaves that are either diminished or modified into spines, thereby aiding in the minimization of water loss. Furthermore, some plant species have evolved to possess folded leaves, which serve to decrease the surface area exposed to sunlight, consequently further curtailing water loss [20].

2.4.2 Root Architecture and the Efficiency with Which Nutrients Are Absorbed

The root system of the plant is its most important organ of absorption of water and nutrients. Hence, it alters structural changes in the roots when subjected to any kind of stressors in order to increase their ability to acquire the required resources. For example, in drought conditions, the plants may have a more extensive and deep root system where they might tap into the deeper parts within the soil where the reserves of water are kept. However, even with such dry conditions in the top layers of the soil, these plants can hydrate because they have this deeper system of roots. In salinized conditions, some plant species begin to possess a shallow root profile to avoid higher salt concentrations in deeper layers of soil [21]. On the other hand, most plant species adapt specific root structures to excrete surplus salt: the prototype of such plants is halophytes, including mangroves.

Plants are known to change their root system architecture to enhance their nutrient uptake capabilities in response to stressors. When nutrient deficiencies are encountered, more root hair production increases the root surface area [22]. This adaptability would allow them to increase the uptake of essential nutrients such as nitrogen, phosphorus, and potassium.

2.4.3 Stem and Shoot Modifications for Structural Support

There are several environmental stressors that lead to brittle plant structures. In such scenarios, plants may modify their stems and shoots in order to increase their rigidity. Some plants increase the amount of lignin deposited in their stems [23]. Lignified plants become stronger and are able to resist physical stress such as wind or drought.

It is also possible for the succulent stems to develop in diffuse-drained situations [24]. Water storage becomes possible, and plants can last longer without requesting external supplies of water. These include desert plants, cacti, and succulents, with their water-storing stems.

Depending on their exact environmental condition, other plants will modify shoot architecture to minimize heat stress through reduced direct sunlight on the critical tissues of the plant. Vertical leaf position and the presence of hairy leaves may reduce solar heat absorption allowing protection for the plant's internal structures during prolonged exposure to elevated temperatures [21, 22]. As a result, these morphological, cellular, and metabolic adaptations enable them to grow in environments with little water, nutrients, and other raw materials. When scientists discover these natural adaptions, the hope is that they can develop crop varieties, which respond to the pressing demands of future environmental conditions, including climate change.

2.5 *Cellular Adaptations*

2.5.1 Membrane Lipid Composition and Fluidity

Cell membranes are the major physical barriers to intrusion by external stressors. The fluidity and integrity of the cell membrane are both crucial for the plant's ability to resist stress, and its proportion of membrane lipids; thus, these factors are interrelated with stress [25]. In terms of lipid metabolism, membranes contain an appropriate amount of saturated and unsaturated acids in the normal state, and an excessive amount of saturation acids can lead to the suppression of membrane fluidity. Consequently, the amount of unsaturation acids is increased to avoid the negative impact of stress on membrane fluidity when the membrane is subjected to stress, such as cold [26]. This makes it possible for the plant to maintain membrane integrity and complete its normal activities even when it is adapting to cold temperatures.

The levels of saturated fatty acids can increase in plants in response to heat stress. This adjustment is necessary to prevent excessive membrane fluidity, which could lead to leakage of cellular contents. Such variations in the lipid composition of membranes play a crucial role in sustaining cellular homeostasis when confronted with challenging environmental factors, including drought, salinity, and fluctuating temperatures.

2.5.2 Ion Transport Mechanisms in Cellular Homeostasis

Ion transport regulation is essential for maintaining plant cellular homeostasis, especially under high metabolic stress. Such ions include sodium (Na^+), potassium (K^+), and calcium (Ca^{2+}), which play a central role in the regulation of osmotic equilibrium, enzymatic activity, and signaling transduction mechanisms [27].

Plants are exposed to saline environments and thus subjected to higher concentrations of sodium ions (Na^+), which is proven detrimental in higher quantities. In order to maintain ionic homeostasis, plants activate vacuolar membrane-bound Na^+/H^+ antiporters. The activity of these antiporters proves to be essential in sequestering sodium ions within the vacuoles in order to reduce the concentration of sodium ions within the cytoplasm. In addition to maintaining the osmotic balance, the process protects cellular activities from salt's harmful effects [28].

Another critical mechanism of ion transport is recognized as a calcium signaling pathway. In some signal transduction pathways, calcium ions play the role of secondary messengers for activation of stress-response gene expression. Alterations in cytosolic concentrations serve to stimulate the expression of stress-responsive genes and proteins that ultimately give rise to physiological responses to abiotic stresses such as salinity, low temperatures, or drought conditions [28].

2.5.3 Role of Organelles (Chloroplasts, Mitochondria) in Stress Response

Chloroplasts and mitochondria are the main energy centers for plant cells, hence providing great contributions toward stress response in plants. Their roles are critical. Specifically, chloroplasts assist in photosynthesis, the process by which light energy is converted into chemical energy. The photosynthetic process often is hindered when plants undergo strenuous environmental factors, either drought or salinity, resulting in the generation of ROS [12]. Mechanisms occur within chloroplasts that could diminish such stress as a result of antioxidant enzymes and activation of superoxide dismutase (SOD) and ascorbate peroxidase. These enzymes assist in the detoxification of reactive oxygen species (ROS) and safeguard the machinery that is responsible for photosynthesis.

Mitochondria, responsible for cellular respiration, are also instrumental in regulating the release of energy under stress conditions. In stress-inducing environments, the functionality of mitochondria changes according to different metabolic pathways; hence, the synthesis of harmful ROS is decreased [29]. Given that they act as signaling centers following stress, mitochondria can also induce programmed cell death (PCD) when damage is regarded as excessively severe to cause harm to the rest of the plant.

2.6 Metabolic Adaptations

2.6.1 Regulation of Photosystem II

Photosystem II (PSII) is a major component in the light-independent reactions occurring during photosynthesis. Photolysis, a part of the light-independent reaction, releases oxygen, electrons, and protons. In cases of excessive light-intensity stress or drought or extreme temperature, PSII is likely to be damaged because this damage decreases the photosynthesis process's functionality [30]. Plants have developed protective mechanisms that would regulate their functionality.

For instance, the mechanism of non-photochemical quenching (NPQ) has created a safe pathway for light energy dissipation as heat and therefore prevents the overexcitation of the power supply interconnection of PSII. Although not so efficient, this protective mechanism ensures that the photosynthetic apparatus is functioning under stress [31].

2.6.2 Alternative Routes for the Transport of Electrons

In addition to the linear chain, stressed plants contain the avenue of electron transport that enables photosynthesis to produce ATP and NADPH. Among these include cyclic electron flow (CEF) around photosystem I [31]. This, therefore, results in the production of more ATP but without the production of NADPH. It saves cellular energy while keeping the ratio of ATP to NADPH as close to each other as possible at all times during stressful conditions, even when carbon fixation is limited by salinity or drought.

Another alternative pathway is the water-water cycle. This cycle involves the movement of electrons from water to oxygen and forms ROS [29], which are neutralized later on through antioxidant mechanisms. This pathway safeguards against oxidative damage of the photosynthetic apparatus caused by stress.

2.6.3 Carbon Partitioning and Sink Strength in Plants that Are Under Stressed Conditions

There is a mechanism known as "carbon partitioning" that is important for plants to make judicious decisions in carbon resource allocation when they get stressed out [32]. During stress, most of the carbon resources in the plant go to protective mechanisms i.e. the production of osmoprotectants (proline and sugars), and antioxidants.

Sink strength is the ability of any plant organ, for example, roots or fruits, to acquire and utilize the photosynthates. Sink strength usually declines in stressed stages. Carbon share taken out from the reproductive organs can be redirected to the roots or shoots to boost chances of survival in case of drought [32]. This carbon distribution helps to maintain the general metabolism of the plant and also to enhance the tolerance of the plant to stress.

2.7 Osmotic Adjustment Mechanisms

Osmotic stress caused by salt or drought leads to the adaptive response in plants by the accumulation of specific solutes referred to as osmolytes [33, 34]. These small organic molecules help maintain osmotic pressure without interfering with cell normal processes, therefore, they are termed as "compatible solutes." Proline, glycine betaine, and soluble sugars are some examples of these.

Proline is very significant in enhancing the stress tolerance of plants through different mechanisms. It is an osmoprotectant that stabilizes proteins, maintains the cellular structure from damage, and reduces the ROS. The accumulation of proline is highly visibly seen in salt- or drought-stressed plants. In similar ways, glycine betaine is also another very important osmolyte that stabilizes the membrane and proteins when such structures are exposed to adverse conditions [35].

2.7.1 The Control of Water Potential and the Maintenance of Osmotic Equilibrium

For a stressed plant, it is quite important that they control the water potential at a precise and fine level so that they retain cellular turgor along with proper physiological functions. The uptake and translocation of water and solutes by the cellular membranes are able to conserve osmotic equilibrium [35]. In drought, water potential in a plant usually drops, risking the onset of wilting and stunted growth.

To counter the said phenomenon, the plant reduces its osmotic potential by accumulating osmolytes, which are chemicals that can enhance the chances for water retention in cells. More importantly, aquaporins constitute forms of water channels in cellular membranes, and their functions significantly contribute to the regulation of the rate of water transport across these membranes to help retain the water balance in plants, particularly under extreme stress [34].

The stomata are small pores situated on the leaf surface, and that is where the regulation of gas exchange and water loss through transpiration occurs. A drought-stressed plant needs to shut down its stomata so as not to lose unnecessary water [22]. This also reduces the intake of carbon dioxide, which is crucial in photosynthesis.

Abscisic acid (ABA) is the most crucial regulator of stomatal opening-closure during drought. Since water stress results in increased ABA concentration, it initiates signaling pathways to cause the stomata apertures to close down [16]. Consequently, this water-conserving mechanism also retains the potential of the plant's water and reduces water loss.

On the other hand, stomatal closure can persist for an extended period and thus retard photosynthesis and further growth. Again, it is important that these plants must know when to regulate their stomata and, simultaneously, require the uptake of carbon dioxide and survive without getting too much photosynthetic activity [22].

With such development, plants have been able to develop various cellular, metabolic, and osmotic adaptations that ensure the survival of these critical physiological processes such as photosynthesis, respiration, and uptake of water. It is due to this understanding that crop varieties are becoming resilient to stresses that are also the resultant effects of global climate change in addition to other stresses.

2.7.2 Antioxidant Defense System

Antioxidants are those chemicals that even in trace amounts prevent oxidation of many molecules. These compounds have a significant role in combating deleterious effects caused by free radicals in biological organisms resulting from oxidative stress. Free radicals are formed with unstable electrons and have a high order of reactivity that may cause harmful reactions to normal metabolic processes. The organism avoids these adverse reactions using particular enzymes that are specifically adapted to free radical detoxification. Superoxide dismutase (SOD), monodehydroascorbate, guaiacol peroxidase, glutathione peroxidase (GPX), glutathione reductase, glutathione S-transferase, dehydroascorbate reductase, and catalase are some of the key enzymatic antioxidants [36] that form a very essential part of an organism's defense machinery, eliminating free radicals generated from metabolic activities of the cells in the body [37]. The focus is on the mechanism of operation of such antioxidants as SOD, GPX, and catalase and, therefore, their importance in mechanisms of disease prevention and stress tolerance.

2.7.3 Enzymatic Antioxidants

Superoxide Dismutase

Superoxide dismutase is an enzyme that is very abundantly distributed throughout the body and catalyzes the decomposition of superoxide radicals into less damaging products. Hydrogen peroxide is a product of this reaction but most likely also one that exacerbated damage initiated by free radicals [36]. Plants generate an enormous quantity of reactive oxygen species, including superoxide, hydrogen peroxide, and hydroxyl radicals with the latter being highly tissue-damaging because it can damage tissues further by attacking surrounding cells. There are three main types of SOD. The most heavily studied is the cytoplasmic type, which contains zinc and copper. The mitochondrial form contains manganese, while the third is extracellular.

Catalase

Catalase is an antioxidant enzyme that facilitates the breakdown of hydrogen peroxide into water and oxygen, effectively neutralizing its harmful effects within cells. The exact concentration of catalase in the cytoplasm is difficult to determine, as much of it is lost during tissue handling [36, 37]. Oxidative stress arises when there is an imbalance between reactive oxygen species and antioxidants, which contributes to the development and progression of various diseases.

2.7.4 Non-enzymatic Antioxidants

This is based on the scavenging or reducing capabilities of glutathione, ascorbate, α-tocopherol, carotenoids, and flavonoids, meaning high levels of non-enzymatic antioxidants improve the plant's tolerance toward different environmental stresses [38]. In most crop plants, stress tolerance is associated with the efficiency of the antioxidant systems of these crops. Hence, it becomes important scientifically to know how far the defense mechanisms of these crops are effective in withstanding challenging environments. Key non-enzymatic antioxidants are ascorbic acid (vitamin C), glutathione (GSH), and α-tocopherol (vitamin E) [38]. Upregulation of such antioxidant pathways during oxidative stress helps reduce the toxic effects of free radicals.

Glutathione

Glutathione peroxidase and glutathione reductase are significant antioxidant enzymes. The reduced form of glutathione (GSH) has a protective role, while its oxidized form (GSSG) does not. Hydrogen peroxide from the intracellular environment is scavenged by reduced glutathione, reducing the possibility of oxidative stress. The cycle of continuous oxidation and reduction makes glutathione very effective in scavenging free radicals; therefore, these enzymes become pivotal in maintaining cellular defense against oxidative damage.

Ascorbate

Being the reduced active form of vitamin C, ascorbate donates its electrons in neutralizing free radicals and ROS by acting as an antioxidant. This donated electron stabilizes and neutralizes them. Finally, ascorbate gets oxidized into its dehydroascorbate form while doing so. Dehydroascorbate can freely diffuse out of the cell, hence allowing movement between different cellular compartments [36]. Once it gets into the cell, it can be enzymatically reduced back to ascorbate, thus restoring its active form. Interconversion involving antioxidant defense between ascorbate and dehydroascorbate is a significant factor. Ascorbate acts as the major antioxidant by neutralizing free radicals directly, while dehydroascorbate acts as a mediator. Due to this, there has to be a balance between the two forms for vitamin C to play its role as an antioxidant effectively. Another function of Vitamin C is its role in recycling other antioxidants, like Vitamin E. A fat-soluble antioxidant, i.e., Vitamin E prevents oxidative damage and protects cell membranes. Once vitamin E counters free radicals, it gets oxidized. Vitamin C can restore the activity of vitamin E by reducing its oxidized form back to its active state so that it can continue protecting cells against oxidative stress [39].

2.8 *Role of Reactive Oxygen Species (ROS) Scavenging in Stress Mitigation*

Major environmental stresses occur during the life cycle of a plant, and ROS plays an important role as regulators in the mechanisms of growth as well as stress response. ROS acts bidirectionally, being essential for normal development within the plant but also increasing tolerance to hostile environmental conditions for growth. Through very precisely regulated signaling pathways, ROS balances the processes related to growth along with the trajectory toward tolerance against adverse conditions for growth. ROS-scavenging enzymes, for instance, superoxide dismutase (SOD), ascorbate peroxidase (APX), and catalase (CAT) are disseminated throughout all cellular compartments in crop plants and are therefore significant in detoxifying ROS thereby protecting against most abiotic stresses [36]. For example, MnSOD transgenic alfalfa has an increased survival rate and yield under drought stress. Like the effects seen on the fertility of rice spikes under cold stress, OsAPX1 has been found to enhance fertility for the transgenic rice plants under stress conditions, OsAPX2 overexpression also results in a relatively drought-tolerant plant compared to wild-type plants.

2.9 Hormonal Regulations

Hormonal regulatory systems constitute a system by which products from one group of cells alter the metabolic processes of another cell type. This applies broadly to cellular differentiation and development, in which cells of a specific developmental class influence the growth of other classes by altering their metabolism. Some intercellular communication is directly from cell-to-cell through direct surface-to-surface contact between immediately adjacent cells, while others involve signaling over somewhat larger distances. In this context, hormonal regulation can be effectively applied to developmental processes. Indeed, evidence for this concept exists in the structural and mechanistic similarities between nerve growth factor (NGF) and insulin.

2.10 Signaling Pathways in Drought Stress Response

Bacterial wilt is a highly destructive plant disease due to the presence of the soil-borne Gram-negative bacterium, *Ralstonia solanacearum* [Betaproteobacteria] [40]. There have been several researches on the pathogen in tropical, subtropical, and temperate regions. These are now recorded to be more than 250 plant species in 54 monocot and dicot families, which include important crops such as tomato, potato, peanut, and tobacco. The bacterium infects the plant through its roots, cracks, or wounds, establishing itself in the root cortex before proceeding into the xylem vessels and penetrating the vascular system toward the aerial parts of the plant. Once established, *Ralstonia solanacearum* grows to high cell densities and, in a quorum-sensing-dependent manner, secretes virulence factors including type III secretion effectors, extracellular polysaccharides, and cell-wall degrading enzymes [40].

The bacterium further regulates the synthesis of virulence factors and cell motility by influencing various host plant metabolites, such as L-glutamic acid, galacturonic acid, tryptophan, and methionine. It also hydrolyzes salicylic acid (SA) signaling components of the plant to inhibit its host immunity, thus causing the roots to wilt due to problems in water transportation and ultimately the plant's death. However, the intricate involvement between *Ralstonia solanacearum* and the host plant is still unknown.

A signaling pathway refers to a series of biochemical events that involve a group of molecules in a cell acting collectively to manage cellular functions, including cell division or apoptosis. Generally, such pathways are an example of a series of reactions wherein intracellular signal transduction, a process that communicates the signals from the surface of the cell to intracellular targets, occurs [41]. The end results of these signaling cascades are often transcription factors, whose function is critical in controlling the expression of genes.

2.10.1 The Signal Transduction Pathway

There are basically three steps to this signal transduction pathway:

Reception: This is the way a cell detects signals produced from its environment.

Transduction: A process of activation of a chain of proteins inside the cell that begins at the cell membrane.

Response: It refers to the alterations in cell behavior that are initiated by the signal detected.

2.10.2 Perception of Stress Signals (Receptors, Sensors)

Environmental stress has impacts on plant growth and development. The plasma membrane is provided with receptor proteins that perceive the stimuli produced by environmental changes, and they transmit the signals to intracellular and intercellular signaling networks within the cells.

One key component of this is receptor-like kinases (RLKs), which help in detecting extracellular ligands and transmitting signals within the cell through phosphorylation of an intracellular serine/threonine kinase domain [42]. Arabidopsis contains more than 600 genes encoding RLKs, which are dedicated to detecting environmental signals throughout the lifecycle of these semi-immobile plants. Other components include histidine kinases that are membrane-localized and play roles in osmotic stress and plant hormone perception.

Plants have developed adaptive mechanisms to live under ever-changing conditions. As sessile organisms, they need to perceive their environment and then adapt their physiology and development in response to adverse and favorable environmental signals. Activation of receptor molecules by external stimuli activates complex downstream signaling networks that interact with other networks to respond appropriately to the wide range of environmental and developmental signals in a coordinated manner. High-yielding promise has been compromised by both abiotic and biotic stresses on both native and economically important crops. This level of growth and productivity impairment necessitates a high interest in the mechanisms related to responses and adaptations of stress, for instance, stress signaling and regulation of gene expression to create more tolerant plants.

2.10.3 Abscisic Acid (ABA) Signaling Pathways in Drought Stress Response

The chemical that is important in plant responses to stress is abscisic acid (ABA), and it is crucial in helping plants adapt to drought conditions [43]. In drought, ABA promotes stomata closure and thus lowers the transpiration rate of water. Explanation of the responses of plants to drought stress is important for understanding mechanisms of the drought signaling network that can subsequently be translated into crop breeding programs. Among the plant hormones, abscisic acid is primary in the transduction of the stress signal.

Abscisic acid regulates the plant stress response through multiple layers of control. This includes, at the most basic level, transcriptional responses mediated through interactions between a set of core transcription factors modulated by ABA and other plant hormones and regulation of ABA metabolism and transport. Of greater importance in the present context, posttranscriptional and posttranslational modification is considered as part of stress signaling that tends to be missed. Effective integration and coordination of ABA metabolism, transport, and the regulation of core signaling elements have become important to maintain tissue- and cell-type-specific hormone concentrations that maximize signaling efficacy for effective growth and developmental responses. Abscisic acid has been known to regulate various life cycle processes including inhibiting seed germination, restricting shoot and root growth, and promoting stomatal closure [43]. In the present scenario, emerging studies on understanding ABA signaling pathways initiated in the plant cell under drought stress and their relation with drought tolerance are of much significance.

2.10.4 Ethylene and Jasmonic Acid Crosstalk in Defense Signaling

The coordination of the plant stress responses involves a complex interplay of both the jasmonic acid (JA) and the ethylene (ET) pathways involved. The responses are collaborative, involving both JAZs-MYC2 and EIN3/EIL1 signaling components, and are generally more defensive against necrotrophic and hemi-biotrophic pathogens [44].

Phytohormones, like jasmonic acid (JA), salicylic acid (SA), and ethylene (ET), form a complex network, supporting different mechanisms against biotic and abiotic stresses [44]. Among the most well-known biotic stress agents in rice (*Oryza sativa*) in Asia is the brown planthopper (BPH) [45]. This is caused by the most destructive piercing-sucking insects brown planthopper, *Nilaparvata lugens*. This sucking insect pest causes damage to the rice plant's growth and vigor, resulting in a broad range of plant death called "hopper burn" [45]. Cloning of endogenous resistance genes in the host plant against BPH has been done under intense effort and research to understand the molecular mechanisms, which will be more pragmatic and environment-friendly than chemical controls. Meanwhile, metabolic pathways of phytohormones involved in resistance mechanisms in rice to BPH have been at the forefront of scientific research.

2.10.5 Role of Salicylic Acid in Systemic Acquired Resistance

Among the most important phytohormones in the activation of SAR, salicylic acid (SA) is actually regarded as one that equips plants with broad-spectrum resistance against a diverse array of both related and unrelated pathogens [44].

It has a critical role in maintaining and controlling various phases of development and growth of plants in addition to the activation of defense mechanisms against both biotic and abiotic stresses [46]. The principle of acquired immunity in plants upon pathogen infection was first conceived way back in 1933, however, actual progress in our understanding of plant immune signaling has really taken place over the last three decades, especially since the discovery of salicylic acid (SA) as an endogenous defense signal [46].

During this period significant research was conducted and has finally advanced the identification of: (i) two clearly separable pathways of SA biosynthesis, (ii) many proteins regulating the biosynthesis and metabolism of SA, and (iii) a number of signaling components operating downstream of SA, including a wide array of SA targets or receptors. Furthermore, it is typically acknowledged that SA is not an activator of the immune response but, rather, the one acting within a network encompassing a good number of other plant hormones. Future efforts to better understand SA-mediated immune signaling will require the identification of gaps in our current knowledge of this pathway. Further research on how the interactions between various hormone signaling pathways shapes an immune response that is robust and optimized for maximum effectiveness, tailored to the characteristics of the invading pathogen.

2.10.6 Signal Transduction Cascades (MAPK Pathways, Ca^{2+} Signaling)

Perhaps one of the most ancient signaling pathways, mitogen-activated protein kinases have been employed through evolution to mediate various physiological processes. For all kinds of eukaryotic cells, many kinds of MAPK pathways lie in their cells that facilitate an integrated function of regulating gene expression, mitosis, metabolism, motility, cell survival, apoptosis, and differentiation.

Plant phosphoinositide/phospholipase C activates the release and signaling of Ca^{2+} [47]. Stimulation of phosphatidylinositol-specific phospholipase C (PI-PLC), most commonly through hormonal signals, results in the hydrolysis of phosphatidylinositol 4,5-bisphosphate (PtdIns(4,5)P_2) into inositol trisphosphate (IP3) and diacylglycerol (DAG). Later, IPKs carry out further phosphorylation on the produced IP3, which results in inositol hexakisphosphate, or IP6, and DAG in a DAGK-catalyzed reaction that results in phosphatidic acid. Both IP3 and IP6 are able to cause the release of intracellular Ca^{2+}, whereas DAG as well as PA activates protein kinase phosphorylation [47]. These responses will lead to the activation of Ca^{2+} sensors and protein kinases (PKs), which may induce target proteins to affect the cell or activate TFs that regulate gene expression.

The creation of adaptive responses in plants results from numerous environmental stresses. A concise pivotal event in the cellular signaling network is the recognition of environmental signals and the transduction of stress signals into intracellular pathways. Ca^{2+} acts as an important second messenger that can relate the environmental stimuli to growth, physiological functions, and detection or response to stress. A widespread feature of many stress-activated pathways is the rise of intracellular calcium concentrations $\{[Ca^{2+}]i\}$.

2.11 Genomic Approaches for Stress-Responsive Gene Identification

The plant root undergoes a complex chain of physiological and molecular responses to osmotic stress, termed as osmotic stress response. Shoots and roots had different responses to osmotic stress as they are different in the aspect of function and anatomy. The first response for shoots is to control stomatal conductance. In response to osmotic stress, plant stomata react by closing to conserve water and keep the cells hydrated. Consequently, transpiration is decreased and loses less water [48]. However, in the case of roots, their response to osmotic stress is mainly through modified uptake and transport of water. Root water uptake is downregulated when the solute concentration in the rhizosphere is high, such as in saline soils or due to drought, thereby causing cellular dehydration. This dehydration is then countered by plants, which absorb more water and retain cellular turgor by several mechanisms, including the accumulation of compatible solutes.

Many responses are similar between the two, yet water stress and salinity stress exert a range of differences in plant responses. Experiments with maize *Zea mays* and tomatoes *Solanum lycopersicum* have established that ABA will enhance root and shoot development by inhibiting ethylene biosynthesis induced following the onset of water stress [43]. Salinity stress affects lateral roots differently. A prior study showed that mild ionic stress stimulates both the initiation and emergence of lateral roots, whereas lateral root emergence was repressed in response to ionic stress in loss-of-function mutants of sos1, sos2, and sos3 but not under osmotic stress [34]. Although some previous studies looked at roots' response to osmotic stress, genome-wide analyses have not been done yet and most were limited to microarray analyses.

High-throughput technologies, such as microarrays and RNA sequencing (RNA-Seq), have been used to identify genes and pathways involved in responses to osmotic stress resulting from high salinity and drought in roots of both model species, such as *Arabidopsis thaliana*, and crop species, such as *Oryza sativa* (rice), maize, and *Medicago sativa* (alfalfa). Much of the earlier studies, especially those conducted using Arabidopsis, relied on microarray analysis. RNA sequencing provides the capability for high-throughput next-generation sequencing, allowing for rapid analysis of large genomic datasets and precise quantification of transcriptomes and splice variants. It has been used extensively to analyze developmental-stage-specific global gene expression profiles and for the study of the effects of diverse stimuli and genotypes on gene expression.

2.12 Transcriptomic Profiling to Understand Dynamic Responses

The RNA transcriptomic profile of an organism is the accumulation of RNA molecules expressed under specific conditions. It varies with the biological process, developmental stage, and environmental factors. There have been several tools like RNA-seq and microarrays developed to measure gene transcription and expression under different states of cells.

2.13 Proteomics and Metabolomics in Elucidating Stress Adaptation Mechanisms

Abiotically tolerant plants have a number of defense mechanisms that alleviate the negative impacts of stressors. The abiotic stresses are saline concentrations, water-deprived and waterlogged issues, high and low temperatures, mineral deficiencies, and toxicity of heavy metals in soil, which significantly alter the normal metabolism, physiology, and growth of plants. Among the other stress management techniques, the reorganization of the metabolite status is one of the majorly adopted stress tolerance mechanisms in a plant [49].

Such stress stimulates a multigene response in the plant, that is, a set of different changes in the accumulation of primary and secondary metabolites. Different intermediates are regulated in response to different abiotic stresses, which include carbohydrates like galactinol, gluconate, raffinose, and trehalose; amino acids such as proline, glutamate, and asparagine; polyamines such as putrescine, spermine, and spermidine; organic acids; flavonoids; and intermediates of the xanthophyll cycle like carotene, zeaxanthin, violaxanthin, and antheraxanthin. Apart from the above, different intermediates of the tricarboxylic acid cycle also get regulated such as citrate and oxoglutarate [49]. These metabolic changes contribute to the adaptive responses of the plant to several environmental stressors. Some metabolites are upregulated and downregulated according to the requirements of the plant in its efforts to cope with prevailing stress conditions. Metabolomic techniques have been employed in investigating mechanisms that account for the tolerance of the plant to various stresses.

Recent developments in metabolomics provide enormous opportunities to broaden our understanding of the mechanisms involved in abiotic stress tolerance in plants.

New metabolomics technologies have enabled the identification, characterization, and quantification of both primary and secondary metabolites vital for abiotic stress tolerance. The resolution and visualization of primary stress-related metabolites in sophisticated metabolic profiles from plants have been improved through modern technologies such as mass spectrometry and nuclear magnetic resonance in combination with chromatographic separation methods. Strengthened by advanced identification and interpretation techniques, metabolomic studies are helping elucidate complex metabolic pathways and their regulation under abiotic stresses. With proper integration of metabolomics with other "omics" approaches, it will be feasible to apply insights gained through metabolomics research toward developing crops that are better equipped to cope with stresses.

2.14 Integrating Multiple-Stress Tolerance Mechanisms

Plants have primarily developed two major adaptation strategies—acclimation and adaptation—to either tolerate or avoid a particular stress (Fig. 3). Many species show conserved responses to similar environmental conditions, thus having evolved mainly through natural selection. Technologies related to genomics, proteomics, metabolomics, transcriptomics, lipidomic, prime-omics, and miRNA omics greatly enhance our understanding at the molecular level of plant responses. The diverse omics approaches and their integration provide integral criteria to push forward with plant systems biology research [50].

These mechanisms include signal transduction, stress perception, synthesis of stress-related proteins, transcriptional activation of stress-responsive genes, and other molecules. This makes the plants thrive in unfavorable environmental conditions through biochemical and physiological changes [49]. Among the significant influences on crop yield are those that result from abiotic and biotic stresses, the causes of abiotic stresses include drought, salinity, cold, and heat. Biotic stress is from a number of living organisms such as fungi, bacteria, viruses, nematodes, and insects.

Recent research has been carried on about the impact on plant physiology and expression of the genes caused by the multiplicity of stresses. Nowadays, the results of such studies are oriented toward model plants, like Arabidopsis, for example, where the effects of six stressors acting simultaneously have been described. In maize, particular interest is paid to the investigation of the double and triple stress response, including the combined effect of drought with heat, heavy metal exposure, or the interaction of drought with insect and fungal infestations. To tackle these problems, researchers are using natural variation in maize and using this in combination with genetic engineering techniques. Molecular breeding and genome-wide association studies (GWAS) have led to the identification of QTL associated with multiple-stress tolerance. These findings would be applied in future breeding programs, leading to more resilient types of maize [51]. In parallel, a transgenic approach has resulted in several hundreds of commercial maize varieties that resist double or triple stresses, most of them have focused on incorporating tolerance for weeds and insect resistance,

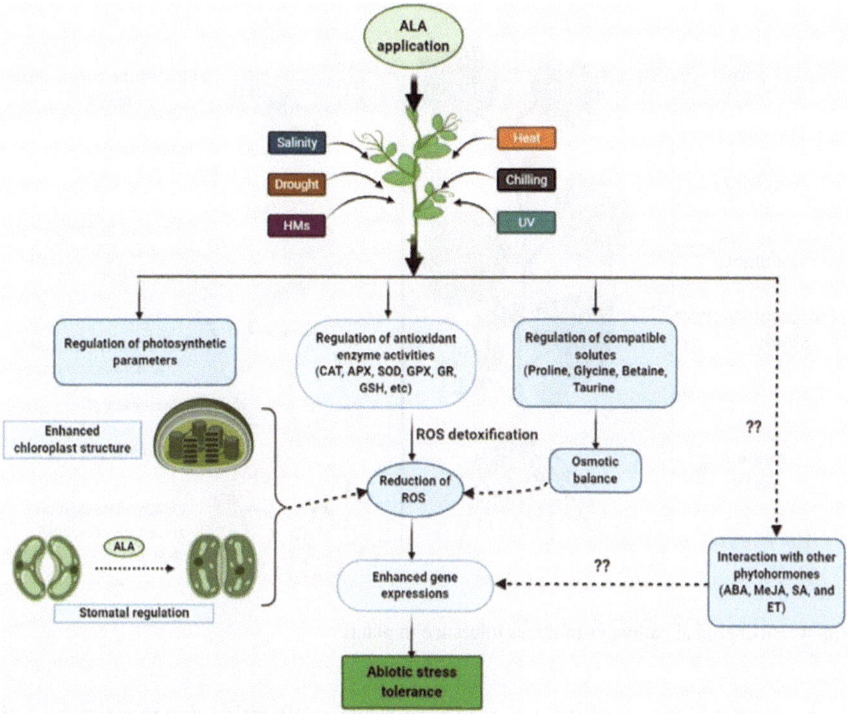

Fig. 3 Multiple stress tolerance mechanism

though some have been shown to have additional increases in drought tolerance [51]. It is expected that the first-generation gene-edited techniques and newly developed base and prime editing applications when combined with routine haploid induction in maize will allow the timely identification of further stress-tolerant alleles into leading varieties.

3 Biochemical Pathways

3.1 The Cell Membrane Stability

Stress imposed on plants causes a reorganization in the cellular membranes, leading to several impairments or overall dysfunction [52]. Though the structural and functional changes provoked by stress are not so clear, the impairment of cellular membranes through stress can be said to result from increased permeability and the leakage of ions (Fig. 4). Quantitatively, this phenomenon may be calculated through the measurement of electrolyte efflux.

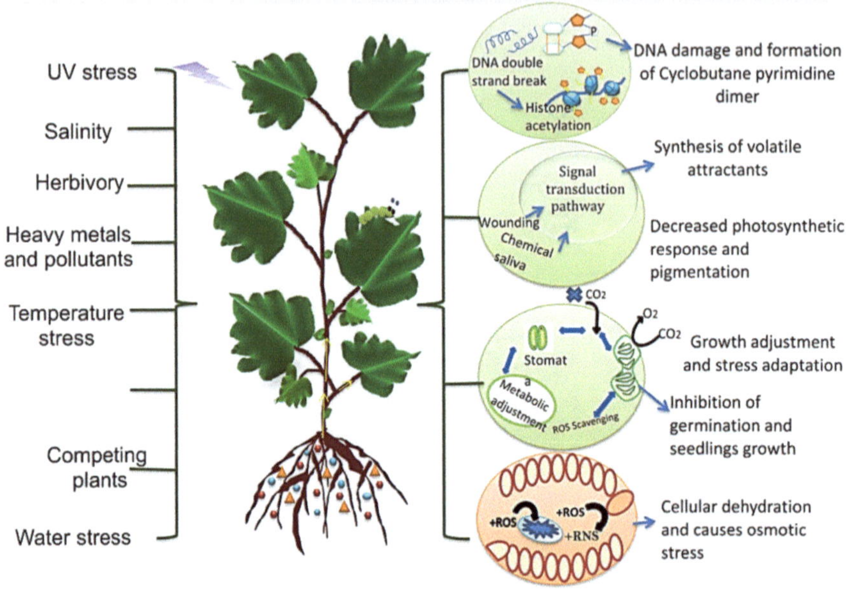

Fig. 4 Biochemical pathways of stress tolerance in plants

As a consequence, quantification of membrane malfunction due to stress by estimating cellular electrolyte leakage from stressed or damaged leaf tissues into an aqueous solution is gaining popularity as a measure of cellular membrane stability, besides having applications as a screening method for resistance to stress [53]. This approach was actually developed in the late 1960s by the late C. Y. Sullivan at the University of Nebraska for testing heat tolerance in sorghum and maize. Later modifications of this method also were developed as tools for determining cold and drought (desiccation) tolerance. Many reports have demonstrated that higher CMS, measured using this test, is positively correlated with higher yields in different genetic materials exposed to stress conditions.

3.2 *Gene Expression and Signal Transduction*

Drought and high salt stress are not the only factors on which the dehydration or water stress of plants takes place; even a low-temperature environment induces dehydration. For all these environmental stresses, plants exhibit different adaptive responses against water stress. The result of water stress brings about several biochemical and physiological reactions leading to a loss in water contents with reductions in turgor pressure within the cell. Abscisic acid is a plant hormone but primarily improves responses to water stress in plants. It also increases tolerance to drought and other conditions such as salinity and cold temperatures.

Recently, several salt, drought, and, cold stress-induced transcriptionally activated genes have been identified [54]. When the stress of water depletion is removed from plants, the mRNA levels of the genes sensitive to water depletion or desiccation decrease, which again supports previous studies that propose these genes are indeed water depletion/desiccation regulated.

A few of the gene products have been predicted to have functions based on sequence homology to known proteins. This means that cells are somehow protected from water deficit [54]. The pattern of expression for dehydration-inducible genes is complex. While some genes respond rapidly, other genes are activated at a slower rate when the ABA level increases [43]. In addition, many genes induced by drought, salt, and cold stress are also stimulated by exogenously applied ABA. Dehydration seems to trigger the biosynthesis of abscisic acid (ABA), which initiates the expression of several genes [54]. Most of the drought or cold-induced genes are not responsive to the exogenous application of ABA. This would not signify an opposition between ABA-dependency and ABA-independency but rather could reflect a kind of parallelism, where stress seems to activate both ABA-dependent and ABA-independent pathways in a way that leads to the specific induction of certain sets of genes. Furthermore, gene promoter studies have also identified several cis-acting elements associated with both ABA-dependent and ABA-independent responses to drought and cold. Even though the nature of molecular mechanisms regulating gene responses to water stress in higher plants has not been well understood, further input at the molecular level is still required to answer many questions. It consists of sensing of water or osmotic stress, transduction of the signal through different cellular activities, transmission of these signals toward the nucleus, regulation of transcription of a stress-inducible gene, and the roles and interactions of these genes involved in promoting water-stress tolerance.

3.3 Metabolic Adjustment

Global warming and climate change are known to directly influence the growth and yield of crops, often exacerbating and frequently enhancing the manifestation of other abiotic stresses like drought, higher temperature, flood, etc. This often leads to various stress factors taking place simultaneously [55]. Increases in soil stresses such as salinity, inorganic nutrient imbalances, heavy metal contamination, and biotic stresses will also complicate the future environmental landscape for plants. Consequently, it becomes much more difficult to predict how plants will respond to these multiple stress challenges.

Besides the induction of specific physiological and molecular responses to stress, plants alter various metabolic pathways that curb the influence of stress on the growth and development of plants. This adaptation enables them to respond to the new energy demands created by diverse climatic and environmental conditions [56, 57]. One of the crucial characteristics of attaining a new balance and successfully adapting to adverse environments is the ability to effectively reconfigure metabolic networks for renewed growth and seed production [56, 57].

The plants synthesize thousands of various metabolites in a spatiotemporal and environment-dependent manner that support diverse processes [58]. That encompasses carbohydrates, phenolics, amino acids, polyamines, and polyols, as well as lipids for which specific biological functions have been established [59]. In general, they are classified as primary metabolites, directly related to the growth of the plant (e.g., carbohydrates, TCA cycle intermediate components, and amino acids),secondary metabolites, which reflect interactions between plants and their environments (e.g., phenolics, terpenes, and nitrogen-containing compounds); and hormonal compounds acting as regulators of varied aspects of plant growth, development, and responses to environmental stress (e.g., ABA, JA, SA, and ethylene) [59, 60].

Recent research shows that some secondary metabolites can perform multiple functions; although they are defense-related compounds against pathogens, they also have the ability to function like 'hormones' in regulating growth and developmental processes in plants. Of these, flavonoids play a pivotal role in modulating growth and development and responding to environmental adaptation in plants [59].

In addition to their roles as downstream targets in various stress-response signaling pathways, some metabolites may also have the ability to function as signals for stress. In this regard, some of the metabolic responses elicited by specific abiotic stresses can condition the plant to preparedness for subsequent environmental stresses. For example, proline, an amino acid-induced by drought stress, is a metabolic stress-activated signal in *Periploca sepium* and tobacco [61, 62]. Other metabolic signals, such as primary metabolites, for example, sugar phosphates, sugars, and many amino acids, and secondary metabolites, have been documented following drought, cold [63], and heat stresses [62, 64]. Accordingly, plants are also able to synthesize numerous regulators that can alter growth and development for survival under different types of stress.

3.4 Epigenetic Regulation

Epigenetic regulation is very crucial for plant growth and development, reproduction, and resistance to pathogens. Moreover, it makes plants more sensitive to temperature fluctuations, salt stress, and nutritional deficiencies. Such mechanisms will serve to enhance crop yield, growth, quality, and production; hence, they will promote sustainable agriculture. These include some of the key epigenetic processes that drive critical agronomic traits during development, by influencing gene expression including histone modifications, DNA methylation, and action of small RNAs that influence some of the most important stages of development: germination, seedling establishment, and fruit formation. They will ultimately affect crop productivity and yields and lead to higher quality [65].

Long-term epigenetic modifications are crucial for the foundation of the evolutionary processes because they create a stable molecular template that drives plasticity at the phenotype level. On the other hand, short-term mechanisms would make the plant survive stressful conditions. Such flexibility in quick adaptation enables the promotion of offspring that are better prepared for changing environmental settings, as seen with natural populations sharing similar genetic makeup, suggesting that those traits are epigenetic. Such improvement in crop yield, therefore, in the context of stressing environmental factors, becomes contingent on a thorough understanding of plant epigenetics and stress-response mechanisms [66].

However, direct evidence that links epigenetic changes with plasticity in phenotypic expression in plants growing under different environmental conditions and stressors is still not available. According to the latest research, by changing DNA methylation patterns, plants are able to achieve a shift in gene expression that helps in the dynamic adjustment of responses to stressors, making the plant more resilient to harmful conditions.

Some stress-induced epigenetic modifications have also been isolated with certain enzymes. For instance, a DNA methyltransferase known as DRM2 was found to be essential for altering DNA methylation in the process of stress-responsive development [65]. Histone modifications and their association with histone acetyltransferases and histone deacetylases have been said to be related to stress responses [66]. A detailed understanding of these mechanisms is necessary for the establishment of plant varieties of enhanced resistance to environmental stress.

3.5 Genetic Engineering

Climate change will have a more catalytic impact than ever before on global food production. Compound drought and heat wave events have rapidly increased over the last decade. [67] found that 49% of wheat-growing regions experienced at least one such event each season and that the frequency and duration in wheat-growing regions increased by 42% since the 1980s. Climate variability is proving to threaten food security due to altered precipitation and temperature patterns and an increase in extreme events across most regions. More importantly, it can cause shifts in the distribution of pests and diseases, making it even more difficult to achieve agricultural resilience.

At the end of the twentieth century when DNA-based markers dramatically hastened the emergence of new agricultural crop lines and varieties [68], streamlined targeted GE techniques replaced the historical processes resulting from this development. Among the earliest GE technologies, one that was successfully used in functional genomics and plant breeding was RNA interference or RNAi. It has disadvantages such as the random insertion of RNAi constructs into the genome and partial suppression of gene functions [69].

Genome editing is the latest crop improvement development, which enables the alteration of the plant genome by deleting unwanted genes or introducing novel functions to the genes themselves [70]. With this, it has become increasingly accessible to breed desirable traits in view of the fact that a lot of crop genomes have already been sequenced and also with advanced genome-editing technologies. To essentially increase food production sustainably, it is of great importance to integrate multiple developed techniques that may incorporate genomics, genome editing (GE), artificial intelligence, and deep learning [68].

Genome-editing tools enable advanced biotechnological approaches that allow the use of precision and efficiency in the modification of the genome of an organism. The innovation and development of a series of new tools in genome and gene editing greatly empower researchers to accurately modify genomic sequences. Techniques in this new tool set provide much insight into functional genomics in plants and facilitate targeted gene expression regulation in defined regions. Technologies, such as CRISPR/Cas9 (clustered regularly interspaced short palindromic repeats/CRISPR-associated systems), TALENs (transcription activator-like effector nucleases), CRISPR/Cas12a (Cpf1, CRISPR from Prevotella and Francisella), and Cas9-derived DNA base editors, have transformed genome engineering [68, 69]. Thus, these tools can potentially put one in a position to be able to make a contribution to ensuring food security across the world.

4 Conclusion

Recent advances in the research have greatly enhanced the understanding of early events in plant stress signaling, especially in terms of the identification of primary sensors that perceive initial stress events. Presently, agriculture is on the threshold of a digital revolution similar to past technological shifts, faced with changing and uncertain environmental conditions, climate change, increasing input costs, the growing population, and market volatility. Advanced data collection, such as through drones and satellites, can have soil conditions, crop growth, and weather patterns that can lead to better food production with minimal environmental impact.

The ability of crops to withstand stress factors depends on an understanding of physiological and biochemical pathways that allow for crop stress tolerance. Resilient varieties in agricultural systems should fight drought, salinity, heat, and diseases. In defense against oxidative damage, the management of ROS is very important and is helped by both enzymatic and non-enzymatic antioxidants. Hormonal regulation, especially via abscisic acid, coordinates responses to stress. New genomic explorations have isolated key genes and transcription factors, which may be used for traditional breeding, as well as biotechnological development, of stress-resistant crops. The integrated approach to breeding must combine physiological, biochemical, and molecular insight with smart agricultural practices including optimal soil management and efficient irrigation.

The creation of broader strategies on how to improve food security in the face of climate change would, therefore, be dependent on the collaboration of scientists, agronomists, and policymakers. A deeper understanding of stress tolerance mechanisms will ensure sustainable agricultural practices that support food security as well as ecosystem health in a changing environment. Innovative research and good management will foster the crops to grow well under stress and lead the way to more stress-tolerant agriculture tomorrow.

Competing Interests The authors declare that they have no competing interest.

Declaration of Generative AI and AI-Assisted Technologies in the Writing Process During the preparation of this work the book chapter author(s) have not taken any assistance from AI, and the book chapter author(s) reviewed and edited the content as needed and take(s) full responsibility for the content of the publication.

References

1. Wang X (2022) Managing land carrying capacity: key to achieving sustainable production systems for food security. Land 11:484
2. Hafeez A, Ali B, Javed MA, Saleem A, Fatima M, Fathi A, Afridi MS, Aydin V, Oral MA, Soudy FA (2023) Plant breeding for harmony between sustainable agriculture, the environment, and global food security: an era of genomics-assisted breeding. Planta 258(5):97
3. Spök A, Sprink T, Allan AC, Yamaguchi T, Dayé C (2022) Towards social acceptability of genome-edited plants in industrialised countries? Emerging evidence from Europe, United States, Canada, Australia, New Zealand, and Japan. Front Genome Edit 4:899331
4. Bhattacharjee S, Bhowmick R, Kant L, Paul K (2023) Strategic transgene-free approaches of CRISPR-based genome editing in plants. Mol Genet Genom 298(3):507–520
5. He Y, Zhao Y (2020) Technological breakthroughs in generating transgene-free and genetically stable CRISPR-edited plants. Abiotech 1(1):88–96
6. Bacelar E, Pinto T, Anjos R, Morais MC, Oliveira I, Vilela A, Cosme F (2024) Impacts of climate change and mitigation strategies for some abiotic and biotic constraints influencing fruit growth and quality. Plants 13(14):1942
7. Kissoudis C, van de Wiel C, Visser RG, van der Linden G (2014) Enhancing crop resilience to combined abiotic and biotic stress through the dissection of physiological and molecular crosstalk. Front Plant Sci 5:207
8. Dos Santos TB, Ribas AF, de Souza SGH, Budzinski IGF, Domingues DS (2022) Physiological responses to drought, salinity, and heat stress in plants: a review. Stresses 2(1):113–135
9. Balasubramaniam T, Shen G, Esmaeili N, Zhang H (2023) Plants' response mechanisms to salinity stress. Plants 12(12):2253
10. Zulfiqar F, Akram NA, Ashraf M (2020) Osmoprotection in plants under abiotic stresses: New insights into a classical phenomenon. Planta 251(1):3
11. Banerjee A, Roychoudhury A (2018) Abiotic stress, generation of reactive oxygen species, and their consequences: an overview. In: Singh VP, Singh S, Tripathi D, Mohan Prasad S, Chauhan DK (eds) Revisiting the role of reactive oxygen species (ROS) in plants: ROS Boon or bane for plants? Wiley, pp 23–50
12. Mittler R, Zandalinas SI, Fichman Y, Van Breusegem F (2022) Reactive oxygen species signalling in plant stress responses. Nat Rev Mol Cell Biol 23(10):663–679
13. Bai H, Chen J, Gao T, Tang Z, Li H, Gong S, Du Y, Yu Y, Wang W (2023) A Na^+/H^+ antiporter localized on the Golgi-to-vacuole transport system from Camellia sinensis, CsNHX6, plays a positive role in salt tolerance. Sci Horticult 309:111704

14. Blumwald E, Aharon GS, Apse MP (2000) Sodium transport in plant cells. Biochim Biophys Acta (BBA) – Biomembr 1465:140–151
15. Fussy A, Papenbrock J (2024) Molecular analysis of the reactions in Salicornia europaea to varying NaCl concentrations at various stages of development to better exploit its potential as a new crop plant. Front Plant Sci 15:1454541
16. Ali MS, Baek KH (2020) Jasmonic acid signaling pathway in response to abiotic stresses in plants. Int J Mol Sci 21(2):621
17. Siddiqi KS, Husen A (2019) Plant response to jasmonates: current developments and their role in changing environment. Bull Natl Res Centre 43(153):1–11. https://doi.org/10.1186/s42269-019-0195-6
18. Yurina NP (2023) Heat shock proteins in plant protection from oxidative stress. Mol Biol 57(6):951–964
19. Sevanto S (2020) Why do plants have waxy leaves? Do we know after all? Tree Physiol 40(7):823–826
20. Roth-Nebelsick A, Krause M (2023) The plant leaf: a biomimetic resource for multifunctional and economic design. Biomimetics 8(2):145
21. Naorem A, Jayaraman S, Dang YP, Dalal RC, Sinha NK, Rao CS, Patra AK (2023) Soil constraints in an arid environment—challenges, prospects, and implications. Agronomy 13(1):220
22. Yetgin A (2024) Exploring the dynamic nature of root plasticity and morphology in the face of changing environments. Ecol Front 44(1):112–119
23. Li Q, Fu C, Liang C, Ni X, Zhao X, Chen M, Ou L (2022) Crop lodging and the roles of lignin, cellulose, and hemicellulose in lodging resistance. Agronomy 12(8):1795
24. Zaib M, Zeeshan A, Aslam S, Bano S, Ilyas A, Abbas Z et al (2023) Drought stress and plants production: a review with future prospects. Int J Sci Res Eng Dev 6(04):1278–1293
25. Yu L, Zhou C, Fan J, Shanklin J, Xu C (2021) Mechanisms and functions of membrane lipid remodeling in plants. Plant J 107(1):37–53
26. Bhattacharya A (2022) Lipid metabolism in plants under low-temperature stress: a review. In: Physiological processes in plants under low temperature stress. Springer, Singapore, pp 409–516. https://doi.org/10.1007/978-981-16-9037-2_5
27. Hasegawa PM (2013) Sodium (Na^+) homeostasis and salt tolerance of plants. Environ Exp Bot 92:19–31
28. Joshi A, Kaur S, Taneja SK, Mandal R (2024) Review article on molecular mechanism of regulation of hypertension by macro-elements (Na, K, Ca and Mg), micro-elements/trace metals (Zn and Cu) and toxic elements (Pb and As). Biol Trace Elem Res 202(4):1477–1502
29. Garcia-Caparros P, De Filippis L, Gul A, Hasanuzzaman M, Ozturk M, Altay V, Lao MT (2021) Oxidative stress and antioxidant metabolism under adverse environmental conditions: a review. Bot Rev 87:421–466
30. Su J, Jiao Q, Jia T, Hu X (2024) The photosystem-II repair cycle: updates and open questions. Planta 259(1):20
31. Han LJ, Fan DY, Wang XP, Xu CY, Xia XL, Chow WS (2023) The protective role of non-photochemical quenching in PSII photo-susceptibility: a case study in the field. Plant Cell Physiol 64(1):43–54
32. Cao H, Ding R, Kang S, Du T, Tong L, Zhang Y, Chen J, Shukla MK (2023) Drought, salt, and combined stresses in plants: effects, tolerance mechanisms, and strategies. Adv Agron 178:107–163
33. Husen A (2025) Roles of osmolytes in plant systems under changing environment. Elsevier Inc., Cambridge
34. Ozturk M, Turkyilmaz Unal B, García-Caparrós P, Khursheed A, Gul A, Hasanuzzaman M (2021) Osmoregulation and its actions during the drought stress in plants. Physiol Plant 172(2):1321–1335
35. Hussain S, Hafeez MB, Azam R, Mehmood K, Aziz M, Ercisli S, Javed T, Raza A, Zahra N, Hussain S, Ren X (2024) Deciphering the role of phytohormones and osmolytes in plant tolerance against salt stress: Implications, possible cross-talk, and prospects. J Plant Growth Regul 43:38–59

36. Mishra N, Jiang C, Chen L, Paul A, Chatterjee A, Shen G (2023) Achieving abiotic stress tolerance in plants through antioxidative defense mechanisms. Front Plant Sci 14:1110622
37. Navyashree R (2024) Antioxidants in Plants. In: Advanced botany
38. Rudenko NN, Vetoshkina DV, Marenkova TV, Borisova-Mubarakshina MM (2023) Antioxidants of non-enzymatic nature: their function in higher plant cells and the ways of boosting their biosynthesis. Antioxidants 12(11):2014.
39. Faizan M, Alam P, Rajput VD, Shareen KK, Faraz A, Minkina T, Maqbool Ahmed S, Rajpal VR, Hayat S (2023) Potential role of tocopherol in protecting crop plants against abiotic stresses. Physiol Mol Biol Plants 29(10):1563–1575
40. Islam T, Haque MA, Barai HR, Istiaq A, Kim JJ (2024) Antibiotic resistance in plant pathogenic bacteria: recent data and environmental impact of unchecked use and the potential of biocontrol agents as an eco-friendly alternative. Plants 13(8):1135
41. Picard M, Shirihai OS (2022) Mitochondrial signal transduction. Cell Metab 34(11):1620–1653
42. Jose J, Ghantasala S, Roy Choudhury S (2020) Arabidopsis transmembrane receptor-like kinases (RLKs): a bridge between extracellular signal and intracellular regulatory machinery. Int J Mol Sci 21(11):4000
43. Zareen S, Ali A, Yun DJ (2024) Significance of ABA biosynthesis in plant adaptation to drought stress. J Plant Biol 67(3):175–184
44. Myers RJ Jr, Fichman Y, Zandalinas SI, Mittler R (2023) Jasmonic acid and salicylic acid modulate systemic reactive oxygen species signaling during stress responses. Plant Physiol 191(2):862–873
45. Bunga AP, Avinashe H, Dubey N, Choudhary S, Rasane P, Sachan S (2024) Dissecting brown planthopper resistance genes in Oryza and its wild relatives: a review. Euphytica 220(3):32
46. Song W, Shao H, Zheng A, Zhao L, Xu Y (2023) Advances in roles of salicylic acid in plant tolerance responses to biotic and abiotic stresses. Plants 12(19):3475
47. Amokrane L, Pokotylo I, Acket S, Ducloy A, Troncoso-Ponce A, Cacas JL, Ruelland E (2024) Phospholipid signaling in crop plants: a field to explore. Plants 13(11):1532
48. Hasanuzzaman M, Zhou M, Shabala S (2023) How does stomatal density and residual transpiration contribute to osmotic stress tolerance? Plants 12(3):494
49. Salam U, Ullah S, Tang ZH, Elateeq AA, Khan Y, Khan J, Khan A, Ali S (2023) Plant metabolomics: an overview of the role of primary and secondary metabolites against different environmental stress factors. Life (Basel) 13(3):706
50. Ijaz S, Iqbal J, Abbasi BA, Yaseen T, Rehman S, Kazi M, Mahmood T (2024) Role of OMICS-based technologies in plant sciences. In: OMICs-based techniques for global food security, pp 45–66
51. Sahito JH, Zhang H, Gishkori ZGN, Ma C, Wang Z, Ding D, Zhang X, Tang J (2024) Advancements and prospects of Genome-Wide Association Studies (GWAS) in Maize. Int J Mol Sci 25(3):1918
52. Ďúranová H, Šimora V, Ďurišová Ľ, Olexiková L, Kovár M, Požgajová M (2023) Modifications in ultrastructural characteristics and redox status of plants under environmental stress: a review. Plants 12(8):1666
53. Alamgir ANM (2023) Cell membrane injury, stability and atpase actibity and their relevance to seed germination, seedling growth tolerance indices of rice under abiotic stresses. Bangladesh J Bot 52(3):889–896
54. Kim JS, Kidokoro S, Yamaguchi-Shinozaki K, Shinozaki K (2024) Regulatory networks in plant responses to drought and cold stress. Plant Physiol 195(1):170–189
55. Rivero RM, Mittler R, Blumwald E, Zandalinas SI (2022) Developing climate-resilient crops: improving plant tolerance to stress combination. Plant J 109(2):373–389
56. Dusenge ME, Duarte AG, Way DA (2019) Plant carbon metabolism and climate change: elevated CO_2 and temperature impacts on photosynthesis, photorespiration and respiration. New Phytol 221(1):32–49
57. Fernie AR, Cavalcanti JHF, Nunes-Nesi A (2020) Metabolic roles of plant mitochondrial carriers. Biomolecules 10(7):1013

58. Fang C, Fernie AR, Luo J (2019) Exploring the diversity of plant metabolism. Trends Plant Sci 24(1):83–98
59. Zandalinas SI, Balfagón D, Gómez-Cadenas A, Mittler R (2022) Plant responses to climate change: metabolic changes under combined abiotic stresses. J Exp Bot 73(11):3339–3354
60. Erb M, Kliebenstein DJ (2020) Plant secondary metabolites as defenses, regulators, and primary metabolites: the blurred functional trichotomy. Plant Physiol 184(1):39–52
61. An Y, Zhang M, Liu G, Han R, Liang Z (2013) Proline accumulation in leaves of *Periploca sepium* via both biosynthesis up-regulation and transport during recovery from severe drought. PLoS ONE 8(7):69942
62. Rather BA, Sehar Z, Majid A, Jahan B, Mir IR, Anjum NA, Masood A, Khan NA (2023) Ethylene and cellular redox management in plants. In: The plant hormone ethylene. Academic, pp. 141–170
63. Jan R, Asaf S, Numan M, Lubna KKM (2021) Plant secondary metabolite biosynthesis and transcriptional regulation in response to biotic and abiotic stress conditions. Agronomy 11(5):968
64. Hemme D, Veyel D, Mühlhaus T, Sommer F, Jüppner J, Unger AK, Sandmann M, Fehrle I, Schönfelder S, Steup M, Geimer S (2014) Systems-wide analysis of acclimation responses to long-term heat stress and recovery in the photosynthetic model organism Chlamydomonas reinhardtii. Plant Cell 26(11):4270–4297
65. Abdulraheem MI, Xiong Y, Moshood AY, Cadenas-Pliego G, Zhang H, Hu J (2024) Mechanisms of plant epigenetic regulation in response to plant stress: Recent discoveries and implications. Plants 13(2):163
66. Hemenway EA, Gehring M (2023) Epigenetic regulation during plant development and the capacity for epigenetic memory. Ann Rev Plant Biol 74(1):87–109
67. He M, He CQ, Ding NZ (2018) Abiotic stresses: general defenses of land plants and chances for engineering multistress tolerance. Front Plant Sci 9:1771
68. Yadav RK, Tripathi MK, Tiwari S, Tripathi N, Asati R, Chauhan S, Tiwari PN, Payasi DK (2023) Genome editing and improvement of abiotic stress tolerance in crop plants. Life 13(7):1456
69. Bowerman AF, Byrt CS, Roy SJ, Whitney SM, Mortimer JC, Ankeny RA, Gilliham M, Zhang D, Millar AA, Rebetzke GJ, Pogson BJ (2023) Potential abiotic stress targets for modern genetic manipulation. Plant Cell 35(1):139–161
70. Wang Y, Demirer GS (2023) Synthetic biology for plant genetic engineering and molecular farming. Trends Biotechnol 41(9):1182–1198

Managing Defense Mechanisms in Crop Plants Using Phytohormones Under Various Contaminants

Vikas Kumar, Ravi Prakash, Nitin Sharma, and Nidhi Gupta

Abstract Exposure of crop plants to broad spectrum of environmental contaminants such as drought, salinity, heavy metals, pesticides, and nanoparticles can impact their growth and productivity. Phytohormones play a crucial role in regulating plant defense mechanism during stress condition. The present chapter offers a comprehensive examination of several phytohormones such as auxin, gibberellins, cytokinins, abscisic acid, ethylene, and brassinosteroids, addressing their role in processes like seed germination, flowering, fruit ripening, and stress resilience. Further, the link between phytohormones and environmental forces is also explored, emphasizing how these relationship influence plant physiology under stress conditions. This chapter explores the application of phytohormones within contemporary plant biotechnology, emphasizing the enhancement of crops and the augmentation of stress resilience, while underscoring the opportunities to leverage hormonal pathways for the sustainability of agriculture. Researchers and practitioners can gain insights into the mechanisms through which phytohormones operate and their interrelations, thereby facilitating the development of innovative strategies to increase plant resistance and productivity.

Keywords Phytohormones · Auxins · Gibberellins · Stress tolerance · Plant biotechnology · Signal transduction · Crop enhancement

---

Vikas Kumar and Ravi Prakash—indicates equal contribution.

V. Kumar (✉) · R. Prakash · N. Gupta
University Institute of Biotechnology, Chandigarh University, Gharuan, Mohali, Punjab, India
e-mail: vikaskmr59@gmail.com

N. Sharma · N. Gupta
Department of Biotechnology, Chandigarh Group of Colleges, Landran, Mohali, Punjab, India

A. Husen (ed.), *Enhanced Crop Production Setup Under Various Micro and Nano Contaminants*, Smart Nanomaterials Technology,
https://doi.org/10.1007/978-981-96-6099-5_15

1 Introduction

Agriculture is the backbone of the worldwide population, and three major cereals are rice, maize, and wheat, which comprise almost half of the world's calorie consumption and enhancing global food security [1, 2]. It is well recognized that tremendous growth of the global population, varying climatic conditions, and the COVID-19 pandemic have highly impacted food production and are highly at risk to food security. Modern farmings, which have become very common, have introduced many new pollutants to the agricultural system through years of urbanization, heavy industrialization, and heavy use of agrochemicals [3]. The rising global population and changing weather conditions, alongside persistent wars, exert an immense pressure on the global supply of food and call for chemicals, high-yielding crops, and an enormous intensification of arable land [4]. Emerging contaminants in agriculture are substances which are not regulated by conventional statutes but have recently been identified as having potential for adverse environmental impact. These include nano- and microplastics, Per- and polyfluoroalkyl substances (PFAS), endocrine-disrupting chemicals (EDCs), antibiotics, hormones, and analgesics. Possible crop, soil, ground-, and surface-water adverse impacts of these emerging contaminants are raising increasing concerns in the environment. These contaminants may originate from various sources, including agricultural practices (application of agrochemicals, pesticides, and herbicides), industrial activities, personal care products, pharmaceuticals, and urban runoff. Modern advancements in pesticide and herbicide formulations are continually developed to respond to evolving insect and weed populations [5, 6].

The prolonged consumption of contaminated irrigation water, fertilizers, pesticides, and soil amendments facilitates the accumulation of cadmium, lead, and arsenic in agricultural soils, significantly impacting the health of plant ecosystems [6–8]. In addition, the use of pharmaceuticals and veternary medicines in animal farming leads to their residues in manure and runoff water, which can be harmful to aquatic systems, soil microorganisms, and agricultural products [9–11]. It is acknowledged that emerging contaminants, such as antibiotics, synthetic hormones, and analgesics, can accumulate in agricultural crops, thus significantly impacting their growth, development, and, ultimately, the quality of the yield. Exposure to antibiotics has, for example, been demonstrated to alter the normal growth and developmental activities in plants [12–14]. Endocrine-disrupting chemicals (EDCs) have the potential to enter agricultural systems via water sources or the application of pesticides, negatively impacting reproductive and developmental processes in both plant and animal species [15–17]. PFAS compounds consist of at least one aliphatic perfluorocarbon moiety. Hence, it has been reported in the soil and water surrounding agricultural areas due to anthropogenic inputs, leading to a growing concern toward their implications for crop production and the food chain [18–20]. High climate patterns globally and rising international trade have promoted the outbreak of new and more virulent plant pathogens, both bacterial and fungal which may significantly impede food production [21]. Over the past 300 years, various countries worldwide have used plants for the decontamination of heavy metals, like lead and zinc [22]. However,

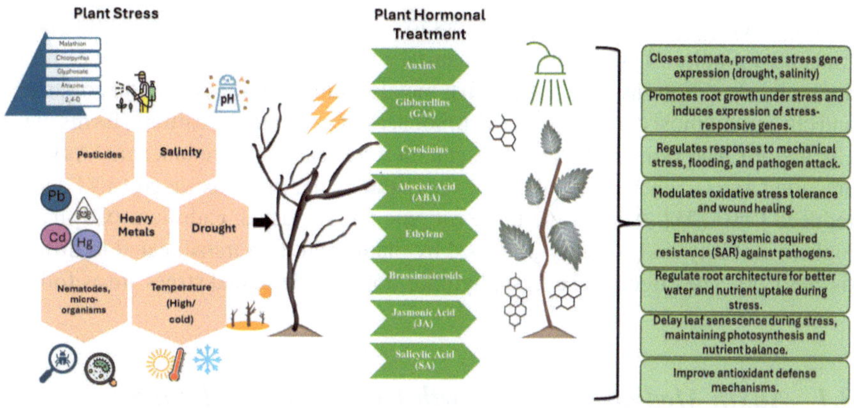

Fig. 1 Schematic depiction of various plant stresses and the impact of phytohormonal treatments in plant stress responses. Several environmental and biological stresses such as salt, drought, temperature variation (high/low), heavy metals, pesticides, and nanoparticles are illustrated alongside the relevant phytohormones that facilitate stress tolerance

for commercial purposes, this technology has been attracting much attention mainly in the last three decades [23]. This chapter attempts to discuss the essential role of phytohormones in modulating defensive mechanisms of crop plants under various environmental pollutants such as heavy metals, pesticides, and organic contaminants. To define the internal regulators that enable the detection, response, and adaptation by plants to stress caused by contaminants, the molecular mechanisms and signaling networks influenced by phytohormones—abscisic acid, jasmonic acid, ethylene, and salicylic acid—are explored in this chapter. The chapter attempts to elucidate possible biotechnological applications of phytohormones for enhancing crop resilience to environmental contaminants and fostering sustainable agricultural practices (Fig. 1).

2 Phytohormonal Regulation Under Heavy Metal Stress

Plants counteract metal-induced abiotic stress by altering their phytohormone concentrations. In response to abiotic stress, root cells can promptly produce phytohormones [24], leading to modifications in root morphology, nutrient absorption, and the expression of metal transporters [25]. Furthermore, phytohormones affect plant growth and development. Phytohormonal application reduces metal accumulation in plants [26]. Phytohormones can improve plant growth under abiotic stress [27], which is essential for increasing agricultural output. Nevertheless, the mechanisms through which they alleviate metal toxicity in plants are still inadequately recognized. Toxic metals include arsenic (As), cadmium (Cd), chromium (Cr), mercury (Hg), and lead (Pb), which are not essential to plant growth and have a harmful effect on plants [28]. They generally thrive well in polluted soils, thus affecting human health adversely [29]. Chief contributors to environmental pollution include industrial and agricultural practices, while natural sources include major rock weathering and volcanic eruptions [30]. Of the five metals examined, the oxyanions of arsenic and chromium most commonly occur in soils, as opposed to the cationic forms of Cd, Hg, and Pb. The primary forms of arsenic in terrestrial soils are arsenate (AsV) and arsenite (AsIII), whereas chromium exists both as the more oxidized CrVI and the reduced CrIII [31]. Among the three cationic metals tested, cadmium (Cd) has the highest mobility and is easily absorbed by plants and then translocated to different tissues [25]. The effects of several phytohormones under metal stress are as follows.

2.1 Auxin Mediates Root Growth Under Heavy Metal Stress

Auxin is a crucial phytohormone in regulating lateral root development [32]. Under metal stress, plants often experience a reduction in auxin accumulation in their meristematic tissue, which in turn inhibits root growth [33]. Root plasticity is a crucial adaptation characteristic, as plants absorb water and nutrients from the soil through their roots [34]. Plants acclimatize to abiotic stress by modifying root development, wherein auxin plays pivotal in regulating root growth [33]. Root cells at the tips undergo frequent division under optimal auxin concentrations, facilitating root elongation and differentiation in *Arabidopsis* thaliana; however, elevated auxin levels induce cellular quiescence [35]. The administration of low auxin concentrations at 200 nM enhances mitotic activity, leading to an enlarged meristematic zone in plants [36, 37]. Metal stress reduces auxin levels, leading to impaired lateral root formation in plants [38], hence, auxin administration enhances root growth [39]. Under metal stress, the expression of auxin production genes in the leaves and transportation genes in lateral roots is also diminished [40], resulting in reduced auxin levels in the lateral roots. Cadmium represses the expression of the auxin efflux genes PIN1, PIN3, and PIN7 in *Arabidopsis thaliana*, thereby lowering the auxin content in the roots, leading to reduced growth in the root meristem [41]. Likewise, the same plant response is linked with the alteration of BjPIN5 and BjPIN3 in the leaves and roots of *Brassica juncea* exposed to arsenic [42]. Furthermore, auxin also regulates lateral root development through interaction with jasmonate. It has been found that jasmonate increases lateral root development, making it possibly mitigate the effects of arsenic and cadmium, especially on lateral roots [38]. Jasmonate does interact with the auxin signaling pathway to improve growth in plant roots by stimulating the root stem cells [32]. The MYB gene family comprises a broad array of proteins that are important for a broad range of biological processes. Cell cycle regulation, morphogenesis, and stress signaling pathways are examples. The transcription factor interacts between auxin and salicylic acid through metal stress through the network that OsMYB-R1 constitutes. Lines that overexpress OsMYB-R1 have a remarkable increase in lateral root development and high levels of auxin [43] and are tolerant of chromium stress. Auxin signaling pathways are mostly downstream effectors, rather than other phytohormones in the plant systems [44]. In addition, several other phytohormones are responsible for governing auxin concentration during metal stress. Among the various phytohormones, brassinosteroids, cytokinins, gibberellic acid, and salicylic acid can increase the auxin concentrations in plants subjected to metal (loid) stress. Conversely, auxin levels are reduced in these organisms in response to decrease in cytokinin levels [45, 46]. The expression of the genes associated with auxin biosynthesis is increased by cytokinin and therefore facilitates the conversion of Indole pyruvic acid to auxin in plants [47, 48]. The expression of auxin biosynthesis genes is stimulated by cytokinin, which stimulates auxin production by converting indole pyruvic acid to auxin in plants [47, 48]. In addition, plants also control root development through the interaction of auxin with cytokinin/ salicylic acid [43]. In roots, auxin stimulates meristematic cell division, while cytokinin inhibits auxin signaling to facilitate differentiation of cells from a meristematic to a differentiated state [49]. In addition, brassinosteroid treatment of radish seedlings elevates auxin levels during Cr (VI) stress, while salicylic acid application to wheat increases auxin levels under Cd stress [50, 51]. Semeradova et al. [52] examined the polar auxin transport under the control of various hormonal mechanisms. The interactions between auxin and other phytohormones with metal stress are not known yet and therefore need to be explored further.

2.2 *Brassinosteroid Enhances Antioxidant System Under Metal Stress*

In metal-induced stress, in addition to direct toxicity, the excessive generation of reactive oxygen species (ROS) triggers oxidative stress in plant systems [27]. The antioxidant system of plants is crucial for coping with oxidative stress through the limiting of the accumulation of ROS and lipid peroxidation as illustrated through proteomic analysis [53]. Flora contains a variety of essential enzymatic antioxidants, such as superoxide dismutase (SOD), catalase (CAT), ascorbate peroxidase (APX), glutathione reductase (GR), and glutathione peroxidase (GPX) [54]. Among these antioxidants, SOD mediates the dismutation of free radical O_2 to produce H_2O_2, whereas CAT and APX can both convert H_2O_2 into H_2O. GR catalyzes the reduction of oxidized glutathione (GSSG) to its reduced form, glutathione (GSH), while GPX catalyzes the reduction of hydrogen peroxide (H_2O_2) to water (H_2O), using GSH as a reducing agent. In addition to enzymatic antioxidants, the antioxidant pool is supplemented by non-enzymatic antioxidants, including GSH and ascorbate.

The water-soluble ascorbate acts as a substrate for ascorbate peroxidase, donating electrons to ROS, can act as an electron donor to ROS and, therefore, regulate both the redox state of amino acids and of proteins by relieving oxidative stress, and GSH can also regulate both the redox state of amino acids and proteins by relieving oxidative stress. Brassinosteroids promote the activity of the antioxidant enzymes and help plants survive metal stress (oxidative stress) [55]. Previous studies have shown that brassinosteroids can enhance stress tolerance to metals via increase in the antioxidant defense capability in plants [54, 56, 57]. For example, a 0.01 μM brassinosteroid in combination with 100 μM Cr (VI) enhances rice resistance by induction of SOD, POD, CAT and APX activities [58]. In bean plants [59], treatment with 5 μM brassinosteroid suppresses 1 mM Cd stress-induced lipid peroxidation by increasing activities of enzymes SOD, CAT, and GR. Brassinosteroid-induced enhanced antioxidant enzyme activity is primarily due to induction of their gene expression [54]. Rice plants (0.5 mM Cr) express the SOD, CAT, APX, and GR genes more at endogenous levels when treated with brassinosteroid (0.01–1 μM), [54]. Induction of temporary apoplastic H2O2 production in brassinosteroid in the external plant application stimulates relative expression of stress proteins and defensive enzymes which might help in buffering metal-induced stress in plants (Kour et al. 2021). For example, in tomato exposed to 0.15–1.5 mM Cu stresses, 0.01 μM brassinosteroid treatment, or 0.1 mM H2O2 administration, SOD, POD, and CAT activities were increased [60]. The beneficial effect of brassinosteroid was counteracted by nitrate reductase, which was shown to reduce the levels of nitrous oxide via nitrate reductase, indicating a potential role for nitrate reductase in the brassinosteroid mediated nitrous oxide production. Kaya et al. [61] demonstrated acceleration of antioxidant enzymes of pepper plants in Cd stress.

2.3 *Salicylic Acid Promotes Cell Wall Synthesis and Nutrient Uptake in Plants*

The cell wall is the first line of defense against metal entry into plants and is essential for plant detoxification [62]. Plants uptake metal and the metal uptake depends on the composition of the cell wall. Their negative charges [63, 64] can diminish their uptake by plants through the binding of pectin, a cell wall component. In plants, pectin methyl esterase bound to free carboxyl groups catalyzes demethylation of pectin by binding metals such as Pb [65, 66]. Hg and Pb can also bind with low-methyl-esterified pectin in the cell wall [65, 67]. In addition, it should also be prohibited to permit the metal to get through the cell membranes and reach the cytoplasm. One of the glucans, callose, accumulates in the cell wall, preventing metal access to the cells and therefore reducing the amount of metal in plants [68]. Furthermore, the plant can enhance the cell wall thickness by precipitating celluloses, hemicelluloses, and lignin to form a complex with metals [69]. Consequently, phytohormones alter cell wall composition leading to the alteration of the metal deposition on the cell wall. It was known that salicylic acid improved the plant resistance to cadmium (Cd). For example, treatment with 100 μM of salicylic acid creates the formation of cell walls within the root of tomato and thus lowers the level of Cd accumulation within the plant [70]. Xu et al. [71] also showed that 100 μM salicylic acid enhanced cell-bound Cd in peanuts grown under 200 μM of Cd stress. Hemicellulose and pectin are the major constituents of the cell wall, crucial for cadmium accumulation because it is negatively charged [72]. Energy-dispersive X-ray microanalysis shows direct visualization of cadmium attachment to plant cell walls [73]. Salicylic acid, in turn, alters the cell wall composition of *Brachypodium distachyon* with increased levels of ferulic and p-coumaric acids [74]. Ferulic and p-coumaric acids were known to make cross-links with hemicellulose and lignin, thereby enhancing cell wall resistance to digestibility [75, 76]. Besides this, ferulic and p-coumaric acids also act to reduce the oxidative stress induced by arsenic in plants [77]. Apart from its function in the cell wall, glutathione is necessary for the defense of plants against oxidative stress elicited by cadmium by acting as a constituent component of phytochelatin that aids in the chelation of cadmium [78]. SA also links up with the biogenesis of glutathione.

The glutathione level is lower in Cd-stressed A. thaliana leaves than that of the wild-type of the SA-mutant [79]. Furthermore, there was an increase in glutathione levels after SA treatment in Cd-stressed peppermint plants [80]. Above induction of glutathione due to SA under cadmium stress can be ascribed to the high transcription of serine acetyltransferase—the precursor gene responsible for cysteine formation [81], better uptake of sulfur, which is an essential component in the synthesis of glutathione [79], and enhanced activities of glutathione synthetase and glutathione reductase to maintain glutathione in its reduced form [79]. To reduce Cd stress, PC-Cd complexes can be sequestered into vacuoles through ATP-binding cassette (ABC) transporters [82]. An ABC transporter was recently identified in soybean triggered by salicylic acid treatment [83]. However, the role of ABC transporters in SA-induced Cd tolerance needs further exploration. Nutrients are necessary for plant growth, although their concentrations in plants are affected by metals (Li et al. 2016). In plants, Cd interferes with C, N, and S metabolism, leading to nutritional deficiencies such as Zn, Ca, Mg, Mn, P, K, and Fe [84]. SA can enhance the nutritional composition in plants subjected to metal stress. For instance, 100 μM SA elevates the iron, calcium, and magnesium levels in soybean subjected to 53 μM cadmium [85], whereas 10 μM SA improves nitrogen fixation and assimilation efficiency in chickpea plants under 223–893 μM Cd stress [86]. SA modifies plasma membrane characteristics to enhance H^+-ATPase activity, hence promoting nutrient absorption in plants under Cd stress [63, 87] (Fig. 2).

2.4 Abscisic Acid Decreases Metal Accumulation in Plants

Various phytohormones, such as auxin, brassinosteroids, gibberellin, jasmonic acid, abscisic acid (ABA), and cytokinin, can reduce metal uptake in plants to mitigate toxicity (Fig. 2). Among these, ABA is the most efficient [88, 89]. ABA effectively reduces Cd accumulation in plants [90, 91]. ABA also decreases Cd accumulation in rice shoots by changing stomatal closure and reducing transpiration rate (Fig. 2; [92, 93]). In addition, application of ABA decreases the transpiration rate and cadmium concentration in rice [94]. ABA biosynthesis inhibitor fluridone reduces ABA accumulation in rice [94] and increases the transpiration rate and cadmium concentration in rice. In rice, transpiration rate is reduced by ABA to reduce the Cd transport to shoots [95]. Downregulation of the ABA inhibited expression of iron-regulated

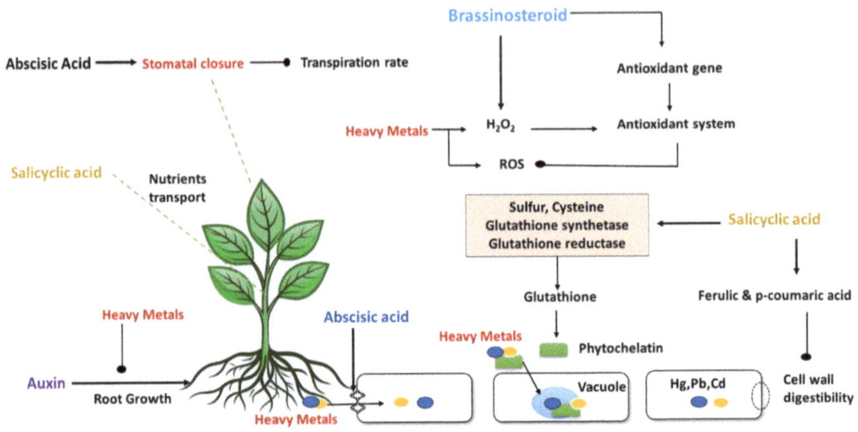

Fig. 2 Mechanism of phytohormone-mediated heavy metal stress tolerance in plants. The roles of various phytohormones such as auxin, abscisic acid, brassinosteroids, and salicylic acid in mitigating the effect of heavy metal stress on plants are shown in illustration

transporter 1 (IRT1) in the roots of *A. thaliana* is shown by [90] to be associated with diminished cadmium accumulation in plants under ABA treatment [96, 97]. Importantly, Cd uptake in an IRT1 mutant is not modulated by ABA treatment in *A. thaliana*, suggesting an essential role of ABA in regulating Cd uptake in *A. thaliana* [90]. Further, the basic region/leucine zipper transcription factor abscisic acid insensitive 5 (ABI5) contributes to cadmium accumulation. ABI5 mutants have higher Cd accumulation in A. thaliana but overexpressed mutants have lower Cd accumulation [98].

3 Interaction of Phytohormones with Organic Contaminants

In traditional agriculture, the overuse of pesticides creates significant threats to agroecosystems, altering soil–plant interactions and hindering physiological processes such as photosynthesis. Although application of pesticide and herbicide are essential for food security and crop protection, they cause plants to experience oxidative stress, which in turn reduces their development and productivity [99, 100]. Pesticides induce oxidative stress by generating ROS and inhibiting essential metabolic pathways. Fomesafen, a commonly utilized herbicide in soybean agriculture, presents difficulties owing to its prolonged soil persistence and adverse impacts on future crops. Pot tests have demonstrated that elevated fomesafen residues (0–0.05 mg/kg) diminished sugar beetroot growth, photosynthetic efficiency, and antioxidant defense, while raising oxidative stress indicators such as malondialdehyde (MDA). Fomesafen can also affect non-target crops such as sugar beets even when applied at as low as

0.025 mg/kg, as shown by Li et al. [101, 102]. Heavy metal stress (HMS) reduces agricultural output, which in turn decreases growth, and food supply, and poses a threat to global food security by reducing the food supply and increasing the likelihood of shortages [103, 104]. These factors stimulate the plants to develop such plant defense mechanisms in response which include syntheses of plant defense-related genes, antioxidant enzymes, metabolites, etc. Auxins, gibberellins, cytokinins, abscisic acid (ABA), salicylic acid (SA), and jasmonic acid (JA) used as plant growth regulators (PGRs) decrease pesticide toxicity by inducing antioxidant protection and reducing reactive oxygen species (ROS) levels and adding to the plant's resilience [105]. Pesticides have been mitigated by plant growth regulators (PGRs) to reduce pesticide toxicity and to promote the degradation of pesticides in plant, resulting in reduced residue accumulation. Phytohormones are critical in regulating the interaction between phytohormones and organic contaminants that govern plant stress tolerance and detoxification, as well as adaptation to environmental stresses.

3.1 Salicylic Acid (SA)

Salicylic acid exogenous administration increased germination rates, radicle length, nutritional content, reduced lipid peroxidation, and total antioxidants in *Fritillaria hupehensis* [106]. However, physiological adaptability of alfalfa plants under Zn stress (1.0 mM, $ZnSO^4$) is improved with 0.5 mM SA exogenous application through foliar spray. SA treatment increases chlorophyll content, improves photosynthesis, elevates proline and soluble sugar levels, and stimulates the antioxidant system and glutathione metabolism. It also boosts the activities of key antioxidant enzymes such as catalase (CAT), peroxidase (POD), and ascorbate peroxidase (APX), collectively reducing oxidative stress, limiting Zn accumulation, and enhancing the plants' tolerance to Zn toxicity [12, 13]. Treatment of young bean seedlings (*Phaseolus vulgaris* L.) with 100 μM prometryne herbicide inhibited plant growth, reduced leaf pigments, decrease in glutathione and elevated oxidative stress markers and increase in activities of antioxidant enzymes (CAT, APX, GST) and proline levels. Application of 1 mM SA counteracted these effects, improving growth, increasing leaf pigment content, regulating antioxidant enzyme activity, and reducing oxidative stress [107]. Fomesafen herbicide negatively impacts sugar beet growth and photosynthesis, while increasing oxidative stress and membrane damage. However, exogenous SA can alleviate these effects, improving plant height, biomass, photosynthetic performance, and antioxidant enzyme activity, with 1 mM SA showing the most pronounced benefits. This suggests SA could be useful in managing herbicide toxicity in crops [101, 102]. Exogenously applied SA mitigates oxidative damage caused by the herbicide mesosulfuron-methyl + iodosulfuron-methyl in wheat by reducing H_2O_2 and malondialdehyde (MDA) levels, enhancing catalase (CAT) activity, and regulating other antioxidant enzymes (SOD, PPO, APX, and GST). SA pre-treatment also reduced proline accumulation, helping to alleviate herbicide-induced stress [108]. Pre-sowing seed treatment with 0.1 mM SA, either alone or combined with kinetin (30 ppm) or calcium chloride (40 mM), can effectively alleviate nickel and lead toxicity in *Phaseolus vulgaris.* This treatment enhances the activities of key enzymes like carbonic anhydrase, nitrate reductase, catalase, peroxidase, and superoxide dismutase, while also improving plant growth, photosynthetic pigments, carbohydrate content, and nitrogen compounds [109]. Glyphosate, the most widely used herbicide globally, poses environmental risks and harms non-target plant growth, necessitating strategies to minimize its impact and improve crop tolerance. Glyphosate (GLY) causes significant growth inhibition, oxidative stress, and lipid peroxidation in *Hordeum vulgare* (barley), along with increased H_2O_2, superoxide anion, and non-protein thiols, and a reduction in ascorbate (AsA) content. Foliar application of SA (100 μM) partially mitigates GLY-induced stress by enhancing antioxidant (AOX) enzyme activity (SOD, CAT, APX, GST) thereby reducing oxidative damage [110]. Exposure to 6.6 mM thiram in tomatoes (*Solanum lycopersicum* Mill.) causes oxidative stress, marked by increased H_2O_2 and MDA levels and a reduction in photosynthetic pigments. SA treatment reduces H_2O_2 and MDA levels, enhances chlorophyll and carotenoid content, and regulated antioxidant enzyme activity. Pre-treatment with SA (1 mM) in thiram-treated leaves boosts the activity of detoxification enzymes, significantly upregulating GST-1, GST2, GST3, and P450 expression [111]. Foliar application of 600 μM SA in potato plants alleviates cadmium (Cd) toxicity by boosting relative water content (RWC), chlorophyll, proline, and endogenous SA levels and activating key enzymes, including superoxide dismutase (SOD), catalase (CAT), ascorbate peroxidase (APX), and glutathione reductase of antioxidant defense system [112].

3.2 *Brassinosteroids (BR)*

Foliar application of 10 μM JA and 10 μM BR, alone or combined, significantly improved bamboo growth and tolerance under Pb stress (0–150 μM) by reducing oxidative stress, membrane damage, and electrolyte leakage, while enhancing antioxidant enzyme activities (SOD, POD, CAT, APX, and GR), non-antioxidants, glyoxylate activity, proline content, metal chelation, photosynthesis, and biomass. This treatment boosted Pb tolerance by enhancing antioxidant and glyoxalase systems and reducing metal translocation [113]. An herbicide Sigma Broad at 3.37 g/ha induces oxidative stress in foxtail millet (*Setaria italica* L.), leading to lipid peroxidation, reduced pigment (chlorophyll) content, and damage to photosystems I and II, ultimately decreasing the photosynthetic rate. However, foliar application of 0.1 mg/L BR significantly enhances plant height, leaf area, and fresh mass, the net photosynthetic rate and pigment content along with activities of antioxidant enzymes SOD, POD, and CAT and decreases malondialdehyde (MDA) levels [114]. A foliar spray of 10^{-8} M BR also enhances the yield and quality of *Solanum lycopersicum* (tomato) under cadmium stress by improving growth, reducing oxidative stress, and boosting antioxidant enzyme activity [115].

3.2.1 24-Epibrassinolide (EBL)

Treatment with 24-epibrassinolide, a plant hormone, can improve plant tolerance to pesticide toxicity through an increase in antioxidant activity and a decrease in oxidative damage. Isoproturon is an effective selective weedicide for controlling weeds but due to its high solubility, low absorption in the soil and its persistence, it can pollute water and is potentially toxic to human beings. Application of 1.0 μmol/L 24-epibrassinolide improves the IPU (0.25–4.0 mg/L) degradation and detoxification in rice (Japonica cv., wild-type) through the induction of the activity of detoxifying enzymes and the upregulation of key Phase I-III genes including GST, CYP450, GTs, and ABC transporters, thereby decreasing the negative effect of herbicide on growth, chlorophyll content, and membrane permeability of rice plant [116]. EBL reduces oxidative damage in maize caused by diazinon insecticide through enhancement of antioxidant systems. It promotes photosynthesis, stomatal conductance, proline and protein, and increases free radical scavenging and reduces lipid peroxidation. In this case, EBL may modulate the activity of antioxidant genes by phosphorylating the promoters of BZR1 and BES1 and by controlling their phosphorylation states. EBL increases the ability of glutathione to bind with pesticide residues and their subsequent transport to vacuoles or out of the cell [117]. 24-epibrassinolide (EBL) along with PGPR (*Pseudomonas aeruginosa, Burkholderia gladioli*) alleviates chlorpyrifos (CP) toxicity in *Brassica juncea* (mustard) by enhancing pigment levels, stomatal function, and upregulating key photosynthetic (psbA and psbB) and stress-related genes (CHS, PSY, and PAL) [118]. Application of the fungicide thiuram (tetramethyl thiuram disulfide to tomatoes (Solanum lycopersicum Mill. causes oxidative stress, however, by lowering levels of hydrogen peroxide and malondialdehyde, treatment with 24-EBL can ameliorate this. It improves the plants' general health while helping in their recovery from challenges by increasing their total protein content and photosynthetic pigments [119]. 24-epibrassinolide (EBL) and PGPRs alleviate chlorpyrifos-induced oxidative stress in *Brassica juncea* seedlings by enhancing antioxidant defense systems, reducing reactive oxygen species, and improving nitric oxide levels, leading to improved growth and stress recovery [120]. Use of glyphosate herbicide reduce phenolic compounds and elevates stress markers, such as proline and malondialdehyde (MDA), indicating oxidative damage in winter wheat (*Triticum aestivum* L.). However, foliar treatment with 1 μM 24-epibrassinolide (EBL) in winter wheat (*Triticum aestivum* L.) led to a moderate activation of the antioxidant defense system (including POX, GR, GST, and CAT) and increase in phenolic compounds, along with detoxifying enzyme GST activity, resulting in decreased MDA and proline levels [121]. Boscalid, used to extend postharvest storage of cherry tomatoes, leaves pesticide residues, posing a potential food safety risk for consumers. Exogenous application of 24-epibrassinolide (24-EBL) in cherry tomatoes (*Lycopersicon esculentum*) activated detoxification enzymes, including glutathione reductase (GR), peroxidase (POD), and glutathione S-transferase (GST), leading to a significant reduction in boscalid pesticide residues—71.2% in core, 44.8% in peel, and 54.0% in pulp and increase in soluble protein content [122]. Pre-treatment of grapevine leaves with 24-EBL significantly increases the expression of antioxidant and defense-related genes when exposed to chlorothalonil (CHT). Furthermore, 24-EBL promotes the transcription of chlorothalonil-degrading genes, including P450 monooxygenase (P450), glutathione S-transferase (GST), glutathione reductase (GR), and MRP-type transporter (MRP) [123]. Pre-soaking *Brassica juncea L.* seeds with 24-EBL significantly reduced imidacloprid pesticide residues in green pods by nearly 30%, with no residues found in mature seeds. This treatment also boosted the activities of key antioxidative enzymes (such as glutathione S-transferase, glutathione reductase, glutathione peroxidase, ascorbate peroxidase, and guaiacol peroxidase) and increased glutathione content [124].

3.2.2 Castasterone (CS)

Castasterone (CS), a bioactive brassinosteroid, plays a significant role in mitigating both heavy metal and pesticide phytotoxicity in plants. CS-seed treatment of *Brassica juncea* enhances seedling growth, boosts pigment content, reduces oxidative stress, and activates the antioxidant defense system under imidacloprid (IMI) insecticide stress. It also promotes the production of Krebs cycle intermediates and phenolics, while modulating gene expression related to antioxidative enzymes (SOD, CAT, POD, DHAR, GR, GST), ROS generation, pigment metabolism, carbon fixation, and pesticide detoxification, leading to lower IMI residues [125]. Exposure to Cu can cause significant oxidative stress in *Brassica juncea* plants, but pre-treatment of seeds with CS enhances the antioxidative defense system by modulating the ascorbate–glutathione cycle and amino acid metabolism. Additionally, CS pre-treatment results in increased gene expression of key antioxidant enzymes, dehydroascorbate reductase (DHAR), glutathione reductase (GR), glutathione S-transferase (GST-1), and glutathione synthetase (GSH-S) [126]. Seed treatment with CS effectively mitigates cadmium (Cd (II)) toxicity in *Brassica juncea.* It reduces the expression of chlorophyllase (CHLASE) and enhances the expression of phytoene synthase (PSY) and chalcone synthase (CHS) genes with CS treatments compared to Cd (II) only. This treatment also increases the accumulation of compatible solutes [127].

3.3 Cytokinins (CK)

Cytokinin, which is an antisenescing phytohormone, was found to have the capacity to improve heavy metal resistance. Pb inhibits cytokinins, auxins, and gibberellins and increases abscisic acid in *Acutodesmus obliquus.* Pb stress is reduced by exogenous application of auxins and cytokinins due to the modulation of phytohormone concentrations. While auxins stimulate the activity of phytochelatin synthase to increase phytochelatin content for Pb immobilization, cytokinin treatment decreases phytochelatin synthesis. Changes in metal-chelating metabolites and glutathione pathways were also established and auxins and cytokinins were identified as factors that algal adaptation to Pb stress [128]. *Kosteletzkya pentacarpos* seedlings exposed to Cd and Zn showed improved growth, photosynthesis, stomatal conductance, and reduced oxidative stress when treated with the cytokinin trans-zeatine riboside. Exposure to zinc reduced CK levels in plant, but exogenous cytokinin increased resistance to Zn. Salinity helped plants cope with heavy metal stress. However, cytokinin's effectiveness was lower in saline conditions [129].

3.4 Strigolactone (SL)

Strigolactones help mitigate arsenic stress in rice by regulating uptake, biosynthesis, and antioxidant defenses [130]. A comparison between wild-type (WT) rice with strigolactone-deficient mutants (d10 and d17) shows that WT rice exhibits less growth inhibition and oxidative stress under sodium arsenate exposure, with lower arsenic accumulation and higher antioxidant activity. Gene expression analysis has revealed reduced phosphate transporter levels in WT roots, explaining decreased arsenate uptake [131]. Herbicides penoxsulam (PXL) and bensulfuron-methyl (BSM) both negatively affects root growth and photosynthetic pigments in watermelon (*Citrullus lanatus*) plant possibly causing ultrastructural damage and cell death in leaf and root tissues. The activation of SL-related gene expression indicates that SL might help mitigate herbicide toxicity. Additionally, exogenous SL improves antioxidant activity and upregulated detoxification-related genes, contributing to its protective effects against herbicide stress [132]. Synthetic strigolactone GR24 can alleviate cadmium (Cd) stress in switchgrass (*Panicum virgatum*) by improving growth, root development, and photosynthesis while enhancing nutrient uptake and reducing Cd accumulation [133].

3.5 Jasmonic Acid (JA)

Jasmonic acid (JA) is an essential plant hormone that regulates the response of plants to heavy metals, influencing their accumulation and distribution within plant tissues. Priming by Jasmonic acid (JA) g (100 nmol) and foliar application of spermidine (Spd) (1 mM), both individually and in combination, mitigate Cd stress (100 μM) in wheat by improving plant height, dry mass, and photosynthetic parameters. These treatments enhance antioxidant enzyme activity, increase levels of secondary metabolites and osmolytes, and reduce Cd accumulation in seedlings [134]. JA enhances the degradation of chlorothalonil (CHT) in tomatoes regulated by glutathione (GSH), reducing CHT residues by 28%. This process involves nitric oxide signaling and the upregulation of antioxidant and detoxification enzymes [135].

Methyl jasmonate (Me-JA) enhances the tolerance of *Cajanus cajan* seeds to the HM stress, particularly Cu and Cd. The Me-JA treatment raised the proline, glutathione (41.95% increase at 5 mM Cu), and ascorbic acid (160.4% increase) contents thus improving the oxidative status of the plants [136]. The presence of Cd stress induces jasmonic acid synthesis that in turn governs the JA signaling system and regulates nitric oxide quantity to hinder Cd uptake and thereby lessen its toxic effect in *Arabidopsis*. External use of methyl jasmonate (Me-JA) reduces chlorosis in young leaves under Cd stress through downregulating the effect of genes such as AtIRT1, AtHMA2, and AtHMA4 that are involved in the absorption and transportation of Cd in plants as well as reducing the concentration of cadmium in the root cell sap and shoots [137]. When exposed to 100 μM Cu, JA treatment at 1-, 5-, or 10-mM significantly decreased Cu concentrations in both roots and leaves of alfalfa plants. This shows that external application of JA is critical in reducing Cu toxicity in alfalfa plants. In addition to that, the treatment lessened the photocatalytic reduction of heavy metals, leading to increased root and leaf biomass, improved chlorophyll contents, and high antioxidant enzyme activities with low MDA and H_2O_2 levels [138]. JA at 100 nM when applied to Pb stressed tomato (*Lycopersicon esculentum*) seedlings at 0.25, 0.50 and 0.75 mM reduced the toxic effects of Pb by increasing chlorophyll content through the modulation of chlorophyllase and chalcone synthase, phenyl alanine ammonia lyase, succinate dehydrogenase, and fumarate hydratase. In addition, the JA treatment increased the accumulation of in vitro polyamines, osmolytes, and organic acids in Pb treated seedlings [139]. The combined treatment of JA, SA, and proline reduced the negative impact of Pb on the growth of maize plants through enhancing growth, increasing biosynthesis of the pigment, reducing EL, MDA, and the Pb content in plants [140].

3.6 *Auxins*

It is well-known that auxin plays an essential function in plant development and growth, especially in controlling cell elongation, division, and differentiation [141]. Nevertheless, there is growing interest in its role in reducing metal stress. A pot experiment showed that foliar application of 10^{-8} M indole-3-acetic acid (IAA) effectively enhanced growth, photosynthesis, and carbohydrate metabolism in *B. juncea* under copper (Cu) stress (30 or 60 mg kg^{-1}). This concentration minimized oxidative stress, improved gas exchange, and maintained chloroplast structure while reducing malondialdehyde (MDA) and electrolyte leakage. Moreover, IAA promoted Cu transport from roots to shoot to improve phytoremediation capacity in contaminated soils [142]. The interrelationship between Se, Cd, auxin, and P is examined in a study focusing on protecting tobacco plants from Cd stress. Selenium reduces the effects of Cd toxicity through the promotion of biomass, the improvement of the antioxidant enzyme activity and the decrease of Cd accumulation. Low Se (10 μM) increases auxin accumulation by activating auxin biosynthesis-related genes and NtPT2 overexpression enhance Cd tolerance by enhancing P and Se content and decreasing the content of Cd in plants [143].

3.7 Gibberellin (GA)

Gibberellins (GA) facilitate plants in managing trace metal toxicity by reducing oxidative stress, which is induced by the accumulation of detrimental reactive oxygen species (ROS) within plant cells [131]. Environmental factors such as Cd stress can greatly limit plant development by interfering with physiological processes, decreasing nutrient absorption, and inflicting oxidative damage. Corn is protected from Cd-induced stress by combining 0.75% biochar (BC), a carbon-rich byproduct of organic matter pyrolysis, with 10 ppm gibberellic acid (GA3). This combination improves corn's growth, protein content, phenolics, chlorophyll levels, and stress tolerance. This approach has the potential to enhance the resistance of maize to soil contaminated with heavy metals [144]. By increasing osmolyte levels (proline and glycine betaine), decreasing Cr uptake and oxidative damage, and improving growth, photosynthesis, and antioxidant enzyme activities (SOD, CAT, APX, and GR), exogenous kinetin and gibberellic acid (GA3) reduced chromium (Cr) stress in faba bean (*Vicia faba*) [145]. The application of 1 nM JA and 10^{-6} M GA3, individually or in combination, to chickpea (*Cicer arietinum*) plants exposed to 150 μM cadmium sulfate reduced reactive oxygen species, malondialdehyde, and electrolyte leakage. It also increased relative water content, chlorophyll and carotenoid content, improved nutrient uptake, enhanced enzymatic activities, and lowered metal accumulation in the plants [146].

3.8 Ethylene

Ethylene is a key plant hormone that plays various roles in plant growth, development, and responses to environmental stimuli. A study investigated the role of ethylene in mustard (*B. juncea*) plants exposed to high Zn concentrations (200 mg kg^{-1} soil). The application of 200 μL L^{-1} of ethylene as ethephon restored photosynthesis and growth by mitigating oxidative stress through enhanced antioxidant activity by upregulating genes (SOD, APX, GR, GPX, GST), inhibited ROS production, proline metabolism (GK and PROX), glyoxalase system (Gly I and Gly II), and nutrient homeostasis [147].

3.9 Abscisic Acid (ABA)

Priming maize (*Zea mays* L.) seedlings subjected to different concentrations of the herbicide 2,4-dichlorophenoxyacetic acid (2,4-D) resulted in an increase in osmolytes (proline and glycine betaine) levels, a decrease in the Na^+/K^+ ratio, and enhanced growth, pigment levels, and tissue hydration. Treatment with ABA restored photosynthesis and fluorescence activity, decreased reactive oxygen species, and lowered oxidative stress markers like as malondialdehyde and carbonyl content. Furthermore, ABA priming augmented antioxidant enzyme activity, contributing to the maintenance of redox homeostasis in response to herbicidal toxicity [148]. A hydroponic study investigated the impacts of ABA and 6-benzylaminopurine (BAP) on tomato (*Solanum lycopersicum* L.) plants subjected to cobalt (Co) stress (400 μM). Co treatment alone decreased growth, relative water content, and chlorophyll concentrations while elevating oxidative stress. The simultaneous administration of ABA and BAP markedly enhanced plant biomass, root shape, and chlorophyll content, while diminishing oxidative stress indicators. BAP, especially at 10 μM in conjunction with ABA, significantly reduced Co buildup in roots and shoots by 48.4 and 70.2%, respectively [149].

4 Phytohormones and Nanocontaminants: Their Mutual Interactions

It has been eluted that nanocontaminants such as nanoparticles (NPs) have a significant impact on plant hormones causing changes in their physiological status, growth and health. The increase in the application of nanoparticles (NPs) in many sectors such as agriculture is inherently linked to their release into the environment. Since NPs are incorporated into fertilizers, insecticides, and other agricultural inputs, they can enter soil, water, and air systems. Although these nanocontaminants can be useful to promote growth in plants and help the plants to withstand stress, they have been found to cause hormonal changes in the plants affecting the plant's physiological and pathological conditions. Cerium dioxide (CeO_2), gamma irons oxide (γFe_2O_3), silver (AgNPs), zinc oxide (ZnO), and lanthanum oxide (La_2O_3) have been shown to prevent nutrient concentration and hormone regulation in plants. CeO_2 (cerium dioxide) nanoparticles at 0, 100 and 500 mg·L^{-1} significantly reduced the nutrient content (Zn, Mg, Fe, and P) and IAA and ABA in the roots of traditional cotton. Using TEM, [150] found that CeO_2 nanoparticles entered the roots and moved to the stems and leaves causing the swelling and bursting of chloroplasts, particularly in the Bt-transgenic cotton, as well as disrupting the vascular bundles. Transmission electron microscopy also showed that γFe_2O_3 NP altered the rice roots by affecting the production of hormones such as ABA and IAA. The differences in hormone levels were more pronounced in transgenic rice compared to non-transgenic varieties. In the shoots, no significant differences in IAA and ABA levels were observed across treatments, suggesting that root hormones may be more sensitive to nanoparticle exposure [151]. Exposure to silver nanoparticles (AgNPs) can elicit a response in plants that can resemble their reaction to mechanical stress. Treatment with silver nanoparticles (AgNPs) and silver ions (Ag^+) has significantly increased levels of stress phytohormones—jasmonic, abscisic, and salicylic acids—in both wild-type *Arabidopsis* seedlings and especially in the glabra 1 mutant [152]. Another study examined the effects of zinc oxide nanoparticles (ZnO NPs) at concentrations ranging from 0.16 to 100 mg L^{-1} on five phytohormones: cytokinins, auxins, abscisic acid, salicylic acid, and jasmonic acid in apices, leaves, and roots of *Arabidopsis thaliana*. Higher concentrations of ZnO NPs suppressed the production of growth-promoting hormones in shoot apical meristems while increasing stress-related hormones like abscisic acid and cis-zeatin, particularly in roots. The findings indicate that plants recognize ZnO nanoparticles as serious stress, resulting in a multifaceted hormonal response [153]. Exposure of pepper (*Capsicum annuum* L.) to silver nanoparticles (AgNPs) and Ag^+ ions significantly elevated the levels of isoprenoid cytokinins (CK) in the leaves, while negatively impacting plant growth by diminishing height and biomass [154]. Peanut seeds growth and development were found to be stimulated and impaired by Ag NPs at 50, 500, and 2000 mg·kg^{-1} during 98 days of exposure The nanoparticles were detected in the edible part of the peanuts [155]. These outcomes propose prospective risks associated with AgNPs in plant systems following their release into the environment. Another type of nanoparticles, La_2O_3, has also been known to affect the level of plant hormones. A study showed that La_2O_3 NPs at concentrations of 100 and 300 mg/L greatly affected the radish plants with root breaking and reduced storage root biomass by 38 and 60%, respectively. This treatment altered the root cell wall architecture through the down-regulation of some important genes which are involved in cell wall pectin and IAA synthesis which consequently affected the photosynthesis efficiency parameter [156]. Control treatment, 5 and 50 mg/L of La_2O_3 was tested for 72 and 144 h and observed that the La_2O_3 nanoparticles affected the plant growth, reduced water uptake, and increased abscisic acid accumulation was due to the enhanced formation of apoplastic barrier in roots of juvenile maize plants [157]. The nanoparticles affect plants hormonally, and these plants can deploy JA, SA, and ethylene to manage the impact of nanoparticles, proving the sturdiness and flexibility of the plant world [158].

5 Future Perspectives and Conclusions

In response to this, there is a possibility of making the crops immune to pollution by regulating the protective mechanisms of crops with phytohormones. However, although the complexity of phytohormonal interactions and the wide range of contaminants involved have been well established, the detailed analysis of these pathways is still insufficient to effectively utilize them in agricultural production. Genetic manipulation offers prospects for producing crops having improved resistance to contaminants within the phytohormonal system. Manipulation of crops may be made to enhance the ability to tolerate heavy metals, organic pollutants or any other toxin in the environment by either silencing or overexpressing the genes that are involved in hormone biosynthesis and signalization or their perception. CRISPR-Cas9 and other tools of genome modification are expected to play a critical role in manipulating hormonal reactions to certain stimuli. The idea of using precision agriculture by applying the technologies of remote sensing and automated systems in combination with the controlled application of phytohormones may be considered as an innovative concept of eliminating plant stress. Thus, designating the early signs of stress, particular hormone therapy will help to strengthen the defense systems without relying on chemical fertilizers and pesticides so necessary for present agriculture. However, while much has been learned about the particular action of phytohormones, more work must be done to determine how these hormones respond to multi-contaminant stress. Comprehensive analyses of hormone signaling networks and the transcriptional and epigenetic modifications they provoke will be crucial for clarifying their synergistic effects. A deeper comprehension of how plants prioritize hormonal signals in response to simultaneous abiotic stressors may enhance their application in agricultural contexts. There is growing interest in non-conventional phytohormones, such as strigolactones and brassinosteroids, because of their functions in stress alleviation. New approaches to improving crop tolerance, especially in challenging environments, may result from examining their ability to reduce contaminant-induced stress. Future research should concentrate on the application of phytohormones to enhance phytoremediation strategies, particularly in areas with significant heavy metal pollution. Enhancing hormone therapy to improve root growth, metal sequestration, and detoxification processes enables plants to function as more effective agents for environmental cleanup. Incorporating biotechnological techniques might improve the effectiveness and sustainability of phytoremediation.

The use of phytohormones to manage crop plant defense mechanisms against various contaminants is one of the most important and promising innovations in the sphere of safe agriculture. Signaling phytohormone networks are vital in regulation of plant stress response to contaminated environment including heavy metals, herbicides, and organic pollutants. By the incorporation of these natural compounds, plants can improve their protective mechanisms and continue with production even under unfavorable conditions. The studies show that abscisic acid, ethylene, jasmonic acid, and salicylic acid are phytohormones that have unique but interacting responsibilities in the regulation of plant defense mechanisms. The inclusion of these hormones into the agricultural systems presents potential for enhancing crop resistance and minimizing on the use of synthetic chemicals. The intricate pattern of hormonal connections, especially when using multiple stress models, is a challenge and a potential for further research. Subsequent studies on crops with higher resistance to environmental pollution may be useful with phytohormone strategies in combination with genetic engineering and precision agriculture. The ongoing study of phytohormonal processes especially in such unconventional fields as phytoremediation and biotechnological enhancement is crucial for developing better and sustainable agricultural systems that can effectively grow in contaminated environment.

Competing Interests The authors declare that they have no competing interest.

Declaration of Generative AI and AI-Assisted Technologies in the Writing Process During the preparation of this work, the book chapter author(s) have not taken any assistance from AI, and the book chapter author(s) reviewed and edited the content as needed and take(s) full responsibility for the content of the publication.

References

1. Grote U, Fasse A, Nguyen TT, Erenstein O (2021) Food security and the dynamics of wheat and maize value chains in Africa and Asia. Front Sustain Food Syst 4:617009. https://doi.org/10.3389/fsufs.2020.617009
2. Erenstein O, Jaleta M, Sonder K, Mottaleb K, Prasanna BM (2022) Global maize production, consumption and trade: trends and R&D implications. Food Secur 14(5):1295–1319. http://dx.doi.org/10.1007/s12571-022-01288-7
3. Bayabil HK, Teshome FT, Li YC (2022) Emerging contaminants in soil and water. Front Environ Sci 10:873499. https://doi.org/10.3389/fenvs.2022.873499
4. Galanakis CM (2023) The "Vertigo" of the food sector within the triangle of climate change, the post-pandemic world, and the Russian-Ukrainian war. Foods 12(4):721. https://doi.org/10.3390/foods12040721
5. Kilani-Morakchi S, Morakchi-Goudjil H, Sifi K (2021) Azadirachtin-based insecticide: overview, risk assessments, and future directions. Front Agron 3:676208. https://doi.org/10.3389/fagro.2021.676208
6. Sun N, Thompson RB, Xu J, Liao S, Suo L, Peng Y et al (2021) Arsenic and cadmium accumulation in soil as affected by continuous organic fertilizer application: implications for clean production. Agronomy 11(11):2272. https://doi.org/10.3390/agronomy11112272
7. Mohanty B, Das A (2023) Heavy metals in agricultural cultivated products irrigated with wastewater in India: a review. AQUA Water Infrastruct Ecosyst Soc 72(6):851–867. https://doi.org/10.2166/aqua.2023.122
8. Rashid A, Schutte BJ, Ulery A, Deyholos MK, Sanogo S, Lehnhoff EA, Beck L (2023) Heavy metal contamination in agricultural soil: environmental pollutants affecting crop health. Agronomy 13(6):1521. https://doi.org/10.3390/agronomy13061521
9. Charuaud L, Jarde E, Jaffrezic A, Thomas M-L, Bot BL (2019) Veterinary pharmaceutical residues from natural water to tap water: sales, occurrence and fate. J Hazard Mater 361:169–186. https://doi.org/10.1016/j.jhazmat.2018.08.075
10. Khan AHA, Barros R (2023) Pharmaceuticals in water: risks to aquatic life and remediation strategies. Hydrobiol 2(2):395–409. https://doi.org/10.3390/hydrobiology2020026
11. Nightingale J, Carter L, Sinclair CJ, Rooney P, Kay P (2023) Influence of manure application method on veterinary medicine losses to water. J Environ Manag 334:117361. https://doi.org/10.1016/j.jenvman.2023.117361
12. Li L, Li T, Liu Y, Li L, Huang X, Xie J (2023) Effects of antibiotics stress on root development, seedling growth, antioxidant status and abscisic acid level in wheat (Triticum aestivum L.). Ecotoxicol Environ Saf 252:114621. https://doi.org/10.1016/j.ecoenv.2023.114621
13. Li Q, Guan C, Zhao Y, Duan X, Yang Z, Zhu J (2023) Salicylic acid alleviates Zn-induced inhibition of growth via enhancing antioxidant system and glutathione metabolism in alfalfa. Ecotoxicol Environ Saf 265:115500. https://doi.org/10.1016/j.ecoenv.2023.115500
14. Zhang H, Li X, Yang Q, Sun L, Yang X, Zhou M et al (2017) Plant growth, antibiotic uptake, and prevalence of antibiotic resistance in an endophytic system of pakchoi under antibiotic exposure. Int J Environ Res Public Health 14(11):1336. https://doi.org/10.3390/ijerph14111336
15. Pridemore CM (2021) The effects of endocrine disrupting chemicals on plants. Theses, Dissertations and Capstones. https://mds.marshall.edu/etd/1412
16. Sharma BM, Scheringer M, Chakraborty P, Bharat GK, Steindal EH, Trasande L et al (2022) Unlocking India's potential in managing endocrine-disrupting chemicals (EDCs): importance, challenges, and opportunities. Expo Health. https://doi.org/10.1007/s12403-022-00519-8
17. Thacharodi A, Hassan S, Hegde TA, Thacharodi DD, Brindhadevi K, Pugazhendhi A (2023) Water a major source of endocrine-disrupting chemicals: an overview on the occurrence, implications on human health and bioremediation strategies. Environ Res 231:116097. https://doi.org/10.1016/j.envres.2023.116097

18. Adu O, Ma X, Sharma VK (2023) Bioavailability, phytotoxicity and plant uptake of per-and polyfluoroalkyl substances (PFAS): a review. J Hazard Mater 447:130805. https://doi.org/10.1016/j.jhazmat.2023.130805
19. Karamat A, Tehrani R, Foster GD, Van Aken B (2023) Plant responses to per- and polyfluoroalkyl substances (PFAS): a molecular perspective. Int J Phytoremed 18:1–9. https://doi.org/10.1080/15226514.2023.2232874
20. Nassazzi W, Wu T-C, Jass J, Lai FL, Ahrens L (2023) Phytoextraction of per- and polyfluoroalkyl substances (PFAS) and the influence of supplements on the performance of short–rotation crops. Environ pollut 333:122038. https://doi.org/10.1016/j.envpol.2023.122038
21. Singh BK, Delgado-Baquerizo M, Egidi E, Guirado E, Leach JE, Liu H et al (2023) Climate change impacts on plant pathogens, food security and paths forward. Nat Rev Microbiol 21:640–656. https://doi.org/10.1038/s41579-023-00900-7
22. Cunningham SD, Berti WR (1993) Remediation of contaminated soils with green plants—an overview. Vitro Cellul & Dev Biol—Plant 29:207–212. https://doi.org/10.1007/BF02632036
23. Sharma P, Pandey S (2014) Status of phytoremediation in world scenario. Int J Environ Bioremediat Biodegrad 2(4):178–191. https://doi.org/10.12691/ijebb-2-4-5
24. Zhang QM, Gong MG, Liu KY, Chen YL, Yuan JF, Chang QS (2020) Rhizoglomus intraradices improves plant growth, root morphology and phytohormone balance of Robinia pseudoacacia in arsenic-contaminated soils. Front Microbiol 11:1428–1440. https://doi.org/10.3389/fmicb.2020.01428
25. Rasafi TE, Oukarroum A, Haddioui A, Song H, Kwon EE, Bolan N, Tack FMG, Sebastian A, Prasad MNV, Rinklebe J (2022) Cadmium stress in plants: a critical review of the effects, mechanisms, and tolerance strategies. Crit Rev Environ Sci Technol 52(5):675–726. https://doi.org/10.1080/10643389.2020.1835435
26. Verma G, Srivastava D, Narayan S, Shirke PA, Chakrabarty D (2020) Exogenous application of methyl jasmonate alleviates arsenic toxicity by modulating its uptake and translocation in rice (Oryza sativa L.). Ecotoxicol Environ Saf 201:110735. https://doi.org/10.1016/j.ecoenv.2020.110735
27. Sytar O, Kumari P, Yadav S, Brestic M, Rastogi A (2019) Phytohormone priming: regulator for heavy metal stress in plants. J Plant Growth Regul 38(2):739–752. https://doi.org/10.1007/s00344-018-9886-8
28. Bandara T, Franks A, Xu J, Bolan N, Wang H, Tang C (2020) Chemical and biological immobilization mechanisms of potentially toxic elements in biochar-amended soils. Crit Rev Environ Sci Technol 50(9):903–978. https://doi.org/10.1080/10643389.2019.1642832
29. Balali-Mood M, Naseri K, Tahergorabi Z, Khazdair MR, Sadeghi M (2021) Toxic mechanisms of five heavy metals: mercury, lead, chromium, cadmium, and arsenic. Front Phamacol 12:643972. https://doi.org/10.3389/fphar.2021.643972
30. da Silva EB, Gao P, Xu M, Guan DX, Tang XJ, Ma LQ (2020) Background concentrations of trace metals As, Ba, Cd Co, Cu, Ni, Pb, Se, and Zn in 214 Florida urban soils: different cities and land uses. Environ Pollut (Barking, Essex: 1987) 264:114737. https://doi.org/10.1016/j.envpol.2020.114737
31. de Oliveira LM, Lessl JT, Gress J, Tisarum R, Guilherme LR, Ma LQ (2015) Chromate and phosphate inhibited each other's uptake and translocation in arsenic hyperaccumulator *Pteris vittata* L. Environ Pollut 197:240–246. https://doi.org/10.1016/j.envpol.2014.11.009
32. Betti C, Della Rovere F, Piacentini D, Fattorini L, Falasca G, Altamura MM (2021) Jasmonates, ethylene and brassinosteroids control adventitious and lateral rooting as stress avoidance responses to heavy metals and metalloids. Biomolecules 11(1):77
33. Tu T, Zheng S, Ren P, Meng X, Zhao J, Chen Q, Li C (2021) Coordinated cytokinin signaling and auxin biosynthesis mediates arsenate-induced root growth inhibition. Plant Physiol 185(3):1166–1181. https://doi.org/10.1093/plphys/kiaa072
34. Ogura T, Goeschl C, Filiault D, Mirea M, Slovak R, Wolhrab B, Satbhai SB, Busch W (2019) Root system depth in Arabidopsis is shaped by *EXOCYST70A3* via the dynamic modulation of auxin transport. Cell 178(2):400–412. https://doi.org/10.1016/j.cell.2019.06.021

35. Petersson SV, Johansson AI, Kowalczyk M, Makoveychuk A, Wang JY, Moritz T, Grebe M, Benfey PN, Sandberg G, Ljung K (2009) An auxin gradient and maximum in the *Arabidopsis* root apex shown by high-resolution cell-specific analysis of IAA distribution and synthesis. Plant Cell 21(6):1659–1668. https://doi.org/10.1105/tpc.109.066480
36. Ruzicka K, Simaskova M, Duclercq J, Petrasek J, Zazimalova E, Simon S, Friml J, Van Montagu MCE, Benkova E (2009) Cytokinin regulates root meristem activity via modulation of the polar auxin transport. Proc Natl Acad Sci USA 106(11):4284–4289. https://www.jstor.org/stable/40441801, https://doi.org/10.1073/pnas.0900060106
37. Ruzicka K, Simaskova M, Duclercq J, Petrasek J, Zazimalova E, Simon S, Friml J, Van Montagu MCE, Benkova E (2009) Cytokinin regulates root meristem activity via modulation of the polar auxin transport. Proc Natl Acad Sci USA 106(11):4284–4289. https://doi.org/10.1073/pnas.0900060106
38. Ronzan M, Piacentini D, Fattorini L, Federica D, Caboni E, Eiche E, Ziegler J, Hause B, Riemann M, Betti C, Altamura MM, Falasca G (2019) Auxin jasmonate crosstalk in Oryza sativa L. root system formation after cadmium and/or arsenic exposure. Environ Exp Bot 165:59–69. https://doi.org/10.1016/j.envexpbot.2019.05.013
39. He H, Yamamuro C (2022) Interplays between auxin and GA signaling coordinate early fruit development. Hortic Res 9:uhab078. https://doi.org/10.1093/hr/uhab078
40. Fattorini L, Ronzan M, Piacentini D, Della Rovere F, De Virgilio C, Sofo A, Altamura MM, Falasca G (2017) Cadmium and arsenic affect quiescent centre formation and maintenance in *Arabidopsis thaliana* post-embryonic roots disrupting auxin biosynthesis and transport. Environ Exp Bot 144:37–48. https://doi.org/10.1016/j.envexpbot.2017.10.005
41. Yuan HM, Huang X (2016) Inhibition of root meristem growth by cadmium involves nitric oxide-mediated repression of auxin accumulation and signalling in Arabidopsis. Plant Cell Environ 39(1):120–135. https://doi.org/10.1111/pce.12597
42. Praveen A, Pandey A, Gupta M (2019) Nitric oxide alters nitrogen metabolism and *PIN* gene expressions by playing protective role in arsenic challenged *Brassica juncea* L. Ecotoxicol Environ Saf 176:95–107. https://doi.org/10.1016/j.ecoenv.2019.03.054
43. Tiwari P, Indoliya Y, Chauhan AS, Singh P, Singh PK, Singh PC, Srivastava S, Pande V, Chakrabarty D (2020) Auxin-salicylic acid cross-talk ameliorates *OsMYB–R1* mediated defense towards heavy metal, drought and fungal stress. J Hazard Mater 399:122811. https://doi.org/10.1016/j.jhazmat.2020.122811
44. Mathur P, Roy S, Nasir Khan M, Mukherjee S (2022) Hydrogen sulphide (H_2S) in the hidden half: role in root growth, stress signalling and rhizospheric interactions. Plant Biol 24(4):559–568. https://doi.org/10.1111/plb.13417
45. Ronzan M, Piacentini D, Fattorini L, Della Rovere F, Eiche E, Rieman M, Altamura MM, Falasca G (2018) Cadmium and arsenic affect root development in Oryza sativa L. negatively interacting with auxin. Environ Exp Bot 151:64–75. https://doi.org/10.1016/j.envexpbot.2018.04.008
46. Mohan TC, Castrillo G, Navarro C, Zarco-Fernandez S, Ramireddy E, Mateo C, Zamarreno AM, Paz-Ares J, Munoz R, Garcia-Mina JM, Hernandez LE, Schmulling T, Leyva A (2016) Cytokinin determines thiol-mediated arsenic tolerance and accumulation. Plant Physiol 171(2):1418–1426. https://doi.org/10.1104/pp.16.00372
47. Di DW, Wu L, Zhang L, An CW, Zhang TZ, Luo P, Gao HH, Kriechbaumer V, Guo GQ (2016) Functional roles of ArabidopsisCKRC2/YUCCA8 gene and the involvement of PIF4 in the regulation of auxin biosynthesis by cytokinin. Sci Rep 6:36866. https://doi.org/10.1038/srep36866
48. Zhou ZY, Zhang CG, Wu L, Zhang CG, Chai J, Wang M, Jha A, Jia PF, Cui SJ, Yang M, Chen R (2011) Functional characterization of the *CKRC1/TAA1* gene and dissection of hormonal actions in the Arabidopsis root. Plant J 66(3):516–527. https://doi.org/10.1111/j.1365-313X.2011.04509.x
49. Su YH, Liu YB, Zhang XS (2011) Auxin–cytokinin interaction regulates meristem development. Mol Plant 4(4):616–625. http://dx.doi.org/10.1093/mp/ssr007

50. Choudhary SP, Kanwar M, Bhardwaj R, Yu J-Q, Lam-Son Phan T (2012) Chromium stress mitigation by polyamine-brassinosteroid application involves phytohormonal and physiological strategies in Raphanus sati- vus L. PLoS ONE 7(3):e33210. https://doi.org/10.1371/journal.pone.0033210
51. Shakirova FM, Allagulova CR, Maslennikova DR, Klyuchnikova EO, Avalbaev AM, Bezrukova MV (2016) Salicylic acid-induced protection against cadmium toxicity in wheat plants. Environ Exp Bot 122:19–28. https://doi.org/10.1016/j.envexpbot.2015.08.002
52. Semeradova H, Montesinos JC, Benkova E (2020) All roads lead to auxin: Post-translational regulation of auxin transport by multiple hormonal pathways. Plant Commun 1(3):100048. https://doi.org/10.1016/j.xplc.2020.100048
53. Ovečka M, Takáč TJ (2014) Managing heavy metal toxicity stress in plants: biological and biotechnological tools. Biotechnol Adv 32(1):73–86. https://doi.org/10.1016/j.biotechadv.2013.11.011
54. Sharma A, Kumar V, Thukral AK, Bhardwaj R (2016) Epibrassinolide–imidacloprid interaction enhances non-enzymatic antioxidants in *Brassica juncea* L. Indian J Plant Physiol 21(1):70–75. https://doi.org/10.1007/s40502-016-0203-x
55. Guedes F, Maia CF, Silva B, Batista BL, Alyemeni MN, Ahmad P, Lobato A (2021) Exogenous 24-epibrassinolide stimulates root protection, and leaf antioxidant enzymes in lead stressed rice plants: central rolesto minimize Pb content and oxidative stress. Environ Pollut (Barking, Essex: 1987) 280:116992. https://doi.org/10.1016/j.envpol.2021.116992
56. Siddiqi KS, Husen A (2021) Significance of brassinosteroids and their derivatives in the development and protection of plants under abiotic stress. Biologia 76:2837–2857. https://doi.org/10.1007/s11756-021-00853-3
57. Jan S, Alyemeni MN, Wijaya L, Alam P, Siddique KH, Ahmad P (2018) Interactive effect of 24-epibrassinolide and silicon alleviates cadmium stress via the modulation of antioxidant defense and glyoxalase systems and macronutrient content in *Pisum sativum* L. seedlings. BMC Plant Biol 18(1):146. https://doi.org/10.1186/s12870-018-1359-5
58. Basit F, Bhat JA, Dong Z, Mou QS, Zhu XB, Wang Y, Hu J, Jan BL, Shakoor A, Guan YJ, Ahmad P (2022) Chromium toxicity induced oxidative damage in two rice cultivars and its mitigation through external supplementation of brassinosteroids and spermine. Chemosphere 302:134423. https://doi.org/10.1016/j.chemosphere.2022.134423
59. Rady MM (2011) Effect of 24-epibrassinolide on growth, yield, antioxidant system and cadmium content of bean (*Phaseolus vulgaris* L.) plants under salinity and cadmium stress. Sci Hortic 129(2):232–237. https://doi.org/10.1016/j.scienta.2011.03.035
60. Nazir F, Fariduddin Q, Hussain A, Khan TA (2021) Brassinosteroid and hydrogen peroxide improve photosynthetic machinery, stomatal movement, root morphology and cell viability and reduce Cu-triggered oxidative burst in tomato. Ecotoxicol Environ Saf 207:111081. https://doi.org/10.1016/j.ecoenv.2020.111081
61. Kaya C, Ashraf M, Alyemeni MN, Ahmad P (2020) The role of nitrate reductase in brassinosteroid-induced endogenous nitric oxide generation to improve cadmium stress tolerance of pepper plants by upregulating the ascorbate–glutathione cycle. Ecotoxicol Environ Saf 196:110483. https://doi.org/10.1016/j.ecoenv.2020.110483
62. Parrotta L, Guerriero G, Sergeant K, Cai G, Hausman J-F (2015) Target or barrier? The cell wall of early- and later-diverging plants vs cadmium toxicity: differences in the response mechanisms. Front Plant Sci 6:133. https://doi.org/10.3389/fpls.2015.00133
63. Guo B, Liu C, Liang Y, Li N, Fu Q (2019) Salicylic acid signals plant defence against cadmium toxicity. Int J Mol Sci 20(12):2960. https://doi.org/10.3390/ijms20122960
64. Hossain AK, Hossain MA, Asgar MA, Tosaki T, Koyama H, Hara T (2006) Changes in cell wall polysaccharides and hydroxycinnamates in wheat roots by aluminum stress at higher calcium supply. J Plant Nutr 29(4):601–613. https://doi.org/10.1080/01904160600564303
65. Krzesłowska M, Rabęda I, Basińska A, Lewandowski M, Mellerowicz EJ, Napieralska A, Samardakiewicz S, Woźny A (2016) Pectinous cell wall thickenings formation – a common defense strategy of plants to cope with Pb. Environ Pollut (Barking, Essex: 1987) 214:354–361. https://doi.org/10.1016/j.envpol.2016.04.019

66. Sheshukova EV, Komarova TV, Pozdyshev DV, Ershova NM, Shindyapina AV, Tashlitsky VN, Sheval EV, Dorokhov YL (2017) The intergenic interplay between *aldose 1-epimerase-like protein* and *pectin* methylesterase in abiotic and biotic stress control. Front Plant Sci 8:1646. https://doi.org/10.3389/fpls.2017.01646
67. Carrasco-Gil S, Alvarez-Fernandez A, Sobrino-Plata J, Millan R, Carpena-Ruiz RO, Leduc DL, Andrews JC, Abadia J, Hernandez LE (2011) Complexation of Hg with phytochelatins is important for plant Hg tolerance. Plant Cell Environ 34(5):778–791. https://doi.org/10.1111/j.1365-3040.2011.02281.x
68. O'Lexy R, Kasai K, Clark N, Fujiwara T, Sozzani R, Gallagher KL (2018) Exposure to heavy metal stress triggers changes in plasmodesmatal permeability via deposition and breakdown of callose. J Exp Bot 69(15):3715–3728. https://doi.org/10.1093/jxb/ery171
69. Kovac J, Lux A, Vaculik M (2018) Formation of a subero-lignified apical deposit in root tip of radish (Raphanus sativus) as a response to copper stress. Ann Bot 122(5):823–831. https://doi.org/10.1093/aob/mcy013
70. Jia H, Wang X, Wei T, Wang M, Liu X, Hua L, Ren X, Guo J, Li J (2021) Exogenous salicylic acid regulates cell wall polysaccharides synthesis and pectin methylation to reduce Cd accumulation of tomato. Ecotoxicol Environ Saf 207:111550. https://doi.org/10.1016/j.ecoenv.2020.111550
71. Xu LL, Fan ZY, Dong YJ, Kong J, Bai XY (2015) Effects of exogenous salicylic acid and nitric oxide on physiological characteristics of two peanut cultivars under cadmium stress. Biol Plant 59(1):171–182. https://doi.org/10.1007/s10535-014-0475-9
72. Zhu XF, Lei GJ, Jiang T, Liu Y, Li GX, Zheng SJ (2012) Cell wall polysaccharides are involved in P-deficiency-induced Cd exclusion in *Arabidopsis thaliana*. Planta 236(4):989–997. https://doi.org/10.1007/s00425-012-1652-8
73. Wójcik M, Vangronsveld J, D'Haen J, Tukiendorf A (2005) Cadmium tolerance in *Thlaspi caerulescens*: II. Localization of cadmium in *Thlaspi caerulescens*. Environ Exp Bot 53(2):163–171. https://doi.org/10.1016/j.envexpbot.2004.03.011
74. Napoleão TA, Soares G, Vital CE, Bastos C, Castro R, Loureiro ME, Giordano A (2017) Methyl jasmonate and salicylic acid are able to modify cell wall but only salicylic acid alters biomass digestibility in the model grass Brachypodium distachyon. Plant Sci Int J Exp Plant Biol 263:46–54. https://doi.org/10.1016/j.plantsci.2017.06.014
75. Buanafina MMD (2009) Feruloylation in grasses: current and future perspectives. Mol Plant 2(5):861–872. https://doi.org/10.1093/mp/ssp067
76. Jung HJ, Samac DA, Sarath G (2012) Modifying crops to increase cell wall digestibility. Plant Sci 185:65–77. https://doi.org/10.1016/j.plantsci.2011.10.014
77. Zulfiqar F, Ashraf M (2022) Antioxidants as modulators of arsenic-induced oxidative stress tolerance in plants: an overview. J Hazard Mater 427:127891. https://doi.org/10.1016/j.jhazmat.2021.127891
78. Szalai G, Kellős T, Galiba G, Kocsy G (2009) Glutathione as an antioxidant and regulatory molecule in plants under abiotic stress conditions. J Plant Growth Regul 28(1):66–80. https://doi.org/10.1007/s00344-008-9075-2
79. Guo B, Liu C, Li H, Yi K, Ding N, Li N, Lin Y, Fu Q (2016) Endogenous salicylic acid is required for promoting cadmium tolerance of *Arabidopsis* by modulating glutathione metabolisms. J Hazard Mater 316:77–86. https://doi.org/10.1016/j.jhazmat.2016.05.032
80. Ahmad B, Jaleel H, Sadiq Y, Khan MM, Shabbir A (2018) Response of exogenous salicylic acid on cadmium-induced photosynthetic damage, antioxidant metabolism and essential oil production in peppermint. Plant Growth Regul 86(2):273–286. https://doi.org/10.1007/s10725-018-0427-z
81. Freeman JL, Garcia D, Kim D, Hopf A, Salt DE (2005) Constitutively elevated salicylic acid signals glutathione-mediated nickel tolerance in Thlaspi nickel hyperaccumulators. Plant Physiol 137(3):1082–1091. https://doi.org/10.1104/pp.104.055293
82. Bovet L, Eggmann T, Meylan-Bettex M, Polier J, Kammer P, Stark M, Zažímalová E, Amrhein N, Martinoia E (2005) Transcript levels of AtMRP1, an Arabidopsis ATP binding cassette transporter gene, are regulated by cadmium and the glutathione status. FEBS Lett 581(3):328–332. https://doi.org/10.1016/j.febslet.2005.01.001

83. Eichhorn H, Klinghammer M, Becht P, Tenhaken R (2006) Isolation of a novel ABC-transporter gene from soybean induced by salicylic acid. J Exp Bot 57(10):2193–2201. https://doi.org/10.1093/jxb/erj179
84. Kupper H, Kochian LV (2010) Transcriptional regulation of metal transport genes and mineral nutrition during acclimatization to cadmium and zinc in the Cd/Zn hyperaccumulator, Thlaspi caerulescens (Ganges population). New Phytol 185(1):114–129. https://doi.org/10.1111/j.1469-8137.2009.03051.x
85. Drazic G, Mihailovic N (2005) Modification of cadmium toxicity in soybean seedlings by salicylic acid. Plant Sci 168(2):511–517. https://doi.org/10.1016/j.plantsci.2004.09.019
86. Hayat S, Hayat Q, Alyemeni MN, Ahmad A (2014) Salicylic acid enhances the efficiency of nitrogen fixation and assimilation in *Cicer arietinum* plants grown under cadmium stress. J Plant Interact 9(1):35–42. https://doi.org/10.1080/17429145.2012.751635
87. Gordon LK, Minibayeva FV, Rakhmatullina DF, Alyabyev AJ, Ogorodnikova TI, Loseva NL, Valitova YN (2004) Heat production of wheat roots induced by the disruption of proton gradient by salicylic acid. Thermochim Acta 422(1–2):101–104. https://doi.org/10.1016/j.tca.2004.04.032
88. Saha I, Hasanuzzaman M, Adak MK (2021) Abscisic acid priming regulates arsenite toxicity in two contrasting rice (*Oryza sativa* L.) genotypes through differential functioning of *sub1A* quantitative trait loci. Environ Pollut 287:117586. https://doi.org/10.1016/j.envpol.2021.117586
89. Vishwakarma NK, Patel VK, Mitra P, Ramesh K, Mitra K, Vishwakarma S, Acharya K, Misra N, Maiti P, Ray B (2021) Synthesis of ABA-type double hydrophilic amphiphilic PU-based block copolymers of poly(*N*-vinylpyrrolidone) and poly(*N*-isopropylacrylamide) *via* click chemistry. J Macromol Sci Part A 58(3):192–205. https://doi.org/10.1080/10601325.2020.1840920
90. Fan SK, Fang XZ, Guan MY, Ye YQ, Lin XY, Du ST, Jin CW (2014) Exogenous abscisic acid application decreases cadmium accumulation in Arabidopsis plants, which is associated with the inhibition of IRT1-mediated cadmium uptake. Front Plant Sci 5:721. https://doi.org/10.3389/fpls.2014.00721
91. Shen G, Niu J, Deng Z (2017) Abscisic acid treatment alleviates cadmium toxicity in purple flowering stalk (*Brassica campestris* L. *ssp. chinensis var. purpurea* Hort.) seedlings. Plant Physiol Biochem 118:471–478. https://doi.org/10.1016/j.plaphy.2017.07.018
92. Hauser F, Li Z, Waadt R, Schroeder JI (2017) SnapShot: abscisic acid signaling. Cell 171(7):1708. https://doi.org/10.1016/j.cell.2017.11.045
93. Uraguchi S, Mori S, Kuramata M, Kawasaki A, Arao T, Ishikawa S (2009) Root-to-shoot Cd translocation via the xylem is the major process determining shoot and grain cadmium accumulation in rice. J Exp Bot 60(9):2677–2688. https://doi.org/10.1093/jxb/erp119
94. Hsu YT, Kao CH (2003) Role of abscisic acid in cadmium tolerance of rice (*Oryza sativa* L.) seedlings. Plant Cell Environ 26(6):867–874. https://doi.org/10.1046/j.1365-3040.2003.01018.x
95. Hsu YT, Kao CH (2005) Abscisic acid accumulation and cadmium tolerance in rice seedlings. Physiol Plant 124(1):71–80. https://doi.org/10.1111/j.1399-3054.2005.00490.x
96. Kobayashi T, Nishizawa NK (2012) Iron uptake, translocation and regulation in higher plants. Annu Rev Plant Biol 63:131–152. https://doi.org/10.1146/annurev-arplant-042811-105522
97. Vert G, Grotz N, Dedaldechamp F, Gaymard F, Guerinot ML, Briat JF, Curie C (2002) IRT1, an Arabidopsis transporter essential for iron uptake from the soil and for plant growth. Plant Cell 14(6):1223–1233. https://doi.org/10.1105/tpc.001388
98. Zhang C, He ML, Wang SL, Chu LY, Wang C, Yang NM, Ding GD, Cai HM, Shi L, Xu FS (2021) Boron deficiency-induced root growth inhibition is mediated by brassinosteroid signalling regulation in Arabidopsis. Plant J: Cell Mol Biol 107(2):564–578. https://doi.org/10.1111/tpj.15311
99. Hatamleh AA, Danish M, Al-Dosary MA, El-Zaidy M, Ali S (2022) Physiological and oxidative stress responses of Solanum lycopersicum (L.) (tomato) when exposed to different chemical pesticides. RSC Adv 12(12):7237–7252. https://doi.org/10.1039/D1RA09440H

100. Pathak VM, Verma VK, Rawat BS, Kaur B, Babu N, Sharma A, Dewali S, Yadav M, Kumari R, Singh S, Mohapatra A (2022) Current status of pesticide effects on environment, human health and it's eco-friendly management as bioremediation: a comprehensive review. Front Microbiol 13:962619. https://doi.org/10.3389/fmicb.2022.962619
101. Li X, Riaz M, Song B, Liu H (2022) Phytotoxicity response of sugar beet (Beta vulgaris L.) seedlings to herbicide fomesafen in soil. Ecotoxicol Environ Saf 239:113628. https://doi.org/10.1016/j.ecoenv.2022.113628
102. Li X, Riaz M, Song B, Liang X, Liu H (2022) Exogenous salicylic acid alleviates fomesafen toxicity by improving photosynthetic characteristics and antioxidant defense system in sugar beet. Ecotoxicol Environ Saf 238:113587. https://doi.org/10.1016/j.ecoenv.2022.113587
103. Babula P, Adam V, Opatrilova R, Zehnalek J, Havel L, Kizek R (2010) Uncommon heavy metals, metalloids and their plant toxicity: a review. In: Organic farming, pest control and remediation of soil pollutants: organic farming, pest control and remediation of soil pollutants, pp 275–317. https://doi.org/10.1007/978-1-4020-9654-9_14
104. Vasilachi IC, Stoleru V, Gavrilescu M (2023) Analysis of heavy metal impacts on cereal crop growth and development in contaminated soils. Agriculture 13(10):1983. https://doi.org/10.3390/agriculture13101983
105. Jan S, Singh R, Bhardwaj R, Ahmad P, Kapoor D (2020) Plant growth regulators: a sustainable approach to combat pesticide toxicity. 3 Biotech 10(11):466. https://doi.org/10.1007/s13205-020-02454-4
106. Cheng W, Luo X, Jiang L, Du L, Wang S, Zhang L, Hong J (2023) Relief effects of salicylic acid against the stress of heavy metal in fritillaria hupehensis hsiao et KC Hsia. Bangladesh J Bot 601–606. https://doi.org/10.3329/bjb.v52i20.68230
107. Boulahia K, Ouldsaid C, Abrous-Belbachir O (2023) Exogenous application of salicylic acid improve growth and some physio-biochemical parameters in herbicide stressed Phaseolus vulgaris L. Gesunde Pflanzen 75(6):2301–2318. https://doi.org/10.1007/s10343-023-00878-5
108. Khatami SA, Barmaki M, Alebrahim MT, Bajwa AA (2022) Salicylic acid pre-treatment reduces the physiological damage caused by the herbicide mesosulfuron-methyl+ iodosulfuron-methyl in wheat (Triticum aestivum). Agronomy 12(12):3053. https://doi.org/10.3390/agronomy12123053
109. Khalil R, Haroun S, Bassyoini F, Nagah A, Yusuf M (2021) Salicylic acid in combination with kinetin or calcium ameliorates heavy metal stress in Phaseolus vulgaris plant. J Agric Food Res 5:100182. https://doi.org/10.1016/j.jafr.2021.100182
110. Spormann S, Soares C, Fidalgo F (2019) Salicylic acid alleviates glyphosate-induced oxidative stress in Hordeum vulgare L. J Environ Manag 241:226–234. https://doi.org/10.1016/j.jenvman.2019.04.035
111. Yüzbaşıoğlu E, Dalyan E (2019) Salicylic acid alleviates thiram toxicity by modulating antioxidant enzyme capacity and pesticide detoxification systems in the tomato (Solanum lycopersicum Mill.). Plant Physiol Biochem 135:322–330. https://doi.org/10.1016/j.plaphy.2018.12.023
112. Li Q, Wang G, Wang Y, Yang D, Guan C, Ji J (2019) Foliar application of salicylic acid alleviate the cadmium toxicity by modulation the reactive oxygen species in potato. Ecotoxicol Environ Saf 172:317–325. https://doi.org/10.1016/j.ecoenv.2019.01.078
113. Emamverdian A, Khalofah A, Pehlivan N, Zia-ur-Rehman M, Li Y, Zargar M (2024) Exogenous application of jasmonates and brassinosteroids alleviates lead toxicity in bamboo by altering biochemical and physiological attributes. Environ Sci Pollut Res 31(5):7008–7026. https://doi.org/10.1007/s11356-023-31549-7
114. Yuan XY, Zhang LG, Huang L, Yang HJ, Zhong YT, Ning N, Wen YY, Dong SQ, Song XE, Wang HF, Guo PY (2017) Spraying Brassinolide improves Sigma Broad tolerance in foxtail millet (Setaria italica L.) through modulation of antioxidant activity and photosynthetic capacity. Sci Rep 7(1):11232. https://doi.org/10.1038/s41598-017-11867-w
115. Hayat S, Alyemeni MN, Hasan SA (2012) Foliar spray of brassinosteroid enhances yield and quality of Solanum lycopersicum under cadmium stress. Saudi J Biol Sci 19(3):325–335. https://doi.org/10.1016/j.sjbs.2012.03.005

116. Su X, Liu X, Li C, Zhang Y (2024) 24-epibrassinolide as a multidimensional regulator of rice (*Oryza sativa*) physiological and molecular responses under isoproturon stress. Ecotoxicol Environ Saf 281:116575. https://doi.org/10.1016/j.ecoenv.2024.116575
117. Mehrian SK, Karimi N, Rahmani F (2023) 24-Epibrassinolide alleviates diazinon oxidative damage by escalating activities of antioxidant defense systems in maize plants. Sci Rep 13(1):19631. https://doi.org/10.1038/s41598-023-46764-y
118. Bakshi P, Sharma P, Chouhan R, Mir BA, Gandhi SG, Bhardwaj R, Alam P, Ahmad P (2023) Interactive effect of 24-epibrassinolide and plant growth promoting rhizobacteria inoculation restores photosynthetic attributes in Brassica juncea L. under chlorpyrifos toxicity. Environ Pollut 320:120760. https://doi.org/10.1016/j.envpol.2022.120760
119. Dalyan E (2023) Amelioration of the adverse effects of thiram by 24-epibrassinolide in tomato (Solanum lycopersicum Mill.). Arch Biol Sci 00:15–15. https://doi.org/10.2298/ABS230201015D
120. Bakshi P, Chouhan R, Sharma P, Mir BA, Gandhi SG, Landi M, Zheng B, Sharma A, Bhardwaj R (2021) Amelioration of chlorpyrifos-induced toxicity in Brassica juncea L. by combination of 24-epibrassinolide and plant-growth-promoting rhizobacteria. Biomolecules 11(6):877. https://doi.org/10.3390/biom11060877
121. Shopova E, Katerova Z, Brankova L, Dimitrova L, Sergiev I, Todorova D, Talaat NB (2021) Modulation of physiological stress response of Triticum aestivum L. to glyphosate by brassinosteroid application. Life 11(11):1156. https://doi.org/10.3390/life11111156
122. Yang Y, Mao L, Guan W, Wei X, Shao Y, Luo Z, Lin X, Li L (2021) Exogenous 24-epibrassinolide activates detoxification enzymes to promote degradation of boscalid in cherry tomatoes. J Sci Food Agric 101(6):2210–2217
123. Wang Z, Jiang Y, Peng X, Xu S, Zhang H, Gao J, Xi Z (2017) Exogenous 24-epibrassinolide regulates antioxidant and pesticide detoxification systems in grapevine after chlorothalonil treatment. Plant Growth Regul 81:455–466. https://doi.org/10.1007/s10725-016-0223-6
124. Sharma A, Kumar V, Bhardwaj R, Thukral AK (2017) Seed pre-soaking with 24-epibrassinolide reduces the imidacloprid pesticide residues in green pods of Brassica juncea L. Toxicol Environ Chem 99(1):95–103. https://doi.org/10.1080/02772248.2016.1146955
125. Sharma A, Yuan H, Kumar V, Ramakrishnan M, Kohli SK, Kaur R, Thukral AK, Bhardwaj R, Zheng B (2019) Castasterone attenuates insecticide induced phytotoxicity in mustard. Ecotoxicol Environ Saf 179:50–61. https://doi.org/10.1016/j.ecoenv.2019.03.120
126. Yadav P, Kaur R, Kanwar MK, Sharma A, Verma V, Sirhindi G, Bhardwaj R (2018) Castasterone confers copper stress tolerance by regulating antioxidant enzyme responses, antioxidants, and amino acid balance in B. juncea seedlings. Ecotoxicol Environ Saf 147:725–734. https://doi.org/10.1016/j.ecoenv.2017.09.035
127. Kaur R, Yadav P, Sharma A, Thukral AK, Kumar V, Kohli SK, Bhardwaj R (2017) Castasterone and citric acid treatment restores photosynthetic attributes in Brassica juncea L. under Cd (II) toxicity. Ecotoxicol Environ Saf 145:466–475. https://doi.org/10.1016/j.ecoenv.2017.07.067
128. Piotrowska-Niczyporuk A, Bajguz A, Kotowska U, Zambrzycka-Szelewa E, Sienkiewicz A (2020) Auxins and cytokinins regulate phytohormone homeostasis and thiol-mediated detoxification in the green alga Acutodesmus obliquus exposed to lead stress. Sci Rep 10(1):10193. https://doi.org/10.1038/s41598-020-67085-4
129. Zhou M, Ghnaya T, Dailly H, Cui G, Vanpee B, Han R, Lutts S (2019) The cytokinin trans-zeatine riboside increased resistance to heavy metals in the halophyte plant species Kosteletzkya pentacarpos in the absence but not in the presence of NaCl. Chemosphere 233:954–965. https://doi.org/10.1016/j.chemosphere.2019.06.02
130. Siddiqi KS, Husen A (2017) Plant response to strigolactones: current developments and emerging trends. Appl Soil Ecol 120:247–253. https://doi.org/10.1016/j.apsoil.2017.08.020
131. Mostofa MG, Rahman MM, Nguyen KH, Li W, Watanabe Y, Tran CD, Zhang M, Itouga M, Fujita M, Tran LSP (2021) Strigolactones regulate arsenate uptake, vacuolar-sequestration and antioxidant defense responses to resist arsenic toxicity in rice roots. J Hazard Mater 415:125589. https://doi.org/10.1016/j.jhazmat.2021.125589

132. Ali A, Malangisha GK, Yang H, Li C, Wang C, Yang Y, Mahmoud A, Khan J, Yang J, Hu Z, Zhang M (2021) Strigolactone alleviates herbicide toxicity via maintaining antioxidant homeostasis in watermelon (Citrullus lanatus). Agriculture 11(5):419. https://doi.org/10.3390/agriculture11050419
133. Tai Z, Yin X, Fang Z, Shi G, Lou L, Cai Q (2017) Exogenous GR24 alleviates cadmium toxicity by reducing cadmium uptake in switchgrass (Panicum virgatum) seedlings. Int J Environ Res Public Health 14(8):852. https://doi.org/10.3390/ijerph14080852
134. Alharbi K, Alghanem SM, Begum N, Alrashidi AA, Alaklabi A, Alnusairi GS, Alsudays IM, Imtiaz KHAN, Soliman MH (2024) Jasmonic acid priming and foliar application of spermidine up-regulates the tolerance mechanisms to alleviate the damaging effects of cadmium stress on growth and photosynthesis in wheat. Notulae Botanicae Horti Agrobotanici Cluj-Napoca 52(3):13614. https://doi.org/10.15835/nbha52313614
135. Yu GB, Chen RN, Chen QS, Chen FQ, Liu HL, Ren CY, Zhang YX, Yang FJ, Wei JP (2022) Jasmonic acid promotes glutathione assisted degradation of chlorothalonil during tomato growth. Ecotoxicol Environ Saf 233:113296. https://doi.org/10.1016/j.ecoenv.2022.113296
136. Kaushik S, Sharma P, Kaur G, Singh AK, Al-Misned FA, Shafik HM, Sirhindi G (2022) Seed priming with methyl jasmonate mitigates copper and cadmium toxicity by modifying biochemical attributes and antioxidants in Cajanus cajan. Saudi J Biol Sci 29(2):721–729. https://doi.org/10.1016/j.sjbs.2021.12.014
137. Lei GJ, Sun L, Sun Y, Zhu XF, Li GX, Zheng SJ (2020) Jasmonic acid alleviates cadmium toxicity in Arabidopsis via suppression of cadmium uptake and translocation. J Integr Plant Biol 62(2):218–227. https://doi.org/10.1111/jipb.12801
138. Dai H, Wei S, Pogrzeba M, Rusinowski S, Krzyżak J, Jia G (2020) Exogenous jasmonic acid decreased Cu accumulation by alfalfa and improved its photosynthetic pigments and antioxidant system. Ecotoxicol Environ Saf 190:110176. https://doi.org/10.1016/j.ecoenv.2020.110176
139. Bali S, Jamwal VL, Kohli SK, Kaur P, Tejpal R, Bhalla V, Ohri P, Gandhi SG, Bhardwaj R, Al-Huqail AA, Siddiqui MH (2019) Jasmonic acid application triggers detoxification of lead (Pb) toxicity in tomato through the modifications of secondary metabolites and gene expression. Chemosphere 235:734–748. https://doi.org/10.1016/j.chemosphere.2019.06.188
140. Sofy MR, Seleiman MF, Alhammad BA, Alharbi BM, Mohamed HI (2020) Minimizing adverse effects of pb on maize plants by combined treatment with jasmonic, salicylic acids and proline. Agronomy 10(5):699. https://doi.org/10.3390/agronomy10050699
141. Husen A (2022) Environmental, physiological and chemical controls of adventitious rooting in cuttings. Elsevier Inc., Cambridge. https://doi.org/10.1016/C2020-0-04484-1, ISBN 978-0-323-90636-4
142. Mir AR, Alam P, Hayat S (2022) Auxin regulates growth, photosynthetic efficiency and mitigates copper induced toxicity via modulation of nutrient status, sugar metabolism and antioxidant potential in Brassica juncea. Plant Physiol Biochem 185:244–259. https://doi.org/10.1016/j.plaphy.2022.06.006
143. Luo Y, Wei Y, Sun S, Wang J, Wang W, Han D, Shao H, Jia H, Fu Y (2019) Selenium modulates the level of auxin to alleviate the toxicity of cadmium in tobacco. Int J Mol Sci 20(15):3772. https://doi.org/10.3390/ijms20153772
144. Anwar T, Qureshi H, Jabeen M, Zaman W, Ali HM (2024) Mitigation of cadmium-induced stress in maize via synergistic application of biochar and gibberellic acid to enhance morpho-physiological and biochemical traits. BMC Plant Biol 24(1):192. https://doi.org/10.1186/s12870-024-04805-2
145. Alam P, Azzam MA, Balawi TA, Raja V, Bhat JA, Ahmad P (2022) Mitigation of negative effects of chromium (VI) toxicity in Faba bean (Vicia faba) plants through the supplementation of Kinetin (KN) and gibberellic Acid (GA3). Plants 11(23):3302. https://doi.org/10.3390/plants11233302
146. Ahmad P, Raja V, Ashraf M, Wijaya L, Bajguz A, Alyemeni MN (2021) Jasmonic acid (JA) and gibberellic acid (GA3) mitigated Cd-toxicity in chickpea plants through restricted cd uptake and oxidative stress management. Sci Rep 11(1):19768. https://doi.org/10.1038/s41598-021-98753-8

147. Khan MIR, Jahan B, Alajmi MF, Rehman MT, Khan NA (2019) Exogenously-sourced ethylene modulates defense mechanisms and promotes tolerance to zinc stress in mustard (Brassica juncea L.). Plants 8(12):540. https://doi.org/10.3390/plants8120540
148. Sarkar B, Das A, Pal S, Sil SK, Hasanuzzaman M, Adak MK (2024) Insights into the mechanism of abscisic acid-induced 2, 4-D tolerance in maize (Zea mays L.) seedlings: Study on the regulation of osmolytes, fluorescence, photosynthetic carbon reactions, and antioxidant metabolism. Plant Stress 11:100396. https://doi.org/10.1016/j.stress.2024.100396
149. Kamran M, Danish M, Saleem MH, Malik Z, Parveen A, Abbasi GH, Jamil M, Ali S, Afzal S, Riaz M, Rizwan M (2021) Application of abscisic acid and 6-benzylaminopurine modulated morpho-physiological and antioxidative defense responses of tomato (Solanum lycopersicum L.) by minimizing cobalt uptake. Chemosphere 263:128169. https://doi.org/10.1016/j.chemosphere.2020.128169
150. Nhan LV, Ma C, Rui Y, Liu S, Li X, Xing B, Liu L (2015) Phytotoxic mechanism of nanoparticles: destruction of chloroplasts and vascular bundles and alteration of nutrient absorption. Sci Rep 5:11618. https://doi.org/10.1038/srep11618
151. Gui X, Deng Y, Rui Y, Gao B, Luo W, Chen S, Van Nhan L, Li X, Liu S, Han Y, Liu L (2015) Response difference of transgenic and conventional rice (Oryza sativa) to nanoparticles (γFe_2O_3). Environ Sci Pollut Res 22:17716–17723. https://doi.org/10.1007/s11356-015-4976-7
152. Angelini J, Klassen R, Široká J, Novák O, Záruba K, Siegel J, Novotná Z et al (2022) Silver nanoparticles alter microtubule arrangement, dynamics and stress phytohormone levels. Plants 11(3):313. https://doi.org/10.3390/plants11030313
153. Vankova R, Landa P, Podlipna R, Dobrev PI, Prerostova S, Langhansova L et al (2017) ZnO nanoparticle effects on hormonal pools in Arabidopsis thaliana. Sci Total Environ 593:535–542. https://doi.org/10.1016/j.scitotenv.2017.03.160
154. Vinković T, Novák O, Strnad M, Goessler W, Jurašin DD, Parađiković N, Vrček IV (2017) Cytokinin response in pepper plants (Capsicum annuum L.) exposed to silver nanoparticles. Environ Res 156:10–18. https://doi.org/10.1016/j.envres.2017.03.015
155. Rui M, Ma C, Tang X, Yang J, Jiang F, Pan Y et al (2017) Phytotoxicity of silver nanoparticles to peanut (Arachis hypogaea L.): physiological responses and food safety. ACS Sustain Chem Eng. 5(8):6557–6567
156. Xiao Z, Yue L, Wang C, Chen F, Ding Y, Liu Y, Cao X, Chen Z, Rasmann S, Wang Z (2021) Downregulation of the photosynthetic machinery and carbon storage signaling pathways mediate La2O3 nanoparticle toxicity on radish taproot formation. J Hazard Mater 411:124971. https://doi.org/10.1016/j.jhazmat.2020.124971
157. Yue L, Chen F, Yu K, Xiao Z, Yu X, Wang Z, Xing B (2019) Early development of apoplastic barriers and molecular mechanisms in juvenile maize roots in response to La2O3 nanoparticles. Sci Total Environ 653:675–683. https://doi.org/10.1016/j.scitotenv.2018.10.320
158. Gayathiri E, Prakash P, Pandiaraj S, Ramasubburayan R, Gaur A, Sekar M, Viswanathan D, Govindasamy R (2023) Investigating the ecological implications of nanomaterials: unveiling plants' notable responses to nano-pollution. Plant Physiol Biochem 206:108261. https://doi.org/10.1016/j.plaphy.2023.108261

Managing Defense Mechanisms in Crop Plants Using Nanomaterials Under Contaminants

Iram Gul, Syeda Irsa Mazhar, Muhammad Arif Ali, and Muhammad Arshad

Abstract Agriculture is facing the challenge of feeding of growing population globally. Moreover, crop productivity is steadily being threatened by environmental contaminants including heavy metals, pesticides, and industrial chemicals. These pollutants interfere with nutrient uptake and plant physiological processes that reduce yields. Nevertheless, even high levels of heavy metals causing impairment to growth and oxidative stress continue unabated among the constitutive and inducible mechanisms developed in plants to cope with stress. Nanotechnology is very promising in solving these problems because nanomaterials enhance stress tolerance, nutrient uptake, and resistance against contaminants. Nanomaterials enhance the activity of antioxidants, photosynthesis, and assimilation of nutrients, which enable plants to recover from stress conditions. Nanomaterials help target fertilizer and pesticide delivery to root systems, reducing environmental impacts and increasing agricultural efficiency. Nanomaterials promote microbial growth, which is good for soil health. However, there are concerns about the long-term health and environmental impacts of nanomaterials that require risk assessments and regulatory regulation before they can be widely used. This chapter aims to discuss nanomaterials that might be used to combat soil contamination and enhance plant defensive mechanisms.

Keywords Soil contamination · Heavy metals · Defense mechanism · Nanomaterial · Crops

I. Gul · S. I. Mazhar
Institute of Environmental Sciences and Engineering, School of Civil and Environmental Engineering, National University of Sciences and Technology, Islamabad, Pakistan

M. A. Ali · M. Arshad (✉)
Department of Environmental Sciences, Bahauddin Zakariya University, Multan, Pakistan
e-mail: marshad@iese.nust.edu.pk

A. Husen (ed.), *Enhanced Crop Production Setup Under Various Micro and Nano Contaminants*, Smart Nanomaterials Technology,
https://doi.org/10.1007/978-981-96-6099-5_16

1 Introduction

Agriculture is considered one of the main sectors of global food security feeding an increasingly growing population of the globe. However, with increased soil, water, and air contamination, there are a lot of productivity challenges with crops. Environmental toxic substances such as heavy metals, pesticides, industrial chemicals, and many other pollutants have become ubiquitous and seriously threatened to be hazardous to the health and development of plants [1]. These harmful contaminants interfere with nutrient absorption, disrupt physiological functions, and contribute to crop yield reduction, thus, priority emphasis is assigned to the development of sustainable approaches to mitigating their effects. As dependence on agriculture increases to provide food for the world's population, it is critically important to search for innovative strategies to protect crops from damaging contaminants.

Interestingly, the defense mechanisms of plants against biotic stressors and abiotic environmental toxins are complex and multi-faceted. Defense mechanisms have been broadly grouped into two classes referred to as constitutive and inducible responses. Constitutive defenses relate to the structural barriers, such as cuticles, trichomes, and cell walls that present the first defense lines against environmental stress. Moreover, plants constantly release small amounts of defense-related enzymes and secondary metabolites that neutralize contaminants before they cause much damage. Inducible defense reactions in plants can be activated against stressful conditions; they are "induced" and are characterized by a chain reaction of molecular and physiological responses, including ROS production, activation of antioxidant enzymes, synthesis of stress-related proteins, and alterations in hormonal signaling pathways [2]. This should relate to the roles played by these responses, inhibition of damage, detoxification of harmful substances, and recovery of cellular homeostasis.

However, the protective mechanisms from the plants are generally not strong enough to counteract the levels of pollutants, especially in largely polluted areas. For instance, some of the heavy metals such as cadmium (Cd), lead (Pb), and arsenic (As) can be trapped within the tissues of the plant causing oxidative stress and interference with photosynthesis and plant growth and development within the tissues of the plant [3]. Overexposure to pesticides and herbicides may also lead to the building up of toxic residues in the soil that further weakens plant defenses, hence increasing susceptibility to stressors. Thus, it requires some unconventional approaches in order to enhance the resistance of plants against environmental contaminants. Nanotechnology is evolving as a wonderful tool to overcome such problems in agriculture. These materials provide some novel solutions for the management of plant defense mechanisms under contaminant stress. Nanomaterials possess some specific properties such as a high surface area, greater reactivity, and the ability to interact at the molecular level [4, 5]. Over the past years, there has been increased interest in developing nanomaterials to enhance crop stress tolerance, improve nutrient uptake, and decrease the toxic effect of contaminants in plants. The application of nanotechnology to agriculture has proved quite promising as many studies demonstrated improving growth, stress tolerance, and increased crop yields in adverse environmental conditions [6, 7]. Nanomaterials can be synthesized to interact with plants either by their size, composition, or surface chemistry. They can be applied directly onto the soil or sprayed directly on plant surfaces, but also through irrigation. Once inside the plant, nanomaterials can act as carriers for nutrients and growth regulators or other beneficial molecules. It has been shown that metal nanoparticles zinc (Zn), copper (Cu), and iron (Fe) enhance the activity of photosynthesis, assimilation of nutrients, and antioxidant levels in plants, hence making them survive better under oxidative stress resulting from contaminants in the medium [8]. Among the most noted impacts of nanomaterials on helping the plant overcome stress is its ability to regulate the production and scavenging of ROS. ROS is normally produced as a byproduct of cellular metabolism, which is involved in signaling and defense. However, exposure of organisms under stress conditions to heavy metals and other types of contaminants vastly increases ROS production, culminating in oxidative damage to proteins, lipids, and DNA [9]. Nanomaterials, specifically antioxidant materials, could also reduce oxidative stress by activating plant antioxidant enzymes such as superoxide dismutase (SOD), catalase (CAT), and peroxidase (POD) [10].

Nanomaterials can be applied in the targeted delivery of fertilizers, pesticides, and stress-relief agents to plants. These substances, applied using traditional methods, bring about massive losses because of leaching, volatilization, or degradation; thus, they become less effective besides increasing the adverse environmental influence [11]. In contrast, nanomaterials are designed to deliver these substances in precise and efficient manners, thus permitting assured delivery that shall target tissues and ensure controlled release. For example, fertilizer formulations released at a slow rate encapsulated within nanomaterials have proven to increase nutrient absorption, consequently reducing the ecological footprint of application [12]. This not only increases the rate of growth and stress tolerance but also reduces the likelihood of contamination of neighboring ecosystems.

There is a tremendous potential for nanomaterials in governing plant defense under contaminant stress; however, several challenges and concerns remain. Another major fear is that the nanomaterials will be de-positively introduced into the environment and, subsequently, the food chain, which will cause great harm to human health and the ecology. Though there have been so many proven benefits from the introduction of nanomaterials into agriculture, there is still very little known about the long-term health and environmental impacts. This therefore requires high-level safety assessments and regulatory frameworks to deem the use of nanomaterials in agriculture both safe and sustainable. This chapter highlights the contamination of agricultural soil and the use of nanomaterial to restore the contaminated soil. It also focused on the use of nanomaterials to enhance the defense mechanism of crops in contaminated soils.

2 Global Crops Production

Over the last twenty years, global production volumes of primary crops have consistently increased to meet the rising global demand (Fig. 1). From 2015 to 2022, there was a notable growth rate of 56%, driven by advancements in production technologies and intensified agricultural practices, including greater use of irrigation, fertilizers, pesticides, and high-yield crop varieties, as well as the expansion of arable land, despite challenges posed by climate change. Since 2000, the annual growth rate of crop production has experienced fluctuations, with peaks ranging from 4 to 6%. Notably, it recorded negative growth in 2009 (−0.4%) following the 2008 financial crisis and again in 2020 (−0.1%) during the COVID-19 pandemic. However, the downturn in 2009 was followed by a recovery in 2011, which saw the highest growth rate during this period at +5.7%. After 2013, growth rates gradually decreased, fluctuating between −0.1% in 2020 and 2.3% in 2017. The decline in 2020 was primarily attributed to decreased production of sugar beet and sugarcane, influenced by adverse weather, political factors, and the spread of the yellow beet virus, rather than the pandemic's economic impacts. In 2022, crop production growth reached + 0.7%, likely affected by market disruptions following the onset of the war in Ukraine and high inflation rates [13].

Globally and regionally (specifically Pakistan and neighboring countries), following crops are produced every year, wheat, rice, cotton, maize, sugar cane, soya bean, potato, tomato, canola, and sunflower. Over the past decade, global wheat production has steadily increased due to advances in agricultural technology and improved crop varieties. The production ranged from 713 million metric tons (MMTs) in 2015 to around 762 MMTs in 2022 as shown in Fig. 1. Major producers like China, India, and Russia contributed significantly to this growth, though climate change has caused periodic droughts and extreme weather events that led to regional

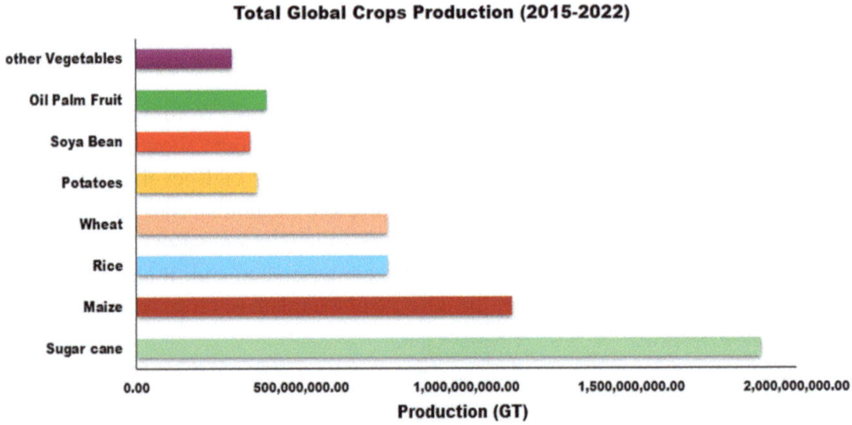

Fig. 1 Histogram of global primary crops production during 2015–2022 years [13]

declines. Global rice production grew steadily from 478 MMTs in 2015 to approximately 760 MMTs in 2022. The top producers, China and India, have maintained their dominance in production, while countries like Vietnam and Indonesia experienced moderate growth. Climate change, specifically rising temperatures and erratic rainfall, has challenged production in some regions. Maize production worldwide has consistently grown over the past decade, rising from 1,016 MMTs in 2015 to about 1,138 MMTs in 2022. The USA remains the largest producer, with Brazil and China also showing significant increases. Technological innovations, including genetically modified (GM) crops, have enhanced yields, but weather disruptions occasionally cause declines. Sugarcane production globally grew from 1.76 MMTs in 2015 to about 1.9 MMTs in 2022. Brazil remains the largest producer, followed by India and China. The sugar industry faced volatility due to fluctuating global sugar prices and changes in biofuel policies in major producing countries. Soybean production experienced substantial growth from 276 MMTs in 2015 to 347 MMTs in 2022, driven largely by increasing demand for livestock feed and vegetable oil. Brazil has emerged as the largest producer, overtaking the USA. The expansion of soybean farming has raised concerns over deforestation, particularly in South America. Potato production worldwide increased from 368 MMTs in 2015 to around 376 MMTs in 2022. China and India are the largest producers, with Europe also playing a significant role in global production. Rising demand for processed foods has contributed to stable production growth. Sunflower seed production grew from 40 MMTs in 2015 to about 56 MMTs in 2022. Ukraine and Russia have historically been the largest producers, but the 2022 Russia-Ukraine conflict caused disruptions in sunflower oil supply chains, impacting global markets. Global canola production has risen from 67 MMTs in 2015 to about 72 MMTs in 2022. Canada remains the top producer, followed by China and the European Union. Canola's importance as a source of biodiesel and vegetable oil has driven its production growth [13, 14]. In Pakistan, wheat production has seen some fluctuations but generally remained stable, ranging from 25 MMTs in 2015 to around 28.17 MMTs in 2023. Rice production in Pakistan has increased from 6.8 MMTs in 2015 to about 9.32 MMTs in 2023. In Pakistan, potato production grew from 3.8 MMTs in 2015 to 4.5 MMTs in 2023, with Punjab leading production. Improved varieties and better disease management practices have supported this growth, making Pakistan a notable producer in South Asia. In Pakistan, sunflower production has seen marginal growth from 8.6 MMTs in 2015 to around 12.40 MMTs in 2023 as shown in Fig. 2. Sunflower is being promoted for its high oil content, with initiatives in place to reduce Pakistan's dependency on imported edible oils [15].

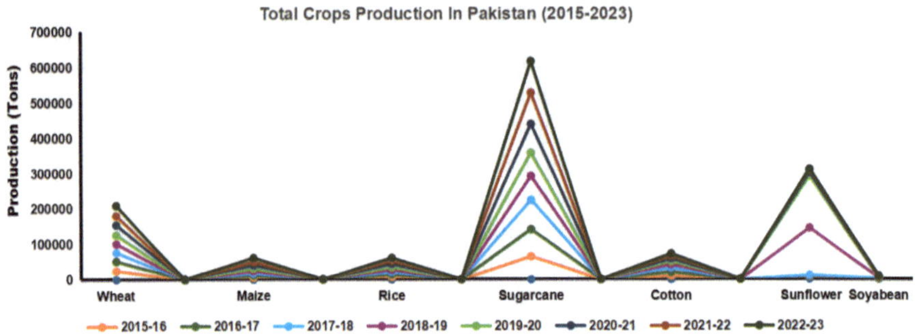

Fig. 2 Pakistan agricultural crops production during the years from 2015 to 2023 ([15]; production is in (000) metric tons and cotton production is in thousand bales of 375 lbs. each)

3 Agricultural Soil Contamination with Heavy Metals (HMs)

3.1 Global Sources of Heavy Metals in Soil

The primary drivers for uncontaminated HM emissions are natural sources, which may be additionally impacted by human activities, leading to increased human-related emissions. In those areas where there is a dominancy of economic growth/ industrial development, heavy metal concentration significantly increases. Markedly, in these regions like Europe (36.2%), Southern China (38.4%), Northern China (34.2%), the USA (33.4%), the Caribbean (31.2%), and the Middle East (30.5%), heavy metals contamination in soil and water bodies is primarily due to industrial emissions (Fig. 3). The proliferation of heavy industry and mining activities results in substantial environmental releases of toxic heavy metals, encompassing lead, mercury, cadmium, and arsenic, via industrial waste streams. Soil pollution in regions like India (32.9%), North China (32.6%), Africa (30.3%), and the Caribbean (30.3%) is significantly attributed to agricultural practices. Chronic irrigation with untreated industrial and domestic wastewater, coupled with improper fertilizer application, leads to the buildup of hazardous materials, including heavy metals (HMs) and other pollutants [16, 17]. This contamination is intensified using fertilizers containing hazardous elements like cadmium, lead, and mercury, pertaining to severe soil health risks, crop safety, and human well-being [18]. Moreover, the characteristic composition of soil mediums in regions for instance Africa (31.7%) and Australia/New Zealand (28.8%) contributes significantly to HM contamination. Heavy metals' high-level contamination in the topsoil natural sources is volcanic activity and rock weathering [19]. Inadequate use of fertilizer and its management further complexes this issue of contamination.

Atmospheric deposition linked to transportation is the main cause of soil heavy metal pollution in these regions; Brazil (33.1%), India (37%), South America (31.3%), and China (30.1%). The uncontrolled urbanization and increased number

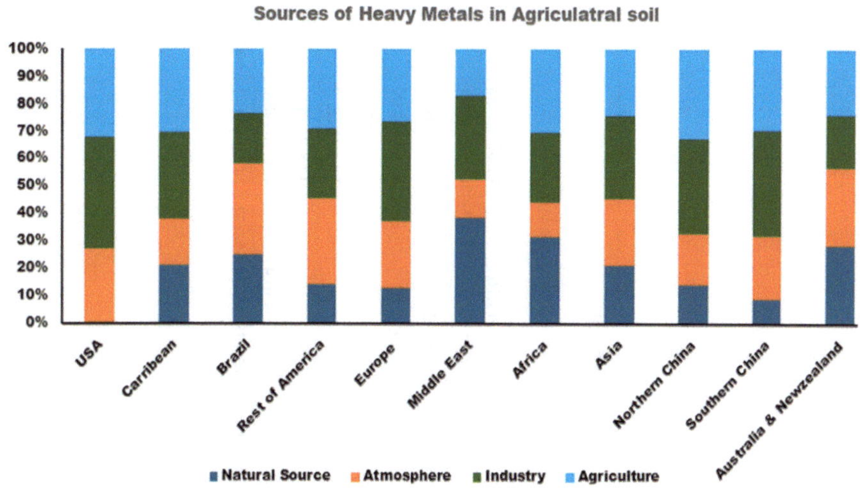

Fig. 3 Distribution of soil contamination through heavy metals in different regions of the world [17]

of vehicles in these regions further add to the problem of pollution [20]. The emission from these sources releases a variety of pollutants including nitrogen oxides and particulate matter, transported through air movement, and finally settles in the soil. This atmospheric transport process significantly contaminates the soil with harmful substances [21]. The startling increase in car emissions highlights the pressing need for focused actions to address this environmental issue.

3.2 *Heavy Metals Contamination in Agricultural Soil*

Table 1 illustrates the geographical distribution of heavy metals (HMs) in soil. Notably, HM levels in the Indian region exceed those in other countries and regions, with remarkably high concentrations of chromium (Cr), copper (Cu), and zinc (Zn) [22]. In contrast, zinc (Zn) displays minimal spatial variation. Manganese (Mn) and nickel (Ni) are also prominently present in Indian soils. While nickel (Ni) and lead (Pb) exhibit limited regional variation, Africa records the highest levels. Conversely, the USA has the highest concentrations of cadmium (Cd) and arsenic (As), consistent with existing literature on HM pollution [23]. Strikingly, US soil HM levels surpass those in other industrialized regions like Europe. The US' extensive industrialization and urbanization, as a global economic leader, have resulted in substantial emissions of metals like Pb, Cd, Cr, and Zn [24]. Indian region, an emerging economy, is experiencing rapid development, which may exacerbate HM pollution.

Indian region waste management infrastructure lags behind certain developed nations, resulting in heightened heavy metal (HM) influx into its soils [26]. Table 1

Table 1 Heavy metals distribution in contaminated soil in different regions of the world [25]

Heavy metals (HMs) concentration mg/kg	Indian region	Russia	Africa	North and South America	China	Australia	Europe
Cr	75.48–107	72.46–75.47	67.32–72.45	64.41–67.31	49.17–64.40	47.39–49.16	46.67–47.38
Cu	58.25–83.29	41.25–52.38	41.25–52.38	41.25–52.38	17.33–30.85	17.32	30.86–38.41
Ni	41.53–56.11	32.15–41.52	41.53–56.11	32.15–41.52	21.22–26.26	19.32–21.21	21.22–26.26
Pb	35.28–40.36	35.28–40.36	44.03–69.17	25.72–29.32	29.33–35.27	18.20	40.37–44.02
Zn	108.50–193.25	55.20–67.31	77.55–108.49	67.32–77.54	77.55–108.49	32.11	47.61–55.19
As	14.82–17.65	7.13–9.12	13.42–14.81	5.81–7.12	9.13–10.17	5.80	7.13–9.12
Cd	1.03–1.42	1.03–1.42	1.03–1.42	0.35–0.47	0.10–0.34	0.09	0.48–0.75
Mn	619.19–785.12	267.55–350.74	350.75–436.47	267.55–350.74	488.25–581.28	350.75–436.47	436.48–488.24
Fe	28,414.35–35,722.37	18,654.25–21,893	28,414.35–35,722.37	21,893.78–24,785.71	24,785.72–28,414.34	16,178.26–18,654.24	14,009.61–14,602.91

demonstrates the elevated HM concentrations in Indian soils, particularly zinc (Zn) and chromium (Cr). This contamination can be partly attributed to the Indian region's inherently high geological background levels, characterized by naturally elevated concentrations of specific HMs in the earth's crust. Additionally, regions rich in mineral resources, such as Africa, face unique challenges. Unregulated mining activities, exacerbated by inadequate local regulations, have led to increased levels of lead (Pb), nickel (Ni), and cadmium (Cd) in some African countries [27]. This highlights the need for effective environmental governance to mitigate HM pollution.

4 Plant Defense Mechanism

4.1 Morphological Defense Mechanism

Constitutive defenses are inherent protective measures that are always active, helping plants ward off threats even before they occur. The cell wall serves as the primary entry point for pathogens attacking plants. The cell wall is made up of cellulose, lignin, and pectin, all of these offer firmness and aid as obstruction against physical damage and microbial attack [28]. Throughout the co-evolution of plants and microbes, plants have developed a multilayered defense system, with the cell wall being a crucial element. Pathogens establish a relationship with host cells by evading the cell wall, which necessitates effective host recognition strategies and the development of infection structures or chemical defenses [29]. Microbes that cannot evolve effective methods to bypass the host plant's cell wall are typically classified as non-pathogens or poorly adapted pathogens. The waxy cuticle on the plant's exterior creates a hydrophobic layer that minimizes water loss and blocks pathogen entry as shown in Fig. 4a. Lignin and cellulose deposition on the plant cell walls and tissue of plants made them hard and tough and resistant to insect pest tearing by mandibles or penetration by ovipositors of insect pests as shown in Fig. 4b [30]. The leaves cell walls are also secured during pest attacks through the use of macromolecules, such as lignin, cellulose, suberin, and callose, coupled with small organic molecules, such as phenols, and sometimes inorganic silica particles. The development of trichomes involves the evolution of intricate plant defense strategies against insect pests [28]. These plant outgrowths have a variety of shapes/ morphology, ranging from straightforward and spiral forms to stellate, hooked, and glandular forms [31]. Insect motion, affection for the plant, shelter on the plant, feeding from the plant, and survival on the plant are predominantly obstructed by trichomes. Many plant species possess intricate vascular networks comprising laticifers and ducts resin, which store latex and resins under pressure. When damaged, these channels release their contents, potentially entrapping or intoxicating herbivores [32]. Plants have evolved unique defense mechanisms against insect pests, including the accumulation of specific minerals in their cuticles that hinder insect feeding. Research has identified deposits of silica and calcium carbonate in various plant species [28]. Mimicry and Camouflage strategic tactics of a plant enable it to prey on the invader easily and launch a surprise attack.

Research by [33] revealed that during infection, oligosaccharide elicitors are released from the host plant cell wall DAMPs/PAMPs as a part of triggering an immune response. These elicitors are recognized by plasma membrane-bound immune receptors, activating signaling cascades that activate various DAMP/PAMP-activated immunities. Defense system, similarly known as Defense Triggered Immunity (DTI) or PTI, respectively as shown in Fig. 5 [34]. Elicitors can induce the thickening of outer layer parenchyma cells and the production of amorphous, fibrillar material to entrap bacteria. Conversely, in response to fungal attacks, plants produce

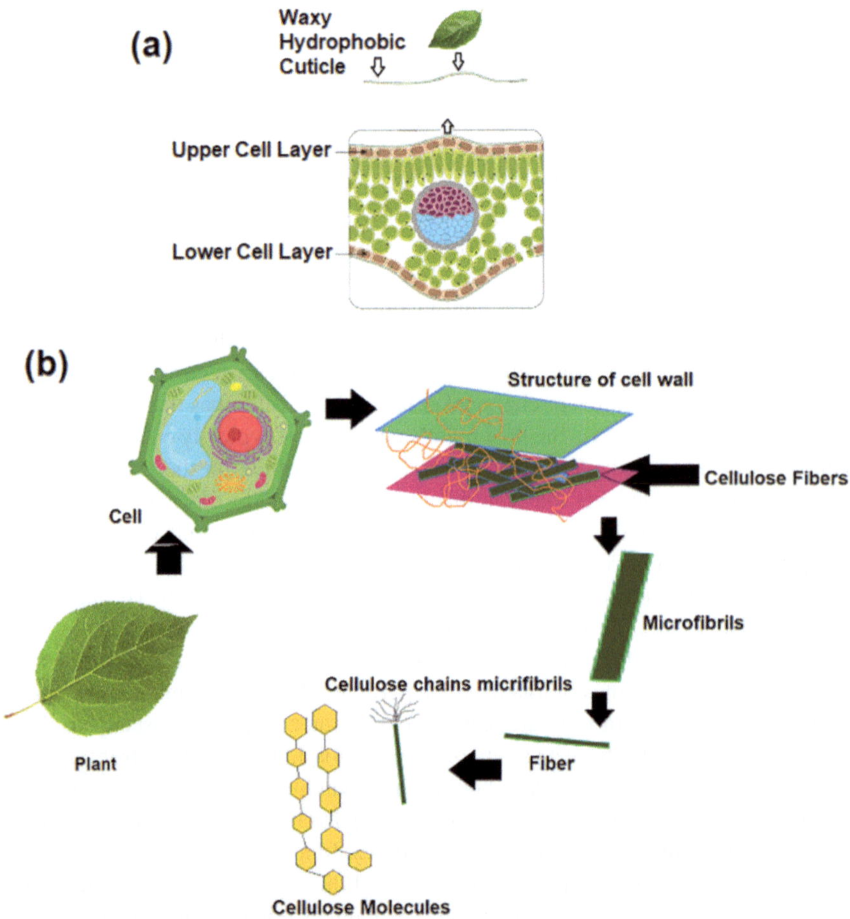

Fig. 4 **a** Illustration showing the structure of leaf tissues, including the lower and the upper cell layers with a waxy cuticle layer at the top, **b** structure of plant cell wall and how they are formed from cellulose

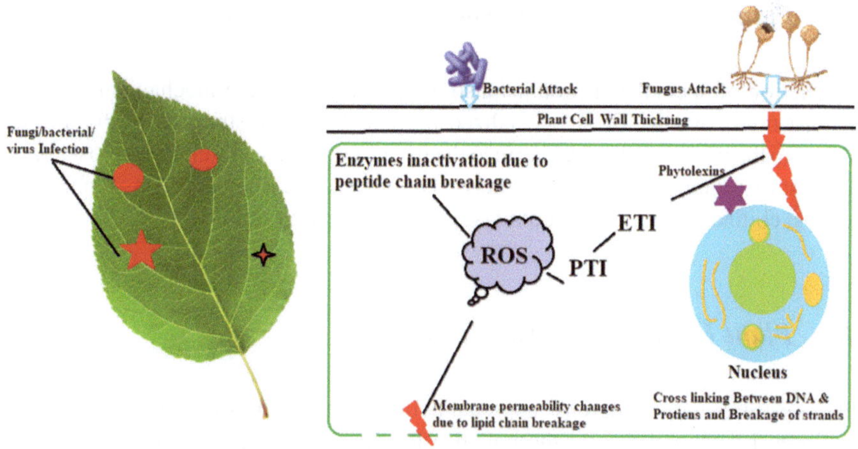

Fig. 5 Schematic representation of molecular and cellular defense mechanism of plant cells in response to pathogens

and deposit defensive substances such as callose, hydroxyproline-rich glycoproteins, phenolic compounds (including lignin and suberin), and mineral elements (like calcium and silicon) into the cell wall to provide protection.

4.2 Biochemical Defense Mechanisms

Plants synthesize a diverse array of biochemical compounds, categorized into primary and secondary metabolites. Primary metabolites play a crucial role in plant growth, development, and reproductive processes. Conversely, secondary metabolites serve as a defense mechanism, safeguarding plants against herbivores and pathogens. These compounds exhibit dual functionality, repelling generalist insects while attracting specialist insects. Notably, toxic secondary metabolites can incapacitate non-specialized herbivores, whereas specialized insects must allocate resources to detoxify these compounds, resulting in impaired growth and development [35]. The list of players that play an important role in biochemical defense mechanisms includes alkaloids, benzoxazinoides, cyanogenic glucosides (CNglcs), glucosinolates (GSLs), phenols, and terpenoids.

4.3 Plant Defense Mechanisms Based on Proteins

Production of specialized proteins also aids in the defensive mechanism of plants against insect attack. These specialized proteins help in restricting the food availability to the insect or physical harm to the insect's digestive system. The most important list of protein players in this defensive process includes α-amylase inhibitors, lectins, chitinases, polyphenol oxidases, and proteinase inhibitors [36].

4.4 Plant Indirect Defense Mechanisms

Plants practice indirect defense approaches by taking benefits of beneficial organisms to diminish herbivore attacks through the following steps like production of volatiles, extra-floral nectar, food bodies, and sheltered sites [37]. Indirect defense plays an essential role in defending plants against herbivore attacks. This defense mechanism involves both the enhancement of attraction and retention of enemies, which prey on herbivores [38] These defenses can either be innate or induced in response to mechanical damage and herbivore-derived elicitors [39].

4.5 Plants Resistance Through Hormones

Plant defense against herbivore attacks relies on complex signaling networks orchestrated by plant hormones, which play a pivotal role in regulating growth, development, and defense mechanisms. Following herbivore damage, various plant hormones facilitate intra- and inter-plant communication, triggering a coordinated response. Key signaling pathways mediated by jasmonic acid (JA), salicylic acid (SA), and ethylene are primarily responsible for activating plant defense responses to insect attacks [9]. Upon wounding or insect feeding, these pathways selectively induce specific defense-related genes, mobilizing the plant's defense machinery as shown in Fig. 6.

5 Nanomaterials in Agriculture

Nanomaterials of different types are used in agriculture to promote crop yields, minimize agrochemical use, and alleviate the effects of environmental pollutants.

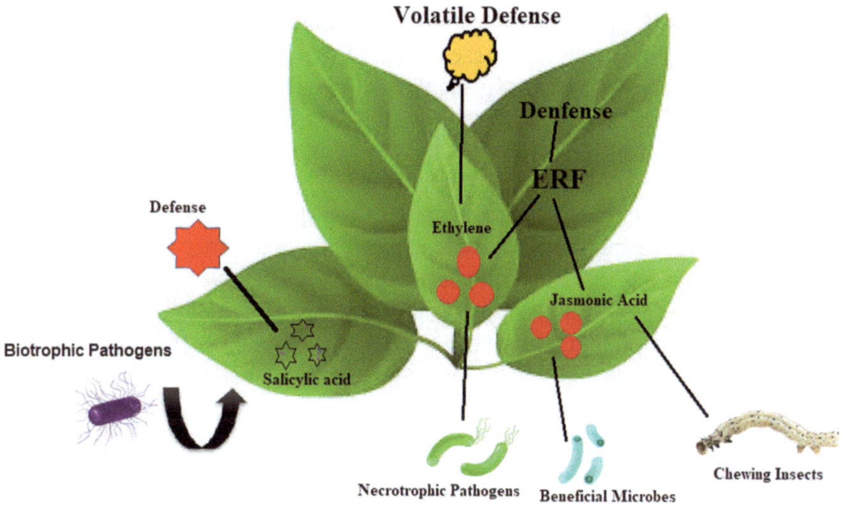

Fig. 6 Schematic representation of plant-induced defense mechanism by releasing ethylene, salicylic acid, and jasmonic acid

5.1 *Metal-Based Nanoparticles*

Zinc-based, iron-based, silver-based, copper-based, and titanium-based nanoparticles are among the most widely studied. These provide increased nutrient uptake, and enhanced growth, and elevate the plant's inherent defense mechanisms against biotic and abiotic stressors.

Zinc oxide (ZnO) nanoparticles have been proven to enhance plant growth and development by enhanced photosynthesis and nutrient uptake. Zinc is an essential micronutrient for plant growth and development; its deficiency leads to stunted growth, apart from lower yields. ZnO nanoparticles are extremely bioavailable zinc forms that plants can easily absorb through roots. Raliya et al. [40] have shown that foliar application of ZnO nanoparticles enhanced the chlorophyll content and improved nutrient uptake for enhanced wheat yield through enhanced water use efficiency. Iron oxide nanoparticles (Fe_3O_4) are used in agricultural practices to overcome iron deficiency in plants, which may cause chlorosis, reducing photosynthesis. The distribution of iron can be made by Fe_3O_4 nanoparticles in an efficient manner compared to that of usual chemical fertilizers. According to [41], peanut plants with enhanced plant growth and higher chlorophyll content exhibited increased yields.

It has been proven that silver nanoparticles (AgNPs) have antimicrobial activities, which leads to their use in controlling plant pathogens. It has been shown the AgNPs-reduced growth of particular bacterium, fungi, and viruses that prevents certain plant diseases. Jo et al. [42] found that AgNPs could be very useful in inhibiting the phytopathogenic fungus, which attacks tomatoes and causes late blight. The incidence of late blight has come down significantly by the application of AgNPs.

CuNPs are reportedly under study against their fungicidal as well as bactericidal action. The investigation proved that the fungicidal activity of CuNPs is more potent than those copper-based fungicides available on the market. It has been demonstrated that CuNPs inhibit the mycelial growth of pathogens in the roots of crops such as *Fusarium oxysporum* and *Aspergillus niger* while reducing root rot as well as other fungal diseases drastically [43].

Another class of metal-based nanomaterials that recently gained attention is titanium dioxide (TiO_2) nanoparticles. These can potentially enhance photosynthesis in crops. Research has indicated that TiO_2 nanoparticles enhance photosynthetic activity through increased light absorbance in plants. The incorporation of TiO_2 nanoparticles into the growth medium improved the growth and yield of spinach as reported by [44]. It recorded an increase in light intake, intensified activity of photosynthetic enzymes, and enhanced plant growth.

5.2 *Carbon-Based Nanomaterials*

Carbon nanomaterials, including carbon nanotubes and graphene oxide, have been studied for agriculture applications. These materials possess some electrical, thermal, and mechanical properties that could be beneficial for increasing crop yield, stress tolerance, and plant growth. CNTs have shown the capability to accelerate seed germination, root growth, and nutrient absorption of several crops. Khodakovskaya et al. [45] found CNTs to enhance tomato seedlings' growth by increasing the absorption of water and nutrient intake by roots. The permeation ability of CNTs in the plant cell wall allows for the plant to absorb water and nutrients at an efficient rate to the different tissues. Other than growth, CNTs increased the resistance of plants to drought stress through increased retention of water in the soil. GO is particularly interesting due to its growth-enhancing properties and protective properties against environmental stresses. GO helps not only improve the retention of water in the soil but also nutrient availability and increases stress-related protein production in plants. Recently, the work conducted by [10] proved that GO nanoparticles are effectual, as wheat plant growth was improved along with its drought tolerance due to an enhancement of water uptake and the expression of stress-related genes.

5.3 *Polymer Nanomaterials*

Polymer nanomaterials have been used in agriculture for the controlled release of fertilizers, pesticides, and other agrochemicals. Chitosan nanoparticles and nanogels are two of them. They encapsulate active ingredients into their matrix, and in due course, these may often gradually release those active ingredients. Hence, there is no need for frequent applications and therefore less chance of polluting the environment. Many studies on chitosan nanoparticles have been performed due to the potential of using these nanoparticles in sustainable agriculture. Chitosan is nontoxic and has antimicrobial activity and elicits plant defense responses, therefore providing a very effective means of protecting crops against pathogens. In addition, chitosan nanoparticles can also be used for controlled release delivery of fertilizers and pesticides, which would minimize the consumption of chemicals. Liu et al. [46] demonstrated that chitosan nanoparticles enhanced growth and disease resistance in tomatoes due to the reduction in the use of synthetic pesticides.

The third classification of the nanomaterials in the family polymer-based is nanogels, or nanoscale hydrogels, that can be used in controlled release formulations of agrochemicals. Fertilizers, pesticides, and growth regulators can be encapsulated by nanogels, hence their gradual release due to environmental conditions. The application of nanogels has brought improvement in efficiency in delivering pesticides to cotton plants and minimized the overall dosage of pesticides needed to control insect pests, according to [3]. Table 2 summarizes the role of different nanoparticles in enhancing crop growth.

6 Managing Abiotic Stress with Nanomaterial

6.1 Nanomaterials in Drought Stress Management

Drought is one of the most severe abiotic stresses occurring in crops. This abiotic stress is caused by water deficiency, thereby reducing photosynthesis and cell turgor pressure and causing oxidative damage in the plant. Nanomaterials have promising applications in modulating drought tolerance in crops through various mechanisms, including better retention of water, stomatal closure, and activation of antioxidant defense systems. ZnO-NPs enhance water conservation in crops by enhancing the synthesis of Osmo protectants, which sustain plant water status when they are under stress. Tripathi et al. [10] reported through their study that foliar spray of ZnO-NPs on wheat enhanced water use efficiency and exhibited reduced oxidative stress conditions when antioxidant enzyme activities, such as SOD and CAT, were enhanced. Silicon nanoparticles (Si-NPs) have gained abundant attention in drought stress management. Silicon was reported to enhance the drought tolerance capability of an organism through an increase in cell wall strength and a decrease in water loss via stomata. Siddiqui et al. [62] have reported that the application of Si-NPs improves water retention and growth in rice under drought conditions. These water uptake and osmotic adjustment stress-responsive genes were regulated in Si-NPs, thus showing an improvement in plant ability to endure long periods of water deficit. Carbon nanotubes have been demonstrated to improve the water uptake of a plant through enhancement of root growth and subsequent intake of water. Khodakovskaya et al. [45] indicated that drought stress could be improved in tomato plants with CNT treatment by developing better root elongation and improved water use efficiency that would enable the plants to respond effectively under drought. Besides, CNTs

Table 2 Nanomaterial application in different crops for enhanced production

Nanomaterial	Application in crop production	Effects on crops	Key references
Silver–silica nanocomposites	Antimicrobial applications for fruit crops (e.g., strawberry)	Reduced microbial contamination, extended shelf life of fruits	De Oliveira et al. [47]
Cerium oxide (CeO_2) nanoparticles	Soil application for stress management in soybean	Reduced oxidative stress, improved yield under abiotic stress conditions	[48]
Gold nanoparticles (AuNPs)	Foliar application to Arabidopsis and maize for enhanced growth	Increased photosynthetic efficiency, enhanced root and shoot development	[49]
Magnesium oxide (MgO) nanoparticles	Fertilizer application in maize	Improved germination rates, enhanced seedling vigor and nutrient uptake	[50]
Titanium dioxide (TiO_2) nanoparticles	Enhancing photosynthesis in spinach	Increased photosynthetic efficiency, improved light absorption	[51]
Chitosan nanoparticles	Foliar application in rice, control of fungal pathogens	Enhanced disease resistance, reduced fungal infections	[52]
Manganese dioxide (MnO_2) nanoparticles	Soil application for stress alleviation in wheat	Enhanced antioxidant defense, improved resistance to environmental stress	[53]
Silicon dioxide (SiO_2) nanoparticles	Application in rice for improved pest resistance	Enhanced resistance to insect pests, improved growth and development	[54]
Boron nitride (BN) nanoparticles	Enhancing nitrogen fixation in leguminous crops	Increased nitrogen fixation, improved productivity in legumes	[55]
Graphene oxide (GO) nanoparticles	Drought stress management in wheat	Improved drought tolerance, enhanced water retention, upregulation of stress-related genes	[56]
Copper nanoparticles (CuNPs)	Antimicrobial treatment for *Phytophthora infestans* in potatoes	Suppressed fungal growth, reduced plant disease severity	[57]

(continued)

Table 2 (continued)

Nanomaterial	Application in crop production	Effects on crops	Key references
Iron oxide (Fe_3O_4) nanoparticles	Soil application to wheat and spinach	Enhanced photosynthesis, improved iron uptake, increased root and shoot biomass	[58]
Zinc oxide (ZnO) nanoparticles	Foliar application to tomato plants	Enhanced growth, improved zinc uptake, increased chlorophyll content	[59]
Silver nanoparticles (AgNPs)	Antimicrobial treatment for rice and wheat crops	Reduced fungal infections, enhanced resistance to bacterial pathogens	[60]
Carbon nanotubes (CNTs)	Seed treatment in maize	Improved seed germination, enhanced root elongation, increased water and nutrient uptake	[61]

enhance aquaporin expression, which is the protein mediating the transport of water across the cell membrane.

6.2 *Nanomaterials in Salinity Stress Management*

Soil salinity is another major abiotic stress that causes devastating effects on crop production, especially in arid and semi-arid regions. High salt concentrations in soil cause ion toxicity, osmotic stress, and oxidative damage in plants. Nanomaterials have been considered for their plant growth-promoting activities that improve nutrient uptake enhance ion homeostasis and reduce the oxidative stress of a plant suffering from salinity stress. Iron oxide nanoparticles (Fe_3O_4-NPs) have been shown to improve salt tolerance in crops by alteration in ion transport and nutritional availability. In the paper presented by [58] on spinach, Fe_3O_4-NPs were found to increase the uptake of necessary nutrients, namely potassium and calcium but decreased the accumulation of toxic sodium ions. This helped in maintaining ion homeostasis within the plant, thus having minimal salt stress that transpired with an impediment in growth and development.

Salinity stress research also has been carried out on the role of MgO-NPs. Mg is crucial for chlorophyll biosynthesis and photosynthesis and can cause serious damage to plant growth in the case of deficiency under salt stresses. It has been reported that

MgO-NPs enhanced the chlorophyll content, resulting in increased photosynthesis efficiency in maize at salt stress conditions [50]. Graphene oxide nanoparticles have recently emerged as the newly discovered material that helps in solving the salinity stress problem in crop plants. GO-NPs are used for carrying nutrients and water because of increased tolerance in plants toward salty environments. Mukherjee et al. [56] reported that GO-NPs improved the germination and growth of wheat at higher salinity levels, this was due to the improvement in water retention through altered expression of stress-related genes.

6.3 Nanomaterials in Temperature Stress Management

Generally, it has been reported that temperature stresses which include both heat and cold stresses have been known to impact positively the growth and yield of plants. New research studies on nanomaterial potential established that these materials could play a profound role in making plants more resistant to temperature stresses through modifications of pathways of stress signaling, protection of cellular structure, and increase of nutrient supply.

AgNPs improve crop thermal tolerance by stimulating the antioxidant defense system. The AgNPs have the ROS scavenging potential, which is generated in response to thermal stress, hence preventing oxidative damage to cellular components. Recently, [63] observed that AgNPs improved the tolerance of wheat plants to heat by increasing the activities of antioxidant enzymes along with lipid peroxidation levels. Gold nanoparticles (AuNPs) have gained much attention and have been proposed as an agent for cold stress alleviation in crops. The suitability of AuNPs is their ability to effectively regulate the stress-activated signal transduction network and enhance photosynthetic efficiency at a low temperature. Tripathi et al. [64] demonstrated that foliar spraying with AuNPs increased chlorophyll content while reducing oxidative stress in maize, leading to enhanced crop cold tolerance.

Chitosan nanoparticles also increase crop tolerance to temperature stress. Chitosan, as a biopolymer of chitin, is widely applied in agriculture due to its biocompatibility and bioactivity. Chitosan nanoparticles can activate the heat shock proteins, which play an important role in protecting plants from damage caused by heat. A study by [52] was important to demonstrate the potential of chitosan nanoparticles for a significant increase in HSPs expression and antioxidant activity to facilitate heat tolerance in tomato plants.

6.4 Nanomaterials in Heavy Metal Stress Management

Heavy metal contamination of soil, mainly due to excessive application of chemical fertilizers and various industrial practices, has been a major environmental hazard throughout the world. Toxic concentrations of heavy metals such as cadmium (Cd), lead (Pb), and arsenic (As) resulted in growth retardation and chlorosis in the soil. Nanomaterials are capable of acting as an antidote to heavy metal stress by immobilizing heavy metals and, consequently, reducing their bioavailability, thus promoting the antioxidant defense mechanism in plants. Heavy metal antidotes applied to crops against the toxicity of heavy metals include silicon dioxide nanoparticles or SiO_2-NPs. They act as neutralizing agents that douse free, ionic forms of heavy metals in the soil and significantly inhibit their uptake by the plants. Cadmium and arsenic reduced uptake in rice consequently improving growth and reducing oxidative stress according to the study of [54].

There has been much research into the nature of cerium oxide nanoparticles (CeO_2-NPs), particularly on their antioxidant property and ROS scavenging. CeO_2-NPs have been reported to alleviate heavy metal stress-induced oxidative damage both by scavenging ROS and the induction activity of antioxidant enzymes. Fan et al. [48] reported that the applications of soybean plants under cadmium stress by CeO_2-NPs brought about diminished oxidative damage and enhanced growth. Magnetite nanoparticles (Fe_3O_4-NPs) also have been shown to decrease the availability of heavy metals in the soil. Fe3O4-NPs bind with heavy metals, which thus reduces mobility and uptake by plants. He et al. [58] observed that the uptake of lead in spinach decreases, improvement in growth conditions with reduced oxidative stress caused by Fe_3O_4-NPs. Table 3 summarizes the application of nanomaterials in reducing heavy metals stress and enhancing crop growth.

7 Defense Mechanism in Plants Using Nanomaterials Under Heavy Metals Stress

Nanomaterials, due to their high surface area, small size, and unique physicochemical properties, have shown promise in mitigating heavy metal stress by immobilizing metals, reducing their bioavailability, enhancing antioxidant defenses, and activating stress-responsive pathways in plants. Below is an overview of how crops defend themselves against heavy metal stress when exposed to nanomaterials, detailing the various steps and processes involved.

7.1 Heavy Metal Uptake and Initial Stress Response

In crop exposure to heavy metals, the results were that the absorbed metal is transferred through the roots into other tissues within the plant. Once inside the plant, heavy metals interfere with cellular processes by binding with essential biomolecules, causing oxidative stress, deactivation of enzymes, and damage to proteins, lipids, and DNA.

Table 3 Application of nanomaterials to enhance crop growth in heavy metals contaminated soil

Nanomaterial	Dose of nanomaterial	Size of nanomaterial	Heavy metal	Crop	Impact on crop	Key references
Silver nanoparticles (AgNPs)	50 mg/L	40 nm	Chromium (Cr)	Cucumber	Enhanced growth and reduced Cr toxicity	[65]
Gold nanoparticles (AuNPs)	10 mg/L	30 nm	Cadmium (Cd)	Arabidopsis	Enhanced growth, increased phytochelatin synthesis	[49]
Cerium oxide nanoparticles (CeO_2)	200 mg/L	15 nm	Cadmium (Cd)	Soybean	Alleviated oxidative stress, improved photosynthesis	[48]
Silicon dioxide nanoparticles (SiO_2)	100 mg/kg	50 nm	Cadmium (Cd)	Rice	Reduced Cd uptake, enhanced growth and antioxidant activity	[54]
Chitosan nanoparticles	200 mg/L	60 nm	Cadmium (Cd)	Tomato	Increased ABA and stress-related gene expression	[52]
Titanium Dioxide nanoparticles (TiO_2)	100 mg/L	25 nm	Cadmium (Cd)	Spinach	Improved photosynthetic efficiency and growth	[51]
Magnetite nanoparticles (Fe_3O_4)	50 mg/kg	20 nm	Lead (Pb)	Spinach	Improved growth, reduced Pb accumulation	[58]
Titanium dioxide nanoparticles (TiO_2)	100 mg/L	25 nm	Lead (Pb)	Maize	Improved photosynthetic efficiency and growth	[62]

Nanomaterials and regulation of heavy metal uptake: Nanomaterials can be a key player at this level by reducing the uptake of toxic heavy metals. For example, silicon dioxide nanoparticles have been found to reduce the bioavailability of heavy metals such as cadmium and arsenic in rice by immobilizing these metals into the soil, thereby excluding their entry into the plant [54]. For instance, magnetite nanoparticles or Fe3O4-NPs could immobilize heavy metals within the soil, resulting in limited uptake of such crops such as spinach and wheat [58]. Chelation and compartmentalization: Once heavy metals penetrate into the plant, they become chelated by organic molecules, which include both phytochelatins and glutathione. The metal-binding chelating molecules counteract the toxic effect of heavy metals by making complexes, which are then compartmentalized in vacuoles to keep them from interfering with other cellular functions. It has been shown by studies that nanomaterials such as cerium oxide nanoparticles (CeO_2-NPs) enhance the synthesis efficacy of these chelators, which in turn enhances the sequestration of the heavy metals inside the vacuoles, and this is the earliest defense mechanism during metal detoxification processes [48].

7.2 *Oxidative Stress and ROS Scavenging*

One of the most lethal effects of heavy metal stress on plants is the formation of ROS such as superoxide anions (O^{2-}), hydrogen peroxide (H_2O_2), and hydroxyl radicals (OH•). ROS is highly reactive and causes oxidative damage to cellular constituents such as proteins, lipids, and nucleic acids. The buildup of ROS reflects oxidative stress, which hinders plant development and productivity.

Nanomaterials as antioxidants: Nanomaterials showed a pronounced enhancement in the antioxidant defense system of plants that scavenge excess ROS and mitigate oxidative damage. For example, the scavenging activity for ROS is recognized in cerium oxide nanoparticles (CeO_2-NPs) due to its dual oxidation state, such as Ce^{3+} and Ce^{4+}. Thus, CeO_2-NPs behave as antioxidant enzymes, such as superoxide dismutase (SOD) and catalase (CAT), by decreasing the concentrations of ROS thus protecting plant cells from oxidative stress damage [48].

Soliman et al. [63] demonstrated that AgNPs increased the production of antioxidant enzymes in a wheat plant treated with heavy metals, which lowered the content of ROS and lipid peroxide. In addition, nanomaterials assist in enhancing the antioxidant defense of the plant by maintaining cellular redox balance, which is critical for survival under heavy metal stress.

7.3 *Activation of Stress-Responsive Signaling Pathways*

The activation of diverse stress-responsive signaling pathways in plants involved modulation of gene expression, hormone production, and synthesis of protective proteins that would ultimately cope with heavy metal stress. Nanomaterials can interact with the signaling pathway to better help plants cope with heavy metal stress.

These include hormones such as abscisic acid (ABA), ethylene, and jasmonic acid (JA) that activate the expression of genes related to stress and induce the production of protective proteins—HSPs and metallothioneins that bind and detoxify metals.

Chitosan nanoparticles represent a polymer-based nanomaterial, which is also reported to produce ABA and activate the stress-responsive genes in crops upon the stress induced by heavy metals. Chitosan nanoparticles increased the expression of a metal detoxification-related gene and tolerance to the stress in a tomato treated with cadmium, which implies that chitosan nanoparticles could influence the hormonal signaling pathways associated with the improvement of the adaptive response of the plant toward the heavy metal stress.

7.4 Enhancement of Photosynthesis and Growth

Heavy metals will severely inhibit photosynthesis through chloroplast damage and interfere with the synthesis of chlorophyll, thereby weakening the electron transport chain. This affects the photosynthetic efficiency of the plant and, subsequently, its growth and yield. Nanomaterials have been found capable of alleviating the photosynthetic inhibition of heavy metals as they protect the chloroplasts and enhance the photosynthetic machinery.

Silicon nanoparticles (Si-NPs) and photosynthesis: Silicon nanoparticles have been extensively explored to enhance photosynthesis in heavy metal-stressed conditions. Si-NPs in lead-exposed maize plants enhanced chlorophyll content, and photosystem II efficiency, and suppressed chloroplast damage [62]. The leaf accumulation of heavy metals reduced by Si-NPs has stopped direct toxicity to the photosynthetic apparatus. Si-NPs also increase photosynthetic efficiency and assist in maintaining growth and productivity even under metal stress.

Titanium dioxide nanoparticles (TiO_2-NPs) have also been shown to promote photosynthesis in plants that are under heavy metal stress. In an experiment, [51] proved that treatment of cadmium-stressed spinach plants with TiO_2-NPs increases the rate of photosynthesis and the chlorophyll content, in turn increasing growth and biomass yield.

7.5 Cellular Detoxification and Metal Sequestration

Detoxification and sequestration of metals to prevent interference in essential cellular activities mark the final phase of plant protection mechanisms against heavy metal stress. This occurs through the production of metal-binding proteins and through the further compartmentalization of metals into vacuoles with reduced toxicity of metal complexes to cells.

Nanomaterials and detoxification pathways: Nanomaterials will fortify these detoxification pathways by increasing the synthesis of metal-binding proteins, especially metallothioneins and phytochelatins. AuNPs have been shown to stimulate upregulation of the expression of metal detoxification-related genes. In Arabidopsis, AuNPs induced the synthesis of phytochelatins, which were shown to bind cadmium and lead, thereby sequestering them in vacuoles and minimizing their toxicity [49]. For the past couple of years, magnetite nanoparticles have been used to sequester heavy metals into vacuoles, hence controlling the toxicity inside the cytoplasm. To confirm this, [58] showed that Fe_3O_4-NPs enhanced the vacuolar sequestration of lead in spinach which was followed by securing the cytoplasm against oxidative damage, most importantly enhancing the status of the plant.

7.6 Long-Term Adaptation and Cross-Tolerance

The most interesting feature of plant responses to heavy metal stress is cross-tolerance, which is the phenomenon in which exposure to one type of stress increases tolerance to other types of abiotic stress. Nanomaterials have been shown to induce cross-tolerance in plants through the modulation of common pathways of stress response. For example, CNTs enhanced the resistance of maize to heavy metals and drought by water uptake and the activation of stress-related genes for ROS scavenging as well as metal detoxification, according to [61]. The nature of such evidence supports the view that nanomaterials could bring a broad-spectrum protective effect to prevent plants from fighting many stresses simultaneously.

8 Risks, Challenges, and Future Perspectives

The integration of nanomaterials in agricultural practices raises concerns regarding their potential toxicity to plants and humans upon ingestion or absorption, potentially perpetuating accumulation in the food chain, especially when employed to remediate contaminants like lead, mercury, and arsenic. Moreover, nanomaterials may harm beneficial soil microorganisms and disrupt the ecological balance necessary for mitigating heavy metal contamination. Furthermore, nanomaterials' environmental persistence poses risks to ecosystems and may exacerbate the detrimental effects of heavy metals through interactions. Current regulatory frameworks and standards governing nanomaterial use in agriculture are inadequate, particularly regarding applications in heavy metal-contaminated soils. Key research priorities include: Developing cost-effective and scalable nanomaterial production methods, ensuring targeted delivery of nanomaterials to specific plant tissues or cells, elucidating interactions between nanomaterials, plants, and heavy metals, establishing guidelines for safe nanomaterial use in contaminated agricultural settings. Future research directions may focus on: targeted heavy metal removal and plant defense stimulation using nanomaterials, enhancing plant resilience to heavy metal stress and reducing chemical fertilizer/pesticide reliance, integrating nanomaterials with technologies like genetic engineering for enhanced crop protection, and developing nanomaterial-based sensors for heavy metal detection and plant health monitoring.

9 Conclusion

Nanotechnology is offering solutions to the challenges of agriculture through environmental contaminants that threaten crop productivity and food security. Although plants naturally defend themselves, this mechanism is ineffective at high contaminant levels. Nanotechnology is a promising intervention as it can make the plant defense stronger or at least tolerate a greater amount of contamination. Nanotechnology enhances nutrient uptake but reduces environmental impacts through increased efficiency in the delivery of fertilizers and pesticides. More investigation is, however, needed for long-term effects on health and ecosystems from nanomaterials. As a result, future research will be focused on safety assessments, regulatory frameworks, and the integration of nanotechnology with sustainable practices to have a more resilient agricultural system in the future.

Competing Interests The authors declare that they have no competing interest.

Declaration of Generative AI and AI-Assisted Technologies in the Writing Process During the preparation of this work, the book chapter author(s) have not taken any assistance from AI, and the book chapter author(s) reviewed and edited the content as needed and take(s) full responsibility for the content of the publication.

References

1. Gul I, Manzoor M, Hashim N, Shah GM, Waani SPT, Shahid M, Antoniadis V, Rinklebe J, Arshad M (2021) Challenges in microbially and chelate-assisted phytoextraction of cadmium and lead–a review. Environ Pollut 287:117667
2. Mittler R (2002) Oxidative stress, antioxidants and stress tolerance. Trends Plant Sci 7(9):405–410
3. Singh P, Kim YJ, Zhang D, Yang DC (2016) Biological synthesis of nanoparticles from plants and microorganisms. Trends Biotechnol 34(7):588–599
4. Arshad M, Nisar S, Gul I, Nawaz U, Irum S, Ahmad S, Sadat H, Mian IA, Ali S, Rizwan M, Alsahli AA, Alyemeni MN. 2021. Multi-element uptake and growth responses of Rice (*Oryza sativa* L.) to TiO_2 nanoparticles applied in different textured soils. Ecotoxicol Environ Saf 215:112149.
5. Waani SPT, Irum S, Gul I, Yaqoob K, Khalid MU, Ali MA, Manzoor U, Noor T, Ali S, Rizwan M, Arshad M (2021) TiO_2 nanoparticles dose, application method and phosphorous levels influence genotoxicity in rice (*Oryza sativa* L.), soil enzymatic activities and plant growth. Ecotoxicol Environ Saf 213:111977
6. Husen A (2022) Engineered nanomaterials for sustainable agricultural production, soil improvement and stress management. Elsevier Inc., USA. https://doi.org/10.1016/C2021-0-00054-7
7. Husen A (2023) Nanomaterials and nanocomposites exposures to plants (Response, interaction, phytotoxicity and defense mechanisms). Springer Nature Singapore Pte Ltd., Singapore. https://doi.org/10.1007/978-981-99-2419-6
8. Rico CM, Majumdar S, Duarte-Gardea M, Peralta-Videa JR, Gardea-Torresdey JL (2011) Interaction of nanoparticles with edible plants and their possible implications in the food chain. J Agric Food Chem 59(8):3485–3498

9. Gill SS, Tuteja N (2010) Reactive oxygen species and antioxidant machinery in abiotic stress tolerance in crop plants. Plant Physiol Biochem 48:909–930
10. Tripathi DK, Singh S, Singh VP, Prasad SM, Dubey NK, Chauhan DK (2017) Silicon nanoparticles more efficiently alleviate arsenic toxicity than silicon in maize cultivers. Front Environ Sci 5:1–10
11. Fraceto LF, Grillo R, de Medeiros GA, Scognamiglio V, Rea G, Bartolucci C (2016) Nanotechnology in agriculture: which innovation potential does it have? Front Environ Sci 4:20
12. DeRosa MC, Monreal C, Schnitzer M, Walsh R, Sultan Y (2010) Nanotechnology in fertilizers. Nat Nanotechnol 5:91–93
13. FAO (2023) Food and Agriculture Organization of the United Nations: Crop Production Data
14. USDA (2023) United States Department of Agriculture: World Agricultural Supply and Demand Estimates
15. Pakistan Bureau of Statistics (2023) Pakistan agricultural statistics
16. Dong B, Zhang R, Gan Y, Cai L, Freidenreich A, Wang K, Guo T, Wang H (2019) Multiple methods for the identification of heavy metal sources in cropland soils from a resource-based region. Sci Total Environ 651:3127–3138
17. Zhou L, Zhao X, Meng Y, Fei Y, Teng M, Song F, Wu F (2022) Identification priority source of soil heavy metals pollution based on source-specific ecological and human health risk analysis in a typical smelting and mining region of South China. Ecotoxicol Environ Saf 242:113864
18. Sakizadeh M, Zhang C (2021) Source identification and contribution of land uses to the observed values of heavy metals in soil samples of the border between the Northern Ireland and Republic of Ireland by receptor models and redundancy analysis. Geoderma 404:115313
19. Liu D, Ma J, Sun Y, Li Y (2016) Spatial distribution of soil magnetic susceptibility and correlation with heavy metal pollution in Kaifeng City, China. CATENA 139:53–60
20. Declercq Y, Samson R, Castanheiro A, Spassov S, Tack FMG, Van De Vijver E, De Smedt P (2019) Evaluating the potential of topsoil magnetic pollution mapping across different land use classes. Sci Total Environ 685:345–356
21. Kabata-Pendias A (2011) Trace Elements in Soils and Plants, 4th edn. CRC Press, Boca Raton, MA, USA ISBN 978-1-4200-9368-1
22. Yuan X, Xue N, Han ZA (2021) Meta-analysis of heavy metals pollution in farmland and urban soils in China over the past 20 years. J Environ Sci 101:217–226
23. Han I, Whitworth KW, Christensen B, Afshar M, An Han H, Rammah A, Oluwadairo T, Symanski E (2022) Heavy metal pollution of soils and risk assessment in Houston, Texas following hurricane Harvey. Environ Pollut 296:118717
24. Rafique N, Tariq SR (2016) Distribution and source apportionment studies of heavy metals in soil of cotton/wheat fields. Environ Monit Assess 188(5):309
25. Zhao X, Zang J, Ma R, Luo H, Wan T, Yu D, Hong Y (2024) Worldwide examination of magnetic responses to heavy metals pollution in agricultural soils. Agriculture 14:702
26. Adimalla N, Qian H, Wang H (2019) Assessment of heavy metal (HM) contamination in agricultural soil lands in northern Telangana, India: an approach of spatial distribution and multivariate statistical analysis. Environ Monit Assess 191:246
27. Jiang B, Adebayo A, Jia J, Xing, Y, Deng S, Guo L, Liang Y, Zhang D (2019) Impacts of heavy
28. Abebe W (2021) Review on plant defense mechanisms against insect pests. Int J Novel Res Interdiscip Stud 8:15–39
29. Turra D, ElGhalid M, Rossi F, DiPietro A (2015) Fungal pathogen uses sex pheromone receptor for chemotropic sensing of host plant signals. Nature 527:521–524
30. Raupp MJ (2008) Effects of leaf toughness on mandibular wear of the leaf beetle, *Plagiodera versicolora*. Ecol Entomol 10:73–79
31. Hanley ME, Lamont BB, Fairbanks MM, Rafferty CM (2007) Plant structural traits and their role in antiherbivore defense. Perspec Plant Ecol Evol Syst 8:157–178
32. Agrawal AA, Konno K (2009) Latex: a model for understanding mechanisms, ecology, and evolution of plant defense against herbivory. Annu Rev Ecol Syst 40:311–331

33. Boller T, Felix G (2009) A renaissance of elicitors: perception of microbe—associated molecular patterns and danger signals by pattern-recognition receptors. Annu Rev Plant Biol 60:379–406
34. Jones JDG, Dangl JL (2006) The plant immune system. Nature 444:323–329
35. Kessler A, Baldwin IT (2002) Plant responses to insect herbivory: the emerging molecular analysis. Ann Rev Plant Biol 53:299–328
36. Rodrigues Macedo ML, Das Graças Machado Freire M (2011) Insect digestive enzymes as a target for pest control. Invertebr Surv J 8:190–198
37. Dicke M, Sabelis MW, de Jong M (1988) Analysis of prey preference in phytoseiid mites by using an olfactometer, predation models and electrophoresis. Exp Appl Acarol 5:225–241
38. Tibebu B (2018) Defense mechanisms of plants to insect pests: From morphological to biochemical approach. Trends Tech Sci Res 2(2):555584
39. War AR, Paulraj MG, Ahmad T, Buhroo AA, Hussain B, Ignacimuthu S, Sharma HC (2012) Mechanisms of plant defense against insect herbivores. Plant Signal Behav 7:1306–1320
40. Raliya R, Nair R, Chavalmane S, Wang WN, Biswas P (2015) Synergistic improvement organic matter content and nutrient uptake in wheat through the use of ZnO nanoparticles. Agric Food Chem 63(18):3821–3831
41. Khot LR, Sankaran S, Maja JM, Ehsani R, Schuster EW (2012) Applications of nanomaterials in agricultural production and crop protection: a review. Crop Protect 35:64–70
42. Jo YK, Kim BH, Jung G (2012) Antifungal activity of silver ions and nanoparticles on *Phytophthora infestans* causing late blight of potato. Plant Dis 93:1037–1043
43. Ramyadevi J, Jeyasubramanian K, Marikani A, Rajakumar G, Rahuman AA (2012) Synthesis and antimicrobial activity of copper nanoparticles. Mater Lett 71:114–116
44. Hong F, Yang F, Liu C, Gao Q, Wan Z, Gu F, Wu C, Yang P (2005) Influence of nano-TiO_2 on the chloroplast aging of spinach under light. Biol Trace Elem Res 104:249–260
45. Khodakovskaya MV, de Silva K, Biris AS, Dervishi E, Villagarcia H (2009) Carbon nanotubes induce growth enhancement of tobacco cells. ACS Nano 3:3221–3227
46. Liu Y, Huang W, Wang Y, Du Y, Li J (2017) Chitosan nanoparticles enhance the growth and disease resistance of tomato plants. Carbohydr Polym 172:327–334
47. de Oliveira Lima AM, Fragal EH, Caldas BS, Nakamura TU, Rubira AF, Silva R (2022) Functional mesoporous silica decorated with Ag nanoparticles as chemo-photothermal agents. Micropor Mesopor Mat 341:112097
48. Fan Y, Song Y, Xia X (2021) Cerium oxide nanoparticles enhance oxidative stress resistance in soybean under abiotic stress. J Plant Physiol 265:153480
49. Ferrari E, Barbero F, Busquets-Fité M, Franz-Wachtel M, Köhler HR, Puntes V, Kemmerling B (2021) Growth-promoting gold nanoparticles decrease stress responses in Arabidopsis seedlings. Nanomaterials (Basel) 11(12):3161
50. Kalteh M, Alipour ZT, Ashraf S, Aliabadi MM, Mosavi M (2021) Application of magnesium oxide nanoparticles improves seedling vigor and nutrient uptake in maize. Nanomaterials 11:852
51. Ma X, Geisler-Lee J, Lombi E (2020) Titanium dioxide nanoparticles in the soil-plant system. Environ Pollut 263:114281
52. Ghormade V, Deshpande MV, Paknikar KM (2020) Perspectives for nano-biotechnology enabled protection and nutrition of plants. Biotechnol Adv 40:107520
53. Adhikari T, Kundu S, Rao AS, Tarafdar JC (2020) Effect of manganese oxide nanoparticles on wheat: a case study on soil enzymes and stress tolerance. Agricult Res 9:280–286
54. Ahmed B, Chaudhary P, Singh PN, Rani P (2020) Silicon dioxide nanoparticles improve pest resistance in rice. Environ Nanotechnol Monit Manag 14:100346
55. Pereira AES, Oliveira HC, Fraceto LF (2019) Polymeric nanoparticles as an alternative for application of gibberellic acid in sustainable agriculture: a field study. Sci Rep 9(1):7135
56. Mukherjee A, Majumder D, Dhara S (2019) Graphene oxide enhances drought tolerance in wheat via improved water retention. Agronomy 9(8):411
57. Abdel-Aziz MS, Shaheen MS, El-Nekeety AA, Abdel-Wahhab MA (2019) Antifungal activity of copper nanoparticles against phytopathogenic fungi. J Plant Protect Res 59:1–13

58. He S, Yang X, Wang L, Zhu S (2019) Magnetite nanoparticles reduce lead uptake and improve growth in spinach. Sci Rep 9:1–9
59. Kumar P, Raliya R, Saharan V, Biswas P, Bhargava R (2019) Zinc nanoparticles promote growth in tomato plants via an increase in chlorophyll content. J Plant Physiol 234:30–36
60. Raliya R, Biswas P, Saharan V (2018) Silver nanoparticles reduce bacterial pathogens in wheat and rice plants. Plant Dis 102(2):254–260
61. Yan A, Cheng F, Sharma P, Chen X (2018) Carbon nanotubes boost seed germination and enhance nutrient uptake in maize. Plant Physiol 176(1):1216–1224
62. Siddiqui MH, Al-Whaibi MH, Faisal M (2018) Role of silicon in enhancing photosynthetic performance and mitigating lead toxicity in maize. J Plant Physiol 226:104–113
63. Soliman H, Hammad A, Hanafi A (2020) Silver nanoparticles mitigate heat and heavy metal stress in wheat. Environ Sci Pollut Res 27(19):23324–23335
64. Tripathi S, Sharma S, Suri S, Tiwari K, Tripathi DK, Sharma S (2024) Insights into physiological and molecular responses of plants under metal-nanoparticle stresses. In: Molecular and physiological insights into plant stress tolerance and applications in agriculture-part 2. Bentham Science Publishers, pp 147–173
65. Zia MH, Ali S, Khatak Z (2023) Mitigating cadmium accumulation and toxicity in plants: the promising role of nanoparticles. J Hazard Mater 445:130084

Safety Issues Associated with Micro and Nanocontaminants

Dipanjan Das, Tapas Pal, Rittick Mondal, Pankaj Mandal, and Amit Kumar Mandal

Abstract Micro and nanoplastic pollution poses significant environmental and health risks worldwide. Effective policies are crucial to mitigate this issue. This study evaluates policy effectiveness, highlighting successes and challenges. Extended Producer Responsibility, single-use plastic bans, and waste management regulations demonstrate potential. However, lack of standardization, inadequate enforcement, and limited scope hinder progress. Best practices, including integrated approaches, stakeholder engagement, and continuous monitoring, enhance policy impact. Case studies from Sweden, Rwanda, and California illustrate successful policy implementation. Recommendations for harmonized global regulations, robust enforcement, and international cooperation are presented. This research informs policymakers and stakeholders on effective strategies to combat micro and nanoplastic pollution.

Keywords Microplastics · Nanoplastics · Policy effectiveness · Sustainability · Environmental governance

1 Introduction

In recent years, the escalating occurrence of micro- and nanocontaminants in our ecosystems, food supplies, and user products has raised substantial concerns on human health and ecosystems, despite the myriad benefits they provide. The rapid growth of industrial activities, technological advancements, and changing consumer behaviors have contributed to the widespread presence of these tiny pollutants, often invisible to naked eye, have been associated to various health issues, environmental

D. Das · R. Mondal · P. Mandal · A. K. Mandal (✉)
Department of Sericulture, Raiganj University, Raiganj, West Bengal, India
e-mail: amitmandal08@gmail.com

T. Pal
Department of Geography, Raiganj University, Raiganj, West Bengal, India

A. Husen (ed.), *Enhanced Crop Production Setup Under Various Micro and Nano Contaminants*, Smart Nanomaterials Technology,
https://doi.org/10.1007/978-981-96-6099-5_17

degradation, and economic losses. Micro and nanocontaminants pose unique challenges owing to their small size, making them problematic to identify and eliminate [1]. Similarly, environmental pollution from plastics, particularly microplastics (MPs), has developed as an important community problem. Upon disposal, larger plastic items undergo degradation into smaller fragments, classified as micro- and non-sized plastic particles [2]. MPs are specifically known as particles with diameters ranging from 1 μm to 5 mm, while nanoplastics (NPs) are characterized by sizes between 1 nm and <1 μm [3]. Research has detected MPs in both bottled mineral water—regardless of whether contained in plastic or glass bottles [4]—and in raw and drinking water sources [5]. The prevalence of MPs in sampled studies varied significantly, with occurrences reported between 24 and 100%, and concentrations fluctuating from 0 to 1,247 MPs/L. The predominant morphologies observed were fragments, followed by fibers [6]. Notably, there exists a paucity of reliable data regarding smaller NPs (20 μm); nonetheless, these particles are postulated to possess the most pronounced toxicological effects [7]. Human health risks associated with microplastics (MPs) arise from exposure via ingestion, inhalation, and dermal contact, with toxicity mechanisms encompassing oxidative stress, inflammation, and DNA damage, particularly impacting vulnerable populations [8]. Therefore, there exists an urgent imperative to evaluate the potential adverse effects of MPs on human health. Following exposure, MPs may exert effects locally within tissues or translocate to other sites, as inflammation typically enhances the permeability of epithelial barriers. Additionally, circulating MPs have been linked to a range of cardiovascular issues, including hypertension [9], thrombus formation [10], increased coagulability [11], and cytotoxicity to blood cells [12]. Due to their considerable surface area, MPs may function as carriers for oxidizing species adsorbed onto their surfaces. Economic burdens encompass healthcare costs, environmental remediation, and product recalls, while social challenges include public trust, confidence in regulatory frameworks, and disproportionate impacts on marginalized communities. Addressing these challenges requires a comprehensive approach integrating scientific research, policy development, industrial collaboration, and public engagement. This chapter explores the safety issues associated with micro and nanocontaminants, highlighting potential risks to human and environment.

2 Human Health Risks

Micro and nanocontaminants have become a pressing concern for human health, with far-reaching implications for our well-being [13]. These tiny particles, often measuring < 100 nm in size, can infiltrate our bodies through various means, including ingestion, inhalation, and skin contact [14]. Once inside, they can wreak havoc on our biological systems, triggering a cascade of adverse health effects (Fig. 1). One of the most alarming health risks associated with micro and nanocontaminants is their potential to cause cancer [15]. Previous study revealed that experience of these particles can result in DNA damage, disturbance of cellular signaling pathways, eventually increasing the risk of carcinogenesis [16]. For instance, microplastics have been found in tap water samples worldwide, while nanoparticles in personal care stuffs have been linked to an increased cancer risk [17]. Beyond cancer, micro and nanocontaminants also pose significant neurological and cardiovascular risks. These particles can induce neuroinflammation and oxidative stress. Moreover, they possess the potential to contribute to the development of Alzheimer's and Parkinson's diseases [18]. Moreover, MPs in drinking water and nanoparticles in air pollution have been connected to cardiovascular disease and mortality. Reproductive and developmental health is also vulnerable to micro and nanocontaminants. Hormone disruption, embryonic development impairment, and fetal toxicity have been reported [19]. Experience to microplastics during fetal development has been related to birth defects and reproductive issues, while nanoparticles have been shown to increase reproductive toxicity by 20% [1]. These results underscore the serious requisite for protective measures to safeguard human health. Moreover, the pervasive occurrence of micro and nanocontaminants in the environment intensifies their health effects. Air pollution, water contamination, and food exposure routes all contribute to our cumulative exposure. To mitigate these risks, it is essential to adopt a multifaceted approach, encompassing regulatory frameworks, risk assessments, labeling, and research. By understanding the scope of this problem and working together, we can minimize exposure to micro and nanocontaminants and protect human health. Current research has revealed concerning health implications associated with micro and nanocontaminants. Earlier research found that exposure to microplastics increased cancer risk [20]. Furthermore, researchers discovered that nanoparticles can penetrate the blood–brain barrier, potentially leads brain cancer. The micro and nanocontaminants have been linked to neurological effects. Reproductive and developmental health are also vulnerable to micro and nanocontaminants. An earlier study suggests that exposure to microplastics during pregnancy led to birth defects and reproductive issues in mice [21]. The environmental implications of micro and nanocontaminants are equally concerning. A 2020 study published in Science found that microplastics contaminated 90% of the world's rivers [22]. Moreover, researchers discovered that nanoparticles in air pollution traveled long distances, contaminating remote regions. Overall, these findings underscore the urgent need for protective measures to protect human health and the environment from micro and nanocontaminants.

3 Environmental Risks

Micro and nanoplastics (MNPs) contaminate waterways via several pathways, including atmospheric deposition, agricultural runoff, wastewater treatment plants, and industrial effluent [23]. Once in the water, these particles can cause harm to aquatic life through ingestion, entanglement, physical harm, toxicity, bioaccumulation, and biomagnification. Research has shown that microplastics have been found in 90% of the world's rivers, and nanoparticles have been noticed in drinking water sources worldwide [24]. Micro and nanoplastics also contaminate soil ecosystems

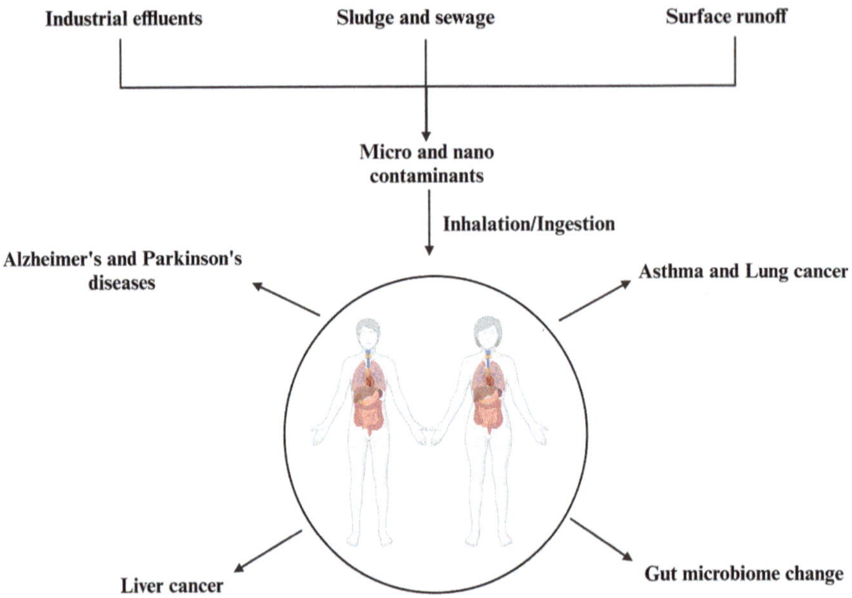

Fig. 1 Effect of micro and nanocontaminants on human health

through agricultural practices, waste disposal, and atmospheric deposition [25]. These particles can alter soil structure and fertility, reduce microbial diversity, increase soil pollution, and impact plant growth and development [26]. For instance, results have found that microplastics were able to reduce water holding capacity of soil and affect microbial communities [27]. MNPs become airborne through industrial processes, waste burning, and vehicle emissions. Once in the air, these particles can mobile extensive distances, contaminating remote regions and affecting human health. Experience to airborne microplastics has been connected to respiratory problems and cardiovascular disease [28]. These particles can be consumed by animals, cause bioaccumulation and biomagnifications [29]. This can cause changes in species composition, impacts on ecosystem services, and even extinction. For example, microplastics have been observed in the stomachs of marine animals, including whales and sea turtles [30]. Micro and nanoplastics disrupt ecosystems by altering nutrient cycling, impacting decomposition, and changing food webs [17]. This can lead to a loss of ecosystem resilience, making it vulnerable to other disturbances. Research has demonstrated that microplastics can adversely impact the growth and development of aquatic plants, as well as modify the composition of microbial communities [31]. The environmental risks of micro and nanoplastics highlight the need for urgent action to mitigate their impacts. By understanding the sources, pathways, and effects of these particles, we can establish an advanced approach to reduce their presence in our environment. The persistence of MNPs in the environment is a major trouble. These particles can remain intact for centuries, slowly breaking down into smaller pieces that can be consumed by smaller creatures

[32]. This persistence can lead to chronic exposure and bioaccumulation, potentially causing irreversible harm to ecosystems [33]. Micro and nanoplastics exert indirect effects on ecosystems by altering the structure of microbial communities, which are vital for nutrient cycling and decomposition [34]. Such changes can lead to cascading impacts on ecosystem function and resilience. Furthermore, climate change intensifies the environmental risks associated with micro and nanoplastics, as rising temperatures and shifting precipitation patterns facilitate their dispersion via atmospheric deposition and runoff [35]. Additionally, climate-driven changes in ocean currents and circulation can transport microplastics to remote regions. The economic impacts of micro and nanoplastic pollution are substantial. Earlier research projected that marine plastic pollution imposes an annual cost of ~$2.5 trillion on the global economy [36]. This includes losses to fisheries, tourism, and coastal protection. Effective mitigation strategies are needed to address the environmental risks of micro and nanoplastics. Collaboration among governments, industries, and individuals is essential to mitigate plastic production, enhance recycling efforts, and enact policies aimed at preventing pollution. Biodegradable alternatives to traditional plastics offer promise in reducing micro and nanoplastic pollution. However, more research is needed to ensure these alternatives are truly biodegradable and do not pose additional environmental risks. Here are enlisted some examples of successful implementations of micro and nanoplastic mitigation strategies: The European Union's Single-Use Plastics Directive has reduced single-use plastic consumption by 40% in member states. Similarly, Kenya's plastic bag ban has led to a 80% reduction in plastic litter [37]. The Sweden's waste-to-energy program converts 99% of household waste into energy, minimizing landfill waste [38]. Singapore's waste management system, which includes advanced recycling facilities, has achieved a recycling rate of 70% [39]. The companies like Lanza Tech and Novamont are developing biodegradable plastics from renewable resources. Researchers have created a microplastic-filtering membrane for wastewater treatment. The Surfrider Foundation's Beach Cleanup program has collected over 300,000 pounds of trash worldwide [40]. The city of Vancouver's "Zero Waste" strategy aims to eliminate single-use plastics by 2025. A conducted study by the Ellen MacArthur Foundation found that implementing circular economy practices can save businesses $1 trillion annually [41]. The recycling industry in the USA generates over $117 billion in economic activity each year [42].

4 Exposure Routes

MPs and NPs infiltrate the human system via ingestion, inhalation, and skin contact [8, 43]. MPs, which are commonly found in rubber tires and synthetic textiles, can be aerosolized and inhaled as part of urban dust [28]. They also come in the human system through food, and water [44]. Although MPs and NPs are unable to penetrate the skin barrier directly, they can gain entry through hair follicles, sweat glands, wounds [45]. Consequently, both MPs and NPs are detectable within the human body through all three exposure routes, with environmental sources and seafood representing the highest risk factors. These environments facilitate the extended degradation of polymers, the leaching of chemical additives, residual monomers, experience to various contaminants, and the proliferation of harmful microbes [46–50].

4.1 Exposure by Ingestion

Humans are frequently vulnerable to MNPs through oral routes, primarily via contaminated food, water, and cosmetic products. The pervasive utilization of plastic materials within drinking water systems, combined with the gathering of particles in long-standing water pipes, can lead to significant exposure to elevated levels of MNPs, which may adversely affect health [51]. Numerous reports have illustrated that individuals may ingest MNPs particles from various sources, including table salt, seafood, bottled beverages, tea bags, honey, and beer. These contaminants can bioaccumulate in affected organisms, thereby posing additional risks to human health [52, 53].

Moreover, MNPs may be introduced into food during production or packaging processes [54–57]). Research by [58] on fluorescent polystyrene microplastics (5 and 20 μm) in murine models revealed accretion in the kidneys, liver, and intestines leading to disruptions in metabolism of lipid and inducing oxidative stress. They further noted that particles up to 150 μm can traverse the intestinal barrier, albeit with a low absorption rate of less than 0.3% [59]. Campanale et al. [60] found that particles as small as 10 μm can infiltrate all organs, including the brain, although the full implications of this penetration remain unclear.

4.2 Inhalation Exposure

Inhalation represents one of the primary pathways through which humans are exposed to MNPs. MNPs are prevalent in the atmosphere due to sources such as synthetic textiles, tire wear, and various polymeric fibers. Individuals are estimated to inhale approximately 26,130 airborne plastic particles daily [28]. Upon inhalation, MNPs penetrate the respiratory epithelium through diffusion, endocytosis, or phagocytosis, with particles between 1 and 3 μm primarily cleared via phagocytosis in the alveoli [61]. Research indicates that indoor environments contain higher concentrations of MNPs than outdoor air, suggesting greater human exposure indoors [62]. While inhaled particles may be cleared through mucociliary transport or macrophage activity, some can accumulate in the lungs and enter the bloodstream. Additionally, inhaled nanoparticles may reach the central nervous system via the olfactory bulb. Studies indicate that microplastics can be transferred from pregnant mothers to fetuses, potentially impacting fetal development and increasing the risk of neurological disorders.

4.3 Skin Contact Exposure

MNPs present in personal hygiene products and contaminated water can adversely impact skin health. Laboratory studies have shown that epidermal cells exhibit stress due to oxidative effects from these particles. Synthetic fibers smaller than 25 μm can penetrate skin pores, typically measuring 40–80 μm, thereby bypassing the stratum corneum, the outermost skin layer. Notably, nanoparticles smaller than 100 nm can be absorbed by the stratum corneum, highlighting the varying susceptibility among individuals due to differences in skin pore characteristics. Research in nanotoxicology indicates that nanoparticles under 40 nm can breach the skin barrier [45, 63]. This exposure often occurs during activities such as showering or using products containing nanopolystyrene [64, 65]. In response to these concerns, the FDA has enacted regulations to limit microbeads in certain products, aiming to reduce dermal exposure and environmental residues. Additionally, microplastics are commonly used in medical applications for drug delivery, potentially exposing individuals to significant quantities. Given the growing severity of plastic pollution, further research is essential to evaluate the risks of MNP absorption through skin exposure, including their accumulation in hair follicles and interaction with immune cells like Langerhans cells [66].

5 Mitigation Strategies

To mitigate micro and nanoplastic pollution, reducing plastic production is crucial. This can be achieved through implementing Extended Producer Responsibility (EPR) policies, promoting circular economy practices, increasing recycling rates, and encouraging biodegradable alternatives [67]. Governments and industries must work together to minimize plastic waste. Improving waste management is also essential. Upgrading wastewater treatment plants, implementing effective waste collection and disposal, reducing litter and illegal dumping, and implementing Deposit Refund Systems (DRS) can significantly reduce microplastic pollution [68]. Effective regulatory frameworks are necessary to combat microplastic pollution. Establishing national and international regulations, setting standards for microplastic-free products, implementing bans on single-use plastics, and enforcing pollution prevention measures can help minimize plastic waste [69]. Research and development play a vital role in addressing microplastic pollution. Investigating biodegradable materials, developing effective microplastic removal technologies, studying environmental impacts, and improving waste management infrastructure are critical areas of focus [70]. Raising awareness and educating the public are also crucial strategies. Conducting public awareness campaigns, educating consumers on sustainable choices, training waste management professionals, and integrating environmental education in school curricula can inspire behavioral change. International cooperation is necessary to tackle the global issue of microplastic pollution. Establishing global agreements, sharing best practices, collaborating on research, and developing harmonized regulations can facilitate collective action. Notable initiatives and policies, such as the European Union's Single-Use Plastics Directive, the United Nations' Clean Seas Campaign, Ocean Conservancy's Trash Free Seas Program, and Japan's Extended Producer Responsibility Law, demonstrate effective mitigation strategies.

6 Conclusion

Governments worldwide have implemented policies to mitigate micro and nanoplastic pollution, but assessing their effectiveness is crucial. Successful policies include Extended Producer Responsibility (EPR), single-use plastic bans, and waste management regulations. These measures encourage manufacturers to design sustainable products and recycling systems, reduce plastic consumption and litter, and improve waste collection and recycling infrastructure. However, challenges and limitations hinder policy effectiveness. Lack of standardization across countries and regions makes international cooperation difficult. Inadequate enforcement, insufficient monitoring, and lenient penalties undermine policy impact. Moreover, policies often focus on specific products or industries, overlooking broader systemic issues. Best practices can enhance policy effectiveness. Integrated approaches combining policy, education, and technological innovation are essential. Stakeholder engagement involving industries, communities, and NGOs in policy development ensures inclusivity. Continuous monitoring and adaptation of strategies enable policymakers to refine their approaches. Case studies demonstrate policy success. Sweden's waste-to-energy program has achieved 99% household waste conversion to energy. Rwanda's plastic bag ban reduced plastic litter by 90%. California's microbead ban eliminated microbeads in personal care products. These examples highlight the potential of effective policy-making. To improve policy effectiveness, key recommendations include harmonizing global regulations, enhancing enforcement mechanisms, and fostering international cooperation. Standardized policies facilitate collaboration and knowledge-sharing. Robust enforcement ensures compliance, while cooperation accelerates progress. In conclusion, evaluating policy effectiveness is vital to combating micro and nanoplastic pollution. By understanding successes and challenges, governments and stakeholders can refine strategies, foster cooperation, and create a more sustainable future.

Competing Interests The authors declare that they have no competing interest.

Declaration of Generative AI and AI-Assisted Technologies in the Writing Process During the preparation of this work, the book chapter author(s) have not taken any assistance from AI, and the book chapter author(s) reviewed and edited the content as needed and take(s) full responsibility for the content of the publication.

References

1. Sharma RK, Kumari U, Kumar S (2024) Impact of microplastics on pregnancy and fetal development: a systematic review. Cureus 16(5)
2. Zhang K, Hamidian AH, Tubić A, Zhang Y, Fang JK, Wu C, Lam PK (2021) Understanding plastic degradation and microplastic formation in the environment: a review. Environ Pollut 274:116554
3. Anderson JC, Park BJ, Palace VP (2016) Microplastics in aquatic environments: implications for Canadian ecosystems. Environ Pollut 218:269–280
4. Welle F, Franz R (2018) Microplastic in bottled natural mineral water – literature review and considerations on exposure and risk assessment. Food Addit & Contam: Part A 35(12):2482–2492
5. Koelmans AA, Nor NHM, Hermsen E, Kooi M, Mintenig SM, De France J (2019) Microplastics in freshwaters and drinking water: Critical review and assessment of data quality. Water Res 155:410–422
6. Pivokonsky M, Cermakova L, Novotna K, Peer P, Cajthaml T, Janda V (2018) Occurrence of microplastics in raw and treated drinking water. Sci Total Environ 643:1644–1651
7. Triebskorn R, Braunbeck T, Grummt T, Hanslik L, Huppertsberg S, Jekel M, Knepper TP, Krais S, Müller YK, Pittroff M (2019) Relevance of nano-and microplastics for freshwater ecosystems: a critical review. TrAC Trends Anal Chem 110:375–392
8. Rahman A, Sarkar A, Yadav OP, Achari G, Slobodnik J (2021) Potential human health risks due to environmental exposure to nano-and microplastics and knowledge gaps: a scoping review. Sci Total Environ 757:143872
9. Amabile N, Heiss C, Real WM, Minasi P, McGlothlin D, Rame EJ, Grossman W, De Marco T, Yeghiazarians Y (2008) Circulating endothelial microparticle levels predict hemodynamic severity of pulmonary hypertension. Am J Respir Crit Care Med 177(11):1268–1275
10. Jones AE, Watts JA, Debelak JP, Thornton LR, Younger JG, Kline JA (2003) Inhibition of prostaglandin synthesis during polystyrene microsphere-induced pulmonary embolism in the rat. Am J Physiol-Lung Cellul Mol Physiol 284(6):L1072–L1081
11. Dong M, Zhang Q, Xing X, Chen W, She Z, Luo Z (2020) Raman spectra and surface changes of microplastics weathered under natural environments. Sci Total Environ 739:139990
12. Çobanoğlu H, Belivermis M, Sıkdokur E, Kılıç Ö, Çayır A (2021) Genotoxic and cytotoxic effects of polyethylene microplastics on human peripheral blood lymphocytes. Chemosphere 272:129805
13. Anetor GO, Nwobi NL, Igharo GO, Sonuga OO, Anetor JI (2022) Environmental pollutants and oxidative stress in terrestrial and aquatic organisms: examination of the total picture and implications for human health. Front Physiol 13:931386
14. Oberdörster G, Maynard A, Donaldson K, Castranova V, Fitzpatrick J, Ausman K, Carter J, Karn B, Kreyling W, Lai D, Olin S (2005) Principles for characterizing the potential human health effects from exposure to nanomaterials: elements of a screening strategy. Part Fibre Toxicol 2:1–35.
15. Kumar R, Manna C, Padha S, Verma A, Sharma P, Dhar A, Ghosh A, Bhattacharya P (2022) Micro (nano) plastics pollution and human health: how plastics can induce carcinogenesis to humans? Chemosphere 298:134267
16. Langie SA, Koppen G, Desaulniers D, Al-Mulla F, Al-Temaimi R, Amedei A, Azqueta A, Bisson WH, Brown D, Brunborg G, Charles AK (2015) Causes of genome instability: the effect of low dose chemical exposures in modern society. Carcinogenesis 36(Suppl_1):S61–S88
17. Ghosh S, Sinha JK, Ghosh S, Vashisth K, Han S, Bhaskar R (2023) Microplastics as an emerging threat to the global environment and human health. Sustainability 15(14):10821
18. Calderón-Garcidueñas L, Torres-Jardón R, Greenough GP, Kulesza R, González-Maciel A, Reynoso-Robles R, García-Alonso G, Chávez-Franco DA, García-Rojas E, Brito-Aguilar R, Silva-Pereyra HG (2023) Sleep matters: neurodegeneration spectrum heterogeneity, combustion and friction ultrafine particles, industrial nanoparticle pollution, and sleep disorders—Denial is not an option. Front Neurol 14:1117695

19. Choi SM, Yoo SD, Lee BM (2004) Toxicological characteristics of endocrine-disrupting chemicals: developmental toxicity, carcinogenicity, and mutagenicity. J Toxicol Environ Health Part B 7(1):1–23
20. Segovia-Mendoza M, Nava-Castro KE, Palacios-Arreola MI, Garay-Canales C, Morales-Montor J (2020) How microplastic components influence the immune system and impact on children health: focus on cancer. Birth Defects Res 112(17):1341–1361
21. Dou Y, Zhang M, Zhang H, Zhang C, Feng L, Hu J, Gao Y, Yuan XZ, Zhao Y, Zhao H, Chen ZJ (2024) Lactating exposure to microplastics at the dose of infants ingested during artificial feeding induced reproductive toxicity in female mice and their offspring. Sci Total Environ 949:174972
22. Sarijan S, Azman S, Said MI, Jamal MH (2021) Microplastics in freshwater ecosystems: a recent review of occurrence, analysis, potential impacts, and research needs. Environ Sci Pollut Res 28:1341–1356
23. Talukdar A, Kundu P, Bhattacharya S, Dutta N (2024) Microplastic contamination in wastewater: Sources, distribution, detection and remediation through physical and chemical-biological methods. Sci Total Environ 170254
24. Shen M, Song B, Zhu Y, Zeng G, Zhang Y, Yang Y, Wen X, Chen M, Yi H (2020) Removal of microplastics via drinking water treatment: current knowledge and future directions. Chemosphere 251:126612
25. Nath S, Enerijiofi KE, Astapati AD, Guha A (2024) Microplastics and nanoplastics in soil: sources, impacts, and solutions for soil health and environmental sustainability. J Environ Qual
26. Seo Y, Zhou Z, Lai Y, Chen G, Pembleton K, Wang S, He JZ, Song P.(2024) Micro-and nanoplastics in agricultural soils: assessing impacts and navigating mitigation. Sci Total Environ 172951
27. Sajjad M, Huang Q, Khan S, Khan MA, Liu Y, Wang J, Lian F, Wang Q, Guo G (2022) Microplastics in the soil environment: a critical review. Environ Technol & Innov 27:102408
28. Correia-Prata J (2018) Airborne microplastics: consequences to human health? Environ Pollut 234:115–126. https://doi.org/10.1016/J.SOILBIO.2017.10.027
29. Miller ME, Hamann M, Kroon FJ (2020) Bioaccumulation and biomagnification of microplastics in marine organisms: a review and meta-analysis of current data. PLoS ONE 15(10):e0240792
30. Park B, Kim SK, Joo S, Kim JS, Jo K, Song NS, Im J, Lee HJ, Kim SW, Lee SB, Kim S (2023) Microplastics in large marine animals stranded in the Republic of Korea. Mar Pollut Bull 189:114734
31. Yang GL, Zheng MM, Liao HM, Tan AJ, Feng D, Lv SM (2022) Influence of cadmium and microplastics on physiological responses, ultrastructure and rhizosphere microbial community of duckweed. Ecotoxicol Environ Saf 243:114011
32. Bajt O (2021) From plastics to microplastics and organisms. FEBS Open Bio 11(4):954–966
33. Paasivirta J (2001) Long-term effects of bioaccumulation in ecosystems. In: Bioaccumulation–new aspects and developments. Springer Berlin Heidelberg, Berlin, Heidelberg, pp 201–233
34. Naylor D, McClure R, Jansson J (2022) Trends in microbial community composition and function by soil depth. Microorganisms 10(3):540
35. Haque F, Fan C (2023) Fate of microplastics under the influence of climate change. Iscience
36. Walker TR, Fequet L (2023) Current trends of unsustainable plastic production and micro (nano) plastic pollution. TrAC Trends Anal Chem 160:116984
37. Kasznik D, Łapniewska Z (2023) The end of plastic? The EU's directive on single-use plastics and its implementation in Poland. Environ Sci Policy 145:151–163
38. Malinauskaite J, Jouhara H, Czajczyńska D, Stanchev P, Katsou E, Rostkowski P, Thorne RJ, Colon J, Ponsá S, Al-Mansour F, Anguilano L (2017) Municipal solid waste management and waste-to-energy in the context of a circular economy and energy recycling in Europe. Energy 141:2013–2044
39. Vuk A, Szűcs I, Bauerné Gáthy A (2024) Waste management and plastic waste recycling in Japan, China, Singapore and South Korea–What trends can be observed under different regulations. Int Rev Appl Sci Eng

40. Stickel BH, Jahn A, Kier B (2012) The cost to West Coast communities of dealing with trash, reducing marine debris
41. Macarthur EL, Heading HE.(2019) How the circular economy tackles climate change. Ellen MacArthur Found 1–71.
42. Golden JS, Handfield RB, Daystar J, McConnell TE (2015) An economic impact analysis of the US biobased products industry: a report to the Congress of the United States of America. Ind Biotechnol 11(4):201–209
43. Prata JC, da Costa JP, Lopes I, Duarte AC, Rocha-Santos T (2020) Environmental exposure to microplastics: an overview on possible human health effects. Sci Total Environ 702:134455
44. Carbery M, O'Connor W, Palanisami T (2018) Trophic transfer of microplastics and mixed contaminants in the marine food web and implications for human health. Environ Int 115:400–409
45. Schneider M, Stracke F, Hansen S, Schaefer UF (2009) Nanoparticles and their interactions with the dermal barrier. Dermato-Endocrinology 1(4):197–206
46. Brennecke D, Duarte B, Paiva F, Caçador I, Canning-Clode J (2016) Microplastics as vector for heavy metal contamination from the marine environment. Estuar Coast Shelf Sci 178:189–195
47. Camacho M, Herrera A, Gómez M, Acosta-Dacal A, Martínez I, Henríquez-Hernández LA, Luzardo OP (2019) Organic pollutants in marine plastic debris from Canary Islands beaches. Sci Total Environ 662:22–31
48. Li J, Zhang K, Zhang H (2018) Adsorption of antibiotics on microplastics. Environ Pollut 237:460–467
49. Rochman CM, Kurobe T, Flores I, Teh SJ (2014) Early warning signs of endocrine disruption in adult fish from the ingestion of polyethylene with and without sorbed chemical pollutants from the marine environment. Sci Total Environ 493:656–661
50. Viršek MK, Lovšin MN, Koren Š, Kržan A, Peterlin M (2017) Microplastics as a vector for the transport of the bacterial fish pathogen species Aeromonas salmonicida. Mar Pollut Bull 125(1–2):301–309
51. Cox KD, Covernton GA, Davies HL, Dower JF, Juanes F, Dudas SE (2019) Human consumption of microplastics. Environ Sci Technol 53(12):7068–7074
52. Oßmann BE (2021) Microplastics in drinking water? Present state of knowledge and open questions. Curr Opin Food Sci 41:44–51
53. Yee MS, Hii LW, Looi CK, Lim WM, Wong SF, Kok YY, Tan BK, Wong CY, Leong CO (2021) Impact of microplastics and nanoplastics on human health. Nanomaterials 11(2):496
54. Hernandez LM, Xu EG, Larsson HC, Tahara R, Maisuria VB, Tufenkji N (2019) Plastic teabags release billions of microparticles and nanoparticles into tea. Environ Sci Technol 53(21):12300–12310
55. Mason SA, Welch VG, Neratko J (2018) Synthetic polymer contamination in bottled water. Front Chem 6:389699
56. Santillo D, Miller K, Johnston P (2017) Microplastics as contaminants in commercially important seafood species. Integr Environ Assess Manag 13(3):516–521
57. Du F, Cai H, Zhang Q, Chen Q, Shi H (2020) Microplastics in take-out food containers. J Hazard Mater 399:122969
58. Deng Y, Zhang Y, Lemos B, Ren H (2017) Tissue accumulation of microplastics in mice and biomarker responses suggest widespread health risks of exposure. Sci Rep 7(1):46687
59. Revel M, Châtel A, Mouneyrac C (2018) Micro (nano) plastics: a threat to human health? Curr Opin Environ Sci & Health 1:17–23
60. Campanale C, Massarelli C, Savino I, Locaputo V, Uricchio VF (2020) A detailed review study on potential effects of microplastics and additives of concern on human health. Int J Environ Res Public Health 17(4):1212
61. Wright SL, Kelly FJ (2017) Plastic and human health: a micro issue? Environ Sci Technol 51(12):6634–6647
62. Dris R, Gasperi J, Mirande C, Mandin C, Guerrouache M, Langlois V, Tassin B (2017) A first overview of textile fibers, including microplastics, in indoor and outdoor environments. Environ Pollut 221:453–458

63. Vogt A, Rancan F, Ahlberg S, Nazemi B, Choe CS, Darvin ME, Hadam S, Blume-Peytavi U, Loza K, Diendorf J, Epple M (2014) Interaction of dermatologically relevant nanoparticles with skin cells and skin. Beilstein J Nanotechnol 5(1):2363–2373
64. Galloway TS (2015) Micro-and nano-plastics and human health. In: Marine anthropogenic litter, pp 343–366
65. Lai H, Liu X, Qu M (2022) Nanoplastics and human health: hazard identification and biointerface. Nanomaterials 12(8):1298
66. Lademann J, Richter H, Schanzer S, Knorr F, Meinke M, Sterry W, Patzelt A (2011) Penetration and storage of particles in human skin: perspectives and safety aspects. Eur J Pharm Biopharm 77(3):465–468
67. Leal Filho W, Saari U, Fedoruk M, Iital A, Moora H, Klöga M, Voronova V (2019) An overview of the problems posed by plastic products and the role of extended producer responsibility in Europe. J Clean Prod 214:550–558
68. Usman S, Abdull Razis AF, Shaari K, Azmai MN, Saad MZ, Mat Isa N, Nazarudin MF (2022) The burden of microplastics pollution and contending policies and regulations. Int J Environ Res Public Health 19(11):6773
69. Knoblauch D, Mederake L (2021) Government policies combatting plastic pollution. Curr Opin Toxicol 28:87–96
70. Moshood TD, Nawanir G, Mahmud F, Mohamad F, Ahmad MH, AbdulGhani (2022) A. Sustainability of biodegradable plastics: new problem or solution to solve the global plastic pollution?. Curr Res Green Sustain Chem 100273

MIX
Papier aus verantwortungsvollen Quellen
Paper from responsible sources
FSC® C105338

If you have any concerns about our products,
you can contact us on
ProductSafety@springernature.com

In case Publisher is established outside the EU,
the EU authorized representative is:
Springer Nature Customer Service Center GmbH
Europaplatz 3, 69115 Heidelberg, Germany

Printed by Libri Plureos GmbH
in Hamburg, Germany